9th International Frohlich's Symposium: Electrodynamic Activity of Living Cells including Microtubule Coherent Modes and Cancer Cell Physics

Journal of Physics: Conference Series Volume 329

Prague, Czech Republic
1 – 3 July 2011

Editors:

Michal Cifra
Jiri Pokorny
Ondrej Kucera

ISBN: 978-1-61839-352-4
ISSN: 1742-6588

Printed from e-media with permission by:

Curran Associates, Inc.
57 Morehouse Lane
Red Hook, NY 12571

CURRAN ASSOCIATES INC. proceedings.com

Some format issues inherent in the e-media version may also appear in this print version.

Printed by Curran Associates, Inc. (2011)

Institute of Physics
Dirac House, Temple Back
Bristol BS1 6BE UK

Phone: 44 1 17 929 7481
Fax: 44 1 17 920 0979

techtracking@iop.org

Additional copies of this publication are available from:

Curran Associates, Inc.
57 Morehouse Lane
Red Hook, NY 12571 USA
Phone: 845-758-0400
Fax: 845-758-2634
Email: curran@proceedings.com
Web: www.proceedings.com

TABLE OF CONTENTS

The Symposium was organized by

Institute of Photonics and Electronics, Academy of Sciences of the Czech Republic

First Faculty of Medicine, Charles University, Prague

under the aegis of

Professor Tomáš Zima
Dean of the First Faculty of Medicine, Charles University

Professor Otomar Kittnar
Director of the Institute of Physiology, First Faculty of Medicine, Charles University

and
Dr. Vlastimil Matějec
Director of the Institute of Photonics and and Electronics AS CR

Symposium was technically sponsored by

IEEE

Further sponsor was

HEKA Elektronik
Dr. Schulze GmbH

Scientific Committee
Chair: J. Pokorný, Czechia
A. Bandyopadhyay, Japan
I. Belyaev, Slovakia
V.N. Binhi, Russia
L.S. Brizhik, Ukraine
M. Cifra, Czechia
A. Čoček, Czechia
D. Dimitrov, Bulgaria
A. Foletti, Italy
S. Hameroff, USA
R. Hölzel, Germany
G. Hyland, England
A. Jandová, Czechia
F. Jelínek, Czechia
I. Jerman, Slovenia
O. Kittnar, Czechia
A.R. Liboff, USA
L. Mora, France
M. Nedbalová, Czechia
M. Satarić, Serbia
F. Šrobár, Czechia
J. Tuszyński, Canada
E. Tvrzická, Czechia
C. Vedruccio, Italy
G. Vitiello, Italy
J. Vrba, Czechia
J. Zon, Poland

Organizing Committee
Chair: M. Cifra, Czechia
D. Havelka, Czechia
A. Jandová, Czechia
R. Janča, Czechia
F. Jelínek, Czechia
O. Kučera, Czechia
M. Nedbalová, Czechia
J. Pokorný, Czechia
Š. Říhová, Czechia
F. Šrobár, Czechia
B. Vrbová, Czechia

Introductory Remarks

This volume contains papers presented at the International Fröhlich's Symposium entitled "Electrodynamic Activity of Living Cells" (July 1–3, 2011, Prague, Czech Republic). The Symposium was the 9th meeting devoted to physical processes in living matter organized in Prague since 1987. Hypothesis of oscillation systems in living cells featured by non-linear interaction between elastic and electrical polarization fields, non-linear interactions between the system and the heat bath leading to energy downconversion along the frequency scale, energy condensation in the lowest frequency mode and creation of a coherent state was formulated by H. Fröhlich, founder of the theory of dielectric materials. He assumed that biological activity is based not only on biochemical but also on biophysical mechanisms and that their disturbances form basic links along the cancer transformation pathway. Fröhlich outlined general ideas of non-linear physical processes in biological systems. The downconversion and the elastic–polarization interactions should be connected in a unified theory and its solution based on comprehensive non-linear characteristics.

Biochemical and genetic research of biological systems are highly developed and have disclosed a variety of cellular and subcellular structures, chemical reactions, molecular information transfer, and genetic code sequences– including their pathological development. Nevertheless, the cancer problem is still a big challenge. Warburg's discovery of suppressed oxidative metabolism in mitochondria in cancer cells suggested the essential role of physical mechanisms (but his discovery has remained without impact on cancer research and on study of physical properties of biological systems for a long time). Mitochondria, the power plants of the cell, have several areas of activity–oxidative energy production is connected with formation of a strong static electric field around them, water ordering, and liberation of non-utilized energy to their surroundings. Mitochondrial function connected with water ordering and excitation of oscillations in microtubules may play a central role in biological activity, in particular in transport, organization, interactions, and information transfer. Mitochondrial dysfunction results in disturbances of the generated electrodynamic field with bad consequences in biological activity and creation of pathological states. A special issue of the biological activity concerns the brain function (consciousness is not yet adequately understood). Experimental investigation using nanotechnology would supply yet unknown data and parameters of physical mechanisms in living systems. Extremely weak biological signals have to be separated from technical noise under conditions of possible non-linear mutual interactions.

Some authors questioned the validity of the Fröhlich hypothesis. Foster and Baish (J. Biol. Phys. 26, 2000, 255) neglected water ordering and concluded that strong damping by water viscosity effects prevents formation of coherent state. Reimers et al. (PNAS 106, 2009, 4219) and McKemmish et al. (Phys. Rev. E 80, 2009, 021912-1) omitted non-linear elastic-electrical polarization interactions and analyzed a linearized model of downconversion with strong damping that cannot represent the Fröhlich system. Fröhlich assumed high quality non-linear system with energy supply. Some methods used for analysis of linear system (for instance method of superposition) are not valid in non-linear systems. For this reason also experimental analysis based on subtraction of the noise from the measured signal spectrum is not a simple question.

There is another special issue concerning the biological activity. Living state and in particular consciousness are very often connected with an idea of non-material and non-measurable entity entering the biological system from outside. There is a splendid harmony and order in nature. Science

Journal of Physics: Conference Series **329** (2011) 011001 doi:10.1088/1742-6596/329/1/011001

should disclose measurable mechanisms of the harmony and order. But the human knowledge about the electrodynamic and electromagnetic fields in biological systems is still at low level.

The Symposium continued in the series of international scientific meetings devoted to physical processes in living cells organized in Prague. The first meeting was entitled "Biophysical Aspects of Cancer" (July 6–9, 1987). At this occasion the anglo-german physicist H. Fröhlich presented a lecture "Coherence in Biology". The next meeting devoted to the Fröhlich coherent systems, information transfer, and neural activity was in 1993. The role of the Fröhlich coherence in the neural activity was included in the meeting "Biophysical Aspects of Coherence" in 1995 too. The further symposia were entitled "Electromagnetic Fields in Biological Systems" (1998), "Electromagnetic Aspects of Selforganization in Biology" (2000), "Endogenous Physical Fields in Biology" (2002), "Coherence and Electromagnetic Fields in Biological Systems" (2005), and "Biophysical Aspects of Cancer – Electromagnetic Mechanisms" (2008). In 2008 a novel project for research of convergence of physics and oncology was triggered in the USA by the National Cancer Institute and the Institute of Public Health.

This issue contains main part of papers presented at the Symposium. The ideas presented at the Symposium might have impact on future research of physical processes and mechanisms in biological systems. Experimental research may provide a background for understanding of the neglected part of biological activity and reveal the physical mechanisms of the cancer transformation pathway.

The Symposium and this issue were prepared by a scientific team whose members were M. Cifra, D. Havelka, A. Jandová, F. Jelínek, O. Kučera, M. Nedbalová, and F. Šrobár.

Jiří Pokorný

Electromagnetic activity –
essential property of living systems
Generating structure – microtubules
Function of the electromagnetic field in

Journal of Physics: Conference Series **329** (2011) 012001 doi:10.1088/1742-6596/329/1/012001

The interplay of biomolecules and water at the origin of the active behavior of living organisms

E Del Giudice[1], **P Stefanini**[2], **A Tedeschi**[3], **G Vitiello**[4]
[1]IIB, Neuss, Germany
[2]Hadoshiatsu, Florence, Italy
[3]WHITE HB, Milano, Italy
[4]University of Salerno, Italy

E-mail: emilio.delgiudice@mi.infn.it

Abstract. It is shown that the main component of living matter, namely liquid water, is not an ensemble of independent molecules but an ensemble of phase correlated molecules kept in tune by an electromagnetic (e.m) field trapped in the ensemble. This field and the correlated potential govern the interaction among biomolecules suspended in water and are in turn affected by the chemical interactions of molecules. In particular, the phase of the coherent fields appears to play an important role in this dynamics. Recent experiments reported by the Montagnier group seem to corroborate this theory. Some features of the dynamics of human organisms, as reported by psychotherapy, holistic medicine and Eastern traditions, are analyzed in this frame and could find a rationale in this context.

1. Introduction

Modern molecular biology has been able to produce a scheme of the living organism where most biological events are traced back to sequences of chemical reactions, ordered in space and time [1]. Such chemical reactions don't seem to occur at random, since there is no evidence of the presence of the results of molecular encounters not obeying specific chemical codes [2,3]. However the mainstream molecular biology has not been able so far to show a rationale for the appearance of such codes.

On the other hand an increasing evidence has been accumulated along the years about the role of electromagnetic fields in the biological dynamics [4,5]. Recently a group led by Luc Montagnier [6] has reported that some bacterial and viral DNA sequences have been found to induce low frequency electromagnetic waves in high aqueous dilutions. These waves are able to imprint pure water contained in nearby vessels so that similar sequences are able to grow up when the appropriate component molecules are poured in the liquid .

The problem arises of finding a dynamics including molecular chemical activity and electromagnetic fields. An intriguing feature of the electrodynamics of living organisms is the property of such organisms to respond to very weak fields, quite unable to transfer significant amounts of energy. This property has been detected for the first time in the microwave experiments. For instance in the 70's the group led by Devyatkov [7] was able to show that unicellular organisms had significant changes when irradiated with microwave fields at selected frequencies and very low intensities. The biological response didn't depend on intensity (above a very low threshold) but on frequency; each organism had its own spectrum of frequencies, possibly dependent on time, at which a microwave field was able to elicit a response. The interaction therefore seemed to imply not an exchange of energy but a correlation mediated by the phase. The possibility of a phase interaction in biological organisms could open the way to the understanding of the class of the so called synchronic phenomena [8] occurring in living organisms, investigated in holistic therapeutic approaches and in psychotherapy where biological events are induced without the presence of significant exchange of energy.

The aim of the present paper is to try to shed some light on the above questions. Our conceptual framework is Quantum Electrodynamics (QED), which is the theory describing the interaction between the electromagnetic field and an ensemble of molecules. It has been shown recently [9,10,11,12] that an ensemble of molecules, put initially in the empty space in absence of externally applied fields, becomes unstable above a density threshold and below a temperature threshold. The system in the above conditions decays into a lower energy state where all the molecules are phase correlated among them in tune with an electromagnetic field grown up from their vacuum quantum fluctuations. This theorem could provide the rationale for establishing the interplay between chemistry and electrodynamics we have asked for above. Moreover this scheme would implement the requirements formulated more than 40 years ago by Herbert Fröhlich [13].

The article is organized as follows. In section 2. we will provide the theoretical scheme of the interplay between molecules and e.m.f.; in section 3. we will apply this scheme to an analysis of Montagnier experiments. In section 4. we will analyze in some depth the phase relationships emerging from such scheme, whereas in section 5. we will apply the above concepts to understand the emergence of an active behavior in living organisms and of phenomena which cannot be reduced to exchanges of energy.

2. Electrodynamics of water and living matter

In this section we summarize shortly the quantum field theoretical mechanism able to produce the spontaneous emergence of a coherent electromagnetic (e.m.) field within an ensemble of molecules which become in turn coherent among them because of the interaction with the self-produced field. This phenomenon depends on the basic tenet of quantum physics that all physical objects (either matter or fields) fluctuate spontaneously. As far back as in 1916, Walter Nernst [14] wondered whether such spontaneous fluctuations of microscopic components could tune together giving rise to a macroscopic oscillation of the ensemble of components, which would become in this way phase correlated. Many decades later [9÷13] the above question has found a positive answer through the introduction of the concept of coherence, which means the property of an ensemble of components of having a well defined value of the phase in tune with a field present in the ensemble.

The physical entities Quantum Physics deals with include of course the vacuum, which is the minimum energy state of the system; vacuum too fluctuates and its fluctuations couple with the fluctuations of the components. In particular, let us consider an ensemble of molecules/atoms, which are composite systems made up of electrically charged particles (nuclei and electrons), able to show a spectrum of configurations, having each one a defined energy.

Let us consider a state q belonging to the spectrum. Its excitation demands an energy E_q

$$E_q = h\,\nu_q = h\,\omega_q/2\,\pi \qquad (1)$$

where h is the Planck constant and ν is the frequency of the e.m. mode, whose wavelength is λ, which is able to raise the atom/molecule from the ground state to the excited state by resonating with it. The probability of its excitation is governed by the so called oscillator's strength f_{0q} [15]. The ensemble of the oscillator's strengths for all possible excitations obeys to the Thomas-Reiche-Kühn sum rule [15].

$$\Sigma_q\, f_{0q} = n \qquad (2)$$

Journal of Physics: Conference Series **329** (2011) 012001 doi:10.1088/1742-6596/329/1/012001

where n is the number of electrons of the single molecule. At the lower order in the equations (see below) the state q can be excited in two ways:

1) by direct excitation via a photon of frequency ν_q

2) by a two step excitation where a level $q' \neq q$ is excited and subsequently decays into the state q. The levels q' to be considered are not only the bound states but also the levels belonging to the continuum spectrum where one electron is expelled from the molecule; in this case the excitation energy would be larger than the ionization energy.

We will give now an intuitive argument in the simplified case that the possibility 1) only is at work. Let us assume that N microscopic units, as described above, are present within a volume $V= \lambda^3$ spanned by the e.m. mode resonant with them. According to QED there is a probability P (which can be calculated for instance by the Lamb shift [16]) that a photon emerging from the vacuum could be coupled with an isolated molecule. If there are N molecules in the volume occupied by the photon, the probability P_N that the photon would be coupled with one of them is

$$P_N= P\, N = P\, \lambda^3 (N/V)= P\, \lambda^3 d \qquad (3)$$

It is apparent that when the density d becomes large enough, P_N reaches the value 1, namely each e.m. vacuum fluctuation is transformed into a molecular excitation. In the volume V (which from now on we will term Coherence Domain (CD)) a self-trapped e.m. field makes molecules oscillating in unison. Let us address the general problem now in mathematical terms. We single out from the spectrum a particular level q, whereas the other levels are present as mediators, as in the way 2) of excitations described above. We describe the e.m. field by its potential $\mathcal{A}$ (averaged on the directions) and the ensemble of molecules by a two-component matter field χ describing the space-time distribution of molecules; the field χ_0 describes the molecules in the ground state, whereas the field χ_q describes the molecules in the level q. It has been shown in refs. [10,11] that the interaction between the matter field and the e.m. field is governed by the equations :

$$\begin{aligned} i\dot{\chi}_0(\tau) &= g\chi_q(\tau)\mathcal{A}^*(\tau) \\ i\dot{\chi}_q(\tau) &= g\chi_0(\tau)\mathcal{A}(\tau) \\ -\frac{1}{2}\ddot{\mathcal{A}}(\tau)+i\dot{\mathcal{A}}(\tau)-\mu\mathcal{A}(\tau) &= g\chi_0^*(\tau)\chi_q(\tau) \end{aligned} \qquad (4)$$

In eqs. (4) the dot means a differentiation with respect to the "adimensional" time $\tau= \omega_q t$; moreover we have neglected the space dependence of the fields which can be assumed not varying too much in the CD. The coupling constants which appear in eqs. (4) are connected with the two ways of excitation. The constant g

$$g = \frac{\omega_p}{\omega_q}\sqrt{2\pi f_{0q}} \qquad (5)$$

where ω_p

$$\omega_p = e\sqrt{\frac{N}{Vm_e}} \qquad (6)$$

is the plasma frequency of the ensemble of electrons whose mass is m_e , is governing the direct excitation of the level q (possibility 1)). The constant

$$\mu = \mu_0 + \mu_{cont} \qquad (7)$$

where

$$\mu_0 = -\frac{3}{2}\omega_p^2 \sum_{n \neq q} \frac{f_{0n}}{\omega_n^2 - \omega_q^2} \qquad (8)$$

$$\mu_{cont} = -\frac{3}{2}\omega_p^2 \int_{\omega_{cont}}^{+\infty} \frac{f_0(\omega)d\omega}{\omega^2 - \omega_q^2} \qquad (9)$$

is governing the two step process (possibility 2)) of the excitation of the level q. Finally eqs. (4) admit three constants of motion :

$$\chi_0^* \chi_0 + \chi_q^* \chi_q = 1 \qquad (10)$$

$$Q = \mathcal{A}\mathcal{A}^* + \frac{i}{2}\left(\mathcal{A}^*\dot{\mathcal{A}} - \dot{\mathcal{A}}^*\mathcal{A}\right) + \chi_0^*\chi_0 \qquad (11)$$

$$H = Q + \frac{1}{2}\dot{\mathcal{A}}^*\dot{\mathcal{A}} + \mu\mathcal{A}^*\mathcal{A} + g\left(\mathcal{A}^*\chi_0^*\chi_q + \mathcal{A}\chi_0\chi_q^*\right) \qquad (12)$$

Eq. (10) expresses the conservation of the total number of molecules spread between the two levels. Eq. (11) is connected with the conservation of the total momentum Q whereas eq. (12) gives the energy per particle of the system. A deeper understanding of the above mathematics can be achieved in refs. [10÷12]. It is important to point out that the above quantities (in particular *H*) are conserved until the system keeps the same ground state; when the ground state changes as a consequence of the phase transition the conserved quantities could have a sudden jump. This is not the case of eqs. (10) and (11) but is the case of eqs. (12) where *H* will be shown to decrease.

We will show now the mechanism of the transition of the system to the coherent state. Let us assume as initial condition at τ=0 :

$$A(0) = \chi_q(0) = O(q.\ f.) \sim 0$$
$$\chi_0(0) = 1 \qquad (13)$$

which means that all molecules are in their individual ground state and there is no e.m. field, apart from the quantum fluctuations. This situation holds only at zero temperature T. The case T≠0 is discussed in ref. [10]. Let us now differentiate once more the third of equations (4) , introduce into it the first two eqs. (4) and eqs. (13). We get finally the third order linear equation describing the evolution of the system just after the initial configuration:

$$-\frac{1}{2}\dddot{\mathcal{A}}(\tau)+i\ddot{\mathcal{A}}(\tau)-\mu\dot{\mathcal{A}}(\tau)+ig^2\mathcal{A}(\tau)=0. \quad (14)$$

We look for solutions like exp(ipτ) which express the usual oscillatory behavior of the e.m. field in the quantum fluctuations. By replacing the exponential in eq. (14) we get the following algebraic equation for the parameter p :

$$\frac{1}{2}p^3-p^2-\mu p+g^2=0. \quad (15)$$

which admits three solutions. When these solutions are all three real, the system is stuck in an oscillatory behavior, whose amplitude is in the order of quantum fluctuations. The most interesting case is when there is one real solution and two complex conjugate ones. It is possible to realize that this occurs when

$$g^2 \geq g_c^2 \quad (16)$$

$$g_c^2=\frac{8}{27}+\frac{2}{3}\mu+\left(\frac{4}{9}+\frac{2}{3}\mu\right)^{3/2} \quad (17)$$

From eqs. (5), (6), (7), (8) and (9) we can see that both g^2 and μ are proportional to the density d so that condition (16) is a condition on the molecule density. Equation (14) would admit a runaway exploding behavior (p having an imaginary part) when density exceeds a threshold as suggested by our initial intuitive argument. The e.m. field therefore grows up evolving toward a new configuration. It is possible to check by direct inspection [10,11] that eqs. (4) admit a nontrivial limit cycle. The physical process can be therefore summarized as follows. The system has initially the configuration where all particles are independent and the e.m. field is absent; subsequently it runs away from this configuration, driven by the exploding e.m. field and finally settles in the configuration described by the limit cycle. This limit cycle describes just a coherent situation since the fields, both e.m. and matter, have defined phases and moreover the frequency of the e.m. field happens to be equal to the difference of the frequencies of χ_0 and χ_q , which is the frequency of oscillation of the molecular system. Therefore molecules and e.m. field oscillate in tune.

Finally it is possible to show that H , after inserting the limit cycle fields, assumes a negative value, which means that the phase transition is a spontaneous process.

The e.m. field is trapped in the CD because in the limit cycle its frequency decreases sharply producing on the internal border a situation of total reflection. The field therefore falls off exponentially out of the CD, giving rise to an evanescent field. This field acts as a trap for surrounding molecules so that the molecule density in the coherent state increases until reaching a saturation value when the intermolecular distance reaches the value of the radius of the molecular hard core.

The above mathematical treatment could be applied to all levels of the molecule spectrum and the level should be selected for which the density threshold is the lowest, namely the first one to be encountered by the system when evolving from a low density situation (gaseous state).
Water has been analyzed [10,11] in the above QED framework and the coherent state has been shown to exhibit the following features:

1. the critical density has been found to be 0.31 g/cc [11] , in striking agreement with experiment
2. the selected level appears to be the 5d state at 12.06 eV , just below the ionization threshold at 12.60 eV. Consequently the size of the CD is the wavelength of the e.m. mode of 12.06 eV, namely 0.1 microns.
3. the coherent state is the superposition of the ground state with a weight of .87 and the excited state with a weight of .13 ; this means that in the CD there are .13 almost free electrons per molecule. Since in each CD, at the density of liquid water, there are several millions of molecules, we get about 1 million of almost free electrons per CD, a quite large number indeed! Coherent water becomes therefore a semiconductor, whereas noncoherent water, where all electrons are tightly bound to their parent molecules, is an insulator. Therefore the pair coherent water-noncoherent water would be an electric pile.
4. the situation at $T \neq 0$ has been examined and appears to be similar to the one predicted for superfluid Helium [17]. Thermal fluctuations are able to push out of tune some molecules which would therefore give rise to a gas like ensemble filling the interstices among CDs. Consequently the whole liquid is a mixture of coherent and noncoherent phases. The dependence of the relative abundances upon T is shown in ref. [18]. At room temperature the two fractions are almost equal.
5. near a wall the attraction molecules wall could decrease the impact of thermal fluctuations so that interfacial water should be more coherent (or much more coherent , like in the case of biological surfaces) than bulk water.
6. the interface between CDs and noncoherent fraction exhibits highly interesting properties. Solute molecules which are unable to penetrate the CDs because of the close packing of coherent molecules are confined within the noncoherent fraction. Those of them which are able to approach the CD surface could feel the evanescent field whose phase is well defined. According to a general theorem of electrodynamics, whose validity has been proven in Laser physics [19] , molecules able to resonate with an extended e.m. field where they are embedded attract strongly each other on a range having the size of the e.m. field. A selective long-range attraction able to account for the existence of the organic codes mentioned in the introduction therefore appears. Moreover the CD surface is a place able to release electrons coming out either by quantum tunneling or by mild external excitations. This feature could account for the huge discharges of electric charge from water droplets which give rise to lightning. Recently evidence for charge release in liquid water has been achieved [20]. This property could give a rationale for the dominance of redox chemical reactions among biochemical reactions [21]. The well known fact that this dominance decreases sharply in degenerative diseases could be considered as an evidence that these diseases imply a loss of coherence.
7. the existence of a field gradient at the interface between CDs and noncoherent fraction implies the appearance of a difference of electric potential across the interface. This difference has been calculated in ref. 22 the existence of a field gradient at the interface between CDs and noncoherent fraction implies the appearance of a difference of electric potential across the interface. This difference has been calculated in ref. [22] and found to be enclosed in the interval between 50 and 100 mV, in striking agreement with the observed cell membrane electric potential. Consequently we are forced to conclude that the membrane is the effect of the potential difference and not vice-versa as usually believed! The decrease of the value of the membrane potential should then be interpreted as a loss of coherence.
8. the presence of a plasma of almost free electrons within the CDs makes possible the excitation of CDs into higher states. The existence of excited coherent states of CDs , produced by the

formation of electron vortices, has been discussed in ref. [23]. The presence of a spectrum of CDs makes possible in principle an iteration of the procedure producing the emergence of coherence among molecules. Emergence of coherence among Coherence Domains, or supercoherence, has been discussed in refs. [24, 25] . Since the CD spectrum is not a permanent feature of the CD, like a molecule spectrum but depends on the dynamics of excitations, a large number of coherences is made possible to arise. Moreover the procedure could be further iterated and a hierarchy of coherences would appear bridging the gap between the initial CD as large as one tenth of a micron up to the largest biological organisms, passing through all the intermediate steps.

9. the phase φ of a coherent state is connected with the e.m. potential , not with the field, according to the equations [11]:

$$\vec{A} = \frac{h}{2\pi e}\,\mathrm{grad}\,\phi$$
$$V = -\frac{h}{2\pi e}\frac{d\phi}{dt} \qquad (18)$$

Moreover, contrary to the velocity of propagation of energy, the phase velocity does not have an upper bound. In a coherent state, contrary to the usual noncoherent situation, a change of phase produces observable consequences since could change the vorticity of the electrons and the CD spectrum. This change of phase according to eqs. (18) could be produced by the e.m. potential, without any role of e.m. fields and any exchange of energy.

3. Molecular dynamics and e.m. fields in living organisms

In this section we sketch an outline of the dynamics of living matter in the framework described in the above section.

First of all, we are now in a position to understand the physical consequence of the predominance of water (60 ÷80 % of the mass and more than 99% of the molar concentration) among the components of a living organism. The role of water molecules appears to be that of the producer of a background e.m. field able to govern the dynamics of the other biomolecules according to codes determined by the interplay of their frequencies with the frequency of the water CDs. Coherence is therefore the basic request that biological dynamics demands to water. Actually there is no point in a living organism which is farther than a fraction of a micron from a surface (either membrane or molecular backbone). Consequently we can assume that the whole biological water is interfacial water , which , as said in the last section, is almost totally coherent. Moreover, as discussed in ref. [24], this water is usually supercoherent in the sense that the water CDs are coherent among them. As a matter of fact there is experimental evidence [26] that water within cells is in a seemingly glassy state , meaning that is totally coherent, such as water at low temperature (T<220 K) [18] . As discussed in depth in refs. [25, 27], water CDs are devices able to store large amounts of energy, collected from the environment, and transform them in coherent energy; in a nutshell CDs are able to transform n quanta having a frequency ν into one quantum of frequency n ν . This happens since lifetimes of the coherent excited levels of CDs are very long [23] so that many excitations produced by the external noise could pile up in the CDs producing higher and higher coherent excitations. Starting from a very low frequency noise it is therefore possible [25] to produce high energy excitations able to resonate with the electronic degrees of freedom of molecules, activating them chemically. The following picture therefore emerges: at a given time coherent water oscillates with a particular frequency, all the molecules able to resonate with this frequency get attracted and become able to interact chemically, the energy output of the chemical reaction is assumed by the CDs which consequently change their frequency, becoming able to attract new molecules. Electron transfer from CDs to biomolecules plays an important role in this biochemical scheme. More details can be found in ref. [21].

A corroboration of the scheme outlined above is provided by the experiments performed by the Montagnier group [6]. An analysis of these experiments in the framework of QED has been presented in ref. [28]. In these experiments highly diluted aqueous solutions containing bacterial DNA sequences are found to produce e.m. fields in the low frequency (500÷3000 Hz) interval, provided that an e.m. very low frequency (a few Hertz) noise is present; in absence of this noise nothing happens. Experimental analysis gives evidence that those fields are coming from water surrounding DNA; since water is at the origin of the observed e.m. fields it is clear that by diluting the solutions fields get enhanced.

The fields emitted by the culture are supplied to pure water contained in nearby vessels for a suitable time (about one day). When the appropriate biomolecules are poured in the imprinted water, original microorganisms appear in the solution. The fields emitted by the organisms in the first solution have induced in the irradiated pure water a spacetime distribution of the phase such that it drives the formation of similar organisms, when the appropriate biomolecules become available. These experiments appear to be not inconsistent with the theoretical framework outlined above.

4. Dynamics of living organisms governed by the phase.

The outcome of the above section implies that in a biological system water CDs are kept coherent among them on a large scale (supercoherent) by the interplay of the frequency of their trapped e.m. fields with the energy output of chemical reactions. In this way a macroscopic phase emerges, which in general depends on space and time. This phase function is the continuation on a larger scale of the phases of the single CDs, which in principle don't depend on space in their inside. According to Quantum Field Theory (QFT) [12] an uncertainty relationship connects the phase and the number of elementary components:

$$\Delta N \, \Delta \Phi \geq \tfrac{1}{2} \qquad (19).$$

Since the optimal requirement for a living organism is the vanishing of $\Delta\Phi$, ΔN is compelled to become as large as possible. However the number of components of a living organism, although quite large, is finite so that the organism is forced, say, to go out of himself, to share its oscillation with the oscillation of the largest possible part of the environment. In this way the need of being an healthy individual compels the organism to have a dialogue with other individuals, species, natural objects. The deep meaning of each organism is therefore the dialogue with a large part of the external world, which then becomes its own Double [29]. Contrary to the objects described by Classical Physics, a coherent quantum system is not defined in isolation, but gets defined by the array of its relationships.

In conclusion the biological organism, being coherent, can interact with environment in two basically different ways:

1) through the conventional exchanges of energy which amount to the application of mutual forces. This mechanism of interaction obeys of course to the causality principle, since energy cannot travel faster than light.

2) Through the sharing of the phase with other coherent systems (biological organisms), which amounts to the establishment of a resonance with them. The phase velocity is not bounded above and can be larger than c. As we will discuss in next section, this relationship could provide a rationale to the property of synchronicity introduced by Jung [8].

Eqs. (18) tell us that the phase is connected with the e.m. potential in a mutual relationship so that we could be able to change the phase of a biological organism by applying an e.m. potential [30, 31, 32].

It is conceivable that the e.m. potential could coincide, as suggested by Mae-Wan Ho [33], with the morphogenetic field which is produced by each growing organism and govern its further development. We could interpret now the biological effectiveness of very weak e.m. and magnetic fields , mentioned in the introduction [7], just by assuming that the agent at work in the interaction is not the field but the potential and the mechanism of interaction is the phase-sharing.

Following this line of thought, it is possible now to understand the extraordinary impact of specially prepared waters [34, 35] . Let us describe the procedure of preparation reported in ref. [34]. We quote verbatim:

The following operations are performed:

1) *1 litre of carbonated water enclosed in a 2 litre container is strongly successed and agitated manually.....Hydrated algae and triturated leaves are dropped in the successed water.*
2) *The mixture of water and vegetables is successed again 100 times whereas constantly illuminated by the light of the bulb lamp. Subsequently it rests for 45 minutes, still illuminated. This is the necessary time for the photosynthetic system to reach the peak of self-adaptation.*
3) *The mixture is poured in the transparent vessel. This system becomes the environment able to imprint the plate of glass or plastic. The imprinting is performed as follows. The plate is heated in an oven (up to 200° centigrades in the case of glass and 80° in the case of plastic). The heated plate is hit by the lamp light beam, which has traversed the vessel containing the mixture. The exposition is kept until the plate cools down to the room temperature. This procedure is assumed to produce the imprinting, namely to induce the transition of the inner configuration of the plate to a coherent excited state.*

It is important to stress the following important points in this procedure:

1) a large number of vegetal species is necessary in order to have a highly nontrivial phase function; this accounts for the important role of biodiversity for the health of each individual
2) the irritation of the vegetal leaves is meant to increase as much as possible what could be termed "the instinct for survival" of each species so that their coherent activity is strongly enhanced
3) water receives all these phase signals and modifies correspondingly its own phase distribution
4) because of the long life-time [23] of the coherent excitations of water, this imprinted phase lasts a long time
5) when this phased water is put in contact with another coherent system, in particular a biological organism, shares its phase with it.

In order to store the phase of this special water for a long time glassy materials are imprinted which in turn are used to imprint water and biological organisms when necessary. The mechanism of such phase sharing has been checked experimentally. We quote again from ref. [34]:

The qualitative measurements of the heats of mixing have been performed in the Department of Chemistry of Naples University in the laboratory of Prof. V. Elia by using a flow-mix calorimetry (LKB standard flow microcalorimeter at 25°C with a sensitivity of 0.1 μW). Two vessels, one containing a solution of Sodium Hydroxide and the other containing demineralised water, are connected by a two way peristaltic pump with the calorimeter where the two fluids mix in a cell whose temperature is carefully measured. The heat of mixing is derived by the observed variation of temperature. This procedure allows to have the value of the heat of mixing in real time. By observing this power-time graph we can see that in a normal case we have a transient phase and a stationary phase. If the instrument parameters don't change with time, the stationary phase is represented by an horizontal line, that could last a very long time. A change in the nature of the tested liquid would imply a new stationary phase that would correspond to the heat of mixing of the new fluid. Should the two mixed fluids be identical, the power-time graph would coincide with the baseline.

In Figure 1 the power-time graph shows the outcome of our experiment. We start with normal demineralised water. The arrow 1 denotes the start-up of the experiment. As expected the first part of the graph shows the reaching of the stationary value of the heat of mixing of normal water. At the time marked by the arrow 2 an imprinted plate is slipped below the vessel containing normal water, without the occurrence of any exchange of energy between the plate and the vessel. The graph shows an instant decrease of the heat of mixing, as if water had changed nature! At the time denoted by the arrow 3 part of the water in the vessel is extracted by reducing its volume to 20 ml. A sharper decrease of the heat of mixing is got, indicating a new reorganisation of water. This result is surprising since an intensive quantity as the molar heat of mixing seems to become an extensive one. In conclusion the observation can be summarized as follows: the heat of mixing decreases when water is in contact with the "informed material", which means that presumably a flow of information and hence of negentropy has occurred. The amount of the decrease depends on the volume of water….

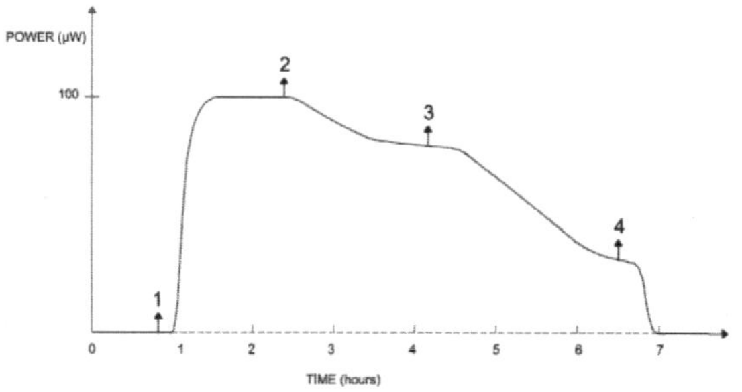

Figure 1

As a control at the time marked by the arrow 4 the plate has been removed and the vessel of water has been replaced by a vessel containing the same titrated alkaline solution. The graph recovers the initial baseline showing that no variation has occurred in the sensitivity of the detector. It can be observed also that the sensitivity of the calorimetric device (0.1 μW) is very high in relation to the differences of the power values between normal and treated water (in the scale of 100 μW), as can be seen in the curve of the calorimetric graph; hence it can be assumed that the phenomenon is real.

This experiment suggests that the phase relationship between coherent systems is a physically sound concept.

5. Evidence of a biological interaction depending on phase

In this section we will exhibit some evidence of phase relationships between living organisms, extracting them from the tradition of healing techniques belonging mainly to the Eastern medicine, but widely spread today in the West, too. In the Japanese tradition Hado is the innate transforming power of each thing and each living being. "Shiatsu" means "finger pressure" and indicates a discipline connected with traditional Chinese medicine. Hado Shiatsu is a therapeutic approach based on "empathetic connections", that means "resonate with" namely the onset of a phase resonance between the therapist and the patient. In this approach results happen without any physical or mental exertion. A sort of "meditative state", where phase is no more chaotic, tunes the phases of the therapist and the patient, between them and with healthy environment. A better life quality is then supported by a sharper definition of the phase of each organism. In this interaction the connectedness of the organism is essential, not the amount of exchanged energy, which could on the contrary be harmful when its amount exceeds the threshold of the "energy gap", namely the amount of energy necessary to make the coherent system boiling and therefore losing coherence.

Important changes are consequently induced in the organism by very gentle touch and very small stimuli. This fact is in agreement with the Weber and Fechner law [36], which states the proportionality of the response of a living organism to the logarithm of the stimulus. Consequently very small stimuli induce very large negative responses, namely responses directed inwards. This law can be understood in the frame of a quantum interaction based on a phase exchange instead of an energy exchange.

It is worthwhile to observe that coherent fields, and hence the phase, are trapped in the coherent regions of water which are present mainly around the long chains of macromolecules; these chains are the microtubules in the cellular medium whereas in the extracellular medium are the chains of the collagen. Should we identify the phase of the field travelling along the shells of water surrounding the chains with what the Eastern tradition calls Qi, which is considered the acting principle of life, then the above water shells could be coincident with the meridians of the Eastern tradition. This point of view has been recently suggested in refs [37, 38].

In the framework of a biology based on coherence, the health of an organism depends on his ability of constructing a well defined phase during the mainly unpredictable vicissitudes of life; the phase has of course to change during the experiences of life but the organism should always be able to have a well defined phase. The external stimuli should always give rise to well defined changes of phase. When this happens, we say that this organism is able to self repair. The capability of self repair of an organism is enhanced by the presence of a well defined phase in what we have called his Double, namely the part of the environment he is connected to. This property coincides just with the Eastern concept of Hado. The connection of the organism to this phase, which could come from loving fellows, from plants, animals, Nature as a whole, but also by dreams and ideas which are the coherent manifestation of a coherent organ, the brain, would give him the ability to self repair.

In a therapeutic setting [39, 40], the therapist acts as a main part of the Double of the patient (transfer), and in some sense also the opposite occurs (counter transfer). They should give rise to a phase relationship, so no exchange of energy is needed, but only a touch expressing the mutual connection of the two. Should the two phases tune together, the phase of the patient, which is disturbed by the loss of coherence produced by the illness, becomes more clearly defined and a process of recovering starts. However, in case of severe diseases, the chemical structure of the organism could become so modified to be unable to resonate with his background water; in these cases of course a chemical repair seems to be necessary. The therapist, for example through a gentle touch, supports his patient to resonate with the environment, namely to dialogue with it. Being a resonance, both partners are on an equal ground, so that the therapist could feel in his own body the phase of the patient, disturbed by the disease. Without any talking or "objective" test, the therapist is able to feel the disease of his patient and to direct his attention to it. He is therefore able to talk to the disease of his patient, who in turn can reply to the therapist in their established correlation. The human strategy outlined above could be always supplemented, and in extreme cases replaced, by artificial means able to deliver a defined phase, such as weak magnetic fields (Magnetotherapy), the phased water described in the last section, low intensity microwave irradiation and so on.

Physics of coherent systems suggests the opening of new horizons to Medicine.

References

[1] Alberts B., Bray D., Lewis J.,Raff M., Roberts K., and Watson J. D., 1994, *Molecular Biology of the Cell*, 3rd edition, Garland Science, New York
[2] Edelman G. M., 1984, Cell adhesion molecules: a molecular basis for animal form. *Scientific American*, April 1984, 80
[3] Barbieri M., 2004, *The Organic Codes*, The University of Cambridge Press.
[4] Fröhlich H., 1980, *Advances in Electronics and Electron Physics*, **53**, edited by L.Marton and C.Marton, New York: Academic Press, 85
[5] Popp F.-A., 2002, Biophotonics, A Powerful Tool for Investigating and Understanding Life, in Dürr H. P., Popp F.-A., Schommers W., (eds) *What is life? Scientific approaches and Philosophical Positions*, World Scientific, Singapore, New Jersey, London
[6] Montagnier L., Aïssa J., Ferris S., Montagnier J.-L., Lavallée C., 2009, *Interdiscip. Sci. Comput. Life Sci.*,**1** , 81
[7] Devyatkov N. D., 1974, *Sov. Phys. Usp*, **16**, 568
[8] Jung C. G., 1952, *Synchronicity. An Acausal Connecting Principle*. Published in 1972 by Routledge and Kegan Paul
[9] Del Giudice E., Preparata G., Vitiello G., 1988, *Phys. Rev. Lett.*, **61**, 1085
[10] Arani R., Bono I.,Del Giudice E., Preparata G., 1995, *Internat. J. Mod. Phys. B*, **9**, 1813
[11] Bono I., Del Giudice E., Gamberale L., Henry M., 2011, Emergence of the coherent structure of liquid water, *Water*, in press
[12] Preparata G., 1995, QED Coherence in Matter, *World Scientific*, Singapore, New Jersey, London
[13] Fröhlich H., 1968, *Int. J. Quantum Chem.*, **2**, 641
[14] Nernst W., *Uber einem versuch von quantentheoretischen betrachtungen zur annahme stetiger energieanderungen zuruckzukehren*. Verh. Deutsche Physikalische Gesellschaft **18**, 83-116 (in German)
[15] Hilborn R. C., 1982, Einstein coefficents, cross sections, f values, dipole moments, and all that, *Am. J. Phys*, **50**, 982-986
[16] Lamb W. E., Retherford R .C., 1947, *Phys, Rev.*, **72**, 241
[17] Tisza L., 1947, *Phys. Rev.*, **72**, 838
[18] Buzzacchi M., Del Giudice E., Preparata G., 2002, *Int. J. Mod. Phys. B*, **16**, 3771
[19] Askar'yan G. A., 1974, *Sov. Phys. Usp*. **16**, 680
[20] Germano R., Tontodonato V., Hison V.C., Cirillo D., Tuccinardi F.P., 2011, *Oxhydroelectric Effect: Electricity from Water by Twin Electrodes*, Key Engineering Materials, in press.
[21] Voeikov V.L., Del Giudice E., 2009, *WaterJournal.org*, **1**, 52
[22] Marchettini N., Del Giudice, E., Voeikov V.L., Tiezzi E. , 2010, *J. Theo. Bio.*, **265**, 511
[23] Del Giudice E., Preparata G., 1998, A new QED picture of water: Understanding a few fascinating phenomena, in Sassaroli E., Srivastava Y., Swain J., Widom A., (Eds.) "*Macroscopic Quantum Coherence*", World Scientific, Singapore, New Jersey, London, 108
[24] Del Giudice E., Tedeschi A., 2009,*Electromagn. Biol. Med.*, **28**, 46
[25] Del Giudice E., Spinetti P.R., Tedeschi A., 2010, *Water*, **2**, 566
[26] Trepat X., Deng L., An S. S., Navajas D., Tschumperlin D. J., Gerthoffer W.T., Butler J. P., Fredberg, J.J, 2007, *Nature*, **447**, 592
[27] Del Giudice E., Pulselli R.M., Tiezzi E., 2009, *Ecol. Model.*, **220**, 1874
[28] Montagnier L., Aissa J., Del Giudice E., Lavallée C., Tedeschi A., Vitiello G. , 2011, *Journal of Physics: Conference Series* **306**, 012007

[29] Vitiello G., 2001. *My Double Unveiled.* John Benjamins, Amsterdam.

[30] Brizhik L., Del Giudice E., Jorgensen S.E., Marchettini N.,Tiezzi E., *Ecological Modelling* **220** (2009) 1865

[31] Brizhik L., Del Giudice E., Tedeschi A., Voeikov V., 2011, *Ecol. Mod.*, doi:10.1016/j.ecolmodel.2011.05.017

[32] Smith C.,1994, *Electromagnetic and magnetic potential bio-information and water, in Ultra high dilution: Physiology and Physics.*, Klüwer Academic, Dordrecht, Netherlands, 187

[33] Ho M.W., French A., Haffegee J., Saunders P.T., 1994. Can weak magnetic fields (or potentials) affect pattern formation? In: Ho M.W., Popp F.-A., Warnke U.,(Eds.), *Bioelectrodynamics and Biocommunication.* World Scientific, NewJersey, London, Singapore, 204

[34] Tedeschi A.,2010, Is the living dynamics able to change the properties of water? *Int. J. Des. Nat. Ecodyn.*, **5**, 60

[35] Katsir Y., Miller L., Aharonov Y., Ben Jacob E., 2007, The effect of rf-irradiation on electrochemical deposition and its stabilization by nanoparticle doping. *J. Electrochem. Soc.*, **154**, D249

[36] Chisholm H. (Ed.), 1911, Weber's Law, in *Encyclopaedia Britannica*, 11th Ed.-Cambridge University Press

[37] Brizhik L.S., Del Giudice E., Maric-Oehler W., Popp F.-A., Schlebusch K.-P. , 2009, *Electromagnetic Biology and Medicine*, **28**, 28

[38] Bajpai R., Brizhik L., Del Giudice E., Finelli F., Popp F.-A., Schlebusch K.-P., 2010, *J Acupunct Meridian Stud* **3**, 291

[39] Stefanini P., 2011, Qi in Shiatsu, in "*Energy Medicine East and West: a natural history of qi*", Churchill Livingstone, Elsevier, pages 211

[40] Stefanini P., 2011, "*Quantum Physics sheds some light on Hado Shiatsu*" Poster for EQBS - International Symposium "Electromagnetic fields and Quantum phenomena in the Biological Systems" - May 20, Poznan University of Medical Sciences, Poland: <http://www.eqbs.ump.edu.pl/ctb/stefanini_eqbs_poster.pdf>

Journal of Physics: Conference Series **329** (2011) 012002 doi:10.1088/1742-6596/329/1/012002

A semi-classical approach of the relationship between simple cells' size and their living temperature limits based on number fluctuations of water coherence domains

E A Preoteasa[1,3] and **C Negoita**[2]

[1] Department of Life and Environment Physics, Horia Hulubei National Institute of Physics and Nuclear Engineering, P.O. Box MG-6, 077125 Bucharest-Magurele, Romania

[2] Laboratory of Microbiology and Immunology, Faculty of Veterinary Medicine, 105 Spl. Independentei, 050097 Bucharest 5, Romania

[1] E-mail: eugenpreoteasa@gmail.com

Abstract. Starting from the concepts of the quantum electrodynamics (QED) theory of coherence domains (CD) in water we propose a model aimed to evaluate the relationship between the size and the living temperature limits for simple, small cells. Cells are described as spherical potential wells with impenetrable walls, with CDs moving inside. The radius of the spherical potential well was estimated for physiological temperatures and the results match to bacteria and yeasts cells' size. As a CD in the spherical cell exerts a force upon the membrane, a 'gas' formed by CDs bears a pressure on the walls. A classical statistical stability condition relates this pressure to cell volume and to the relative fluctuations of the CD number, allowing the evaluation of an upper temperature limit as a function of cellular volume. Assuming further that the CDs in the living cell form together a coherent state, the number-phase incertitude relationship (Heisenberg limit) applies. The maximum coherence between CDs is found in the ground state, a picture consistent also to Fröhlich's postulate. For a given phase dispersion, a lower temperature limit as a function of the cell volume is found. Although we neglected the rod-like shape of certain bacteria and the presence of nucleus in yeasts, the biological data of volume and optimal living temperature intervals fit well to our model's predictions. Moreover the larger the cell volume, the higher are the number of CDs *and* the coherence of their system. In addition we suggest a new classification criterion for small cells based on model's parameters, which show discontinuities between Gram negative and positive microorganisms as well as between prokaryotes and the smallest eukaryotes.

[3] To whom any correspondence should be addressed.

1. The problem of the cell size. An unexplained simple biological fact

Dimensions represent perhaps the most elementary characteristic of the living cells. They are apparently macroscopic objects, with typical sizes of 1 – 100 μm [1]. Smaller biological objects like viruses, proteins or nucleic acids are not actually living objects. Cells are far larger than objects of the microscopic, quantum world such as atoms and molecules, and also larger than nanoparticles, which define the mesoscopic scale (Figure 1).

The dimension of the cells is basically an empirical fact from the standpoint of traditional biology. Molecular biology sensibly postulates that a lower size limit is explained by a minimum number of $5.10^2 – 5.10^3$ different enzymes, while an upper size should be limited by the efficiency of metabolism, by the surface to volume ratio and by the dimensions of vacuoles. But this approach, though significant, is also based indirectly on certain empirical facts and therefore only displaces the problem, failing at least in part to look at the cell size as a basic feature of relevance for the phenomenon of life.

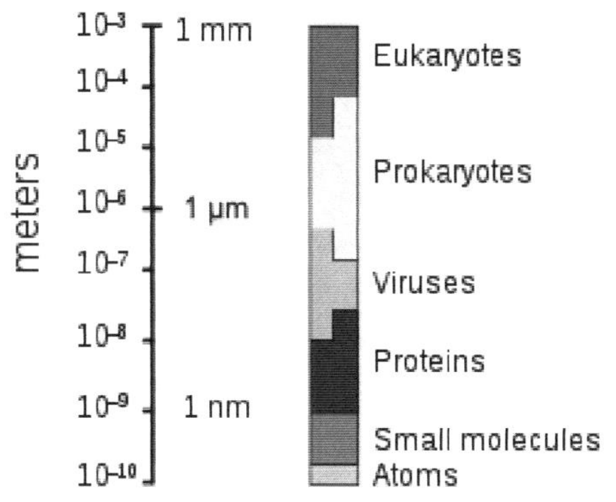

Figure 1. Size scale of biological objects.

2. Physical views on the cell size and the quantum biological approach

From the physical point of view, cell size is not a parameter reducible to the empirical data of molecular and cell biology, which in the large frame of Nature appear elementary, particular and casual, but instead it should be an essential feature of the living matter related to a specific kind of dynamics. Schrödinger [2] pointed out that in a living organism molecules must cooperate, and this condition requires a volume large enough to ensure the cooperation of a sufficient number of molecules against thermal agitation. Along this line the dissipative structures theory [3] is looking up to the cell as a giant density fluctuation and shows that its dimensions must exceed the Brownian diffusion which takes place during the lifetime of the cell. These two views look at the cell as a whole, without taking into account the molecular details inside the cell.

The same perspective is assumed by quantum biology, which proposes to understand the characteristics of the cell and multicellular organisms based on their integrative features, such as the emergence of self-organization and hierarchies of collective order above the molecular level (e.g., [4]). This approach tries to answer whether quantum mechanics plays a non-trivial role in understanding life or not, beyond the quantum chemistry calculations of molecular structure and properties. An answer based on quantum mechanics to the cell size problem was given in a model proposed by

Demetrius [5], which connects size to metabolism, probably the most essential feature of life [6]. The model proved itself capable to derive the empirical allometric relationship, $P = \alpha W^{\beta}$, where W is the weight of a cell or organism and P the metabolic rate, and $\beta < 1$ is an dimensionless scaling exponent which depends on the metabolic efficiency. The model applies Planck's energy quantization rule and statistics for the electron and proton transfer taking place in the coupled cell respiration and oxidative phosphorylation according to the chemiosmotic theory.

However, the dependence of the cell size on metabolism in the quantum model of Demetrius leaves unanswered the question of the nature of metabolic rate (which may be interpreted as an observable of physical origin specific to the uniquely complex system which is the living cell or, alternately, a primary fact of the biological realm).

Another valuable physical approach of the cell size and geometry problem has been proposed recently ([7], to be published in this volume). These authors studied the oscillations of microtubules' network in the dividing and non-dividing eukaryotic cell. The tubulin heterodimer was approximated as an elementary electric dipole performing coherent or quasi-coherent electromechanical longitudinal oscillations with low damping in the surrounding organized water. The size and shape of the cell was related to the electric field calculated around the oscillating microtubule network.

We do not intend to search here an answer to the question of the nature of metabolic rate, a challenge which is neither simple nor obvious. Rather, we propose here a new quantum model for the evaluation of cell size without any hypothesis on metabolism, along a line of attack related to our previous studies [8]. And while the microtubules continue to stimulate quantum mechanical models [9-11], our study is focused on the ubiquitous component of living cells, either prokaryotic or eukaryotic: water.

3. The QED theory of coherent domains of water

The philosophy of our approach finds support in a definition of life given by Del Giudice: "the protagonists of the biological process … [are] mesoscopic collectives characterized by millions of molecules acting in unison in wide regions of the space for long intervals" [12]. This view is in agreement to the theory of long-range coherence of dipole oscillations developed by Fröhlich [13-16] which helps understanding certain fundamental integrative features of living organisms. The very long-range character of this theory overcomes the fragmentary picture of molecular biology which admits information to be carried only by chemical messengers involved in short-range specific interactions and moving only at random. An active role for water in these long-range correlations within the cell, as opposed to its mere function of a cellular environment, is sustained by models of water collective dynamics based on H_2O molecule's electrical properties. Moreover this role is related to inhomogeneous structures formed in water by self-organizing phenomena; we also proposed a model of ionic plasma oscillations accounting for density oscillations in liquid water [17]. As characterized by Ben-Jacob, water is „an active substance saturated with life-giving properties" [18], in accordance to the quantum electrodynamics (QED) theory of coherence domains (CD) of water, described as clusters of electrical dipoles self-organized to oscillate in-phase with each other [19-24].

We note in this connection that quantum coherence is a theoretical concept describing a general property of living and non-living matter [24]. Most important, quantum coherence has been already demonstrated experimentally in plant photosynthesis, where light absorbing molecules capture and transfer energy according to quantum-mechanical laws at temperatures up to 180 K [25], and in a model system represented by a wet conjugated polymer where coherent intrachain energy migration takes place at room temperature [26]. Recent results obtained on the *cryptophytae* marine algae by 25 fs laser pulse irradiation and photoecho spectroscopy demonstrated that there exist long-lasting oscillations exciting a coherent superposition of antenna's molecules vibrational-electronic eigenstates, with correlations across 5 nm long distances among dihydrobiliverdin pigments 'entangled' together by quantum coherence, even at 294 K [27]. *In vitro* studies on various water-containing systems evidenced experimentally a good number of phenomena consistent to the QED theory of CDs. These include among others the water bridge formed in ~15 kV static electric field [28], the long-lasting perturbations induced by weak extremely low frequency electromagnetic fields in solutions of glutamic acid [29], the ion cyclotron resonance (ICR) phenomenon in biological and biochemical systems [30, 31] and the sharp changes in conductivity for amino acids solutions exposed to ICR magnetic fields tuned to the *q/m* ratio of amino acid [32], the gel-state water extending up to ~1-10 μm at interfaces [33], and the phenomenon of autothixotropy evidencing the formation of fragile but trully macroscopic (~cm) structures in water by an ‚ephemeric polymerisation' [34].

A low effective mass (excitation energy) of the water CDs was evaluated to about 12.1-13.6 eV, much lower than *e.g.* the electron mass [21, 24]. Accordingly, their wave-like properties are strongly enhanced because their de Broglie wavelength $\lambda = h/mv$ will be much longer and comparable to the cell size. These properties together with the boson nature of the CDs, as well as with the capacity of CDs to interact with oscillating dipoles up to a few μm apart through the attractive r^{-3} Fröhlich potential [35] endowed with resonant character of Askaryan forces' type [36] sustain the view of a "supercoherence" state, a coherence spanning all the CDs throughout the cell water. Thus a supercoherent network of water CDs in the cell is postulated to play the role of command and control for the huge number of molecular interactions which take place at the right time and place through the life cycle of the cell. Long-range forces may provide an explanation for rouleaux formation [37] and anomalous light diffusion [38] of erythrocytes, dielectrophoresis of small dielectric particles by cells [39], and variations of the optical spectra of a nutrient in suspensions of yeast cells as a function of cell density [40]. A certain type of supercoherent state, or a Fröhlich condensate, has been unambiguously evidenced in living cells by a 8.085-MHz resonance by Pokorny's group [41, 42], who identified microtubules' network oscillations as a possible candidate.

4. Previous models and scope of the present study

Taking as a starting point the concept of water CD elaborated in the (relativistic) QED theory of the Milano group, we previously proposed a few different but related models explaining the size of the cell in the frame of non-relativistic quantum mechanics [8, 43]. To this purpose we postulated that the cell membrane is an impenetrable or semi-penetrable wall for the CDs and the shape and size of the cell selected during evolution are such as to represent a potential well or a resonant cavity imposing to the supercoherent CD system a specific type of dynamics, which is consistent to an integrated and unitary informational control of the molecular dynamics inside the living cell. With this heuristic approach and taking advantage of the low effective mass of CDs ($m_{eff} \cong 13.6$ eV $\cong 2.4 \times 10^{-35}$ kg) we evaluated the lower and higher cell size limits of spherical cells by models based on Bose-like condensation, on CD translation in a spherical well, and on an oscillator consisting of two interacting CDs [8, 43]. In our potential wells and harmonic oscillator models we assumed, for the sake of cell stability, the second energy level to be thermally inaccessible from the first (ground) level, where the system remains during cell life. The results matched well the size and shape of bacteria, yeast and erythrocytes, and explained also why the D_2O toxicity is higher for eukaryotes. For instance in a model of water CDs moving in a spherical well (by analogy to the nucleons in the shell model of the atomic nucleus) the estimated upper limit for the cell radius r_0 is:

$$r_0 \leq \frac{\alpha \cdot \hbar}{\sqrt{m_{eff}\, k_B T}} \tag{1}$$

where $\alpha = \pi = 3.1416$ for the cell with impenetrable walls (describing the cell membrane lipid bilayer with its highly hydrophobic inner region) and $\alpha < \pi$ for the cell with semipenetrable walls (which accounts also for the hydrophilic protein pores in the membrane), $\hbar = h/2\pi = 1.05457 \times 10^{-34}$ J·s = 6.58212×10^{-16} eV·s is the reduced Planck's constant, $k_B \cong 1.380\ 10^{-23}$ J·K^{-1} $\cong 8.617\ 10^{-5}$ eV·K^{-1} is the Boltzmann constant, and T is the absolute temperature. For the model of two coupled water CDs forming a harmonic oscillator (with some particular assumptions) the maximum cell radius is given by:

$$r_0 \leq \frac{4}{\sqrt{3}} \cdot \frac{\hbar}{\sqrt{m_{eff}\, k_B T}} \tag{2}$$

Note that in relation (2) the numerical factor before $\hbar\,/(m_{eff} k_B T)^{1/2}$ changes from $4/3^{1/2} = 2.309$ to $(3^{1/2}/2 + 2\ 2^{1/2}) = 3.694$ when the assumptions in the harmonic oscillator model are slightly changed. In general, depending on the model, one obtains similar formulae for the maximum cell radius which differ only by a numerical factor. They all evidence a relationship between the maximum radius of spherical cells and the living temperature of the cell:

$$r_0 \leq \text{const}\ T^{-1/2} \tag{3}$$

However, cells have various shapes, and in order to facilitate comparisons of various cells we may introduce the maximum cell volume as a function of temperature:

$$V \leq \text{const}\ T^{-3/2} \tag{4}$$

Similarly, one may search for models expressing a relationship of minimum cell volume with temperature. Conversely, it is convenient to express the upper and lower living temperature limits in relation to the cell volume:

$$T \leq \text{const}\ V^{-2/3} \tag{5a}$$
$$T \geq \text{const}\ V^{-2/3} \tag{5b}$$

Here we search for such T-V relationships for the lower and upper limits of temperature which, together with temperatures close to the boiling and freezing points of water which restrict the living conditions of organisms:

$$T \leq 383\ \text{K} \tag{6a}$$

$$T \geq 263 \text{ K} \tag{6b}$$

define a close surface, where one can ideally expect to localize all living cells. To this purpose we evaluate the cell size using the spherical potential well, and take advantage of the QED nature of the water CDs due to which their instantaneous number N in the cell is defined only as an average value with associated incertitude ΔN. Then we apply properties of particle number fluctuations derived from classical statistical physics to the CDs' number fluctuations to deduce an expression of the type of equation (5a) of the upper temperature limit, and the number-phase incertitude relationship (Heisenberg limit) which characterizes quantum relativistic systems of particles to derive the lower temperature limit of the form of equation (5b). Because we use both quantum and classical concepts, the present model may be considered a semi-classical one. The theoretical predictions are compared to biological data. Our models are aimed mainly at prokaryote cells, due to their much simpler internal structure which does not contain a nucleus; however, smaller simple eukaryotes are also considered.

5. Upper temperature limit *vs*. volume: A model based on CD number fluctuations and classical stability conditions

We consider an average number N of CDs in a spherical cell with an impenetrable plasmatic membrane and combine this simple model which applies to each CD with a thermodynamic stability condition for biphasic systems in equilibrium from classical statistical physics.

In the simplest case of a CD inside a spherical well of radius r_0 with *impenetrable* walls (namely, *infinite* deep potential well, $U_0 \to \infty$, with the membrane described also as infinitely thick), the translation energy of each CD is purely kinetic and quantized on an infinite number of discrete levels $E_1, E_2, \ldots$:

$$E_n = \pi^2 \cdot \frac{\hbar^2}{2 m_{eff} r_0^2} \cdot n^2 \qquad (n = 1, 2, \ldots) \tag{7}$$

In our previous model of a CD moving in a spherical potential well [8], we postulated $n = 1$; for N coherence domains in the cell, this means that each CD occupies the same ground state, a situation which is allowed due to the boson nature of the coherence domains of water which accordingly are submitted to the Bose-Einstein statistics.

At the same time each CD in the spherical well collides the walls and such a single quasi-particle exerts a force upon the walls given by [44]:

$$f = -\frac{\partial H}{\partial r_0} = -\frac{\partial E_n}{\partial r_0} = \frac{\pi^2 \hbar^2}{m_{eff} r_0^3} \cdot n^2 \tag{8}$$

where H is the Hamilton function of the CD in translation movement. The pressure exerted by a "gas" formed by the N CDs present in the average in the intracellular water on the walls of the spherical cell of area $A = 4\pi r_0^2$, due exclusively to the water CDs, is therefore:

$$p = -\frac{Nf}{A} = \frac{\pi \hbar^2 n^2 N}{4m_{eff}\left(\frac{3Nv}{4\pi}\right)^{\frac{5}{3}}} \tag{9}$$

where we used the reduced volume per CD, namely $\upsilon = V/N$.

We now turn back to the QED nature of CDs and to the N quasi-particles in the cell. Their instantaneous number is not precisely defined, and only their average number N is a real observable, which can be measured with an incertitude (standard deviation) $\Delta N = (\langle N^2\rangle - \langle N\rangle^2)^{1/2}$. In an intuitive view, the CDs number 'flickers' with a fluctuation ΔN by annihilation in the 'vacuum' represented by the surrounding unordered water and by creation from the same environment. This allows an analogy with the situation described in the classical statistical physics for a liquid in thermodynamic equilibrium with its vapors in a closed vessel. The liquid phase corresponds to the 'gas' of CDs within the cell and the vapor phase to the 'vacuum' of disordered water molecules. The number N of molecules in the liquid phase (which now in the classical description is the instantaneous number, but this aspect is overlooked) undergoes fluctuations of size ΔN. The necessary and sufficient condition which ensures the stability of the system and keeps the system far of the critical point, comes up to [45]:

$$\frac{\partial p}{\partial v} < 0 \tag{10}$$

We impose to the CDs gas this stability condition, which is immediately verified by the derivative $\partial p/\partial \upsilon$ calculated using equation (9):

$$\frac{\partial p}{\partial v} = -\frac{5\pi^2}{9}\left(\frac{4\pi}{3}\right)^{\frac{2}{3}} \frac{\hbar^2 n^2}{N^{\frac{2}{3}} v^{\frac{8}{3}}} \tag{11}$$

In the classical statistical physics the relative fluctuation is given by [45]:

$$\frac{\Delta N}{N} = \frac{\sqrt{\langle N^2\rangle - \langle N\rangle^2}}{N} = \sqrt{\frac{k_B T}{v^2\left(-\frac{\partial p}{\partial v}\right)N}} \tag{12}$$

and substituting here the above result of equation (11) we get for the relative fluctuations of the CD number in the spherical cell:

$$\frac{\Delta N}{N} = \frac{3}{\sqrt{5}} \frac{\sqrt{m_{eff} k_B T}}{\pi \hbar} \frac{r_0}{n\sqrt{N}} \tag{13}$$

where we impose $n = 1$ in numerical calculations. Note that implicitly $\Delta N \sim r_o N^{1/2}/n$; this evidences an obvious conclusion, namely that a living cell must be always of finite size, as the absolute fluctuations may never be infinite. At the same time the relative fluctuations will decrease as $N^{1/2}$, pointing that the larger the number of CDs in the cell the higher the degree of order in its dynamics; thus eq. (13) gives a quantitative expression to the assumption of more than one single CD synergically correlated in every living cell. More important, we obtained an expression which connects T and r_o, and the relationship between these two observables depends on the choice of the relative fluctuation $\Delta N/N$. For the stability of the CD system in the cell we impose a higher limit for the relative fluctuations:

$$\frac{\Delta N}{N} < 1 \tag{14}$$

and thus the maximum radius of the spherical cell r_o is defined for a given temperature T. With the formula of the spherical cell volume, $V = (4\pi/3)r_0^3$, we obtain an *upper limit* for the living temperature of a cell as a function of its volume:

$$T < 5\left(\frac{\pi}{3}\right)^{\frac{8}{3}} 2^{\frac{4}{3}} \frac{\hbar^2}{m_{eff} k_B} \frac{n^2 N}{V^{\frac{2}{3}}} \tag{15}$$

which is of the form of inequality (5a) as expected, i.e., of the form of results obtained by previous models (equations 1 and 2). For instance with $n = 1$ as used before and with $N = 2$ – which is the minimum number of CDs which may correlate themselves in a cell and thus the minimum number compatible with life – we get after substituting the numerical values:

$$T < \frac{9468}{V^{\frac{2}{3}}} \tag{16}$$

where V is in μm^3. The corresponding maximum radius of a cell living at 310 K (37 °C) is:

$$r_0 < \frac{\sqrt{N}}{1.314} \tag{17}$$

and for $N = 2$ a maximum radius of 1.08 μm results, as observed for many bacteria. For the smallest eukaryotic cell, the baker's yeast, one obtains $r_0 < 1.87$ μm with $N = 6$, and for larger eukaryotic cells with e.g. $r_0 < 19$ μm, we find $N = 25$.

6. Lower temperature limit *vs.* volume: A model based on the number-phase incertitude relationship in the cell CD system

The CDs are bosons, therefore we may admit as a plausible approximation that they can form a Bose gas, a quantum ideal gas without interactions between the quasi-particles. Below a critical temperature T_c given by the Bose-Einstein equation:

$$T_c = \varsigma(3/2)^{-2/3}\left(\frac{N}{V}\right)^{2/3}\frac{h^2}{2\pi m_{eff} k_B} \tag{18}$$

a Bose-type condensate with all CDs in the same ground state can form, making possible the role of command and control of the CD system in the cell. Here $h = 2\pi\hbar$ is Planck's constant and $\zeta(3/2) = 2.6124$ is the Riemann zeta function. Due to the low effective mass of the CD, the critical temperature T_c is high. Assuming $N = 2$ as the minimum number of CDs compatible with life, $T_c = 273$ K (water freezing point) for $r_0 = 0.50$ μm and $V = 0.51$ μm^3; $T_c = 310$ K (which is the physiological temperature) for $r_0 = 0.46$ μm and $V = 0.42$ μm^3; and $T_c = 373$ K (water boiling point) for $r_0 = 0.42$ μm and $V = 0.32$ μm^3. Obviously for larger N values T_c can be even higher than 100 °C, and for the later critical temperature value the volume can be larger for larger N values. Such dimensions are indeed confirmed by the size of many prokaryotic cells. On the other hand, the condition of non-interacting bosons is probably only a rough approximation due to the Fröhlich long-range Coulombian attracting forces, and in such conditions the condensate is even more stable even when the number of bosons is small, as we assumed in all our models. Note that an accurate description of the state of the Bose-Einstein condensate is in terms of its wavefunction written as a solution of the Gross-Pitaevski equation [46]. This is a non-linear Schrödinger equation postulated by treating the bosons within a mean field theory, and which takes into account the inter-particle interactions as well as an external potential. Also, the 'internal' degree of freedom associated to the coherent dipole oscillation is neglected in our picture.

In the Bose-type or Fröhlich condensate or supercoherence state, the instantaneous number of CDs is 'ill-defined', and one can measure only its average value N with an associated incertitude ΔN. Therefore the number fluctuations ΔN will be subject to the number-phase realtionship or Heisenberg limit for systems with undetermined number of particles (like the number of electrons in a superconductor and the phase of its Ginzburg–Landau order parameter) [47]:

$$\Delta N \cdot \Delta\varphi \geq 1/2. \tag{19}$$

where $\Delta\varphi$ is the dispersion in the phase of the coherent oscillations which the N CDs perform together. It is obvious that the higher is $\Delta\varphi$, the lower is the coherence in the living cell.

From equation (13) – which was obtained by using both the quantum potential well and the fluctuations expression from classical statistical physics – we obtain the absolute value of fluctuations of N:

$$\Delta N = \frac{3}{\sqrt{5}}\frac{\sqrt{m_{eff} k_B T}}{\hbar}\frac{r_0\sqrt{N}}{n}. \tag{20}$$

Substituting this formula in the incertitude relationship (19) one obtains several interesting results. In the first place, a relation giving the dispersion of phase is established. For simplicity, the phase dispersion in radians at T = 310 K is:

$$\Delta\varphi \geq \frac{0.38\,n}{r_0\sqrt{N}}\,. \tag{21}$$

One sees that the functioning of the cell reaches a maximum coherence ($\Delta\varphi$ = min) in the ground state (n = 1) of the CDs confined in the spherical potential well with impenetrable walls. So far we made arbitrarily the choice n = 1, but this result sustains, at least indirectly, the postulate of Fröhlich theory that a driven set of oscillators can condense with nearly all of the supplied energy activating the vibrational mode of lowest frequency, i.e. the ground state.

The relationship (21) shows also the decrease of $\Delta\varphi$ (the increase of coherence) with the size of the cell r_o and with the square root of N. In fact, the highly specialized eukaryotic cells of higher organisms, which perform strongly specialized functions and thus require a higher coherence of the CDs system, are larger than the simpler prokaryotic cells.

At the same time, for a given value of the phase dispersion $\Delta\varphi_o$, equation (20) defines the lower limit of the viability temperature of the cell as a function of its volume. For the ground state (n = 1):

$$T \geq \frac{5(\pi/3)^{\frac{8}{3}}}{2^{\frac{2}{3}}}\frac{\hbar^2}{m_{eff}k_B}\frac{N}{\Delta\varphi_0^2 V^{\frac{2}{3}}} \tag{22}$$

This relationship is of the same form as the inequality (5b) as expected. Together with (15) and with the limits imposed by the freezing and boiling points of water, it defines a closed surface in a V-T plot predicting the allowed values for the volume and temperature of the prokaryotic living cells.

7. Comparison of theoretical predictions with experimental data. Discussions

The dimensions and optimal temperature ranges for a number of prokaryotic microorganisms and of eukaryotic cells are shown in table 1 (after [1, 48, 49]). One notes the distinctive situation of *Bacillus thermophillus*, which lives in the hot waters of geysers, in a temperature range drastically different of all considered cells. This bacterium is representative for the extremophillic microorganisms, which live in very harsh conditions. Other bacteria from the same taxum live in very cold waters. There is no evidence of clear-cut maximum or minimum temperatures for metabolism [50]. However, even if such organisms survive above the boiling point of water or below its freezing point, they may undergo spore formation in such conditions, and their metabolism would be profoundly changed. Therefore we will limit ourselves to a temperature range which is optimal for the functioning of these microorganisms, and the temperatures of the phase transitions of water as given by (6a) and (6b) will not be surpassed.

The temperature ranges of the considered cells are plotted against the cell volume in figure 2. The plot includes small and larger bacteria as well as two of the smallest unicellular eukaryotes. Remarkably, one can see that by appropriate choices of the parameters in inequalities (15) and (22), practically all the cells can be included in some of the distorted trapezes formed by the 4 curves according to their biological characteristics measured experimentally.

Table 1. Shape, dimensions, volume and living temperature limits of different cells

Cell / taxonomic group / species	Shape	Dimensions		V (μm^3)	Temperature limits (°C)
		r_o (μm)	l_o (μm)		
Prokaryotic cells					
Mycoplasma	Spherical, filamentous, helical	0.06–0.15 (spherical)	–	0.001–0.014	20-38
Prochlorococcus of the *Cyanobacteria* taxum	Spherical	0.15–0.3	–	0.01–0.1	18-38
Staphylococcus	Spherical	0.33–0.52	–	0.1–0.5	20-50
Rickettsia	Rod-like	0.15–0.35	1.5–2.0	0.1–0.8	20-38
Salmonella typhimurium	Rod-like	0.25–0.35	1.3–3.2	0.2–0.9	18-37
Pseudomonas aeruginosa	Rod-like	0.2–0.3	1.5–3	0.2–0.9	15-45
Proteus vulgaris	Rod-like	0.2–0.3	1.0–2.5	0.5–2.8	30-38
Escherichia coli	Rod-like	0.2–0.6	1.0–5.4	0.4–3.7	12-52
Bacillus thermophillus	Rod-like	0.35–0.4	3–4	2.4–8	70-100
Bacillus subtilis	Rod-like	0.3–1.3	1.2-7	0.3–17	10-40
Eukaryotic cells					
Saccharomyces cerevisiae (beaker's yeast) and *Rhodotorulla*	Spherical	1.5–2	–	14–28	25-30
Erythrocyte, human	Disk-like, biconcave	3.7	2	85	36-38
Lymphocyte, human	Spherical, with rough surface	3.5	–	150–270	36-38
Macrophage, human	Spherical, deformed	10-11	–	4200–4800	36-38

According to the values of the parameters of inequations (15) and (22) as synthesized in table 2, the prokaryotes plotted in figure 1 divide in three groups, and a fourth group is constituted by the small eukaryotic unicellular organisms (table 2). A constant trend is evidenced: the larger the cellular volume, the larger is the number N of CDs which co-ordinate the molecular dynamics in the cell and the stronger is the coherence within the CD system (the smaller is $\Delta\varphi_o$).

Apparently, the group of the smallest cells shows the lowest degree of coherence between the two CDs inside. These bacteria are rather diverse structurally, biochemically and functionally. *Mycoplasma* does not have a cell wall, in contrast to most bacteria. It is Gram negative just as *Rickettsia*, *Salmonella*, *Pseudomonas* and *Prochlorococcus*, which have a thinner cell wall formed by an inner peptidoglycan layer covered with lipopolysaccharides. On the other hand *Staphylococcus* is Gram positive, with a thicker multilayered peptidoglycan cell wall. The *Rickettsia*, unlike all other bacteria, use the energetic metabolism of host cells for their multiplication; this feature is puzzling, because the

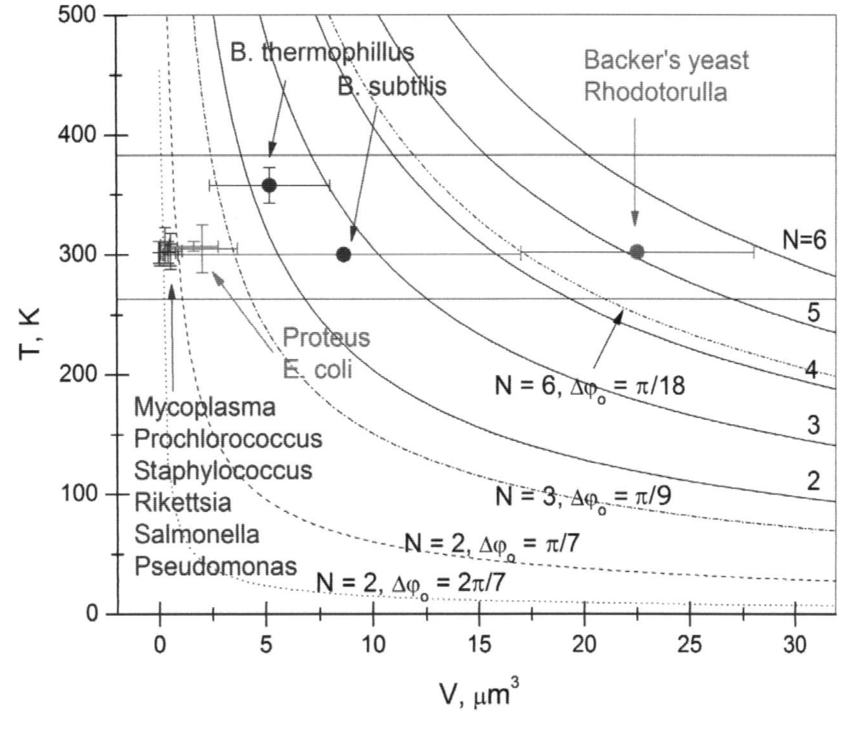

Figure 2. Theoretically predicted upper and lower living temperature limits (solid and broken lines, respectively) vs. cell volume. The horizontal lines are placed slightly below and above the freezing and boiling points of water, respectively. According to the parameter values in equations (15) and (22), the cells group in four distinct clusters represented in different colors: small prokaryotes (N = 2, $\Delta\varphi_o \cong 0.90$ rad $\approx 2\pi/7$), medium-size prokaryotes (N = 2, $\Delta\varphi_o \cong 0.45$ rad $\approx \pi/7$), larger prokaryotes (N = 3, $\Delta\varphi_o \cong 0.35$ rad $\approx \pi/9$) and the smallest eukaryotes (N = 6, $\Delta\varphi_o \cong 0.17$ rad $\approx \pi/18$). Although the clusters of cells are generally well separated, there is some degree of overlapping due to the large spread in volume associated with the high individual variability of some species.

pumping of metabolic energy is essential for the CD's stability. The small size seems to be, perhaps, the most important common observable feature in this group with the lowest N and highest $\Delta\varphi$. The low degree of coherence in this group would mean a poor degree of co-ordination of the molecular dynamics by the CDs.

Or it may point to another fundamental effect, and we would like to comment briefly on a possible hypothesis. In our models based on the spherical well with impenetrable walls we neglected any possible energy exchange between the "moving" CD clusters and the unstructured water around. This is only a rough approximation, and consideration of possible friction of the CD with the "quantum vacuum" constituted by the cytosol – which is a very complex aqueous environment containing dissolved ions, small molecules and large biopolymers – while leaving the energy levels unchanged in the average, would require a description based on the statistical density matrix and the Fokker-Plank equation [51]. This would imply certain non-negligible corrections in the correlations between CDs. But on the other hand, possible frictional effects might be much reduced in the smallest cells, enhancing thus the coherence between CDs and compensating for the high dispersion of phase, due to a property of water confined in very small spaces of up to 1 μm. Using AFM and SEM, Takahashi [52, 53] demonstrated the existence of a special state of water ("super-water") in such spaces which correspond to a "minimum drop" of the liquid; this water looses completely its superficial tension and behaves as an ideal lubricant. The biological significance of this finding was commented [54] starting from the interest shown by Abdus Salam for the living cell and origin of life [55]. One may suppose that in the case of very small cells the dynamics of water CDs may be better described by a frictionless potential well model, in contrast to the larger cells, which would require density matrix treatment and where some coherence might be lost due to dissipation and friction. Although speculative at this moment, this possibility should be kept in mind for further developments.

The second group of cells including common and ubiquitous bacteria like *E. coli* and *Proteus* contains also only two CDs similarly to the microorganisms of the first group, but their coherence is twice better. These Gram negative bacteria have similar metabolism and cell wall and show increased thermotolerance and resistance to putrefaction as compared to the microorganisms of the first group, which are also Gram negative with the only two exceptions mentioned above.

The dispersion of the phase decreases (the coherence increases) even more in the third group of prokaryotes; however, here the cells contain 3 CDs. The larger number of CDs and their increased coherence may be associated to some evolutionary distance of the bacilli belonging to this group and their more simple predecessors. The bacilli in this class, in contrast to the previous two classes of smaller bacteria, are able to form *spores*, which enhances their resistance to the environment.

Finally, the two yeasts representative of small eukaryotic unicellular organisms contain 6 CDs each, a significant jump from the 3 CDs in the previous class. Taking into account the hypothesis that metabolism came first in the origin of life and that the nucleus appeared in the eukaryotes as a cell internalized from outside [6], the fact that the nucleated cells seem to require a stepwise-increase in the number of CDs may be of particular biological significance because it may indicate an *evolutionary discontinuity* with their prokaryotic predecessors. Note also that the bacteria in the first and second class are Gram negative with the only exception of *Staphylococcus* from the first group, while the microorganisms of the third and fourth groups are all Gram positive. Moreover, the Gram-positive *Staphylococcus* may change to Gram-negative when ageing. Therefore the model appears to evidence an evolutionary step by correlating the structural and functional differences between unicellular organisms associated to negative and positive Gram staining of the cell wall, with values of the parameters below and above certain critical values ($N = 3$, $\Delta\phi = 0.35$ rad). At the same time, this correlation underlines the importance of the cell wall, ignored so far in our model.

Altogether, the present model may provide a new classification criterion for cells, able to evidence discontinuity points in the evolution of unicellular organisms. We note finally that our model may accommodate also larger eukaryotic cells if a sufficient number of CDs is included, e.g. some tens in the case of the erythrocyte.

The results of our theoretical model are encouraging beyond expectations. The model succeeds to find reasonable relationships between the upper and lower limits of the living temperatures vs. the cell volume which are consistent with the experimental data, in spite of all its approximations. For instance, the complex composition of cytosol and its properties different of those of pure water may require indeed the density matrix treatment as well as corrections to the classical equation (11). Such corrections may be needed to account also for the higher compressibility of the CDs boson gas as

Table 2. The number of water coherence domains and the phase dispersion of the CD system in relation to some biological characteristics in the four groups of cells discriminated according to the model's predictions.

Species	Shape	Gram staining[1]	*N*	$\Delta\varphi_o$ (rad)	V (μm^3)	*T* domain (°C)
Mycoplasma	Spherical	–				
Prochlorococcus	*ibid.*	–				
Staphylococcus	*ibid.*	+				
Rikettsia	Rod-like	–	2	0.90	0.001-0.9	18-45
Salmonella	*ibid.*	–				
Pseudomonas	*ibid.*	–				
Proteus	Rod-like	–	2	0.45	0.4–3.7	12-52
E. coli	*ibid.*	–				
B. thermophillus	Rod-like	+	3	0.35	0.3–17	10-100
B. subtilis	*ibid.*	+				
S. cerevisiae	Spherical	+	6	0.17	14–28	25-30
Rhodotorulla	*ibid.*	+				

[1]The Gram-positive bacterium *Staphylococcus* may change to Gram-negative when ageing.

compared to an ideal gas [56]. Second, the interspersed classical and quantum character seems epistemologically uncomfortable; but this is not uncommon in many areas of physics and may be not only an expression of the emerging character of quantum biology. We may suppose also that this aspect could be related somehow also to decoherence theory. Considering the cell as a large complex quantum system that collapses to become a classical entity we can expect that the emergence of classicality will involve some quantum *signatures* that cannot be ignored [57] – and size and shape may be among them. Third, the specific numbers of CDs for the considered prokaryotic cells is low, although we use statistical concepts. But given the postulated command and control role of CDs in the cell, their small numbers involving very large relative fluctuations may play a key role in understanding the functioning of the cell.

Fourth and most important, here we did not take into account the shape of the cells except for the simplest spherical cells, and made no distinction between spherical and cylindrical cells (like the rod-like bacteria and the disk-like erythrocyte). A model of cylindrical potential well would lead to energy levels of a more complicated form than equation (7) involving the roots of Bessel functions and, therefore, the two curves given by inequalities (15) and (22) should be replaced by more complicated functions. At the same time, although the two smallest yeast cells may be fitted well by the model, we did not take into consideration the existence of the nuclear membrane, which is expected to divide the spherical potential well in two uneven concentric compartments. However, we expect that these corrections would affect only the 'fine structure' of the curves (15) and (22), and, therefore, to a first approximation we did not restrict the convergence of the theory with the microbiological data by ignoring the cylindrical shape of some bacteria as well as the nucleus in yeasts. This suggests that in our model the size of the cell is more important than shape. The same conclusion has been reached independently, by the completely different approach of [7] (in this volume), who related the cell size and shape to the power of electromagnetic field radiated by (quasi)coherent electromechanical longitudinal oscillations of the microtubules' network in the dividing and non-dividing eukaryotic cell.

The satisfactory fit with microbiology data does not exclude the possibility that other factors be involved, such as decoherence mentioned above. For example one can consider an analogy with superconductors and suggest that the size of the cell is somehow connected to a coherence length ξ describing the size of thermodynamic fluctuations [47]:

$$\xi = \frac{\hbar}{\sqrt{2m_{eff}\,\alpha_o\left|T - T_c\right|}} \tag{23}$$

where T_c should be evaluated by equation (18) to a first approximation. But to make operational such an approach one should perhaps adapt the Ginsburg-Landau theory of superconductors to describe the CD boson system in the cell water before one could define the meaning of the parameter α_o and establish its value. This may be a task for future examination.

8. Conclusions

Our semi-classical approach based on fluctuations of water CDs' number inside cells with impenetrable walls succeeds to accommodate, by appropriate choices of the parameters, the experimentally measured values of biological observables for many typical unicellular organisms within the predicted limits of living temperature and cellular volume. According to the model, the cells cluster in distinct groups; the prokaryotes divide in three classes, and a fourth group is constituted by the small eukaryotic cells. The results suggest not only that the number of CDs which co-ordinate the molecular dynamics in the cell increase with the volume, but also that the coherence within the CD system is stronger in cells with larger volume. By establishing this rule, the model provides a new criterion for the classification cells which is sustained by biological criteria, most notably by evidencing discontinuities in the number of CDs and their coherence at the evolutionary boundary between prokaryotes and eukaryotes as well as between low- and medium-sized bacteria (mostly Gram negative) and larger unicellular organisms (Gram positive). Moreover, one may foresee plausible extensions of the model to larger eukaryotic cells by including a sufficient number of CDs. Finally, the occurrence of maximum coherence predicted in the ground state is consistent to the postulate of

Fröhlich picture stating that a driven set of coupled oscillators can condense in the ground state mode of oscillation.

The convergence of the model with the biological data notwithstanding the present neglect of the geometry of cylindrical and nucleated cells in the model implies that the size of the cell is more important than its shape for the living temperature limits, in agreement to a similar conclusion resulting from a radically different model independently postulated for eukaryotic cells. However, corrections for the shape of the more complex cells using potential wells appropriate for rod-like bacteria and concentric nucleated spherical yeast cells may be foreseen as tasks for further developments. Moreover, a more realistic model should take into account not only the cell membrane as a semi-penetrable wall, but also the more hydrophilic cell wall with its different structures for Gram positive and negative microorganisms, which plays a leading role in maintaining the shape of the cell (especially for rod-like bacilli); this would imply a more complex type of potential well, requiring numerical solutions. Other possible refinements may include consideration of "super-water" for the smallest cells having the size of a "minimum liquid drop" with properties close to an ideal lubricant; taking into account of interactions between CDs and of the internal degree of freedom associated to the coherent dipole oscillation by use of the Gross-Pitaevski equation; corrections associated with the cytosol complex composition including biopolymers which alter the properties of water and may require density matrix treatment and Fokker-Planck equation, as well as decoherence of CDs due to scattering on the intracellular biopolymers and organelles; a possible connection between the size of the cell and other parameters describing the size of thermodynamic fluctuations in the CD system, like for instance a properly defined coherence length; and correlations to energetic metabolism and to various experimental parameters of cytoplasmic water, e.g. pH, ionic strength, colloid-osmotic pressure, water chemical activity, and free vs. 'bound' water content. The evoked possible correlations might open also new prospects to a better understanding of the *Archaea* microorganisms which lived in hostile environments, as well as to the origin on life.

Nevertheless, with all the approximations involved, the model makes successful predictions of direct biological relevance; and therefore warrant efforts for further developments.

Acknowledgments
One of the authors (E.A.P.) expresses his gratitude to Prof. Marian Apostol (Bucharest) for the continuous help and interest as well as for criticism and valuable suggestions throughout this and previous studies. He feels honoured by the interest shown for this study by Prof. Emilio Del Giudice (Milano) and gratefully thanks him for reading the manuscript and for generous encouragement. He thanks also Dr. Cristina Bordeianu (Bucharest) for valuable remarks about the temperature-volume relationship in extremophilic bacteria.

References

[1] Alberts B, Bray D, Lewis J, Raff M, Roberts K and Watson J D 1994 *Molecular Biology of the Cell* (New York – London: Garland Publishing, Inc.).
[2] Schrödinger E 1944 *What is life?* (Cambridge: Cambridge University Press)
[3] Prigogine I 1980 *From Being to Becoming* (San Francisco: Freeman)
[4] Abbot D, Davies P C W and Pati A K 2008 *Quantum Aspects of Life* (London: Imperial College Press)

[5] Demetrius L 2003 *Physica A* **322** 477
[6] Dyson F 2000 *Origins of Life*, 2nd edition (Cambridge: Cambridge University Press)
[7] Havelka D, Cifra M and Vrba J 2011 What is more important for radiated power from cells – size or geometry? *J. Phys. Conference Series* (in press, this volume)
[8] Preoteasa E A and Apostol M V 2008 Collective Dynamics of Water in the Living Cell and in Bulk Liquid. New Physical Models and Biological Inferences *arXiv-0812.0275v2-physics-bio-phys-12-Dec-2008*
[9] Mavromatos N E 1999 *Bioelectrochem. Bioenerg.* **48** 273
[10] Mavromatos N E 2010 Quantum Mechanical Aspects of Cell Microtubules: Science Fiction or Realistic Possibility? *arXiv 1011.6270v1-2010*
[11] Mavromatos N E, Mershin A and Nanopoulos D V 2002 QED-Cavity model of microtubules implies dissipationless energy transfer and biological quantum teleportation *arXiv 0204021v1-2002*
[12] Del Giudice E 2010. *IUMAB International Union of Medical and Applied Bioelectrography Conference*, http://www.iumab.org/index.php/2010conference/818-prometheus-or-the-passionate-soul-of-scientific-reason.
[13] Fröhlich H 1968 *Phys. Lett.* **26A** 402
[14] Fröhlich H 1968 *Int. J. Quant. Chem.* **2** 641
[15] Fröhlich H *Adv. Electron. Electron Phys.* **53** 85
[16] Fröhlich H 1982 *Phys. Lett.* **93A** 105
[17] Apostol M and Preoteasa E 2008 *Phys. Chem. Liquids* **46** 653
[18] Ben-Jacob E 2009 *Water* **2** *Supplement* **1** 8 (*Fourth Annual Conference on the Physics, Chemistry and Biology of Water* October 2009, West Dover, VT, USA)
[19] Del Giudice E and Tedeschi A 2009 *Electromagn. Biol. Med.* **28** 46
[20] Del Giudice E, Doglia S and Milani M 1982 *Phys. Scripta* **26** 232
[21] Del Giudice E, Doglia S, Milani M and Vitiello G 1986 *Nuclear Phys. B* **275** 185
[22] Del Giudice E, Preparata G and Vitiello G 1988 *Phys. Rev. Lett.* **61** 1085
[23] Del Giudice E, Fleischmann M, Preparata G and Talpo G 2002 *Biolectromagnetics* **23** 522
[24] Preparata G 1995 *QED Coherence in Matter*, Singapore–New Jersey: World Scientific
[25] Engel G, Calhoun T, Read E, Ahn T, Mancal T, Cheng Y, Blankenship R and Flemming G 2007 *Nature* **446** 782
[26] Collini E and Scholes G 2009 *Science* **323** 369
[27] Collini E, Wong C Y, Wilk K E, Curmi P M G, Brumer P and Scholes G D 2010 *Nature* **463** 644
[28] Fuchs E C 2010 *Water* **2** 381
[29] De Nino A and Congiu Castellano A 2011, *Bioelectromagnetics* **32** 218
[30] Blackman C F, Benane S G, Rabinowitz J R, House D E and Joines W T 1985 *Bioelectromagnetics* **6** 327
[31] Liboff A R, Rozek R J, Sherman M L, McLeod B R and Smith S D 1987 *J. Bioelectr.* **6** 13
[32] Zhadin M N, Novikov V V, Barnes F S and Pergola N F 1998 *Bioelectromagnetics* **19** 41
[33] Pollack G H 2001 *Cells, Gels and Engines of Life* (Seattle, USA: Ebner and Sons)
[34] Vybiral B and Voracek P 2008 "Autothixotropy" of water - an unknown physical phenomenon, and its possible importance for the cytoskeleton, *Phase Transitions in Cell Biology* eds Pollack G H and Chin W-C (Berlin-Heidelberg-New York: Springer-Verlag)
[35] Tuszynski J A 2004 Fröhlich Theory *Encyclopedia of Nonlinear Science* ed Scott A (New York – Oxon, UK: Taylor & Francis/Routledge) pp. 1-3
[36] Askar'yan G A 1962 *Sov. Phys. JETP* **15** 1088
[37] Rowlands S 1983 Coherent excitations in blood *Coherent Excitations in Biological Systems* eds Fröhlich H and Kremer F (Berlin–Heidelberg: Springer-Verlag) pp. 145-161
[38] Fritz O G 1984 *Biophys. J.* **46** 219

[39] Pohl H A 1983 Natural oscillating fields of cells *Coherent Excitations in Biological Systems* eds Fröhlich H and Kremer F (Berlin–Heidelberg: Springer-Verlag) pp. 199-210
[40] Bercu M, Zhou X, Lee AC, Poenar DP, Heng CK and Tan SN 2006 *Biomed. Microdevices* **8** 177
[41] Pokorny J, Hasek J, Jelínek F, Saroch J and Palan B 2001 *Electro Magnetobiol.* ***20*** *371*
[42] Pokorny J 2004 *Bioelectrochemistry* **63** 321
[43] Preoteasa E A; Apostol M V 2009 *Water* **2** *Supplement* 1 25 (*Fourth Annual Conference on the Physics, Chemistry and Biology of Water*, 2009, West Dover, VT, USA)
[44] Landau L and Lifshitz E 1966 *Mecanique Quantique. Theorie Non Relativiste (Cours de Physique Theorique,* Tome 3*)* eds L Landau and E Lifshitz (Moscou: Editions Mir)
[45] Landau L and Lifchitz E 1965 *Physique Satistique (Cours de Physique Theorique,* Tome 5*)* eds L Landau and E Lifshitz (Moscou: Editions Mir)
[46] Pitaevskii L P and Stringari S 2003 *Bose–Einstein Condensation* (Oxford: Clarendon Press)
[47] de Gennes P G 1995 *Superconductivity of Metals and Alloys*, 2nd Revised Edition (Reading, Mass.-Wokingham: Addison Wesley)
[48] Zarnea G 1983 *A Treatise of General Microbiology,* vol 1 (Bucharest: Romanian Academy Publishing House) (in Romanian).
[49] Merz W G and Hay R J eds 2005 *Medical Micology* 10th edition (*Topley and Wilson's Microbiology and Microbial Infections* series) (London: Hedder and Arnold)
[50] Price P B and Sowers T 2004 *Proc. Natl. Acad. Sci. US* **101** 4631
[51] Risken H 1989 *The Fokker–Planck Equation: Methods of Solutions and Applications*, 2nd edition, Springer Series in Synergetics (Berlin-Heidelberg-New York: Springer-Verlag)
[52] Takahashi R 1989 *Phys. Lett.* A **141** 15
[53] Takahashi R 1982 *Jap. J. Appl. Phys.* **21** 974.
[54] Rosu H 1992 Life and water *ICTP Trieste Internal Report* IC/92/246
[55] Salam A 1991 *J. Mol. Evol.* **33** 105
[56] Hutten H H 1967 *The Ideas of Physics* (Edinburgh: Oliver & Boyd Ltd)
[57] Salari V, Tuszynski J, Rahnama M and G Bernroider G 2011 *J. Phys. Confer. Series* **306** 012075

Methanol, Ethanol and Propanol in EHD liquid bridging

Elmar C. Fuchs[1], Adam D. Wexler[1], Luewton L. F. Agostinho[1], Michael Ramek[2], Jakob Woisetschläger[3]

[1] Wetsus Centre of Excellence for Sustainable Water Technology, Agora 1, 8900CC Leeuwarden, The Netherlands

[2] Institute of Physical and Theoretical Chemistry, Stremayrgasse 9, 8010 Graz, Austria

[3] Experimental Turbomachinery Research and Optical Measurement Group, Institute for Thermal Turbomachinery and Machine Dynamics, Graz University of Technology, Inffeldgasse 25 A, 8010 Graz, Austria

e-mail of corresponding author: elmar.fuchs@wetsus.nl

Abstract. When a high-voltage direct-current is applied to two beakers filled with water or polar liquid dielectrica, a horizontal bridge forms between the two beakers. In this work such bridges made of methanol, ethanol, 1-propanol and 2-propanol are investigated with polarimetry and thermography. Whereas methanol, ethanol and 1-propanol bridges become warm like a water bridge, a 2-propanol bridge cools down relative to the surroundings. It is shown how the different stability of the primary and secondary alcoholate ions and the resulting small difference in conductivity between 1-propanol and 2-propanol is responsible for this novel effect.

1. Introduction

In 1893 Sir William George Armstrong first reported the formation of a horizontal bridge of water forming between two wine glasses filled with chemically pure water when a direct-current (DC) high-voltage was applied [1]. Recently, a number of publications made the broader scientific community aware of this experiment [2-7].

The experiment is stable, easy to reproduce, and grants free experimental access to water inside the bridge under variable high voltage and atmospheric conditions in air, thus giving insights into basic features of dipole-dipole interaction in water [8]. Therefore a number of experiments in the horizontal bridge of water was performed including neutron scattering [4,5], visualization and optical measurement techniques [6], Raman scattering [9], Brillouin scattering [10], reduced gravity experiments [11], and was discussed in relation to molecular vibration coupling [12] and the influence of some electrolytes [13] and non-ionic solvents [14] on the water bridge.

Several authors discussed the electrohydrodynamics of this phenomenon. Widom *et al.* [15] and Saija *et al.* [16] claim that the high dielectric permittivity of water is the reason for a stable bridge forming and used the Maxwell stress tensor to calculate tensions within the bridge. A similar approach was used by Marin and Lohse [17]. It has been shown that a horizontal bridge forms also in methanol [8] and glycerol [17]. Aerov [18] pointed out that a stable equilibrium for the bridge's surface is reached when the axial electric field is strong enough to counteract distortions caused by surface tension.

All electrohydrodynamic discussions of these experiments use previous literature on a phenomenon also named 'liquid bridge', but resembling a vertical column of liquid, pinned at each end between planar electrodes, and surrounded by a non-conducting, dielectric gas. These experiments were performed with both AC and DC fields and led to a detailed analysis of this vertical liquid bridge and its stability [19-23]. It has been shown that the longitudinal field along a liquid jet axis has a stabilizing effect on the liquid – gas interface [24] which was later also discussed for vertical liquid bridges [25]. In this sense the liquid bridge formation can be explained as action of the electrohydrodynamic stress tensor replenishing the liquid within the bridge against any drainage mechanism. Recently, the formation of bridges with acetone, dichloromethane (DCM), dimethylformamide, dimethylsulfoxide, ethanol, glycerol, methanol, 1-propanol, 2-propanol, and tetrahydrofuran have been reported and investigated [26].

In this work the Armstrong experiment is repeated with methanol, ethanol, 1-propanol and 2-propanol. The polarimetric and thermographic properties are reported and discussed.

2. Experimental

For the DC experiments we use two 40 ml glass beakers with spouts, 15 x 5 x 0.25 mm platinum electrodes, a DC high-voltage 0-25 kV power supply Phywe 'Hochspannungs-Netzgerät 25kV' (No 13671.93, Göttingen, Phywe Systeme GmbH, Germany) with a maximum current of 0.5 mA and a 42nF capacitor in parallel to the bridge. For imaging a Canon 300D with a Sigma 105mm 1:2.8 macro lens, a Panasonic 3CCD NV-DX100 camcorder with Raynox DCR-150 or Raynox DCR-250 conversion lenses, a Photron FASTCAM SA1 high-speed camera with a Nikkor 60 mm 1:2.8 macro lens, or a CASIO Exilim High Speed EX-F1 - were used. All liquids are common solvents and were used in a chemically pure grade. Deionized water was provided by a Barnstead NANOpure type I ultrapure water system (Thermo Fisher Scientific Inc., Waltham, MA) with an initial conductivity of 0.056 μS cm-1 recorded with the integrated conductivity meter of the system. This conductivity rose quickly to 0.8 μS cm-1 with a pH value of 5 due to CO2 saturation under atmospheric conditions (Kendall 1916). All images were scaled. Infrared imaging (thermal imaging) was done with a FLIR 620 camera (FLIR systems, MA, USA). The images of the bridges chosen for analysis were taken at a point in time when either a temperature equilibrium was reached, or, if that was not possible, during a time before the bridge broke down when temperature changes were minimal. An emissivity of 0.92 was assumed for all liquids. The lengths of the bridges were the maximum lengths attainable for stable bridges in case of the alcohols. The length of the water bridge was intentionally kept shorter for reasons of better comparison.

For polarometric visualisations a ceramic glass diffuser (50 x 50 cm, 5 mm thick) was placed behind the beakers, illuminated by a halogen lamp from behind with a linear polarizer put on top of the diffuser. A second polarising filter was placed in front of the imaging device and used as the analyzer.

For high voltage conductivity measurements, the set-up consisted of two platinum electrodes dipped 10mm deep into the liquid in a glass cuvette (Groß-Küvette 700.000, Hellma GmbH & Co. KG, Müllheim Germany) with 50 mm inner width, 50 mm inner height and 10 mm inner depth, with 50mm electrode distance under atmospheric conditions. With this set-up, voltage versus current diagrams of the different liquids were recorded. .

3. Results

Fig. 1 shows the alcoholic bridges (b-e) in comparison to a water bridge (a) as well as polarimetric images of such bridges (f-j). Generally, the alcoholic bridges are thinner and shorter than the water bridge, more difficult to create and less stable [26]. The characteristic bright zones on the top and bottom of the bridge initially discovered for aqueous bridges under polarized light [3,6] is present for all alcoholic bridges as well.

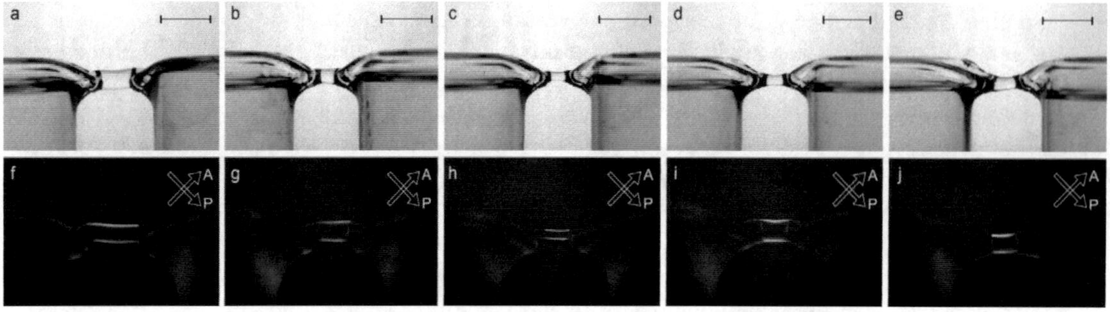

Fig. 1: (a)-(e): Horizontal bridges forming in different polar liquids under atmospheric conditions and DC voltage. The bars are 1 cm. (f)-(j): Liquid bridges between crossed linear polarisers. P indicates the direction of the polarizer, A the direction of the analyzer. (a,f) water 12 kV, (b,g) methanol 9.5 kV, (c,h) ethanol 9.5 kV, (d,i) 1-propanol 10 kV, (e,j) 2-propanol 8 kV

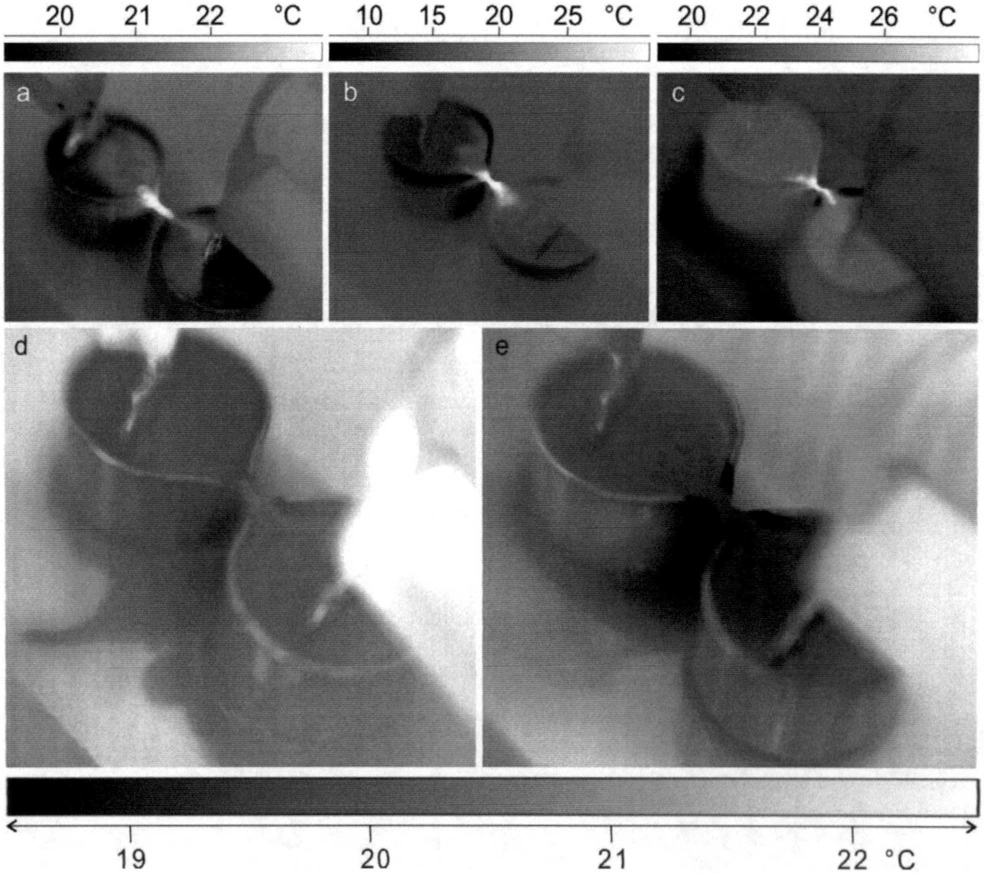

Fig. 2: Thermographic images of horizontal bridges forming in different polar liquids under atmospheric conditions and DC voltage. (a) water 12 kV, (b) methanol 9.5 kV, (c) ethanol 9.5 kV, (d) 1-propanol 10 kV, (e) 2-propanol 8 kV

Journal of Physics: Conference Series **329** (2011) 012003 doi:10.1088/1742-6596/329/1/012003

Fig. 2 shows a thermographic analysis of the alcoholic bridges (b-e) in comparison to a water bridge (a). The scales of the images were adapted to match the observed temperature regions. For the sake of comparison, the same scale was chosen for 1- and 2-propanol, respectively.

The most prominent feature of this image is the fact that the water, methanol, ethanol and 1-propanol bridges are heating up whereas the 2-propanol is slightly cooling down. Next to that, the temperature of the methanol bridge was the highest, followed by ethanol, water, and 1-propanol. These trends are evaluated in Fig. 3.

In addition to the liquid bridge set-up, a single cuvette set-up was used to measure the high voltage conductivities of the liquids investigated under more stable conditions. The measurements showed that these properties are non-Ohmic under the conditions applied, and they are plotted in Fig. 4.

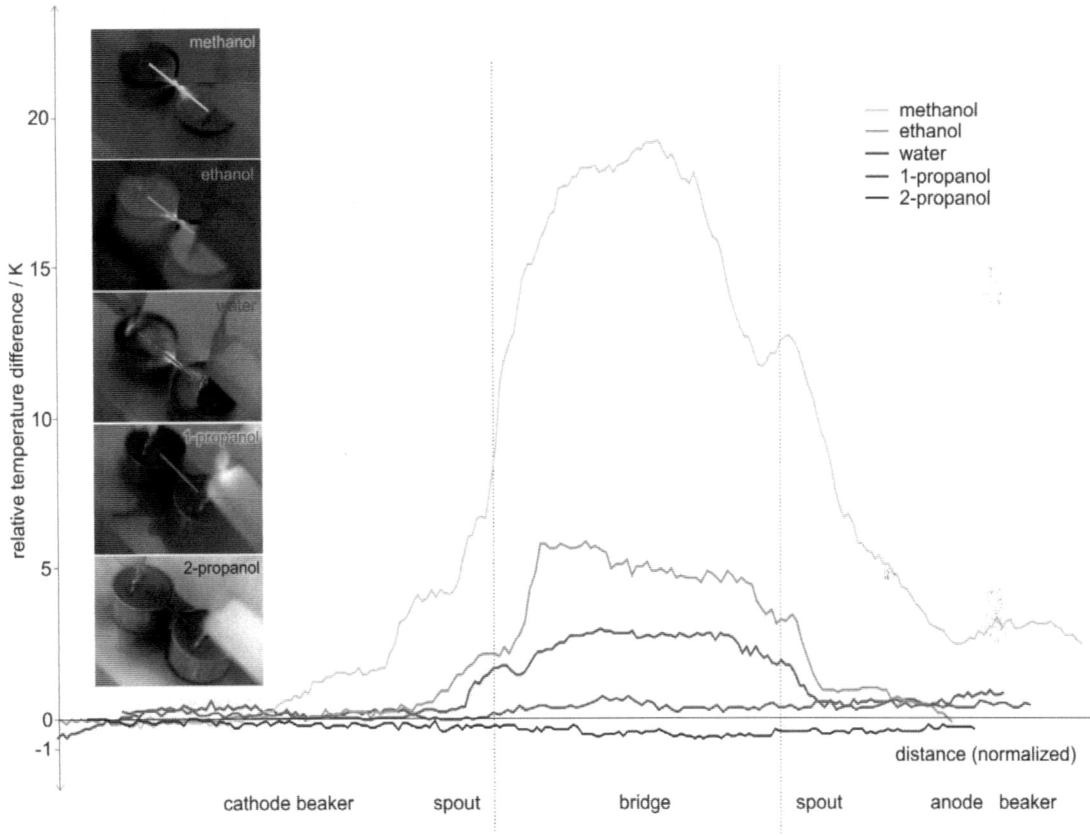

Fig. 3: Relative surface temperature changes across the liquid bridge set-ups. The evaluated paths are shown in the insets on the left. Lengths are normalized for reasons of comparison. The borders between bridge and spout are an approximation.

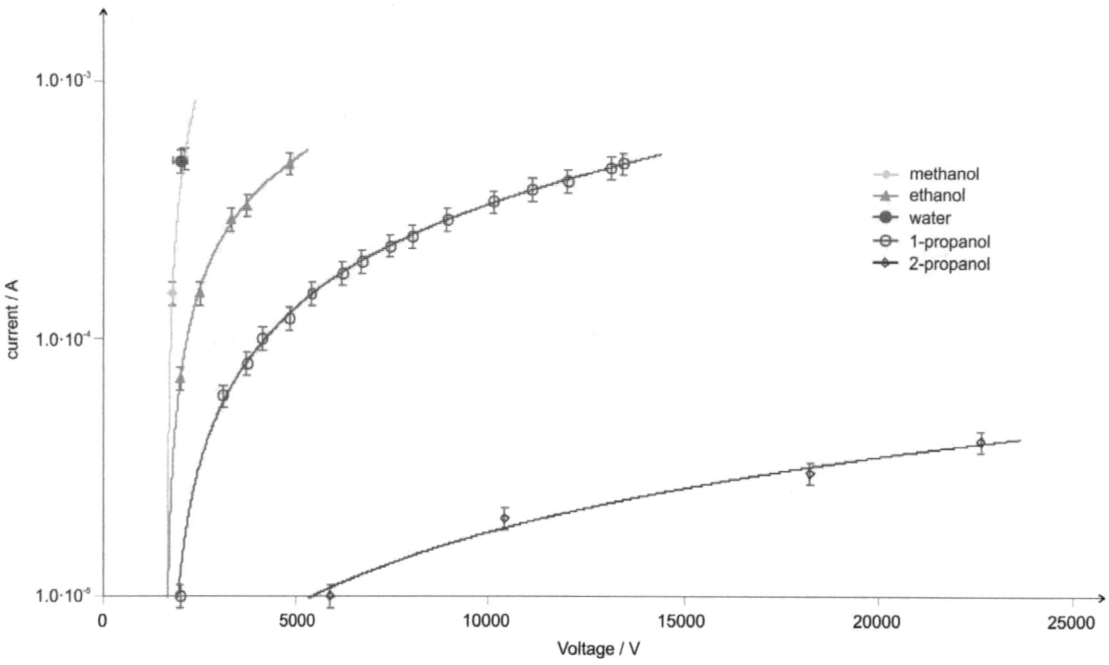

Fig. 4: Non-ohmic behaviour of the liquids used to form a horizontal bridge. The recordings were done up to 25 kV and up to 0.5 mA. The platinum electrodes dipped 10 mm deep into the liquid with 50 mm electrode distance under atmospheric conditions.

4. Discussion

There are two mechanisms at work in all liquids investigated responsible for the temperature changes: Ohmic resistivity causes heating, whereas evaporation causes cooling. In all liquids except for 2-propanol, the Ohmic heating outweighs the evaporation. The following subsections will discuss and explain these facts in detail.

4.1. Alcoholate ion stabilities

It is basic knowledge of organic chemistry that secondary alcoholate anions are less stable than primary ones because a primary carbon atom carries more positive partial charge than a secondary one, thus the negative charge of the oxygen can be distributed more easily in case of a primary alcoholate. This molecular feature becomes measurable as a macroscopic property when looking at the conductivity of the alcohols. Thus, there are far less alcoholate ions in 2-propanol than in 1-propanol, resulting in a lower current (see Fig. 4), and, consequently, in lower Ohmic heating of the liquid.

4.2. Dipole moment and conductivities

The field strength applied in the present experiments does not cause a significant influence on the dipole moment of the single alcohol molecules. Calculations using optimized RHF/6-311++G** structures, which were verified to be local minima via frequency analysis, showed that at the field strengths up to 20kV/cm results do not differ from calculations without external fields. First significant changes of charge distribution and geometry required 10^4 times higher field strengths. This result is in line with a similar calculation for water [27]. However, these calculations do not consider the dissociation of molecules into ion pairs. It is well known that the conductivity of dielectric liquids increases with permittivity due to an accompanying decrease of the electrostatic attraction between ion pairs [28]. Whenever a high-voltage DC field is applied to the fluid, an increasing number of ions escape recombination, causing an increase in conductivity, as was discussed by Onsager and Brière [29,30]. While at low voltage Ohmic behavior is observed due to ions generated by dissociated molecules; at high voltage the dissociation rate is proportional to the electric field but the recombination rate is not and the current approaches saturation [31,32]. These effects are clearly visible for all alcohols in Fig. 4.

4.3. Evaporation

Of all alcohols, methanol has the lowest boiling point (65°C) followed by ethanol (78°C), 2-Propanol (82°C) and 1-Propanol (97°C). According to Hinds [33] a curved surface modifies the attractive forces between surface molecules facilitating diffusion from the liquid surface, and consequently favours evaporation. Thus, when compared to a flat air-liquid interface, an increase on the diffusion of molecules throughout this same interface is expected. Another important point is the presence of surface charge. According to Hartman *et al.* and Darabi *et al.* [34,35] the electrostatic stress created by the presence of these charges counter-acts the surface tension stress on the liquid surface. The net effect is the production of secondary motion that destabilizes the thermal boundary layer near the heat transfer surface, leading to higher effective heat transfer coefficients, and causing the production of vapour along the whole air-liquid interface of the set-up with most of the evaporation taking place where the electric field is highest – across the liquid bridge [26]. This is the reason why the 2-propanol bridge is cooler than the bulk liquid, as can be seen in Figs. 2 and 3.

4.4. Polarimetry

It has been suggested that either bubbles or density gradients [3,6] or birefringence [3,5] are possible sources of the bright zones visible when looking at a water bridge using a polarimetric set-up. Until now, this issue is still not resolved. We suggest that, since this effect is not intrinsic to water (see Fig. 1), what ever physics responsible may apply for the alcohols investigated as well, thereby ruling out water specific origins.

5. Conclusions

The infrared emission of alcoholic EHD bridges made of methanol, ethanol, 1-propanol and 2-propanol were compared to an aqueous EHD bridge. The temperature of the primary alcohol bridges is decreasing with their decreasing conductivity. Water is an exception because of its unusually high heat capacity (4.18 J / g·K) compared to methanol (2.59 J / g·K) and ethanol (2.44 J / g·K), thus aqueous bridges are cooler although the conductivity of water is higher than that of the alcohols. Furthermore, we found that a 2-propanol bridge cools down whereas all other bridges heat up. This effect can be explained by two mechanisms responsible for the temperature changes: Ohmic resistivity causing heating, and evaporation causing cooling. In all liquids except for 2-propanol, the ohmic heating outweighs the evaporation. Moreover, since evaporation is increased by electric fields [34,35] and surface curvatures [33], it is expected to be highest along the bridge, thus explaining why the 2-propanol bridge shows anomalous cooling.

In addition, the bright zones originally found in a polarimetric investigation of the water bridge [3] were reproduced for all liquids, thus showing that they are not a water-intrinsic effect.

6. Acknowledgements

This work was performed in the TTIW-cooperation framework of Wetsus, Centre of Excellence for Sustainable Water Technology (www.wetsus.nl). Wetsus is funded by the Dutch Ministry of Economic Affairs, the European Union Regional Development Fund, the Province of Fryslân, the City of Leeuwarden and the EZ/Kompas program of the "Samenwerkingsverband Noord-Nederland". The authors would like to thank Profs. Eshel Ben-Jacob (Tel Aviv University), Harry Bruning (Wetsus), Cees Buisman (Wetsus), Gert-Jan Euverink (Wetsus), Friedemann Freund (NASA SETI Institute, California), Karl Gatterer (TU Graz), Xingua Guo (TU Graz), Emilio Del Giudice (Universitá di Milano), Gerald H. Pollack (University of Washington), José Teixeira (Laboratoire Léon Brillouin, CEA-CNRS/IRAMIS, CEA/Saclay), as well as Cees Kamp, Astrid H. Paulitsch-Fuchs, Martina Sammer (Wetsus) and especially Mathias Eisenhut (TU Graz), for the ongoing discussion on the water bridge phenomenon (in alphabetic order). The financial support of the participating companies of the Applied Water Physics Theme at Wetsus, Centre of Excellence for Sustainable Water Technology, is acknowledged.

7. References

[1] Armstrong, W.G. The Electrical Engineer **1893**, 10 February 1893, 154-145.
[2] Fuchs, E.C. J. Woisetschläger, K. Gatterer, E. Maier, R. Pecnik, G. Holler and H. Eisenkölbl, J. Phys. D: Appl. Phys. **2007**, 40, 6112-6114.
[3] Fuchs, E.C.; Gatterer, K.; Holler, G.; Woisetschläger, J. J. Phys. D: Appl. Phys. **2008**, 41, 185502-185507.
[4] Fuchs, E.C.; Bitschnau, B.; Woisetschläger, J.; Maier, E.; Beuneu B.; Teixeira, J. J. Phys. D : Appl. Phys. **2009**, 42, 065502:1-065502:4.
[5] Fuchs, E.C.; Baroni, P.; Bitschnau, B.; Noirez, L. J. Phys. D: Appl. Phys. **2010**, 43, 105502:1-105502:5.
[6] Woisetschläger, J.; Gatterer, K.; Fuchs, E.C. Exp. Fluids **2010**, 48-1, 121-131.
[7] Fuchs, E.C.; Agostinho, L.L.F.; Eisenhut, M.; Woisetschläger, J. Proc. SPIE **2010**, 7376, 73761E1-15.
[8] Fuchs, E.C. Water (MDPI) **2010**, 2, 381-410.
[9] Ponterio, R.C.; Pochylski, M.; Aliotta, F.; Vasi, C.; Fontanella, M. E.; Saija, F. J. Phys. D: Appl. Phys. **2010**, 43, 175405:1-175405:8.
[10] Fuchs, E.C.; Bitschnau, B.; Di Fonzo, S.; Gessini, A.; Woisetschläger, J. and Bencivenga, F. *under review*
[11] Fuchs, E.C.; Agostinho, L.L.F.; Wexler, A.; Wagterveld, R.M.; Tuinstra, J.; Woisetschläger, J., J. Phys. D: Appl. Phys. **2011, 44,** 025501:1-025501:8.
[12] Del Giudice, E.; Fuchs, E.C.; Vitiello, G. Water (Seattle) **2010**, 2, 69-82.
[13] Nishiumi, H.; Honda, F. Res. Let. Phys. Chem. **2009**, 2009, 371650:1-371650:3.
[14] Eisenhut, M.; Guo, X.; Paulitsch-Fuchs, A.H.; Fuchs, E.C.; Cent. Eur. J. Chem. **2011** Eisenhut, M.; Guo, X.; Paulitsch-Fuchs, A.H.; Fuchs, E.C.; Cent. Eur. J. Chem. **2011**, 9 , 391-403.
[15] Widom, A.; Swain, J.; Silverberg, J.; Sivasubramanian, S.; Srivastava, Y.N. Phys. Rev. E **2009**, 80, 016301:1-016301:7.
[16] Saija, F.; Aliotta, F.; Fontanella, M.E.; Pochylski, M.; Salvato, G.; Vasi, C.; Ponterio, R.C., J. Chem. Phys. **2010**, 133, 081104
[17] Marin, A. G.; Lohse, D.; Phys. Fluids **2010**, 22, 122104:1-122104:9.
[18] Aerov, A. A.; "Why the Water Bridge does not collapse", arXiv:1012.1592v1, **2010.**
[19] Raco, R.J., Science **1968**, 160, Issue 3825, 311-312.
[20] Gonzalez H.; McCluskey, F.M.J.; Castellanos, A.; Barrero, A., J. Fluid Mech. **1989**, 206, 545.
[21] Saville, D.A., Ann. Rev. Fluid Mech., **1997**, 29, 27-64
[22] Burcham, C.L.; Saville, D.A., J. Fluid Mech.**2000**, 405, 37-56.
[23] Burcham, C.L.; Saville, D.A., J. Fluid Mech. **2002**, 452, 163-187.
[24] Melcher, J.R. and Warren, E.P.; J. Fluid Mech. **1971**, 47, 127 – 143.
[25] Ramos, A.;Castellanos, A., Phys. Fluids **1993,** 6, 207 – 225.
[26] Woisetschläger, J.;Wexler, .A.D; Holler, G.; Eisenhut, M.; Gatterer, K.; Fuchs, E. C., *under*

review

[27] Rai, D.; Kulkami, A. D.; Pathak, R. K. J. Chem. Phys. **2006,** 128, 034310

[28] Watson, P.K.; Conduction, instabilities and breakdown in liquid dielectrics, in: Castellanos, A. (ed) (1998) Electrohydrodynamics, International Centre for Mechanical Sciences, CISM Courses and Lectures No. 380, Springer, Wien, ISBN 3-211-83137-1.

[29] Onsager, L., J. Chem. Phys. **1934,** 2, 599-615.

[30] Briere, G.B., Brit. J. Appl. Phys. **1964**, 15, 413-417.

[31] Jeong, S.I.; Seyed-Yagoobi, J., IEEE Trans. Ind. Appl. **2003,** 39, 355-361.

[32] Zhakin, A.J., Conduction phenomena in dielectric liquids, in: Castellanos, A. (ed) (1998) Electrohydrodynamics, International Centre for Mechanical Sciences, CISM Courses and Lectures No. 380, Springer, Wien, ISBN 3-211-83137-1.

[33] Hinds, W.C., Aerosol Technology. 2 ed. 1998: Wiley Interscience.

[34] Hartman, R.P.A.; Brunner, D.J.; Camelot, D.M.A.; Marijnissen, J.C.M.; Scarlett, B. J. Aerosol Sc. **2000**, 31, 65-95.

[35] Darabi, J.; Ohadi, M.M.; Desiatoun, S.V.; J. Heat Trans. **2000**, 122, 741.

Journal of Physics: Conference Series **329** (2011) 012004 doi:10.1088/1742-6596/329/1/012004

Autothixotropy of Water and its Possible Importance for the Cytoskeletal Structures

Bohumil Vybíral

University of Hradec Králové, Faculty of Science, Department of Physics, Czechia

E-mail: bohumil.vybiral@uhk.cz

Abstract. The article deals with newly observed phenomenon in water that was called 'the autothixotropy of water'. A history of the phenomenon is connected with very fine gravimetric measurements which were made by author in years 1978 till 1986. Autothixotropy is a very weak macroscopic phenomenon that is observed in water that has remained at rest for a certain time (in tens of hours or days). For the experiments distilled water was used, re-boiled before use. The phenomenon is manifested by force (mechanical resistance) to the immersed body in water with a tendency of changing its position. The static and dynamic methods were used for the research. With the static method a moment of force is measured that is necessary for a very distinctive turn of a stainless steel plate hung up on a thin filament and immersed in standing water. With a certain angular torsion of the filament a certain moment of force is achieved when a state of stress reaches a critical value in water which is demonstrated with an expressive changing of angular position of the plate. When the direction of moment of force is changed, it is reflected in the differential rotation of the angle plate (the phenomenon of hysteresis was observed). The phenomenon of autothixotropy expires after a certain time, if water is mechanically mixed or re-boiled. It is significant that the phenomenon of the autothixotropy is not present in deionised distilled water. When using a dynamic method both oscillations of the plate and a very slow fall of a small ball in standing water was observed. The autothixotropy of water can be explained by a hypothesis of a cluster formation by H_2O molecules in standing water. As the phenomenon of autothixotropy is not present in deionised water, it may be primarily caused by a presence of ions in water, as it was demonstrated in the experiment with a salt solution (NaCl). Biophysical applications of the autothixotropy of water remain open. Some possible applications of the phenomenon in an organization of water in cytoskeletal structures are outlined.

1. Introduction – Phenomenon 'Autothixotropy of Water'

Phenomenon in water that was called the autothixotropy of water: it is a very weak macroscopic phenomenon that is observed in distilled water that has remained at rest for a certain time (in about tens of hours or days) i.e. 'standing' water. For experiments distilled water was used, re-boiled before use. The phenomenon is manifested by very small force (definite mechanical resistance) against a body immersed in water and arising when it should change its position. If standing water is vigorously mixed or re-boiled again, the phenomenon disappears and appears again in a matter of about in tens of minutes or hours (or days). Macroscopically therefore standing water acts as a very weak gel, which has approximately tixotropical properties. Because it arises spontaneously a phenomenon was called 'the autothixotropy of water'.

Journal of Physics: Conference Series **329** (2011) 012004 doi:10.1088/1742-6596/329/1/012004

2. History of Discovery

2.1. Problems by Gravitational Measurements

The history of discovery of this phenomenon is connected with very fine gravimetric measurements which I made in the years 1978 till 1986. The subject of research was an experimental verification of gravitational interaction of bodies immersed in fluids – theoretically formulated by Z. Horák [1]. As fluids I used distilled water. First experiments that I proved were made in the year 1978 by an application of Cavendish static torsion – balance method (Figure 1a). The experiments were not successful: during the experiments unknown molecular properties of water were shown disturbingly and superimposed the expected gravitation of the phenomena very strongly. After a period longer than about ten hours it was not possible to control the torsion pendulum with a rotation of the filament hangings – molecular forces of unknown origin took effect against small gravitational forces. That's why in 1986 I decided to use the dynamic Boys method (Figure 1b) which helped to verify the gravitational phenomena in fluids with an accuracy 0.4% [2], [3]. In this case the torsion pendulum was situated in more diluted air and the pendulum oscillated in gravitational field of system of two cast iron spheres immersed in water. Herewith an influence of phenomenon of autohixotropy was excluded.

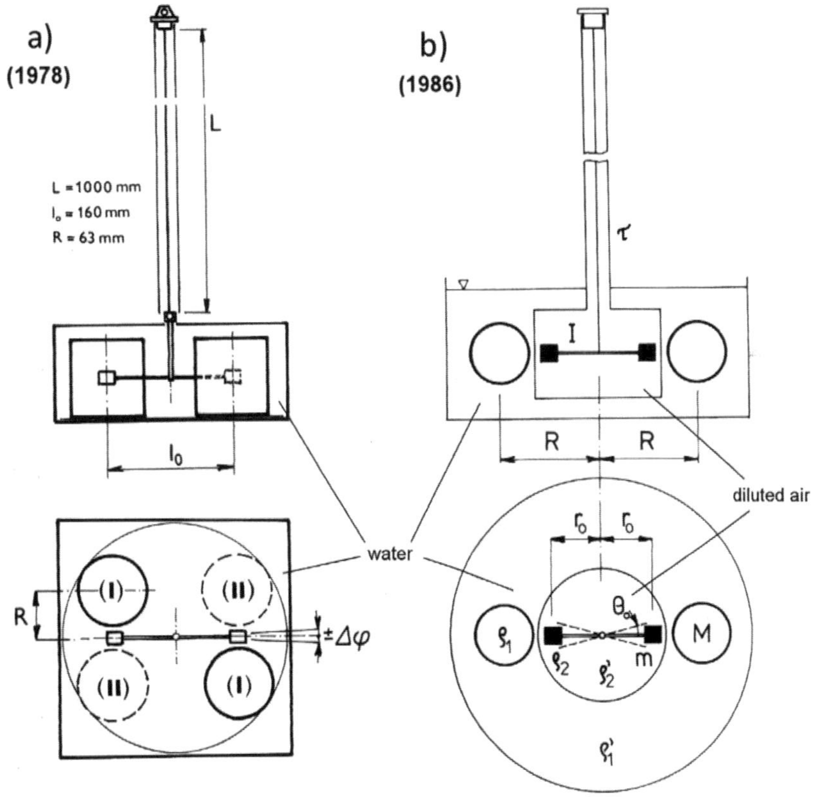

Figure 1. Diagram of apparatus for gravitational measurements: a) Apparatus for static measurement of Cavendish method (1978); b) Apparatus for dynamic measurement of Boys method (1986).

2.2. Qualitative and Quantitative Analyse of Newly Observed Phenomenon in Water

At the recommendation of P. Voráček from University Lund, Sweden, I returned in years 1991 and 1992 to my failed experiments from 1978 and I made series of special experiments leading to an examination of new observed phenomena in water. The results of qualitative analysis were sent in year

1993 for a publication in *Annalen der Physik* (D), *Nature* (GB) and *Science* (USA), but without any positive feedback. Only in 2003 the text was published at http://arxiv.org/abs/physics [4]. The positive acceptance, especially by G. Pollack (USA, WA) and M. Chaplin (GB), brought me to a quantitative measurement of the phenomena in the years 2003 till 2005. The results of this research were published at [5], [6].

3. Static Torsion Method of Experimental Research of Autothixotropy

3.1. Principle of the Method and the Apparatus

Principle of the static torsion method is this ([5], [6]): an experimental body was a flat plate made of stainless steel is hanged up on an elastic filament and immersed into the distilled air-free water (photo of apparatus is on Figure 2). During the experiment it comes to a twisting of the upper filament end and a reaction of the hanged up plate is being watched.

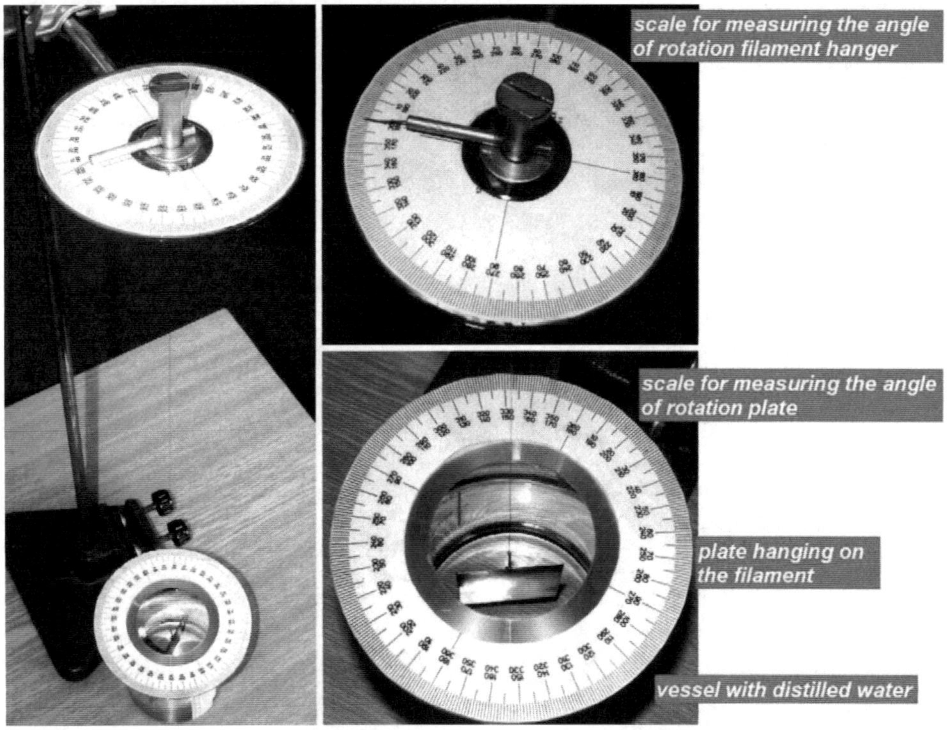

Figure 2. Photo of apparatus for measurements of authotixotropy of water.

Parameters of apparatus:

- A flat stainless steel plate with dimensions: width 38.5 mm, height 60.5 mm, thickness 0.50 mm and mass 8.50 g.
- Phosphor-bronze filament with a cross-section of (0.20×0.025) mm^2 (as used in the clock industry), length L = 465 mm. The torsional rigidity of filament for length L = 465 mm was measured with a dynamic method (method of oscillations) $k_\tau = (1.01 \pm 0.02)\times 10^{-7}$ N.m/rad.
- Distilled air-free water: it was re-boiled for about 3 minutes before the experiment its temperature was kept within the values 24°C – 25°C. Before beginning experiment the water is in the steady state more than about ten hours or days.
- Water with volume of approx. 400 ml was in a glass vessel (beaker) with an inner diameter of 80 mm and a height of 110 mm. The vessel was closed with a paper lid when not in use.
- Accuracy of measurement of angles φ_u , φ_d : ~ 0.5°.

3.2. Measurements of the Phenomenon of Autotixotropy of Water

3.2.1. *Ideal fluid model of water* was assumed first. Then, if we turn the upper end of the filament by angle φ_u, we expect that the plate will follow the rotation of upper, so that approximately is[1] $\varphi_d = \varphi_u$, where φ_d being angle of rotation of the plate (this equality is true for i.e. 'fresh' water).

3.2.2. *Observations.* According to our experiments with 'standing' water, equality $\varphi_d = \varphi_u$ was not achieved. In the experiment, a series of increasing values of angle φ_d is observed, following a very slow – 'step by step' – change of angle φ_u. We measured function $\varphi_d = f(\varphi_u)$. One can specify the action moment of force M_w, arising when the plate influences the water: $M_w = k_\tau(\varphi_u - \varphi_d)$. If angle φ_u reaches a critical value $(\varphi_u)_{crit.}$, the rotation of the plate (i.e. $\dot{\varphi}_d$) becomes relatively quick.

3.2.3. *Quantitative results.* Typically experimental functions $\varphi_d = f(\varphi_u)$ are on Figures 3 and 4, in which there are two examples from many repeatedly measured values [5]. The important points of the graphs are:

- Starting point of equilibrium position $\varphi_u = \varphi_d = 0°$.
- The *critical angle* $(\varphi_u)_{crit}$, i.e. the angle φ_u at which – if reached – the plate began (for a few tens of seconds) to rotate in direction of turning of hanger (angle φ_u), during which time the angle φ_d has been changed considerably.

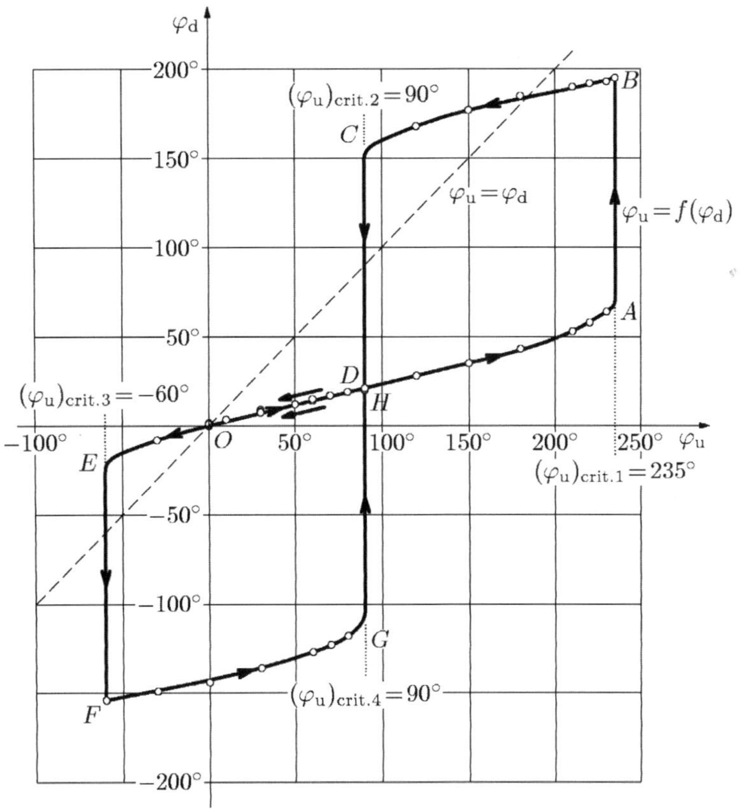

Figure 3. Loop of hysteresis – results of the experiment with the completely immersed plate: loop of the changes of angle $\varphi_d = f(\varphi_u)$; see [5].

[1] See too subsection 5.2 and Figure 5 (the curve ***a***).

3.3. Phenomenon of Hysteresis

When the direction of moment of force is changed, it is reflected in the different rotation of the hanged plate - *the phenomenon of hysteresis* is observed [5]. Measurements for a cyclic change of angle φ_u were carried out – see Figures 3 and 4.

In Figure 4 we see the results of two other experiments: Loop ***a*** is relevant for the experiment with the plate totally immersed in water, while loop ***b*** is related to the experiment with the plate only half-immersed. The effect is clearly more pronounced for the half-immersed plate than for the completely immersed one.

Figure 4 (see loop ***a*** for the experiment with the plate totally immersed in water): we starting in the equilibrium position $\varphi_u = \varphi_d = 0°$ and slowly changing angle φ_u, than the change of angle φ_d did not follow an ideal straight line $\varphi_u = \varphi_d$, but the curve *O-A*. At point *A* the critical value $(\varphi_u)_{crit.1} = 80°$ was reached and then the plate turned into a new equilibrium position – point *B*. The action moment of force is $M_w = k_r(\varphi_u - \varphi_d) \approx 7.1\times10^{-7}$ N.m. With now *decreasing* angle φ_u, than angle φ_d changed according to curve *B-C*, until it reached the second critical value, denoted $(\varphi_u)_{crit.2} = -25°$. Afterwards the plate turned to another equilibrium position – point *D*. When angle φ_u was increasing again, the position of the plate went to origin *O* with the original equilibrium position $\varphi_d \cong 0°$. Herewith the loop of cyclical changes of the angle φ_d of a rotation of the plate is closed. The loops in Figure 4 are simpler than those in Figure 3 and the respective values $(\varphi_u)_{crit.}$ are lower.

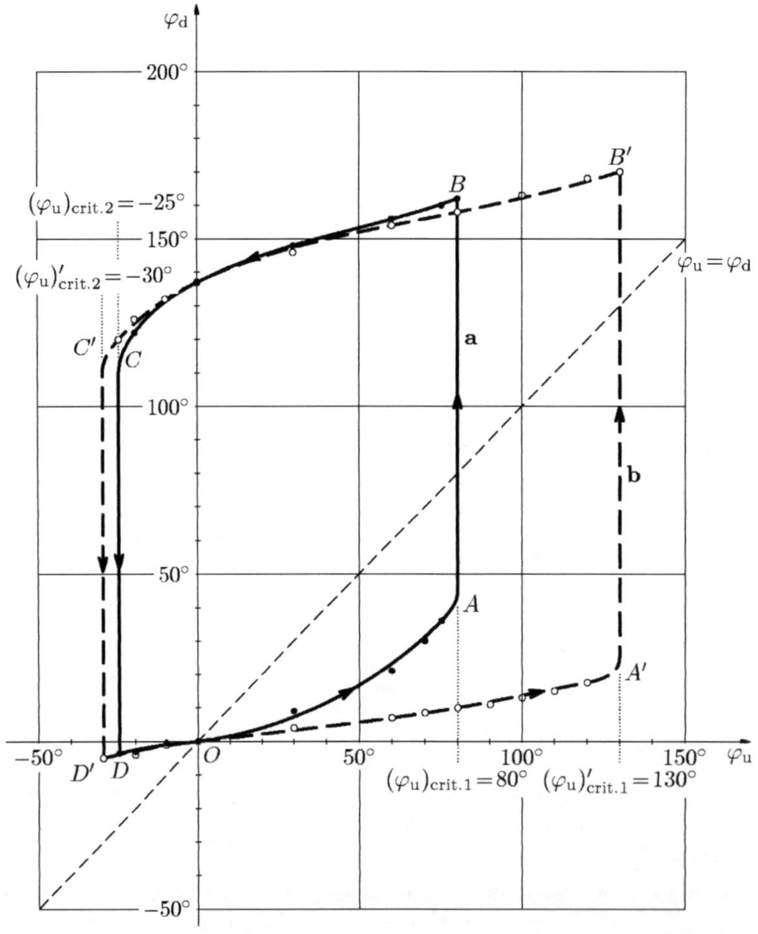

Figure 4. Loops of hysteresis – results of the other two experiments (loops of the changes of angle $\varphi_d = f(\varphi_u)$): with the completely immersed plate (loop ***a***) and with the half-immersed plate (loop ***b***); see [5].

3.4. Additional Measurements

To eliminate other eventual influences on the observed phenomena of autothixotropy, we were realized additional measurements [5].

3.4.1. *Acidity (pH)* of the sample of water was determined by a potentiometric measurement of the pH factor. It did not change considerably during a long period of time (in the range of temperature from 24°C to 25°C the acidity moved in the range 7.1 to 6.9).

3.4.2. *Electric conductivity* of entirely fresh water was 5.6 μS/cm, and after 5 weeks it increased to 30.5 μS/cm at 25°C. A dependence of the observed water properties on this change was not noted.

3.4.3. *Surface tension.* With use of a Du-Noüyhy apparatus (with an accuracy of ~1%), no measurable change of the surface tension was found.

4. Dynamic Method of Torsion Oscillation

4.1. Principle of the Method

We have a plate with an axis of symmetry along with the filament (with torsional rigidity k_τ) on which the plate hangs. We immerse the plate into the water (Figure 2) and consider its torsion oscillations in two situations (see [5]):

- In the 'fresh' water (i.e., with non-exist or negligible autothixotropy); the period of free damped oscillations is T_1.
- In the 'standing' water (i.e., with autothixotropy), we suppose that it is necessary to add, to the quantities related to the elasticity of the filament with torsional rigidity k_τ, the elasticity parameter of putative clusters of water molecules in the considered situation, being represented – for the plate which we used – by a respective torsional rigidity k_w. The period of damped oscillations is $T_2 < T_1$.
- If we measure the periods of oscillation T_1 and T_2, we can calculate [5] the equivalent torsional rigidity:

$$k_w = 4\pi^2 I\left(\frac{1}{T_2^2} - \frac{1}{T_1^2}\right),$$

where I is the moment of inertia of the plate (calculated from the plate dimensions and mass).

4.2. Quantitative Experimental Results of Oscillations

For the measurement (see [5]), an aluminium plate with a thickness of 2.95 mm, width b = 47.59 mm, height h = 50.59 mm and mass 18.70 g, was used. Its moment of inertia was calculated from its dimensions and mass: $I = 7.518 \times 10^{-6}$ kg.m^2. The plate was hung along its longitudinal axis of symmetry on a phosphor-bronze filament with a cross-section of (0.025×0.2) mm^2 and a length of L = 569 mm (the filament had a torsional rigidity $k_\tau = 8.25 \times 10^{-8}$ N.m/rad). The plate was immersed in distilled and re-boiled water in such a way that the upper edge of the plate was 14 mm above the level of the water surface. The experiment was carried out at a temperature of 23°C. The period of the damped torsion oscillations was measured three times – the first one in fresh water. The period of oscillation was $T_1 = (101.7 \pm 1.2)$ s. Then the system was left at rest for seven days, so that a high level of autothixotropy could be reached. Afterwards, the plate was carefully rotated from this new equilibrium position by $\varphi_d \approx 45°$, and at that position it stayed. Then, the plate was given a torsional pulse, by which damped torsion oscillations were initiated. The period of oscillation was measured ten times, the period was distinctively shorter; resulting $T_2 = (5.34 \pm 0.06)$ s. The equivalent torsional rigidity of this system with autothixotropy was determined to be $k_w = (1.04 \pm 0.03)\times 10^{-5}$ N.m/rad. It is possible to say, that the degree of the level of autothixotropy of the system, is ascertainable by means of the measurement of critical angle $(\varphi_u)_{crit.}$. For our system it was determined to be $(\varphi_u)_{crit.} \approx 340°$ (i.e. the action moment of force $M_w \approx 4.0\times 10^{-7}$ N.m).

5. Measurements of the Phenomenon by Changed State

5.1. Deionised Distilled Water

In second phase of experiments [5], water, which was first distilled and then *deionised*, was used. The experiments showed that *in deionised water the phenomenon of autothixotropy and its hysteresis was absent.* The same equipment (Figure 2) was used for the experiment and the plate was immersed both to one half and entirely as well. The period that water was standing still before the measurement was almost ten days. It was found from the measurements that the rotation angle φ_d of the plate, which passed thru the interval $\varphi_d \in (0°, 360°, 0°)$, was equal to angle of torsion φ_u of the upper end of the filament. The ideal equation $\varphi_d = \varphi_u$ was achieved with accuracy of ~1.5°, as evaluated from the repeated measurements. Neither the existence of critical angles $(\varphi_u)_{crit.}$, nor the phenomenon of hysteresis, were found. From the performed experiment, we arrived at the important conclusion that the autothixotropy of water, characterized by a non-zero critical angle, and hysteresis, *is caused by the presence of ions in the water.*

Note: It results from the described experiments with deionised distilled water that in case of an application of deionised distilled water in a gravitational experiment according to Figure 1a, the measurement would be probably successful. This experiment was not made.

5.2. Salt Solution

The observation and newly finding from an experiment in June 2011: During a time period of 10 days deionised distilled water had been observed. The autothixotropy did not appear. Afterwards a salt solution with a concentration of 1 per mille NaCl was prepared from this water (0.400 g salt was dissolved in 400 ml of the examined deionised distilled water). For a period of 7 days no autothixotropy appeared, more to the contrary for angles inequality $\varphi_d > \varphi_u$ was valid (see the curve ***a*** in the Figure 5). Thus the angle φ_d of turning of plate is bigger than the angle φ_u of a primal turning of the upper hanging of filament. It is a remarkable finding because the plate shows a certain additional automotoric movement. The measurement shows that for the certain angle φ_u is the angle of turning φ_d the same (with an accuracy of ~ 0.5%) both for angles φ_u going both up and down too; see the two-way curve ***a*** in the Figure 5. The described situation is valid until autothixotropy appears.

In the observed case with the salt solution the phenomenon of autothixotropy appeared after 7 days from a preparation of the solution. Afterwards, on the contrary, the measurement results $\varphi_d < \varphi_u$. After the critical angle $\varphi_u = (\varphi_u)_{crit.1} = 70°$ is achieved, the angle φ_d of plate turning starts increasing, even if φ_u, is invariable how it is evident from the Figure 5, loop ***b***. If we keep decreasing of the angle φ_u from this status (point A), the angle φ_d of plate turning starts decreasing gradually to the upper part of branch of the loop till the second critical angle $(\varphi_u)_{crit.2} = -40°$ is achieved when the angle φ_d of plate turning decreases spontaneously to negative value of φ_d in comparison with the initial status (point B). During the angle φ_u increasing the loop ***b*** closes up in the initial position $\varphi_d = 0$. A status of hysteresis happened. The status of changes $\varphi_d = f(\varphi_u)$, described with the curves ***a*** and ***b*** in Figure 5, were running in a different way.

5.3. Magnetization

For experiments deionised distilled water was used, partly *not magnetized*, partly *strongly magnetized* (magnetized induction of degree 1 T). The magnetization did not have any influence on the results – the autothixotropy did not appear.

5.4. Ethyl-alcohol

An inspectional measurement was also made when distilled water was substituted with *ethyl-alcohol.* But according to expectations the phenomena of the autothixotropy did not appear with the ethyl-alcohol usage.

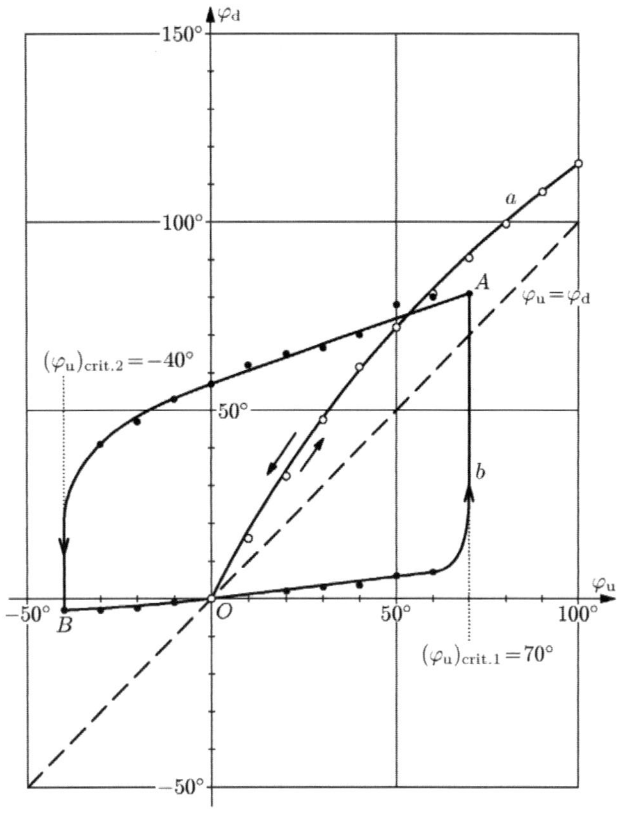

*Figure 5.*The results of experiments with salt solution with the plate totally immersed in solution). The curve ***a*** describes a functional dependence $\varphi_d = f(\varphi_u)$ of a fresh salt solution. The curve ***b*** describes a functional dependence $\varphi_d = f(\varphi_u)$ after arising of autothixotropy. Evidently a phenomenon of hysteresis appeared.

6. Possible Microscopic Interpretation of the Phenomenon

The autothixotropy of water can be explained by a hypothesis of a cluster formation by H_2O molecules in 'standing' water. In the literature about water (collectively see M. Chaplin [7]) we can find various kinds of binding molecules links in a cluster formation on the basis of a hydrogen bridge in case of a water molecule. In case of explication of autothixotropy of water it could deal with a network of icosahedral water super clusters – see [7], [8] and Figure 6. As the phenomenon of autothixotropy is not present in *deionised* water, it may be primarily caused by a presence of ions in water. This hypothesis I was reached by me in 2005 based on the experiments made that time [5]: "From comparison of experiments with natural and deionised distilled water it is possible to deduce that kernels of macroscopic clusters of water molecules are the ions contained in water". In case of the experiment with salt solution NaCl dissociation happens when molecule NaCl breaks up to ions Na^+, Cl^- around which the clusters of molecule H_2O are formed.

Figure 6. Diagram of a super cluster in water (see [7] and [8]). Thirteen complete but overlapping icosahedral clusters form this super-icosahedral structure containing 1820 water molecules (an outer shell of an additional 360 water molecules is shown).

During the experiment with a rotating hanged up plate immersed in water, mechanical characters of a structure of molecules bound in super clusters are probably shown (its elasticity and strength are demonstrated).

7. Possible Importance of the Phenomenon for the Cytoskeletal Structures

Biophysical applications of the autothixotropy of water remain open, but it is expected that the described water structure can be important in biophysics influence on cell characteristics (see eg Pollack [9]). Our observations are consistent with the published results of Wernet *et al.* [10]. Some possible applications of the phenomenon in an organization of water in cytoskeletal structures are outlined.

We thought about this problem in previous work [11] and found the following conclusion: "A manifestation of the autothixotropy of water depends – among others – on the degree of freedom of the transitive motion of its molecules. When the transitive freedom is limited in all possible directions, i.e. in all three space-dimensions, the manifestation of the autothixotropy must – quite logically – become very prominent and influential. Such a situation occurs in small cellular spaces and possibly significantly influences, or even determines, the rigidity of the cytoskeleton of living cell".

8. Conclusions

The experiments with 'standing' distilled water and salt solution proved that after certain time of standing (approximately in about tens of hours or in days) a status of *autothixotrophy of water* is achieved.

- A state of autothixotropy of water is macroscopically demonstrated with very weak mechanical properties analogous to the properties of *solid substances*, such as a certain internal static friction, elasticity and strength. It was shown by means of the experiments with a plate hanged up on an elastic filament and immersed into the exploring water. During a twisting of upper filament end the plate starts rotating progressively when a certain critical angle has been reached. Then it starts spinning and reaches a new relatively equilibrium position.
- During a cycling rotation of the plate (i.e. forth and back) a *phenomenon of hysteresis* was found. It is manifested with a delayed reaction on the filament twisting.
- Water *slightly deviates from an ideal Newtonian viscous fluid*, because the autothixotropy appears in the form of a certain internal static friction, although very weak.
- Experimentally it has been found out that a phenomenon of autothixotropy *does not originate in deionised distilled water*.
- There was proposed a hypothesis that the existence of autothixotropy *is caused by super clusters* that are created in standing water after a certain time and evidently fill up the whole vessel with the exploring water. The kernels of these macroscopic clusters *can be ions* present in water, as it was demonstrated in the experiment with a salt solution (NaCl).
- The super clusters are affected with stirring and shaking of water and are broken with re-boiling over.
- The time interval during which the autothixotropy arises (in case of fresh water or solution or after its re-boiling) has not been definitely determined yet. Obviously it depends on the structure and amount of ions contained in water.
- A hypothesis was expressed that the autothixotropy can have a considerable significance for biophysics because with the aid of autothixotropy *the rigidity of cytoskeleton* of living cell could be explained.

References

[1] Horák Z. Gravitational Interaction of Bodies Immersed in Fluids. *Astrophys. Space Sci.* (1984) **100**: pp 1-11.

Journal of Physics: Conference Series **329** (2011) 012004 doi:10.1088/1742-6596/329/1/012004

[2] Vybíral B. Experimental Verification of Gravitational Interaction of Bodies Immersed in Fluids. *Astrophys. Space Sci.* (1987) **138**: pp 87-98.

[3] Vybíral B. K experimentálnímu ověření gravitační interakce těles ponořených do tekutin. In: *Sborník Pedagogické fakulty*, **54** (*Fyzika*). Praha: SPN, 1989: pp 307-318 (in Czech).

[4] Vybíral B., Voráček P. "Autothixotropy" of Water – an Unknown Physical Phenomenon. Available via ‹http://arxiv.org/abs/physics/0307046›; 2003.

[5] Vybíral B. The Comprehensive Experimental Research on the Autothixotropy of Water. In: Pollack G, *et al.* (eds) *Water and the Cell*. Dordrecht: Springer, 2006, Chap 15: pp 299-314.

[6] Vybíral B., Voráček P. Long term structural effects in water: autothixotropy of water and its hysteresis. *Homeopathy* (2007) **96**: pp 183-188.

[7] Chaplin M. Water as a Network of Icosahedral Water Clusters. Available via ‹http://www.btinternet.com/~martin.chaplin/clusters.html ›, (see on 03. 06. 2011).

[8] Doye, J. P. K., Wales, D. J. and Simdyankin, S. I. Global optimization and the energy landscapes of Dzugutov clusters, *Faraday Disc.* (2001) **118**: pp 159-170.

[9] Pollack G. *Cells, Gels and the Engines of Life*. Seattle WA: Exner and Sons Publisher, 2001.

[10] Wernet Ph., *et al.* The Structure of the First Coordination Shell in Liquid Water. *Science* (2004) **304**: 995-999.

[11] Vybíral B., Voráček P. "Autothixotropy" of Water – an Unknown Physical Phenomenon and its Possible Importance for the Cytoskeleton. In: G.H. Pollack, W.- C. Chin (eds) *Phase Transitions in Cell Biology,* Dordrecht: Springer Science + Bussines Media B.V., 2008: pp 153-157.

Conductivity measurements as a possible means to measure the degree of water ordering

N Verdel[1], **I Jerman**[1], **P Bukovec**[2], **R Krašovec**[1]
[1]Institute Bion, Stegne 21, 1000 Ljubljana, Slovenia
[2]Faculty of Chemistry and Chemical Technology, Aškerčeva c. 12, 1000 Ljubljana, Slovenia

E-mail: igor.jerman@bion.si

Abstract. In our extended experimental work with aqueous solutions of NaHCO3 we noticed higher electrical conductivity in aged solutions in comparison to the conductivity of chemically analogous fresh (one day old) solutions. The phenomenon was found in solutions left undisturbed for longer time (~ 1 year). Most probably the dissolved ions as well as the contact with hydrophilic surfaces play the essential role. We found that higher conductivity is proportional to higher surface vs. volume ratio.

1 Introduction

Water is a simple compound with many peculiar properties. Some of them are well known, while the others are still debated within the scientific community. According to water researchers the extraordinary properties of water represent the reason for the establishment and maintenance of life (Henry 2005, 2009). One of such properties is the so-called water ordering (ordered water, structured water etc.). While it is clear that such phenomenon exists very close to polar (or hydrophilic) surfaces it is not yet resolved whether the order in water can extend considerably far from such surfaces or if it can exist even without such hydrophilic surfaces. Even though the hydrogen bonds that connect water molecules are considered to be responsible for water ordering (Fisenko and Malomuzh 2008) the principles of this ordering are not yet finally settled neither accepted within scientific community.

There are two main streams of concepts that try to model the dynamic order of water. One stems from the quantum field theory and sees the water medium organized in dynamic coherent domains. Namely, an ensemble of molecules interacting with the radiative electromagnetic (EM) field acquires according to Del Giudice above a density threshold and below a critical temperature a new non-trivial minimum energy state, different from the usual one where the oscillations of the molecules are uncorrelated and the EM field is vanishing. The new minimum energy state implies a configuration of the system where all molecules enclosed within an extended region (coherence domain, CD) oscillate in unison and in tune with an EM field trapped within the CD (Montagnier et al. 2010, Del Giudice et al. 2010). The CDs according to Del Giudice characterize every liquid, even the non-polar ones. What could be peculiar in water is that the coherent oscillations are proposed to occur just between the ground and excited state, at 12.06 eV, which is just below the ionization threshold (12.60 eV). In the case of water the dimensions of CDs in normal conditions are proposed to be around 100 nm and they may include an ensemble of almost free electrons that can further accept free energy and convert it to coherent excitations (Del Giudice et al. 2010).

The second stream of concepts trying to explain water ordering seeks various possible cluster formations that would endure more than some ps, perhaps extending into ordinary human time-frame (Chaplin 2004, 2006, Martin Chaplin`s homepage: http://www.lsbu.ac.uk/water/). In any case, it seems that the water ordering is not an all or none phenomenon, but on the contrary, it can have many levels and nobody knows where and what is the supreme level – as far as water still remains a liquid, of course (Stanley et al. 1999).

Regarding its role in maintenance of life, water is mostly treated as a relatively simple, passive solvent for the life`s sustaining molecules, either the small ones (ions, sugars etc.) or the larger ones like proteins, DNA or RNA macromolecules. On the other hand, an increasing number of researchers are coming to the conclusion that water is much more an active ingredient of life as seen conventionally. For instance, Bizzarri and Cannistraro (2002) found the dynamics of the protein and solvent so strongly coupled that they "should be considered as a single entity with a unique rough energy landscape". Therefore, let us look more closely on the influence of hydrophilic and hydrophobic surfaces on the water`s hydrogen bonded network.

Conventionally it is considered that the impact of surfaces on water's hydrogen bonded network extends to a distance of no more than a few molecular layers. Yet, Zheng and Pollack (2003) found that colloidal and molecular solutes are excluded from the vicinity of various hydrophilic surfaces. The width of these solute-free exclusion zones (EZ) may amount to few hundred μm, EZs are physically distinct, less mobile and coexist infinitely with the bulk (Zheng et al. 2006).

These large zones of ordered water are similar to findings of Elia and co-workers that served as a foundation of our empirical work (Verdel et al. Submitted to J. Sol. Chem., Verdel et al.2011). Elia and co-workers explored the physico-chemical properties of extremely dilute aqueous solutions prepared by iterations of dilution and succussion (vigorous shaking). They repeated these two processes until the chemical composition of the final solution was identical to that of the solvent ($NaHCO_3$ (aq.)). They measured electrical conductivities of aged, extremely dilute solutions, and compared the results with electrical conductivities of one day old untreated, chemically analogous solutions and found excess conductivity i.e. higher conductivity values (e.g. Elia et al. 2004, 2005, 2007, 2008, 2009 and 2010). They ascribed their unexpected results to water`s self-organizing abilities that are triggered by the input of kinetic energy during succussion. According to this view, the organised, larger water clusters should accelerate the Grotthuss mechanism of the proton transfer (Agmon 1995) that predominates in the aqueous hydrogen carbonate solutions. Through this mechanism an 'excess' proton or protonic defect diffuses through the network of hydrogen-bonded water molecules through the formation and cleavage of hydrogen and covalent bonds (see Figure 1).

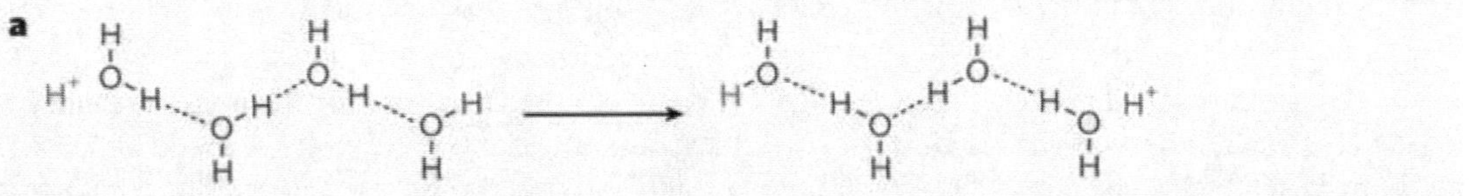

Figure 1. The transport of hydrogen ions (H^+) through water is accomplished by the Grotthuss mechanism whereby hydrogen bonds (dashed lines) and covalent bonds (solid lines) between water molecules are broken and re-formed. The mechanism can involve any number of water molecules, but only four are shown here for simplicity (see also animation on http://en.wikipedia.org/wiki/Grotthuss_mechanism); Source: Vendrell et al. 2009.

In harmony with Elia`s observations, excess conductivities were found only when solutions were left to stand undisturbed for some time (Verdel et al. Submitted to J. Sol. Chem.). However, we found no differences in excess conductivities between diluted and succussed solutions, on the one, and the untreated aged controls, on the other hand. This reminded us on the observations of Vybíral and Voráček (2007), who noticed that distilled water left to stand undisturbed for some time in closed vessels spontaneously developed viscous properties which they ascribed to self-organising abilities of water. The phenomenon was named autothixotropy. Thixotropy is a property of certain gels and liquids that are viscous under normal conditions, but flow (or become less viscous) when shaken or otherwise stressed. In water`s autothixotropic properties, ions, as well as the contact with hydrophilic surfaces could play an essential role. We propose autothixotropy to be the reason for faster proton transfer along the lines of the Grotthuss mechanism found in aged aqueous $NaHCO_3$ solutions in comparison to the freshly prepared solutions with an otherwise identical chemical composition (Verdel et al. submitted to J. Sol. Chem.).

2 The summary of our experiments

Our extended experiments are extensively described in Verdel et al. (submitted to J. Sol. Chem.). We measured the conductivity of aqueous solutions under various conditions and found interesting results wherefrom some important findings can be inferred.

We measured the conductivity of highly dilute fresh (one day old) KCl 10cH and for 310 and 370 days aged KCl 10cH in 0.05 mmol/L $NaHCO_3$. In this, 0.05 mmol/L $NaHCO_3$ was used as starting point solution, solvent. To prepare KCl 1cH, a 1 % mass solution of KCl was prepared in 0.05 mmol/L $NaHCO_3$ and successed for 45 seconds with vigorous mixing. To prepare KCl 2cH, 1 g of KCl 1cH was added to 99 g of 0.05 mmol/L $NaHCO_3$ and successed. Hence, to prepare KCl 10cH the two steps, dilution 1 : 100 and succussion were iterated ten times. The final concentration of KCl in KCl 10cH was identical to the concentration of KCl in the starting point solution. Hence, the final solution (KCl 10cH) was chemically identical to the solvent (0.05 mmol/L $NaHCO_3$).

The actual ionic concentration of aged solutions was determined by ICP-MS. The determination of impurities of aged solutions was namely important because Na^+ ions are constantly leaching from the glass and in aqueous solutions sodium ions react with dissolved CO_2 to form $NaHCO_3$ (Elia et al. 2008).

No differences in conductivity of fresh mechanically treated (successed and diluted) and untreated solutions were found. Basically 2 groups of aged solutions were prepared: in the A group water was aged for 310 days by 2 mL in 2.5-mL flasks and in the B group for 370 days having 2 mL, 5 mL and 10 mL solutions in 20-mL flasks. The solutions in the A group were kept under 3 different conditions (see Table 1).

Table 1. Ageing conditions.

F	S	D
exposed to low temperatures at -20 °C	protected from daylight	exposed to daylight

All solutions in the B group were aged exposed to daylight and other environmental EM disturbances (condition D). The conductivity was measured at 120, 1000, 10,000 and 100,000 Hz 25 and 5 °C. When the frequencies were not in focus conductivity at 1000 Hz was used. For comparison coefficients between measured and theoretical conductivity, σ/σ_{CC}, were used (for detailed information of the method see Verdel et al. submitted to J. Sol. Chem.). From σ/σ_{CC} excess conductivities were deduced by the following equation:

$$\sigma/\sigma_{CC} = \sigma / (\sigma - \sigma^E) \qquad \ldots(1)$$

Conductivity values measured at 1000 Hz were highly reproducible. However, no differences between aged mechanically treated with iteration of dilution and succussion, on the one hand, and aged untreated solutions, on the other hand, were found. Therefore the aged treated and untreated solutions were combined. In Figure 2 the conductivity values of fresh and 370 days aged solutions of the B group are shown.

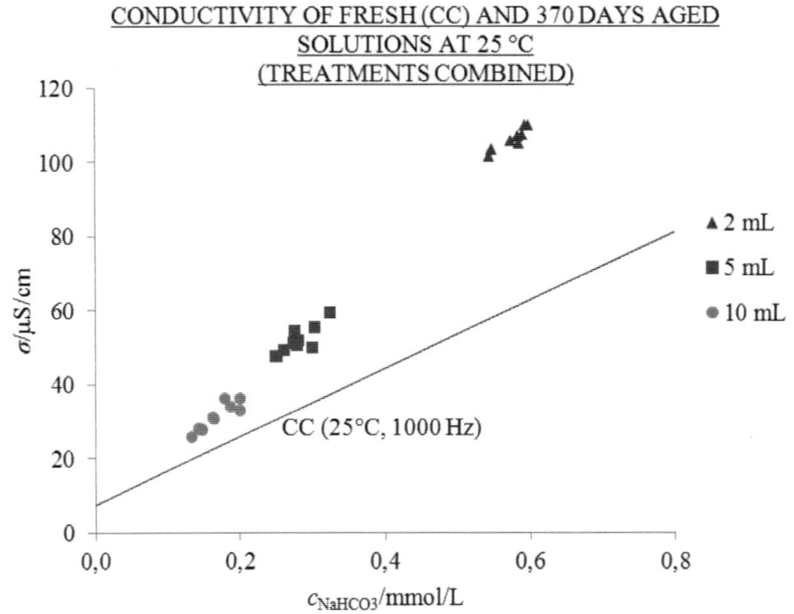

Figure 2. Conductivity of fresh solutions, σ_{CC}, (CC – calibration curve at 25 °C and 1000 Hz) and 370 days aged solutions, σ (the B group), 2 mL (blue), 5 mL (red) and 10 mL (green), as a function of c_{NaHCO3} – all solutions measured at 25 °C and 1000 Hz. Repetitions of 2 mL: 8, 5 mL: 10, 10 mL: 10. Conductivity above CC line is excess.

We can see that with lesser volume the conductivity values are more excess (see Fig. 2) Further calculations uncovered that the excess conductivities are proportional to the ratio of glass surface and solution volume, *S/V*. With increasing *S/V*, the ratio σ/σ_{CC}, increases according to the following equation 2 with $R^2 = 0.928$:

$$\sigma/\sigma_{CC} = 19\ S/V + 100 \qquad \ldots(2)$$

The interesting point was also the condition of ageing (the group A). The frozen samples (condition F) demonstrated even a little subnormal conductivity that we interpret as roughly equal to the theoretical conductivity (see Table 2 and Fig. 3). Hence, no excess conductivity values were found in samples aged frozen. However, the solutions aged under condition S (daylight protected) and D (exposed to daylight) clearly showed excess conductivities. The highest excess conductivities were achieved under the condition S and amounted to 43 % of the measured conductivity σ, the σ^E under condition D amounted to 41.5 % σ, whereas under F σ^E were -17 % σ.

Table 2. Influence of ageing condition on σ/σ_{CC} when frequencies, temperatures and treatments are combined; with standard errors (SE).

condition	D	S	F
σ/σ_{CC}/%	168,6	180,0	88,9
SE/%	2,1	2,5	2,5

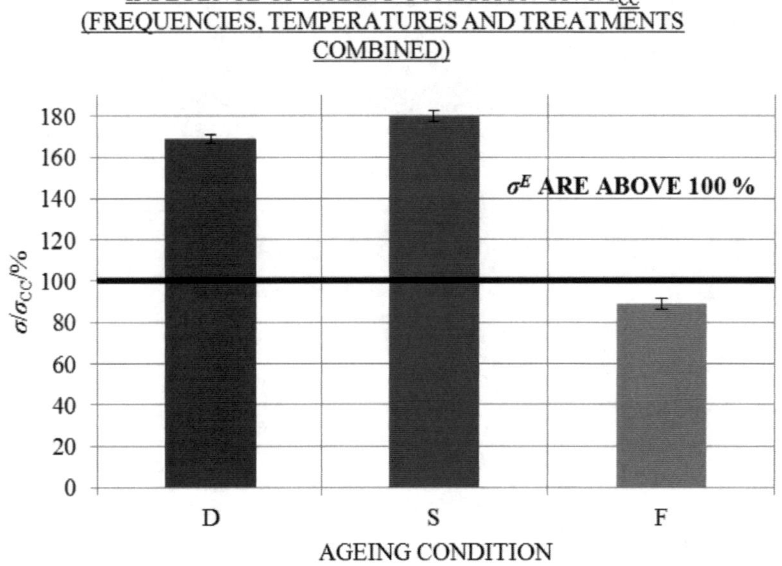

Figure 3. Influence of ageing conditions PR (blue), ST (red) and MD (green) on σ/σ_{cc} with standard error (SE) intervals (treatments, frequencies and temperatures combined). Repetition of D: 28, S: 20, F: 20.

σ/σ_{CC} of solutions aged under condition S were significantly higher than D (p = 0,003), and at the same time σ/σ_{CC} of solutions aged under both conditions (D and S) were significantly higher (p = 0,000) than under F (see Fig. 3).

In general, the results demonstrated anomalies in conductivity of aged solutions in comparison to the chemically analogues freshly prepared controls (CC) except for those samples aged frozen (see excess conductivities in Figs. 2 and 3).

Very interesting, although rather sophisticated, effects were uncovered by experiments concerning temperature and the σ/σ_{CC}. Namely, as shown in Figure 4 with higher frequencies (starting above 1000 Hz and below 10,000 Hz) higher σ/σ_{CC} at lower (5 °C) than at higher (25 °C) temperature are found. The difference is statistically significant at 10,000 Hz and even more at 100,000 Hz.

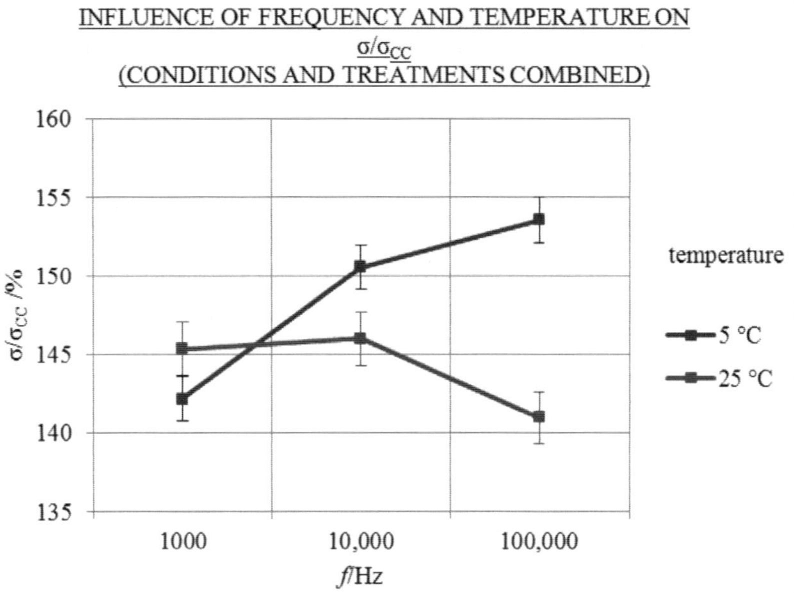

Figure 4. Influence of frequency and temperature on σ/σ_{CC} when conditions and treatments are combined. Repetition of 5 and 25 °C: 68.

3 The most important inferences and further implications

Excess conductivity values (σ^E) in aged aqueous solutions may indicate a highly ordered water system that forms spontaneously with time. Namely the long range and relatively long lasting water structure that should enable a faster transfer of protons – the most mobile ions in aqueous solutions. σ^E – correctly measured and determined – could be regarded as an indicator of water structuring. On its basis we could potentially develop a system of (at least partial) measuring of water ordering. Of course, it is only an idea that should be further thoroughly elaborated and developed in its theoretical as well as empirical aspects.

The second important area of further research following our line of study concerns the living process, especially bioenergetics, mostly founded on the transport of protons.

Let us first examine how some of the most conspicuous results of our research should help in the first task.

3.1. Time effect

Through time of resting the σ^E value in aqueous solutions arises therefore we may conclude that resting time works in favour of spontaneous water structuring. These findings are in harmony with the findings of Vybíral (2006) as well as Vybíal and Voráček (2007) who found the so-called autothixotropic properties of water. The observed phenomenon is relatively weak and appears only on a macroscopic scale, if water is standing still for some time. It causes a force of mechanic resistance against an immersed body, when it should change its position. In their research static and dynamic gravimetric methods were used. The explanation for autothixotropy of water was based on a supposition of cluster formation by H_2O molecules in the aged water. Since the phenomenon does not appear in deionized water, a conclusion was preliminary drawn that the phenomenon is determined by the presence of ions.

Similar time related changes in water characteristics were noticed by Lobyshev and co-workers (1999). They found that distilled water possesses weak luminescence in the near UV and visible regions of the EM spectrum. Its emission spectra possess wide bands at 360 and 410 nm that depend on unusual properties of water. The intensity of luminescence depends on the duration of water storage in a closed vessel. After two to four months of holding the samples in the dark the intensity of the emission spectrum bands increased twofold. Further storage of samples did not change the intensity appreciably.

3.2. **Effect of ions**
Through raising the concentration of ions σ^E values arise (see Fig. 2). Hence, what is the part that ions play in spontaneous structuring of water? Turton et al. (2008) separated rotational and translational motions of the water molecules with the ultra-fast Kerr effect and dielectric relaxation spectroscopic measurements. The results of these two methods showed that salt solutions behave like a supercooled liquid approaching a glass transition, where rotational and translational motions become decoupled. At the glass transition concentration the density of clusters of water molecules and ions is so high that they effectively turn the electrolyte solution into glass.

3.3. **Surface effect**
Higher surface to volume ratio, *S/V* ratio, yields higher σ/σ_{CC}, hence higher σ^E (see Equation 2). Most probably a higher *S/V* ratio promotes higher ordering of water clusters. As already mentioned, the hydrophilic surfaces may have rather thick exclusion zones – zones of highly ordered water (Zheng et al. 2006).

3.4. **Frequency and temperature effect**
As can be seen on Figure 4 at 1000 Hz there is no significant difference between σ/σ_{CC} measured at two different temperatures, 5 and 25 °C. However, at higher frequencies (starting at 10,000 Hz and much more conspicuously at 100,000 Hz) we can clearly see a significantly higher σ/σ_{CC} at *lower* temperature. It seems that even if there is lesser protonic mobility at lower temperature, this is counteracted by higher frequency that somewhat positively interacts with water clusters. And since the clusters are more stable and better organised at lower temperature, a sufficiently high frequency can uncover this order in terms of higher conductivity. It seems that oscillating electric field promotes coherent orientation of water molecules thereby enhancing the formation of ordered chains or Zundel cations (Han et al. 2006).

3.5. **Ageing condition**
We learned that the partially protected flasks under condition S have the highest σ/σ_{CC}, hence the highest σ^E, while no σ^E was found in frozen samples (see Fig. 3). According to our hypothesis ordered water demands liquid phase, the capacity to move that is normally seen as a striking contrast to order. At the same time solutions aged under the condition S (daylight protected) have higher σ^E values than those aged under condition D (exposed to daylight).

As far as the biological implication goes, we know that biological energy supplying electricity is based on proton currents, mostly via mitochondrial membranes. In water the proton current is the quickest and can be raised significantly through water's self-ordering (Han et al. 2006).

If we look on our main results, they indicate the following:
3.6.1. **Aging**
At our measured scale (months) the results would have little biological meaning since biological water is only exceptionally at rest for longer time. But the effect could still have importance on a very different scale, perhaps the one of μs. Further experiments should be devoted to shorter time scales.
3.6.2. **Surface**
This factor should have a profound significance on the ordering of water in biological systems, since the biological (mostly intracellular) surfaces are very large vs. volume. Further experiments should be devoted to explore the σ^E of very small volumes of aqueous solutions in vicinity of very large surfaces.

3.6.3. Ionic composition

As already pointed out by the findings of Turton et al. (2008) as well as the ones of Vybíral and Voráček (2007), the ionic composition of water is very important. In further researches concerning the biological role of enhanced electrical conductivity various ionic solutions should be tested, above all those expected in living tissues.

Acknowledgement
This work was in part financed by the European Union, European Social Fund and was implemented in the framework of the Operational Programme for Human Resources Development for the Period 2007-2013, Priority axis 1: Promoting entrepreneurship and adaptability, Main type of activity 1.1.: Experts and researchers for competitive enterprises.

References

[1] Agmon N. 1995. *Chem. Phys. Lett.*, **244** 456
[2] Bizzarri A. R. and Cannistraro S. 2002. J. Phys. Chem. B, **106** 6617
[3] Chaplin M. F. 2004. *Sci. Society*, **24** 42
[4] Chaplin M. F. 2006. *Nature Rev. Mol. Cell Biol.*, **7** 861
[5] Chaplin M. F. homepage: http://www.lsbu.ac.uk/water/.
[6] Elia V., Napoli E., Niccoli M., Nonatelli L., Ramaglia A., Ventimiglia E. 2004. *J. Therm. Anal. Calorim.*, **78** 331
[7] Elia V., Marchese M., Montanino M., Napoli E., Niccoli M., Nonatelli L., Ramaglia A. 2005. *J. Sol. Chem.*, **34** 947
[8] Elia V., Elia L., Montanino M., Napoli E., Niccoli M., Nonatelli. 2007. *J. Mol. Liq.,* **135** 158
[9] Elia V., Napoli E., Niccoli M., Marchettini N., Tiezzi E. 2008. *J. Sol. Chem.*, **37** 85
[10] Elia V., Napoli E., Niccoli M. 2009. *J. Mol. Liq.,* 148: 45
[11] Elia V., Napoli E., Niccoli M. 2010. *J. Therm. Anal. Calorim.,* **102** 1111
[12] Fisenko A. I. and Malomuzh N. P. 2008. *Chem. Phys.*, **345** 164
[13] Del Giudice E., Spinetti P.R., Tedeschi A. 2010. *Water,* **2** 566
[14] Han J., Zhou X. and Liu H. 2006. *J. Power Sources* **161** 1420
[15] Henry M. 2005. *Cell Mol. Biol.,* **51** 677
[16] Henry M. 2009. *Water,* **1** 3
[17] Lobyshev V. I., Shikhlinskaya R. E., Ryzhikov B. D. 1999. *J. Mol. Liq.,* **82** 73
[18] Montagnier L., Aissa J., Del Giudice E., Lavallee C., Tedeschi A., Vitiello G. 2010. arXiv:1012.5166v1.
[19] Stanley H. E., Buldyrev S. V., Canpolat M., Havlin S., Mishima O., Sadr-Lahijany M. R., Scala A., Starr F. W. 1999. *Physica D,* **133** 453
[20] Turton D.A., Hunger J., Hefter G., Buchner R., Wynne K. 2008. *J. Chem. Phys.,* 128: 161102-1
[21] Vendrell O., Gatti F., Meyer H. – D. 2009. *Angew. Chem. Int. Edit.,* **48** 352
[22] Verdel N., Jerman I., Krašovec R., Bukovec P., Zupančič M. Submitted to *J. Sol. Chem.*
[23] Verdel N., Jerman I., Bukovec P. *Int. J. Mol. Sci.* 2011. Accepted for publication.
[24] Vybíral B. 2006. In: *Water and the Cell.* Ed: Pollack G. H., Cameron I., Wheatley D. Springer, New York: 299
[25] Vybíral B. and Voráček P. 2007. *Homeopathy,* **96** 183
[26] Zheng J. M. and Pollack G. H. 2003. *PRE,* **68** 031408-1
[27] Zheng J.-M., Chin W.-C., Khijniak E., Khijniak Jr. E., Pollack G. H. 2006. *Adv. Colloid Interface Sci.*, **127** 19

Journal of Physics: Conference Series **329** (2011) 012006 doi:10.1088/1742-6596/329/1/012006

On the Photonic Cellular Interaction and the Electric Activity of Neurons in the Human Brain

V Salari[1], J Tuszynski[2,3], I Bokkon[4], M Rahnama[5,6] and M Cifra[7]

[1] Institut de Mineralogie et de Physique des Milieux Condenses, Universite Pierre et Marie Curie-Paris 6, 4 place Jussieu, 75252 Paris cedex 05, France
[2] Department of Experimental Oncology, Cross Cancer Institute, 11560 University Avenue Edmonton, AB T6G 1Z2, Canada
[3] Department of Physics, University of Alberta, Edmonton, AB Canada
[4] Doctoral School of Pharmaceutical and Pharmacological Sciences Semmelweis University, Hungary
[5] Department of Physics, Shahid Bahonar University of Kerman, Kerman, Iran
[6] Afzal Research Institute, Kerman, Iran
[7] Institute of Photonics and Electronics, Academy of Sciences of the Czech Republic, Prague, Czech Republic

E-mail: vahid.salari@impmc.upmc.fr

Abstract. The subject of Ultraweak Photon Emission (UPE) by biological systems is very fascinating, and both evidence of its effects and applications are growing rapidly due to improvements in experimental techniques. Since the relevant equipment should be ultra-sensitive with high quantum efficiencies and very low noise levels, the subject of UPE is still hotly debated and some of the interpretations need stronger empirical evidence to be accepted at face value. In this paper we first review different types of interactions between light and living systems based on recent publications. We then discuss the feasibility of UPE production in the human brain. The subject of UPE in the brain is still in early stages of development and needs more accurate experimental methods for proper analysis. In this work we also discuss a possible role of mitochondria in the production of UPE in the neurons of the brain and the plausibility of their effects on microtubules (MTs). MTs have been implicated as playing an important role in the signal and information processing taking place in the mammalian (especially human) brain. Finally, we provide a short discussion about the feasible effects of MTs on electric neural activity in the human brain.

1. Electromagnetic Radiation and Living Systems

The relation of *biological systems* and *electromagnetic radiation* can be discussed from different points of view. Some of the interesting interactions are as follows:

- *Efficient excitation energy transfer of light by photosynthetic system*

Recently published experimental data in photosynthesis have provided support for the hypothesis that the system uses some nearly100% efficient excitation energy transfer of light (which means almost without dissipation), and it is suggested that quantum coherence plays an important role in this mechanism [65]. This subject has attracted the attention of representing physics and chemistry, especially quantum information theorists who aim to find out how quantum coherence makes the system so efficient. These inquiries resulted in the subject of *quantum biology* becoming a very popular topic in recent years.

- *Response of mammalian cells to near-infrared light*

In a series of studies spanning a period of some 25 years G Albrecht-Buehler (AB) demonstrated that living cells somehow have a molecular analogue of an *eye* which can process light information and react in an *intelligent* manner [66–68]. In his studies microtubular structures especially *centrioles* have been identified as the main candidates for light information processors [68]. He further showed that electromagnetic signals are the triggers for cell repositioning in physical space. It is still largely a mystery how the reception of electromagnetic radiation is accomplished by the centriole. Another mystery related to these observations is the original electromagnetic radiation emitted by a living cell [69]. Using pulsating infra-red signals scattered off plastic beads AB mimicked the effects of the presence of another living cell in the neighbourhood. The question that still remains unanswered and which we address here is the source of infra-red radiation speculated by AB to originate in the mitochondria and later on demonstrated to be correct using quantum mechanical arguments [69].

- *Production of light by living systems*

Photon emission by biological systems can be produced by different mechanisms. In general, light emission can be classified into three groups: (1)Induced light emission, (2)Spontaneous light emission and (3)Black-Body radiation. Here, we discuss a subclass of the second group which is called *Ultraweak Photon Emission* (UPE). All living cells of plants, animals and humans continuously and spontaneously emit ultraweak photons (ultraweak electromagnetic waves) in the optical range of the spectrum, which is associated with their physiological states and can be measured with specific experimental techniques [57]. In different literature sources the UPE is referred to by different names such as ultraweak emission, biophotons, ultraweak bioluminescence, self-bioluminescent emission, photoluminescence, delayed luminescence, ultraweak luminescence, spontaneous chemiluminescence, ultraweak glow, biochemiluminescence, metabolic chemiluminescence, dark photobiochemistry and bioluminescence.

- *Transmission of light by living systems*

Recently, Sun et al. [70] demonstrated that a single neuron can conduct photon signals. Moreover, Wang et al. [71] presented an experimental proof of the existence of spontaneous and visible light induced UPE form freshly isolated rats whole eye, lens, vitreous humor and retina [71].

- *Bio-communication*

There is growing experimental evidence that cells and tissues may interact over distances even when chemically isolated, most likely via electromagnetic fields [51]. Stemming from the pioneering experiments of Gurwitsch in 1920s [52], some researchers confirmed that cellular interactions can be mediated by electromagnetic fields e.g. see [53–56].

Overwhelming majority of the experiments focused on the study of electromagnetic cellular interactions examined in the optical region. For the review of the historical and recent theories and experiments on electromagnetic cellular interactions see [51].

2. UPE emission inside neurons

There are experimental indications that ROS and RNS are responsible for UPE production in living systems [42, 43] and are also necessary for synaptic processes and normal brain functions. Numerous findings have provided evidence of fundamental signal roles of ROS and RNS in cellular processes under physiological conditions. Free radicals and their derivatives act as signaling molecules in cerebral circulation and are necessary for molecular signaling processes in the brain such as synaptic plasticity, neurotransmitter release, hippocampal long-term potentiation, memory formation, etc. [57–63]. Recently Bókkon et al. put forward a molecular hypothesis about biophysical picture representation (intrinsic biophysical virtual visual reality) which states that external photonic signals from an object are converted into electrical signals within the retina and are conveyed to V1 and transformed into regulated UPE via redox processes inside V1 neurons [42,43]. Accordingly, spike-related retinotopic electrical signals - along classical axonal-dendritic pathways can produce synchronized biophotonic signals by redox processes within synchronized retinotopic V1 neurons. In this model, small groups of retinotopic visual neurons can function as visual pixels appropriate to the topological distribution of photonic signals on the retina. As a result, we can get an inherent biophysical picture of the object generated by UPE in early retinotopic V1 during visual perception and imagery [43, 44]. This novel biophysical hypothesis may revive the Kosslyns depictive theory [45] and the *homunculus* (mind's eye) hypothesis [46]. Now the question arises how can this hypothesis be supported experimentally? It should be noted that visual circuits that are normally involved in the detection of visual perception features are also responsible for the generation of the phosphene light perception [43, 48]. Recently Wang et al. presented [49] the first experimental evidence for the existence of spontaneous and visible light induced UPE from in vitro freshly isolated rats whole eye, lens, vitreous humor and retina. In addition, recently, Dotta and Persinger [50] measured significant increases in biophoton emission from near the right hemisphere but not the left for most volunteers when they imagined a white light in a dark room compared to simply casual thinking. These results support the above biophysical picture representation notion [42, 43] and also indicate a more essential role of right hemisphere in visual imagery.

3. Toward coherent states in biological systems?

Biological systems operate within the framework of irreversible thermodynamics and nonlinear kinetic theory of open systems, both of which are based on the principles of non-equilibrium statistical mechanics. The search for physically-based fundamental models in biology that can provide a conceptual bridge between the chemical organization of living organisms and the phenomenal states of life and experience has generated a vigorous and so far unresolved debate [1,2]. Recently published experimental evidence has provided support for the hypothesis that biological systems use some type of quantum coherence in their functions. The nearly 100% efficient excitation energy transfer in photosynthesis is an excellent example [3]. Quantum coherence is a plausible mechanism responsible for the efficiency and co-ordination exhibited by biological systems.

The hypothesis invoking long-range coherence in biological systems was proposed by H. Fröhlich [4–6] and followed by detailed investigations by Tuszynski *et al.* [7–21], Pokorný [22–24], Mesquita *et al.* [25–28] and others for over three decades. The possible role played by coherent states manifested outside low temperature physics has attracted considerable interest in both the physics and biology communities. The original Fröhlich model was very general and did not limit the mechanism of biological coherence to any particular cellular structure. In his model, when the energy supply exceeds a critical level, the dipolar ensemble of biologically relevant molecules populates a steady state of non-linear vibrations characterized by a high degree of structural and functional order [51]. This (electrically polarized) ordered state expresses itself in terms of long-range phase correlations, which are physically similar to such phenomena as superconductivity and superfluidity, where the behaviour of particles is collective and inseparable.

3.1. The Wu-Austin Hamiltonian

There are different approaches possible to be adopted in the analysis of the coherent state generation in biological systems based on Fröhlich coherent states as described in the works of Mesquita et al. [25–27]. The Wu-Austin Hamiltonian [29–31] is the basis of a quantum mechanical approach to Fröhlich coherent states. Bolterauer and Ludwig [72] investigated the thermodynamics of Wu and Austin system quantum mechanically and have shown that even without pumping their Hamiltonian can give rise to Bose condensation. However, the Wu-Austin Hamiltonian has the unphysical property of having no finite ground state. Turcu [32] have obtained a master equation for Fröhlich rate equations. The main aim of his work was to show that there is a rich family of Hamiltonians, modeling differently the pump and the thermal bath, from which the same Fröhlich-like rate equations can be obtained. We believe that the system of neuronal MTs is a good candidate for being described by one of these Hamiltonians. MTs are composed of tubulins which can be considered as biological electric dipoles. Pokorny provided a detailed analysis of the coherent states in MTs. He experimentally observed resonance effects in MTs in the 10 MHz range [22–24].

3.2. Criticism on Coherent states in living systems

Recently, Reimers *et al.* [2] have shown that a very fragile Fröhlich coherent state may occur at sufficiently high temperatures and concluded that there is no possibility for the existence of Fröhlich coherent states in biological systems. Also they provided several diagrams in terms of effective temperature which was defined by the authors as $T_{eff} = \frac{T_s}{T}$, where T_s is the temperature of system and T is the temperature of the thermal bath. Physically, the parameter is wrong because a temperature ratio is a unitless quantity not a quantity with the unit of temperature. They have used the effective temperature parameter for the Wu-Austin Hamiltonian [29–31] and considered it in the high temperature limit. Their diagrams are mostly based on the effective temperature parameter and hence are, in our opinion, not acceptable due to the self-contradictory arguments used in their derivations. For more details see [73]. In fact, the criticism raised by Reimers *et al.* [2, 37] is mainly directed against the so-called Orch OR model which was proposed by Penrose and Hameroff to introduce a physical basis for consciousness. In some formulations of the OrchOR model, a manifestation of quantum coherence involved Fröhlich coherent states in MTs [33–35]. MTs are highly ordered in the neurons of the brain and can indeed be regarded as providing support for Fröhlich coherent states. In this context, the conclusions of our discussion above also apply to MTs. Therefore, we believe that it is still hypothetically possible to generate Fröhlich coherent states in MTs. However, another issue that arises when considering quantum states for MTs is the rapid *decoherence* problem. The question is how it is possible for MTs to be in a coherent state while the environment is relatively hot, wet and noisy? According to decoherence theories, sufficiently strong interactions with the environment cause decoherence, which destroys quantum effects [36]. For macroscopic particles there are two main *natural* ways of experiencing this decoherence: First, decoherence due to collisions with other particles and second the thermal emission of radiation due to the internal heat of an object [38, 41]. Tegmark [39] has calculated decoherence times for MTs based on the scattering between MTs and environmental particles. Hagan *et al.* [40] have shown that Tegmark used wrong assumptions for his investigation of MTs. Another main objection to the estimate in equation (7) is that Tegmarks formula yields decoherence times that increase with temperature contrary to well-established physical laws and the behavior of quantum coherent states. In view of these (and other) problems in Tegmarks estimates, Hagan *et al.* [40] assert that the values of quantities in Tegmarks relation are incorrect and thus the decoherence time should be approximately 10^{10} times larger leading to a ms range of values for typical decoherence times. According to Hagan *et al.*, MTs in neurons could possibly avoid decoherence via several mechanisms for quantum processing to occur there. Tegmark introduced a function for the decoherence rate [47] which is composed of two parts: one for short wavelengths and the other for long wavelengths. Every scattering calculation based on the Coulomb interaction and Tegmarks decoherence rate function leads to decoherence times that are proportional to temperature according to relations such as $\tau_{dec} \propto \sqrt{T}$, $\tau_{dec} \propto \sqrt{T^2}$, $\tau_{dec} \propto \sqrt{T^3}$, etc. Therefore, it can be expected that subsequent calculations based on these criteria are flawed in the high-temperature limit, i.e. as temperature approaches infinity, decoherence time increases too, and if temperature approaches absolute zero, decoherence time approaches zero, a very unphysical conclusion.

4. Microtubules and centrioles

MTs are biological hollow cylinders with a 17-nm inner diameter and a 25-nm outer diameter (see Figure1), composed of units called tubulin dimers, each of which has the dimensions $4nm \times 8nm \times 6nm$ [57]. MTs have been implicated as playing an important role in the signal and information processing taking place in the mammalian, and especially human brain. Earlier, MTs have been considered as optical cavities [74] with quantum properties [75].

Figure 1. Microtubule (MT) is a biological hollow cylinder

It is worth stressing here that centrioles and cilia, which are complex microtubular structures, are involved in photoreceptor functions in single cell organisms and primitive visual systems. Cilia are also found in all retinal rod and cone cells. The dimensions of centrioles and cilia are comparable to the wavelengths of visible and infrared light (see Figure 2) [57]. Albrecht-Buehler has demonstrated that living cells possess a spatial orientation mechanism located in the centriole. This is based on an intricate arrangement of MT filaments in two sets of nine triplets, each of which is perpendicular to the other. This arrangement provides the cell with a primitive *eye* that allows it to locate the position of other cells within two to three degrees of angular accuracy in the azimuthal plane and the same accuracy with respect to the axis perpendicular to it [66].

5. Mitochondria and Microtubules

Both mitochondria and microtubules can form dynamic networks in neurons. Moreover, the refractive index of both mitochondria and microtubules is higher than the surrounding cytoplasm, whose consequence is that mitochondria and microtubules can act as optical waveguides, i.e. electromagnetic radiation (UPE) can propagate within their network [44, 64].

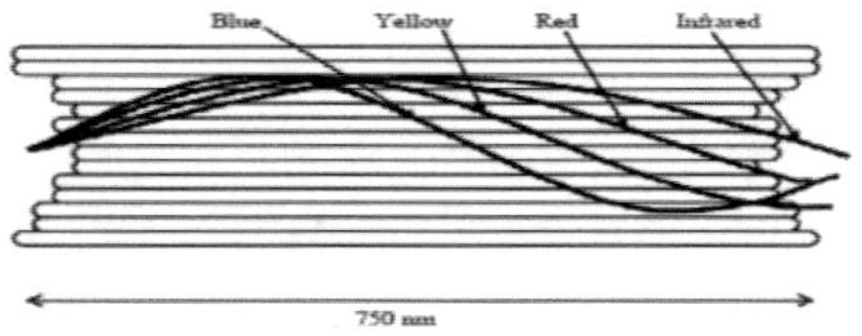

Figure 2. The dimensions of centrioles and cilia are comparable to the wavelengths of visible and infrared light

Regulated UPE (from mitochondrial radicals and excited molecules) can induce polymerization of microtubules. Then, according to the quality of absorbed UPE from mitochondria, microtubules can transport mitochondria in accordance with information processes in cells and neurons. There can be a mutual cross-talk/regulation between mitochondria and microtubules by redox and free radical processes [44].

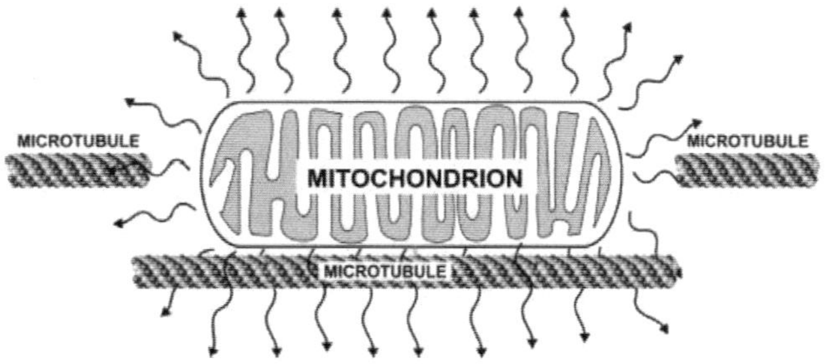

Figure 3. The light production by mitochondria can interact with MTs

6. MTs dynamics and Electric Neural Activity of Neurons

Electrodynamic interactions between various cytoskeletal structures, with MTs playing a central role, and ion channels crucially regulate the neural information-processing mechanism. These interactions involve long-range ionic wave propagation along microtubule networks (MTNs) and actin filaments (AFs), and exhibit subcellular control of ionic channel activity. Hence, they have an impact on the computational capabilities of the entire neural function. Cytoskeletal biopolymers, most importantly AFs and MTs, constitute the basis for wave propagation, and interact with membrane components, leading to a modulation of synaptic connections and membrane ion channels. Association of MTs with AFs in neuronal filopodia guides MT growth and affects neurite initiation [57]. Electric signaling by AFs and MTs may play active roles in coincidence detection and storage of spatiotemporal patterns of inputs, and signaling within the cytoskeleton may be particularly critical to information storage over time scales longer than LTP times. The initial route to the MT network could be through the AFs concentrated in the spines. Inputs to arbitrary sites in the neuron can be transmitted from the neuronal membrane to AFs in spines via scaffolding proteins and signal transduction molecules. Electric signals can then be transmitted, utilizing AF cross-linker proteins to MTs, and subsequently through microtubule associated proteins (MAPs) and signal transduction molecules to other MTs in the network [57].

7. Conclusion

It has been shown that the intensity of UPE is in direct correlation with neural activity, cerebral energy metabolism, EEG (Electroencephalography) activity, cerebral blood flow and oxidative processes [76, 77]. From a theoretical point of view, the interaction of mitochondrial UPE and MTs can take the MTs into coherent states. The synchronous and coherent vibrations of billions of electric dipoles of biomolecules cannot be ignored in the EEG diagrams. MTs are particularly numerous in the brain where they form highly ordered bundles and are the best candidate for long range coherence and large synchrony [57]. In addition to electrical and chemical signals propagating in the neurons of the brain, signal propagation may take place in the form of UPE too. We believe that the role of UPE in the brain merits special attention (see [57]).

References

[1] Abbott D, Gea-Banacloche J, Davies P C W, Hameroff S, Zeilinger A, Eisert J, Wiseman H, Bezrukov S M and Frauenfelder H 2008 *Fluct. Noise Lett.*, **8** c5-c26.
[2] Reimers J R, McKemmish L K, McKenzie R H, Mark A E and Hush NS 2009 *PNAS* **106** 4219-24
[3] Cheng Y C and Fleming G R 2009 *Annu. Rev. Phys. Chem.* **60** 24162
[4] Fröhlich H 1968 *Int. J. Quantum Chem.* **2** 641
[5] Fröhlich H 1970 *Nature* **228** 1093
[6] Fröhlich H 1975 *Proc. Nat. Acad. Sci.* **72** 4211
[7] Tuszynski J A and Dixon J M 2001 *Phys. Rev.* E **64** 051915
[8] Tuszynski J A, Paul R, Chatterjee R and Sreenivasan S R 1984 *Phys. Rev.* A **30** 2666e2675
[9] Tuszynski J A and Paul R 1991 *Phys. Rev.* A **43** 3179e3181
[10] Tuszynski J A, Bolterauer H and Satarić M V 1992 *Nanobiol.* **1** 177e190
[11] Tuszynski J A 1985 *Phys. Lett.* A **107** 225e229
[12] Tuszynski J A 1985 *Int. J. Quantum Chem.* **29** 379e391
[13] Tuszynski J A 1985 *Phys. Lett.* A **108** 177e182
[14] Tuszynski J A 1987 *Nucl. Phys.* B *Proc. Suppl.* **2** 618
[15] Tuszynski J A 1988 *J. Theor. Biol.* **132** 369e373
[16] Bolterauer H, Tuszynski J A 1989 *J. Biol. Phys.* **17** 41e50
[17] Bolterauer H, Tuszynski J A and Satarić M V 1991 *Phys. Rev.* A **44** 1366e1381
[18] Chatterjee R, Tuszynski J A and Paul R 1983 *Int. J. Quantum Chem.* **23** 709e712
[19] Paul R, Chatterjee R, Tuszynski J A and Fritz O G 1983 *J. Theor. Biol.* **104** 169e185
[20] Paul R, Tuszynski J A and Chatterjee R 1984 *Phys. Rev.* A **30** 2676e2685
[21] Portet S, Tuszynski J A, Hogue C W V and Dixon J M 2005 *Eur. Biophys. J.* **34** 912e920
[22] Pokorný J 1982 *J. Theor. Biol.* **98** 21e27
[23] Pokorný J 1999 *Bioelectrochemistry and Bioenergetics* **48** 267e271
[24] Pokorný J 2009 *Electromagnetic Biology and Medicine* **28** 105e123
[25] Mesquita M V, Vasconcellos A R, Luzzi R and Mascarenhas S 2005 *Int. J. Quantum Chem.* **102** 111630
[26] Mesquita M V, Vasconcellos A R and Luzzi R 1996 *Int. J. Quantum Chem.* **60** 68997
[27] Mesquita M V, Vasconcellos A R, Luzzi R and Mascarenhas S 2005 *Brazilian Journal of Physics* **34** 459-88
[28] Marcus V, Mesquita, Aurea R, Vasconcellos, and Luzzi R, 1993 *Phys. Rev. E* 48, 4049-4059.
[29] Wu T M and Austin S 1977 *Phys. Lett.* A **64** 1512
[30] Wu T M and Austin S 1978 *Phys. Lett.* A **65** 746
[31] Wu T M and Austin S 1981 *J. Bio. Phys.* **9** 97107
[32] Turcu I, 1997, *Physics Letters A*, 234, 181-186.
[33] Penrose R and Hameroff S R 1995 *J. Conscious. Stud.* **2** 98
[34] Hameroff S R and Penrose R 1996 *J. Conscious. Stud.* **3** 36
[35] Hameroff S R 1998 *Philos. Trans. R. Soc.* A **356** 1869
[36] Rosa L P and Faber J 2004 *Phys. Rev.* E **70** 031902.
[37] McKemmish L K, Reimers J R, McKenzie R H, Mark A E and Hush N S, 2009 *Phys. Rev.* E **80** 021912
[38] Schlosshauer M 2007 *Decoherence and Quantum-to-Classical Transition* (Berlin: Springer)
[39] Tegmark M 2000 *Phys. Rev.* E **61** 4194-206.
[40] Hagan S, Hameroff S R and Tuszynski J A 2002 *Phys. Rev.* E **65** 061901.
[41] Joos and Zeh, Z. Phys. B, 59, 223-243, (1985).
[42] Bókkon I 2009 *BioSystems* 96, 178-184.
[43] Bókkon I, D'Angiulli A. 2009 *Bioscience Hypotheses* 2, 226-232.
[44] Bókkon I, Salari V, Tuszynski J, Antal I. 2010, *J. Photochem. Photobiol. B Biology* 100, 160-166.

[45] Kosslyn SM. 1994 Image and Brain: The Resolution of the Imagery Debate, MIT Press.
[46] Bókkon I, Salari V, Tuszynski J 2011 *J Integr Neurosci.* 10, 47-64.
[47] Tegmark M 1993 *Found. Phys. Lett.* **6** 571
[48] Bókkon I, 2008 BioSystems 92, 168-174.
[49] Wang C, Bókkon I, Dai J, Antal I. 2011 *Brain Res.* 1369. 1-9.
[50] Dotta BT, Buckner CA, Lafrenie RM, Persinger MA. submitted.
[51] Cifra M, Farhadi A and Fields JZ 2011, *Progress in Biophysics & Molecular Biology*105, 223-246.
[52] Gurwitsch A, 1923, *Archiv für Entwicklungsmechanik der Organismen* 100, 11-40.
[53] Rahn O, 1936, Invisible radiations of organisms, *Verlag von Gebrüder Borntraeger, Berlin.*
[54] Kaznacheev VP, Mikhailova LP 1981 Ultraweak Radiation in Cell Interactions, in Russian, Sverkhslabye izlucheniya v mezhkletochnykh vzaimodeistviyakh, *Novosibirsk: Nauka.*
[55] Albrecht-Buehler G 1992 *PNAS* 89, 8288-8293.
[56] Fels, D 2009 *PLoS ONE* 4, e5086.
[57] Rahnama M, Tuszynski J, Bókkon I, Cifra M, Sardar P, Salari V 2011, *J. Integrative Neuroscience* 10, 65-88.
[58] Kishida KT, Klann E 2007 *Antioxid Redox Signal* 9, 233-244.
[59] Knapp LT, Klann E 2002 *J Neurosci Res* 70, 1-7.
[60] Tejada-Simon MV, Serrano F, Villasana LE, Kanterewicz BI, Wu GY, Quinn MT, Klann E 2005 *Mol Cell Neurosci* 29, 97-106.
[61] Thiels E, Klann E 2002 *Physiol Behav* 77, 601-605.
[62] Thiels E, Urban NN, Gonzalez-Burgos GR, Kanterewicz BI, Barrionuevo G, Chu CT, Oury TD, Klann E 2000 *J Neurosci* 20, 7631-7639.
[63] Volterra A, Trotti D, Tromba C, Floridi S, Racagni G 1994 *J Neurosci* 14, 2924-2932.
[64] Thar R, Khl M 2004 *J Theor. Biol.* 230, 261270.
[65] G.S. Engel, T.R. Calhoun, E.L. Read, T.-K. Ahn, T. Mancal, Y.-C. Cheng, R.E. Blankenship, Fleming GR, 2007 *Nature* 446, 782.
[66] Albrecht-Buehler G, 1994, *Cell Motil. Cytoskeleton* 27, 262-271.
[67] Albrecht-Buehler G, 1995, *Cytoskeleton* 32, 299-304.
[68] Albrecht-Buehler G, 1997, *Exp. Cell Res.* 236, 43-50.
[69] Tuszynski J, Dixon JM, 2001, *Phys. Rev. E* 64, 051915.
[70] Sun Y, Wang C, Dai J, 2010, *Photochem Photobiol Sci* 9, 315-322.
[71] Wang Ch, Bokkon I, Dai J, Antal I, 2011, *Brain Res* 1369,1-9.
[72] Bolterauer H, Ludwig LA, 1993, *Phys. Rev. E* 47, 2122.
[73] Salari V, Tuszynski J, Rahnama M, Bernroider G, 2011, *J. Physics Conf. Ser.*, 306, 012075.
[74] Mavromatos NE, Mershin A, Nanopoulos DV, 2002 *Int J Mod Phys B* 16,3623-3642.
[75] Mavromatos NE,1998 *Int J Mod Phys B* 2, 517-542.
[76] Isojima Y, Isoshima T, Nagai K, Kikuchi K, Nakagawa H, 1995 *Neuroreport* 6, 658-660.
[77] Kobayashi M, Takeda M, Sato T, Yamazaki Y, Kaneko K, Ito K, Kato H, Inaba H, 1999 *Neurosci Res* 34,103-113.

Journal of Physics: Conference Series **329** (2011) 012007 doi:10.1088/1742-6596/329/1/012007

Electrodynamic activity of healthy and cancer cells

J Pokorný

Institute of Photonics and Electronics, Academy of Sciences of the Czech Republic
Chaberská 57, 182 51 Prague 8 – Kobylisy, Czech Republic

E-mail: pokorny@ufe.cz

Abstract. Microtubules in the cell form a structure capable of generating electrodynamic field and mitochondria form their supporting system for physical processes including energy supply. Mitochondria transfer protons from their matrix space into cytosol, create strong static field around them that causes ordering of water and altering it into quasi-elastic medium with reduced viscous damping. Microtubules are composed of heterodimers that are electric dipoles. Microtubule oscillations generate an electrodynamic field. The greatest energy supply may be provided by liberation of non-utilized energy from mitochondria. Microtubules and mitochondria form a unique cooperating system in the cell. Mitochondria form a boundary element whose function depends on chemical-genetic control but their output is essential for physical processes in the cell. Mitochondrial dysfunction in cancer cells results in diminished intensity of the static electric field, disturbed water ordering, increased damping of microtubule oscillations and their shift towards linear region, and decreased energy supply. Power and coherence of oscillations and generated electrodynamic field is weakened. Malignant properties of cancer cell, in particular local invasion and metastasis, may depend on disturbed electrodynamic field. Nanotechnology is promising for investigation of electrodynamic activity in living cells.

1. Introduction

For a long time biological processes and cancer were assumed to be entirely dependent on chemical and genetic changes. In the last decade the attitude has changed. Carew and Huang [1] and Cuezva et al. [2] published papers revisiting the chemical-genetic dominancy and generally accepted opinion that suppression of oxidative metabolism in cancer cells is an irrelevant side effect somehow connected with development of cancer. The oxidative disturbances in cancer cells were disclosed by Warburg et al. [3] and assigned to mitochondrial dysfunction (Warburg [4]). Metabolic imbalances and deficient mitochondrial respiration are described by Kroemer [5] and Brandon et al. [6]. Pedersen [7] claimed that the most striking biochemical phenotype of cancers is their aberrant energy metabolism. Kroemer [5], Brandon et al. [6], and Pedersen [7] papers were published in the special issues of the journal "Oncogene" (2006) and "Journal of Bioenergetics and Biomembranes" (2007) (devoted to mitochondrial dysfunction and the Warburg effect, respectively) which caused the main turn for a real assessment of the Warburg effect. Nevertheless, the mitochondrial function (or dysfunction) was not analyzed in connection with physical processes in living cells (or their changes along the cancer transformation pathway).

Since the first Versailles' conference on "Theoretical Physics and Biology" in 1967 the physical mechanisms (in particular of electrodynamic type) were studied, analyzed, and published. H. Fröhlich [8–12] postulated long-range quantum mechanical phase correlations, existence of electrically polar vibration modes excited far from the thermal equilibrium, and generation of endogenous coherent electrodynamic field in biological systems. The main part of the Fröhlich theory is based on the fundamentally non-linear interactions between elastic and electric polarization fields and non-linear spectral energy transformation. The role of disturbances of endogenous electrodynamic processes in cancer transformation was examined, too (Fröhlich [13]). Theory of synergetic operation in systems far from thermodynamic equilibrium was developed and analyzed by Haken [14].

Attraction of dielectric particles to living cells was observed by Pohl [15], Hölzel and Lamprecht [16], and Hölzel [17]. The dielectric particles move along the gradient of the intensity of the electric field towards the region of the greatest intensity (or vice versa), if the difference of their permittivity and the permittivity of the ambient medium is positive (negative). The number of particles attracted to living cells depends on dielectrophoretic forces. The force acting on the dielectric particle is mainly determined by the value of the permittivity difference, the intensity of the electric field and its gradient, and by the conductivity of the suspending liquid. Geometrical shape of the cell (the tip enhanced field effect), and the phase of the cell cycle affects the number of attracted particles too. The greatest number of particles is attracted in the M phase.

Electromagnetic activity of living cells was disclosed in a wide frequency region. Pelling et al. [18–19] measured elastic oscillations at the cellular membrane with amplitude about 3–4 nm in the acoustic band. Measurement of electric and elastic oscillations in the acoustic frequency range (0.4–2 kHz) was performed by Jelínek et al. [20]. Attachment of tubulin heterodimers to microtubule growing end during polymerization may be inhibited by external electromagnetic field in the frequency range 0.1–0.3 MHz (Kirson et al. [21–22]). Transport and correct orientation of heterodimers may be disturbed due to exerted electrodynamic forces. On the basis of theoretical analysis (Pokorný et al. [23–24]) direct technical measurement of the electrodynamic field was performed at the cellular membrane in the frequency range 8–9 MHz (Pokorný et al. [25]). Sahu et al. [26] claims that microtubules in vitro display resonance at the frequency of about 10 MHz. It might be an important discovery of microtubule properties. Increased damping of the power transmitted from an antenna positioned near a malignant tumor was observed at the frequency 465 MHz and its first and second harmonics (Vedruccio and Meessen [27]). The effect was explained on the basis of resonant interaction between external electrodynamic field and intracellular microtubule oscillations. In cancer cells ordering of water is disturbed that causes increased damping (Pokorný et al. [28]).

Interaction between baby hamster kidney (BHK) cells is mediated by electromagnetic field in the red and near infrared region (Albrecht-Buehler [29]). The Swiss 3T3 cells are able to sense the near infrared radiation and to determine direction of individual sources (Albrecht-Buehler [30]). Formation of aggregates of $3T3_x$ cells (a variant derived from 3T3 cells) is conditioned by electromagnetic interaction in the near infrared range (Albrecht-Buehler [31]).

Microtubules in the eukaryotic cells fulfill conditions for excitation of the electrodynamic oscillations. Mitochondria and microtubules are essential cellular parts for cellular electromagnetic activity. This paper analyzes properties of mitochondria and microtubules as the entities conditioning essential physical processes in cells and their disturbances caused by the cancer transformation. The generating structure in the eukaryotic cells is formed by microtubules (Pokorný [32–34] and Pokorný et al. [35–36]).

2. Water ordering by static electric field

Mammalian cell contains about 70% of water. Jellyfish is alive with more than 99% of water. The high content of water in living systems suggests a question about its role in living state and whether water has any special property that makes life possible. Generally, water exists in three phases–

gaseous, liquid, and solid. The physical parameters of the environment (such as temperature and pressure) determine the water phase. Water state (or phase) may depend on other physical parameters too. Damadian [37] explained different NMR (nuclear magnetic resonance) findings on healthy and cancer cells by different level of water ordering in cells in healthy and tumorous tissues. Regions without solutes were observed around microtubules (Amos [38]). These clear regions were suggested to be connected with the negative charge at the microtubule surface (Stebbings and Hunt [39]). Cleveland et al. [40] evaluated NMR measurement with a conclusion that a large fraction of cellular water is associated with the proteins in such a way that its diffusion coefficient is substantially reduced. Some scientists assumed that water may be capable of forming an "ice-like" or "quasi-crystalline" structure. But the idea that water in living cells may have a different physical structure and that living cells are not containers of a bulk water with macromolecules and various organelles seemed to be impossible and, generally, rejected. But water molecules are electric dipoles and external electric field should have strong effects on water behavior. Nevertheless, theoretical assessment of the effect of the external electric field excluded formation of such physical ordered structure (Booth [41]). Regardless of the negative theoretical analysis experimental investigations disclosed water ordering at the surfaces of hydrophilic materials and inside living cells too. The ordered water may be organized in layers around the surface up to a distance of the order of magnitude 0.1 mm. Ling and Murphy [42] used NMR spectroscopy to assess properties of water from measurement of the spin-lattice (T_1) and the spin-spin (T_2) relaxation time of water protons. The T_1 and T_2 relaxation times depend on rotational relaxation time τ_c (characteristic time for rotation of a molecule or time of its diffusion into its next position). Increase of rotational relaxation time by a factor of about 3 (from about 10^{-12} to 10^{-9} s) results in reduction of solvency and produces water that is non-solvent for Na citrate and sulfate. Zheng and Pollack [43] measured exclusion zones using coated latex microspheres in the vicinity of polyvinyl alcohol gels. Exclusion zones about 0.1 mm thick around the surface were observed. A more detailed measurement of exclusion zones were published by Zheng et al. [44]. Probe measurements disclosed negative potential whose difference across the exclusion zone was about 100 mV. UV–Vis absorption spectrum displays a peak at 270 nm in the exclusion zone. This absorption feature was not measured in the bulk water. Infrared emission from the exclusion zone measured by an infrared camera in the spectral window 3.8–4.6 μm is suppressed. The hydrophilic surfaces play a role more profound than generally assumed.

The ordered water at the interfaces may be measured by AFM (atomic force microscopy) as its physical properties differ from those of the bulk water. The ordered water has higher viscosity (Pollack et al. [45]) and lower thermal motion of molecules (Zheng et al. [44]). Different pH (Chai et al. [46]) and different spectroscopic properties (Chai et al. [47]) were found comparing ordered and bulk water properties. Ordered layers of water are remarkable by separation of charges (Chai et al. [46]) and solvent exclusion (Zheng and Pollack [43] and Zheng et al. [44]). Formation of exclusion zones is promoted by infrared radiation (Chai et al. [46]). A more comprehensive set of ordered water properties was published by Pollack et al. [45, 48]. Water ordering may depend on the phase of the cell cycle. Zimmerman et al. [49] found that increased state of ordered water molecules at metaphase is maintained by non-microtubule factors.

The electric field is a general physical agent ordering water. External field can organize water similarly as the electric field at the hydrophilic surfaces. A strong electric field (about 600–700 kV/m) organizes water and forms a floating water bridge about 1–3 cm long between two glass beakers (Fuks et al. [50–52] and Giuliani et al. [53]).

Quantum electrodynamics theory of creation of ordered water was worked out by Preparata [54] and Del Giudice et al. [55]. They found out that the normal "liquid" water is a mixture of two phases. One phase contains coherent domains that form the ordered water regions. The other phase is composed of gas-like water molecules. The exclusion zones are macroscopic structures corresponding to ordered water domains. The coherent domains are reservoirs of quasi-free electrons that may have fundamental impact on biological activity (Del Giudice et al. [55] and Del Giudice and Tadeschi [56]). The exclusion zone structure may represent a transition structure between a liquid and a solid state material. Protons may diffuse through the exclusion zone and quasi-free electrons may move like electrons in the "conductivity band".

Intramitochondrial water has a high viscosity–30 times larger than the resuspension water (López-Beltrán et al. [57])–that signifies ordered water in the mitochondrial matrix. Bulk solvent water is able to rotate freely, water bound or adsorbed on macromolecular surfaces seems to adopt the correlation time of the host macromolecule. Trombitás et al. [58] observed large water regions near mitochondria and discussed the problem of driving force.

3. Mitochondrial function

Mitochondria are power plants of the cell. Chemical energy stored in pyruvate and fatty acids is parceled out into small parts, transformed into electrochemical proton gradient across the inner mitochondrial membrane, and finally stored into ATP and GTP (adenosine and guanosine triphosphate). Transport of protons into the intermembrane space and their diffusion to the mitochondrial surroundings has essential significance for cellular function. The protons around mitochondria are not only resource of small energy packets but also for formation of a strong static electric field around mitochondria (together with the negative charge in the mitochondrial matrix). Intensity of the static electric field around mitochondria was measured by fluorescence spherical particles 30 nm in diameter (Tyner et al. [59]). The spectrum of the material of the fluorescence particles was dependent on the intensity of the electric field. A zone of a strong electric field is excited around each mitochondrion. At the mitochondrial membrane the intensity of the static electric field is of about 3.5 MV/m. It is measurable even at a distance of several micrometers. As the mitochondria occupy about 22 % of cellular volume the field of different mitochondria may spill over one another. Fluorescent particles may very often measure the field generated by several mitochondria in the vicinity, for instance in the planes below and above the plane of observation. Some experimental results were assumed to be excited only by one mitochondrion and represent the pure cytosolic region. Figure 1 shows the cytosolic intensity of the electric field measured at a mitochondrion at a distance 0.5, 1.0, 1.5, and 2.0 μm (Tyner et al. [59]) and plotted as vertical bars with quadratic regression curve (R). The regression curve has nearly linear dependence on the distance from the mitochondrion (it has a very small quadratic term) which contradicts the theoretical curve (Q) determined from the distribution of the proton space charge around mitochondrion. The intensity values dependent on the proton distribution drop down rapidly near the mitochondrion within a distance of the order of magnitude 10 nm and they are much smaller than the experimental values at larger distances. If the measured intensity actually represents the cytosolic values then water around mitochondrion should be ordered and the proton space charge layer excluded to larger distance. Ordering similar to the interfacial one may be created. The intensity curves of the static electric field of a spherical and cylindrical shape of the exclusion layer model are plotted in figure 1 by short and long dashed curves, respectively. The curve (S2.2) denotes a regression curve of a spherical model for a mitochondrial spherical surface of a radius 2.2 μm.

Protons may be expelled from the exclusion zone and concentrated beyond its distant edge far from mitochondrion membrane. Protons may diffuse into the ordered layer, boundary may be smeared, and the exclusion zone non-uniform. The fact that the exclusion zone of 0.36 mm could be found in 150 mM salt solution argues against a purely electrostatic mechanism (Zheng et al. [44]). Nevertheless, exclusion of protons may depend on the level of ordering. Distribution of the charge depends on its concentration and on the intensity of the static electric field in the exclusion zone. Water ordering alters water from a viscosity liquid to a quasi-elastic gel influencing inner-cellular processes, especially providing low damping of the microtubule vibration system.

But the analyzed static electric field need not be generated only by a mitochondrion. Mitochondria occupy a substantial portion of the cytoplasmic volume of the eucaryotic cells. In a "standard" liver cell (hepatocyte) about 22 % of the total cell volume is occupied by mitochondria. For a spherical cell and spherical mitochondria with a radius 6 μm and 0.3 μm, respectively, the number of mitochondria per cell is $N = 1760$. The full volume of the cell is equivalent to N spherical volumes with a radius about 0.5 μm. The average distance between mitochondria might be about 0.4 μm. Exclusion zones of different mitochondria may overlap. Overlapping may disturb water ordering. Protons may be concentrated in the boundary regions between mitochondria where the intensity of the electric field has the minimum value. Concentration profile of the proton layer and its dynamic equilibrium flow between the inner membrane and the layer remains an open question.

The cytosolic medium around mitochondria is under influence of a strong static electric field polarizing biological molecules and structures. Oscillation processes are shifted into a highly non-linear region.

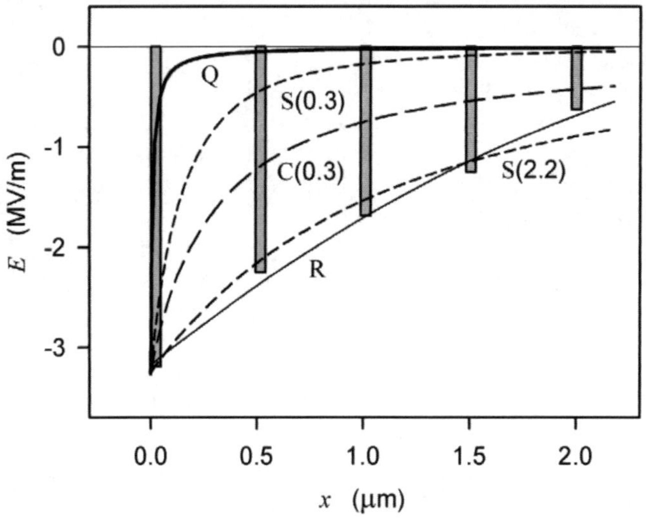

Figure 1. The intensity of the static electric field as a function of the distance from mitochondria,
Vertical bars–measured values (after Tyner et al. [59]),
R–quadratic regression curve,
S(2.2)–regression curve for spherical exclusion layer (radius of mitochondrial surface is 2.2 μm),
Q–a proton layer with no water ordering,
S(0.3) and C(0.3)–a spherical and a cylindrical exclusion layer (radius of mitochondrial surface is 0.3 μm).

Chemical energy of pyruvate and fatty acids supplied to mitochondria is used for production of ATP and GTP. The efficiency of energy conversion is of about 40 % or slightly higher. The non-utilized energy (often denoted the "wasted energy") may be nearly 60%. The non-utilized energy is liberated from mitochondria in the form of photons in the UV (Tilbury and Quickenden [60] and Batyanov [61–62]), visible, and infrared regions, heat and chemical compounds.

4. Microtubule oscillators

The cytoskeleton in eukaryotic cells is a complex network of protein filaments extending through the cytoplasm. Cytoskeleton is a dynamic structure under continuous reorganization connected with shape change, division, coordination of directed movement, organization, response to inner cellular state and environmental conditions. The filamentary structure of the cytoskeleton contains actin filaments, microtubules, intermediate filaments, and accessory proteins. Accessory proteins may have a variety of functions. Some accessory proteins may link filaments to one another. The cytoskeleton forms a compact three-dimensional network in the cytosol conditioning the cell function. Microtubules create the main structure of the network. Microtubules start to polymerize from the centrosome that is positioned in the center of the cell. Microtubules grow in the radial direction. Some microtubules are bound to structures at the membrane. In the M phase microtubules form a mitotic spindle organized by two centrosomes (poles of the mitotic spindle). In the interphase mitochondria are aligned along microtubules. In the M phase their distribution is rather unknown. Organization of microtubules and mitochondria might be a special issue of the cellular static electric and electromagnetic field. It was observed that the microtubule structure organized by a centrosome finds the center of the cell. The cell may form an electromagnetic cavity resonator with dielectric walls (Popp [63] and Jelínek and Pokorný [64]). The position of the microtubules and the centrosome may be controlled by the cavity electromagnetic field. In the interphase organization of mitochondria depends on microtubule structure. The space distribution of the field and the acting forces are different in the interphase and in the M phase (in the one pole and the two poles structure, respectively). Mitochondria may move together with the ordered water and the proton space-charge layer. The final effect of organization may be assessed using real parameters of the system. But this problem has not been solved yet. On the other hand the static electric field might participate in mitochondrial space distribution too. Very often mitochondria remain in position where they cover unusually high ATP consumption. Due to the proton transfer and the strong static electric field mitochondria might provide a long range interactions with the "energy consumption site". The sites of increased energy consumption in the M phase have different locations in comparison with the interphase period.

Microtubule is a hollow cylindrical structure organized in protofilaments composed of tubulin heterodimers. Two globular proteins of a relative mass of about 55000 form a tubulin heterodimer. Each heterodimer is an electric dipole with a dipole moment of about 1000 Debye, i.e. 10^{-26} Cm (Satarić et al. [65] and Tuszynski et al. [66]). The induced dipole moment in a dimer generated only by the motion of mobile electrons or protons was assessed to be 200–400 Debye (Stracke et al. [67]). Polarization changes are connected with elastic oscillations.

Oscillations in microtubules require energy. Energy may be supplied by different mechanisms. Dynamic instability, motion of motor proteins, and liberation of non-utilized energy from mitochondria are the energy supplying processes. Dynamic instability is a special property of microtubules. In the interphase microtubules bound to the structures at the membrane have a turnover of about 18 hours (Pelling et al. [68]) and then they are replaced. The free end microtubules have a short turnover–about 10 min. They continuously polymerize and depolymerize. After polymerization GTP in the β tubulin is hydrolyzed to GDP. A part of the liberated energy is stored in the structure of the microtubule. In the M phase the dynamic instability has a form of treadmilling–microtubules polymerize in the central part and depolymerize at the poles of the mitotic spindle. The described mechanisms of dynamic instability supply energy to microtubules. In the M phase of the cell cycle the rate of exchange of heterodimers is more than one order of magnitude higher than in the interphase and energy supply is correspondingly increased too (Pokorný et al. [69]).

Motor proteins transport particles, vesicles, organelles, etc. along microtubules. Energy for a one "step" of a motor protein "walking" is obtained from hydrolysis of an ATP molecule. A part of the "walking" energy is transferred to microtubules. But the motor protein motion might also disturb and damp microtubule oscillations.

Non-utilized energy liberated from mitochondria may be the greatest contribution to excitation of microtubule oscillations. It may excite vibrations in a wide frequency range.

Biological electromagnetic activity–a novel issue in biology–was observed from acoustic to visible frequency range. Mechanical and electrical oscillations were measured in the acoustic range. Mechanical oscillations were disclosed by AFM (atomic force microscopy) at 1.643 and 0.87 kHz at temperature 30 and 22°C, respectively (Pelling et al. [18–19]). Mechanical and electrical oscillations in the acoustic band were measured by Jelínek et al. [20]. Frequency of oscillations was assessed from experimental result of dielectrophoretic attraction of particles in the band 5 kHz–1 MHz by Pohl [15]. Disturbances of internal cellular electromagnetic field by external electromagnetic field in the frequency range 100–300 kHz were described by Kirson et al. [21–22]. The external field disrupts polymerization of microtubules and can even destroy the cell. Electrodynamic activity of yeast cells and alga cells was measured in the frequency range 1.5–52 MHz by Lamprecht and Hölzel [16] and Hölzel [17]. Pokorný et al. [25] observed and evaluated electrodynamic activity of yeast cells in the frequency band 8–9 MHz. The activity was measured in the M phase. The high activity coincides with the rearrangement of the microtubules into a mitotic spindle, with binding chromatids to kinetochore microtubules, and with elongation of the mitotic spindle during anaphase A and B. Resonant interaction of the electromagnetic field with a cancer tissue at 465 MHz and the first and second harmonics was disclosed by Vedruccio and Meessen [27]. Cells ability to detect electromagnetic signals of other cells in the red and near-infrared range was disclosed by Albrecht-Buehler [29–31]. The observed cells are able to sense the radiation and to determine the direction to the source.

Microtubules in vitro were measured by Sahu et al. [26] (not yet published). A novel method of polymerization was used–polymerization in the oscillating electric field. Electron conductivity in microtubules was investigated. Sahu et al. [26] claims, that electrons move along spiral orbits with different steepness of the helix with no losses (ballistic conductance). The resonant frequencies were around 10 MHz.

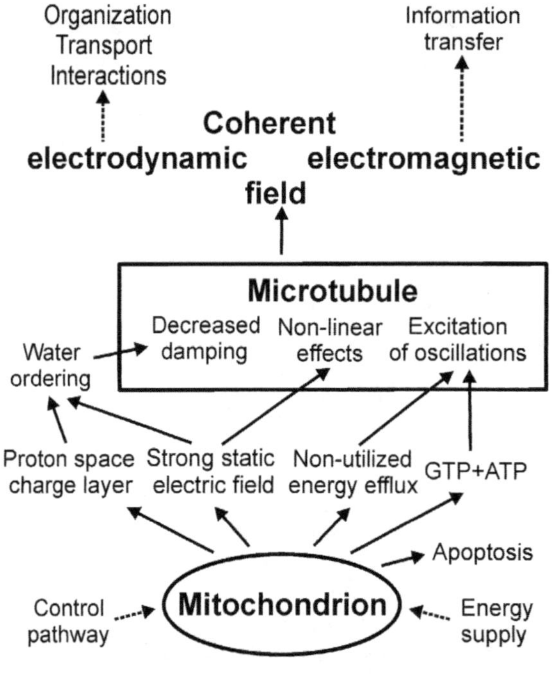

Figure 2. A schematic picture of physical mechanisms in living cells depending on mitochondrial function. Transport of protons into intermembrane space and their diffusion into the cytosol lead to formation of proton space charge layer, strong static electric field, and water ordering around mitochondria resulting in low damping of microtubule oscillations. The static electric field conditions non-linear effects in microtubule. Non-utilized energy efflux and production of ATP and GTP provide energy supply. Microtubule may generate electrodynamic ("static" and "induction"–virtual photon–field in the vicinity of the source) and electromagnetic field (radiation–photon field) at a large distance. Information may be transferred by the photon and the virtual photon field too.

The electrodynamic field measured at the membrane of living cells is assumed to be generated by fundamentally non-linear elastic-electrical oscillations of microtubules. The oscillations may display several resonant frequencies as follows from AFM measurements in the acoustic region and interaction of external electromagnetic fields with living cells, in particular in the 0.1–0.3 MHz and

465 MHz regions. A direct measurement of coherent resonant signals using special nanotechnological systems at physiological temperature may disclose essential spectral properties of oscillations in microtubules and their structures in cells.

5. Cellular physics

Cooperation of mitochondria and microtubules is a unique property in living cells and may form an essential basis for physical processes. The static electric field created by mitochondria may be a fundamental factor for establishment of non-linear properties of microtubules and ordering of water. Non-linear properties make possible special interaction between elastic and polarization fields, water is a quasi-elastic structure whose properties may be at the boundary between a liquid and a solid-state material. Non-utilized efflux of energy from mitochondria seems to be the main source for excitation of oscillation, shifting the system far from thermodynamic equilibrium, and establishment of the coherent state. A considerable part of biological activity may depend on the electrodynamic field which may play a significant role in different cellular processes. It might produce a traction force predicted by Frauenfelder et al. [70] for directed transport of material (Pokorný [71] and Pokorný et al. [35]), morphological force for organization of morphological structures, interaction force for intra and inter-cellular purposes (Pokorný [72]) and may mediate information transfer. Figure 2 shows a schematic picture of physical links of the cellular activity (Pokorný et al. [28]).

6. Physical alterations in cancer cells

Suppression of oxidative metabolism may be one of the biggest differences between healthy and cancer cells. A diverse group of signaling pathways and oncogenes may result in Warburg effect and in resistance to apoptosis. A type of Warburg effect with hyperpolarized mitochondria is in figure 3. The lack of hyperpolarization of the mitochondrial inner membrane in certain types of cancer suggests that mitochondrial dysfunction might be caused by different mechanisms.

Warburg assessed the ratio of the amount of energy produced by fermentation and oxidation (respiration) process in healthy and cancer cells. In healthy cells the oxidation energy production may be more than 10 times greater than the fermentation one (in kidney and liver cells even 100 times). With the exception of the cancer cells with reversed Warburg effect (Pavlides et al. [73]) fermentative production in cancer cells is bigger than in healthy cells. In the cancer cell fermentation may participate in energy production highly approximately by 1/3–2/3 part of the total output and the average value may be of about one half (Warburg [3–4]). If the cancer and the healthy cells have the same amount of the total energy and the same mitochondrial efficiency of utilization of the electrochemical proton gradient (by ATP synthase protein complexes) then only about one half of protons are transported across the inner membrane in comparison with a "healthy" mitochondrion.

Non-utilized energy liberated from mitochondria may be adequately lower too. The intensity of the static electric field and the level of water ordering are diminished, that have a negative impact on microtubule oscillations. Damping of oscillations is increased, excitation is lowered, and non-linear properties are shifted towards linear region. As a consequence power and coherence of the generated electrodynamic field are diminished. Coherent processes in the cell are disturbed, and randomness increased (Pokorný et al. [28,36] and Pokorný [33–34]).

Development of mitochondrial malfunction is produced before the malignant properties of cancer cells may be observed. In the cervical cells the cancer mitochondrial dysfunction turn was observed in the period precancerous–cancer cells (Jandová et al. [74]). The malignant properties, in particular local invasion and metastases, may be caused by disturbances of electromagnetic activity. Local invasion may depend on the interaction mechanisms between cells (Pokorný [72]). The interaction forces between cancer cells may be smaller than interaction forces between healthy and cancer cells. As a result cancer cells are pulled by healthy cells into healthy tissue. Disorganization of the cytoskeleton precedes the metastasis (Beil et al. 75], Suresh et al. 76], and Suresh [77]). Disturbances of the cytoskeleton together with decreased excitation and increased damping may alter the frequency spectrum and the spatial pattern of the generated electromagnetic field with such an impact that cancer

Journal of Physics: Conference Series **329** (2011) 012007 doi:10.1088/1742-6596/329/1/012007

cell does not interact with the surrounding cells, can lose connection with them, can leave the tissue, move freely in the body and create a metastatic nodule even in distant organs.

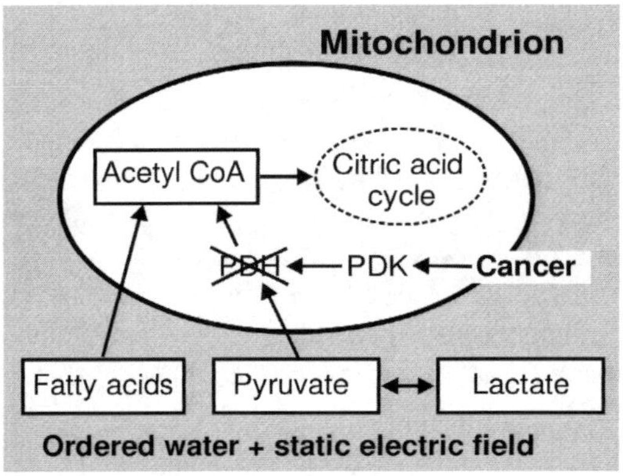

Figure 3. A schematic picture of the glycolytic phenotype of cancer cell (Bonnet et al. [83]). Cancer kinase PDK blocks the pyruvate pathway inhibiting pyruvate dehydrogenase (PDH) enzymes. Only about one half of protons are transferred across the inner membrane into intermembrane space in comparison with fully functional mitochondria. Similar effect of proton transfer inhibition may be caused by defects in the citric acid cycle.

7. Discussion

After elapse of 80 years from the discovery of the suppression of the oxidative mechanism in cancer cells, 40 years of Warburg death, and 10 years of the beginning of the endeavor to revisit the Warburg discovery, the Warburg effect has become an urgent issue. Warburg effect is now understood as one of the most important links along the cancer transformation pathway. Mitochondrial dysfunction may alter processes dependent on the energy supply, strong static electric field, and ordering of water which suggest processes of physical nature. But elucidation of such processes disturbed by the mitochondrial dysfunction is not yet performed. The consequences of the mitochondrial dysfunction may be analyzed on the basis of the Fröhlich hypothesis of electrical polar oscillations in living cells and creation of a coherent state. The Fröhlich hypothetical prediction of oscillation states was branded as impossible due to strong viscosity damping of water (Foster and Baisch [78]) or quite generally by low quality of biological oscillators (Reimers et al. [79] and McKemmish et al. [80]) neglecting interfacial and mitochondrial water ordering in cells. Reimers et al. [79] and Mc Kemmish et al. [80] analyzed linearized system neglecting essentially nonlinear interactions between elastic and polarization fields. But the Fröhlich hypothetical analysis is strongly supported by experimental findings on water ordering, static electric field created around mitochondria, and electrodynamic activity of living cells.

The directions of the future research of electromagnetic processes in living cells should be assessed. Experimental research based on nanotechnological devices makes possible in vivo and in vitro measurements of the electrodynamic field of living cells. The total turnover of the power of the cells is a basic value. Lamprecht [81] measured the yeast cell turnover to be about 10^{-13} W. A similar amount of power might excite electrodynamic oscillations in microtubules that may correspond to the non-utilized power liberated by mitochondria. Assessment of the conversion of the random supply to coherent oscillations, the number of microtubules in the cell, and the power distribution in particular frequency spectral lines, leads to conclusion that the power exciting individual microtubule spectral line may be about 10^{-17} W. The power stored in the oscillating microtubule may be of about 10^{-15} W if the quality factor is high. Measurement may strongly affect the microtubule oscillating system by increased damping and change even the frequency spectrum. In vivo measurement may be performed at the cellular membrane at a patch corresponding to the microtubule binding spot at the inner side of the membrane (Kučera et al. [82]). The linear dimensions of the sensor for detection of the cellular signal should be smaller than about 1 μm. The physiological temperature should be maintained during measurement. Parameters of microtubules for assessment of their electrodynamics characteristic features may be determined by in vitro measurements.

Theoretical research should be bound to experimental research to solve main problems of physical processes in biological activity of living cells. Theoretical treatment should explain the main conditions for water ordering around mitochondria and its essential properties, generation of coherent electrodynamic field by microtubules, interaction inside and between cells mediated by the electrodynamic field, creation of quantum theory of coherence in biological systems, and–very likely–development of a novel fundamentally non-linear quantum (electrodynamic field) theory for biological systems.

Convergence of classical biology, molecular biology, genetics, and biochemistry with the novel biological physics is a highly important issue. The research in the future should bridge the gap between the chemical and genetic processes on the one hand and physical ones on the other hand. Conversion of both parts conditions comprehensive understanding of biological activity and cancer transformation pathway. Biochemical, genetic, and physical processes are mutually dependent and equally important. Building of more sophisticated models of biological activity is required.

A special case of theoretical and experimental investigations of physical processes in living cells concerns the cancer transformation pathway. Fundamentally disturbed electrodynamic field in cancer cells (Pokorný et al. [28,36] and Pokorný [33,34]) results from mitochondrial dysfunction caused by decreased proton pumping from the matrix space and liberation of the non-utilized energy to mitochondrial ambient medium. Mitochondrial dysfunction (that develops before appearance of the cancer malignant properties) and diminished power and coherence of the electrodynamic field may be the most pronounced differences between the healthy and the cancer cells in the clinical phase. Mitochondrial and physical processes should be targeted for cancer treatment. Cancer cell killing is the main target of the standard therapeutic strategy. Nevertheless, such treatment may also damage healthy cells and limit its further application for treatment of recurrent tumors. Restoration of normal (healthy) cell function should be–first of all–the therapeutic strategy. For instance, normal function of mitochondria is restored and the apoptotic pathway unlocked by opening the pyruvate pathway in mitochondria after inhibition of PDK. As a result, the physical processes in the cell and the cell by itself are reversed to normal operation. Consequently, the cell may trigger apoptosis after overturn from cancer to normal if the damage exceeds certain level. Targeting mitochondria acts in the region of the main differences between healthy and cancer cells.

8. Conclusions

Cooperation of mitochondria and microtubules is a unique phenomenon in living cells. Chemical and genetic signals control mitochondrial multifunctional operation. But the main part of mitochondrial activity conditions physical processes. Mitochondria transfer protons across the inner membrane, produce ATP and GTP, and release the non-utilized energy. Formation of a zone of strong static electric field in the cytosol, water ordering, and proton space charge layer are conditioned by proton pumping from the matrix space of mitochondria. The strong static electric field can shift oscillations in microtubules to highly non-linear region that makes possible creation of coherent state. Organized water form a quasi-elastic medium resulting in small damping of microtubule oscillations. A considerable part of energy supply to microtubules may be provided by liberation of non-utilized energy from mitochondria.

Electrodynamic field generated by microtubules might provide essential biological functions. Morphological organization, transport of molecules, vesicles, and reaction components, interactions inside a cell and between cells may depend on generated electrodynamic field. High capacity information transfer between the body organs and the brain may be mediated by the electromagnetic field.

Mitochondrial dysfunction in cancer cells may cause increased damping of microtubule oscillations, diminished energy supply, and shift of the non-linear properties towards linear region. Power and coherence of the cellular electrodynamic field generated in a cancer cell are diminished.

Acknowledgements
The research results presented in this paper were supported by the grant No. 102/11/0649 of the Czech Science Foundation GA CR.

References

[1] Carew JS and Huang P 2002 Mitochondrial defects in cancer *Mol. Cancer* **1** 9–20
[2] Cuezva JM, Krajewska M, López de Heredia M, Krajewski S, Santamaria G, Kim H, Zapata JM, Marusawa H, Chamorro M and Reed J 2002 The bioenergetic signature of cancer: A marker of tumor progression *Cancer Res.* **62** 6674–6681
[3] Warburg O, Posener K and Negelein E 1924 Über den Stoffwechsel der Carcinomzelle *Biochem. Z.* **152** 309–344
[4] Warburg O 1956 On the origin of cancer cells *Science* **123** 309–314
[5] Kroemer G 2006 Mitochondria and cancer *Oncogene* **25** 4630–4632
[6] Brandon M, Baldi P and Wallace DC 2006 Mitochondrial mutations in cancer *Oncogene* **25** 4647–4662
[7] Pedersen PL 2007 Warburg, me and hexokinase 2: multiple discoveries of key molecular events underlying one of cancers' most common phenotypes, the "Warburg Effect", i.e., elevated glycolysis in the presence of oxygen *J. Bioenerg. Biomembr.* **39** 211–222
[8] Fröhlich H 1968 Bose condensation of strongly excited longitudinal electric modes *Phys. Lett.* A **26** 402–403
[9] Fröhlich H 1968 Long-range coherence and energy storage in biological systems *Int. J. Quant. Chem.* **II** 641–649
[10] Fröhlich H 1969 Quantum mechanical concepts in biology *Theoretical Physics and Biology* ed M Marois (Amsterdam: North Holland) (*Proc 1st International Conference on Theoretical Physics and Biology* 1967 Versailles) pp 13–22
[11] Fröhlich H 1973 Collective behaviour of non-linearly coupled oscillating fields (with applications to biological systems) *J. Collect. Phenom.* **1** 101–109
[12] Fröhlich H 1980 The biological effects of microwaves and related questions *Advances Electronics Electron Physics* **53** pp 85–152
[13] Fröhlich H 1978 Coherent electric vibrations in biological systems and cancer problem *IEEE Trans.* MTT **26** 613–617
[14] Haken H 1975 Cooperative phenomena in systems far from thermal equilibrium and in nonphysical systems *Rev. Modern Phys.* **47** 67–121
[15] Pohl HA 1980 Oscillating fields about growing cells *Int. J. Quant. Chem. Quant. Biol. Symp.* **7** 411–431
[16] Hölzel R and Lamprecht I 1994 Electromagnetic fields around biological cells *Neural Net. World* **4** 327–337
[17] Hölzel R 2001 Electric activity of non-excitable biological cells at radio frequencies *Electro-Magnetobiol.* **20** 1–13
[18] Pelling AE, Sehati S, Gralla EB, Valentine JS and Gimzewski JK 2004 Local nano-mechanical motion of the cell wall of *Saccharomyces cerevisiae Science* **305** 1147–1150
[19] Pelling AE, Sehati S, Gralla EB and Gimzewski JK 2005 Time dependence of the frequency and amplitude of the local nanomechanical motion of yeast *Nanomed. Nanotechnol. Biol. Med.* **1** 178–183

[20] Jelínek F, Cifra M, Pokorný J, Vaniš J, Šimša J, Hašek J and Frýdlová I 2009 Measurement of electrical oscillations and mechanical vibrations of yeast cells membrane around 1 kHz *Electromag. Biol. Med.* **28** 223–232

[21] Kirson ED, Gurvich Z, Schneiderman R, Dekel E, Itzhaki A, Wasserman Y, Schatzberger R and Palti Y 2004 Disruption of cancer cell replication by alternating electric fields *Cancer Res.* **64** 3288– 3295

[22] Kirson ED et al. 2007 Alternating electric fields arrest cell proliferation in animal tumor models and human brain tumors *PNAS* **104** 10152–10157

[23] Pokorný J, Jelínek F, Trkal V, Lamprecht I and Hölzel R 1997 Vibrations in microtubules *J. Biol. Phys.* **23** 171–179

[24] Pokorný J, Jelínek F and Trkal V 1998 Electric field around microtubules *Bioelectrochem. Bioenergetics* **45** 239–245

[25] Pokorný J, Hašek J, Jelínek F, Šaroch J and Palán B 2001 Electromagnetic activity of yeast cells in the M phase *Electro- Magnetobiol.* **20** 371–396

[26] Sahu S, Ghosh S, Hirata K, Fujita D and Bandyopadhyay A 2011 Ultra-fast condensation of tubulins into microtubule unravels a generic resonance engineering (personal communication)

[27] Vedruccio C and Meessen A 2004 EM Cancer detection by means of non-linear resonance interaction *Progress in Electromagnetics Research Symposium 2004* (Proc. PIERS, Pisa Italy March 28–31) pp 909–912

[28] Pokorný J, Vedruccio C, Cifra M and Kučera O 2011 Cancer physics: diagnostics based on damped cellular elastoelectrical vibrations in microtubules *Eur. Biophys. J.* **40** 747–759

[29] Albrecht–Buehler G 1992 Rudimentary form of cellular 'Vision' *Proc. Natl. Acad. Sci.* USA **89** 8288–8293

[30] Albrecht–Buehler G 1991 Surface extensions of 3T3 cells towards distant infrared light sources *J. Cell. Biol.* **114** 493–502

[31] Albrecht–Buehler G 2005 A long-range attraction between aggregating 3T3 cells mediated by near-infrared light scattering *PNAS* **102** 5050–5055

[32] Pokorný J 2004 Excitation of vibration in microtubules in living cell *Bioelectrochem.* **63** 321–326

[33] Pokorný J 2009 Biophysical cancer transformation pathway *Electromagn. Biol. Med.* **28** 105–123

[34] Pokorný J 2009 Fröhlich's coherent vibrations in healthy and cancer cells *Neural Netw. World* **19** 369–378

[35] Pokorný J, Hašek J and Jelínek F 2005 Electromagnetic field in microtubules: effects on transfer of mass particles and electrons *J. Biol. Phys.* **31** 501–514

[36] Pokorný J, Hašek J, Vaniš J and Jelínek F 2008 Biophysical aspects of cancer electromagnetic mechanism *Indian J. Exper. Biol.* **46** 310–321

[37] Damadian R 1971 Tumor detection by nuclear magnetic resonance *Science* **171** 1151–1153

[38] Amos LA 1979 Structure of Microtubules *Microtubules* ed K Roberts and JS Hyams (London–New York: Academic Press) pp 1–64

[39] Stebbings H and Hunt C 1982 The nature of the clear zone around microtubules *Cell Tissue Res.* **227** 609–617

[40] Cleveland GG, Chang DC, Hazlewood CF and Rorschach HE 1976 Nuclear Magnetic Resonance Measurement of Skeletal Muscle *Biophys. J.* **16** 1043–1053

[41] Booth F 1951 The dielectric Constant of Water and the Saturation Effect *J. Chem. Phys,* **19** 391–394

[42] Ling GN and Murphy RC 1983 Studies on the physical state of water in living cells and model systems. II NMR relaxation times of water protons in aqueous solutions of gelatin and

oxygen-containing polymers which reduce the solvency of water for Na^+, sugars and free amino acids *Physiol. Chem. Phys. Med. NMR* **15** 137–154

[43] Zheng J and Pollack GH 2003 Long-range forces extending from polymer-gel surfaces *Phys. Rev.* E **68** 031408-1–031408-7

[44] Zheng J, Chin W, Khijniak E, Khijniak E Jr and Pollack GH 2006 Surfaces and interfacial water: evidence that hydrophilic surfaces have long-range impact *Advanc. colloid interface sci.* **127** 19–27

[45] Pollack G, Cameron I and Wheatley D 2006 *Water and the Cell* (Dodrecht: Springer)

[46] Chai B, Yoo H and Pollack GH 2009 Effect of radiant energy on near-surface water *J. Phys. Chem.* B **113** 13953–13958

[47] Chai B, Zheng J, Zhao Q and Pollack GH 2008 Spectroscopic studies of solutes in aqueous solution *J. Phys. Chem.* A **112** 2242–2247

[48] Pollack G, Figueroa X and Zhao Q 2009 Molecules, water, and radiant energy: new clues for the origin of life *Int. J. Molec. Sci.* **10** 1419-1429

[49] Zimmerman S, Zimmerman AM, Fullerton CD, Luduena RF and Cameron IL 1985 Water ordering during the cell cycle: nuclear magnetic resonance studies of the Sea-Urchin egg. *J. Cell Sci.* **79** 247– 257

[50] Fuchs EC, Woisetschläge J, Gatterer K., Maier E, Pecnik R and Eisenkölbl H 2007 "The floating water bridge," *J. Phys. D: Appl. Phys.* **40** 6112–6114

[51] Fuchs EC, Gatterer K, Holler G and Woisetschläger J 2008 Dynamics of the floating water bridge *J. Phys. D: Appl. Phys.* **41** 185502-1–185502-5

[52] Fuchs EC, Bitschnau B, Woisetschläger J, Maier E, Beuneu B and Teixeira J 2009 Neutron scattering of a floating heavy water bridge *J. Phys. D: Appl. Phys.* **42** 065502-1–065502-4

[53] Giuliani L, D'Emilia E, Lisi A, Grimaldi S, Foletti A and Del Giudice E 2009 The floating water bridge under strong electric potential *Neural Netw. World* **19** 393–398

[54] Preparata G 1995 QED *Coherence in Matter* (New Jersey, London, Hong Kong: World Scientific)

[55] Del Giudice E, Elia V and Tedeschi A 2009 The role of water in the living organisms, *Neural Netw. World* **19** 355–360

[56] Del Giudice E and Tedeschi A 2009 Water and autocatalysis in living matter *Electromagn. Biol. Med.* **28** 46–52

[57] López–Beltrán E, Maté M and Cerdán S 1996 Dynamics and environment of mitochondrial water as detected by H NMR *J. Biol. Chem.* **271** 10648–10653

[58] Trombitás K, Baatsen P, Schreuder J and Pollack GH 1993 Contraction-induced movements of water in single fibres of frog skeletal muscle *J. Mus. Res. Cell Mot.* **14** 573–584

[59] Tyner KM, Kopelman R and Philbert MA 2007 'Nanosized voltmeter' enables cellular-wide electric field mapping *Biophys. J.* **93** 1163–1174

[60] Tilbury R and Quickenden T 1992 Luminescence from the yeast Candida utilis and comparisons across three genera *J. Biolum. Chemilum.* **7** 245–253

[61] Batyanov AP 1984 Distant optical interaction of the mitochondria through quartz *Byuleten Exper. Biol. Med.* **97** 675–677

[62] Batyanov AP 1995 Correlation between mitochondria metabolism and the physical characteristics of incubation cells *Biophotonics. Non-equilibrium and Coherent Systems in Biology, Biophysics and Biotechnology* ed L Belousov and F-A Popp (Moscow: Bioinform Services Co) pp 439–446

[63] Popp FA 2006 Coupling of Fröhlich-modes as a basis of biological regulation *Herbert Fröhlich FRS, A physicist ahead of his time* ed GJ Hyland and P Rowlands (Liverpool: The university of Liverpool) pp 139–175

[64] Jelínek F and Pokorný J 2001 Microtubules in biological cells as circular waveguides and resonators. *Electro- Magnetobiol.* **20** 75–80

[65] Satarić M, Tuszyński JA and Žakula RB 1993 Kinklike excitations as an energy transfer mechanism in microtubules *Phys. Rev.* E **48** 589–597

[66] Tuszyński JA, Hameroff S, Satarić M, Trpisova B and Nip MLA 1995 Ferroelectric behavior in microtubule dipole lattices: implications for information processing, signaling and assembly/disassembly *J. theor. Biol.* **174** 371–380

[67] Stracke R, Böhm KJ, Wollweber L, Tuszyński JA and Unger E 2002 Analysis of the migration of single microtubules in electric fields *Biochem. Biophys. Res. Comm.* **293** 602–609

[68] Pelling AE, Dawson DW, Carreon DM, Christiansen JJ, Shen RR, Teitell MA and Gimzewski JK 2007 Distinct contributions of microtubule subtypes to cell membrane shape and stability *Nanomed.: Nanotechnol. Biol. Med.* **3** 43–52

[69] Pokorný J, Hašek J and Jelínek F 2005 Endogenous electric field and organization of living matter *Electromag. Biol. Med.* **24** 185–197

[70] Frauenfelder H, Volynes PG and Austin RH 1999 Biological physics *Mod. Phys.* **71** S419–S430

[71] Pokorný J 2001 Endogenous electromagnetic forces in living cells: implications for transfer of reaction components *Electro- Magnetobiol.* 20 59–73

[72] Pokorný J 2006 The role of Fröhlich's coherent excitations in cancer transformation of cells *Herbert Fröhlich, FRS: A physicist ahead of his time* ed GJ Hyland and P Rowlands (Liverpool: The University of Liverpool) pp 177–207

[73] Pavlides S et al. 2009 The reverse Wartburg effect: aerobic glycolysis in cancer associated fibroblast and the tumor stroma *Cell Cycle* **8** 4984–4001

[74] Jandová A et al. 2009 Cell-mediated immunity in cervical cancer evolution *Electromagn. Biol. Med.* **28** 1–14

[75] Beil M et al., 2003 Sphingosylphosphorylcholine regulates keratin network architecture and visco-elastic properties of human cancer cells *Nature Cell Biol.* **5** 803–811

[76] Suresh S, Spatz J, Mills JP, Micoulet A, Dao M, Lim CT, Beil M and Seufferlein T 2005 Connections between single-cell biomechanics and human disease states: gastrointestinal cancer and malaria *Acta Biomaterialia* **1** 15–30

[77] Suresh S 2007 Biomechanics and biophysics of cancer cells *Acta Materialia* **55** 3989–4014

[78] Foster KR, and Baisch JW 2000 Viscous damping of vibrations in microtubules *J. Biol. Phys.* **26** 255– 260

[79] Reimers JR, McKemmish LK, McKenzie RH, Mark AE and Hush NS 2009 Weak, strong, and coherent regimes of Fröhlich condensation and their applications to terahertz medicine and quantum consciousness *PNAS* **106** 4219–4224

[80] McKemmish LK, Reimers JR, McKenzie RH, Mark AE and Hush NS 2009 Penrose–Hameroff orchestrated objective-reduction proposal for human consciousness is not biologically feasible. *Phys. Rev.* E **80** 021912-1–021912-6

[81] Lamprecht I 1980 Growth and metabolism in yeasts *Biological Microcalorimetry* ed AE Beezer (London: Academic Press) pp 43–112

[82] Kučera O, Cifra M and Pokorný J 2010 Technical aspects of measurement of cellular electromagnetic field *Eur. Biophys. J.* **39** 1465–1470

[83] Bonnet S et al. 2007 A mitochondria-K^+ channel axis is suppressed in cancer and its normalization promotes apoptosis and inhibits cancer growth *Cancer Cell* **11** 37–51

Journal of Physics: Conference Series **329** (2011) 012008 doi:10.1088/1742-6596/329/1/012008

Mechanoelectrical and Photon-Generating Devices in Cells and Organisms: From Molecular Machines to Macroscopic Fields

LV Beloussov
Laboratory of Developmental Biophysics
Faculty of Biology
Moscow State University
Moscow 119991 Russia

E-mail: morphogenesis@yandex.ru

Abstract. The aim of this essay is to review what we know about the transformation of chemical energy into mechanical, electrical and photonic at the different scales of biological organization. We start from the "classical", short-range mechanoelectrical protein machines emphasizing their capacity to slow down the rate of energy relaxation and to concentrate energy onto a restricted number of freedom degrees. Then we pass to the newly described "low entropy machines" and to the macroscopic electromechanical machines which create circuits of the organismal scales. At last, we come to photonic events, paying a special attention to their regular periodicity within several Hz range and to their relations with cytoskeletal structures and their developmental dynamics. We suggest, that this area of investigations should be related with the theory of self-organization and the notion of coherency.

Introduction

All the living beings are known to be far removed from thermodynamic equilibrium, using thus for their purposes and discharging into the outer space measurable amounts of mechanical, electrostatic and electromagnetic energy. Within several last decades a substantial progress was achieved in revealing molecular mechanisms, involved in the generation and transformations of these kinds of energy within a living matter. On the other hand, a simple question remains up to now unanswered – how these micromachines, having characteristic dimensions of about several nanometers and their own oscillations periods within 10^{11} Hz range integrate their activities for providing a coordinated behavior of macroscopic organisms whose linear and temporal dimensions are more than in dozen orders greater?

The aim of this essay is just to attract the interest of investigators to this fundamental problem, rather than to suggest its solution. We would like to start from bridging a gap between our knowledge of the molecular generators and transformers of energy and some biological and physical concepts dealing with macroscopic order in entire organisms. To these concepts belongs, on one hand, the idea of morphogenetic fields and, on the other, the self-organization theory (SOT). The concept of morphogenetic fields postulates that a developing organism is a system of position-dependent forces, regulated from the upper levels and molding successive embryonic shapes. Remarkably, a founder of this theory, Russian biologist Alexander Gurwitsch in his latest version regarded the molecules as the objects of field action, defining a field as a factor vectorizing a certain part of a molecular excitation energy [1]. This makes it reasonable to discuss the relations of this theory with a modern concept of so called protein-machines.

On the other hand, SOT is a well known and a perfectly elaborated theory using a qualitative analysis of non-linear differential equations for explaining how a new macroscopic scale can be generated on the basis of much more local interactions, possessing no linear scale at all [2]. Both morphogenetic field theory and a SOT, in spite of being originated quite independently from each other, became later on to a considerable extent converged.

Accordingly, we start from reviewing what is known about so called "protein machines", that is, elementary transformers of the chemical potentials to mechanical, electrostatic and electromagnetic energy. We shall distinguish two types of such machines: individual protein molecules, creating the "classical", or short range machines, and recently discovered low entropy machines, based upon a coherent behavior of large ensembles of water molecules. Then we shall describe electrical circuits in cells and entire embryos and show that they are extended up to the entire organisms scales and are tightly coupled with the patterns of mechanical stresses. Taking as example lower invertebrates, hydroid polyps, we demonstrate the role of electrical fields in orienting polarized growth.

Next, we come to the photonic events. Here we start from emphasizing the role of collective molecular interactions in photons generation. Passing then towards higher levels, we describe the existence of specific temporal patterns of photon emission with characteristic 10^0 -10^2 Hz range frequencies.

Classical (short range) proteins-machines

By text-book definitions [3] a protein machine consists of a molecule possessing ATPase activity, a fibril along which this molecule is translocated and a load which it transports. This definition fits such constructions as actomyosin and other microfilaments and microtubules associated motors, but it is too narrow and, most important, does not elucidate the basic physical principles of these devices. More broad definition embraces a wider class of the protein-machines, which transform non-vectorized chemical energy to a vectorized mechanical one, even if they do not bear any load and do not split ATP molecule at the given time moment. Several decades ago such devices got a remarkable name of "Brownian ratchets" [4] which meant that they transform chaotic (Brownian) molecular movement into a directed one. Actin microfilaments, pushing the leading edge of a crawling fibroblast, were often taken as an example. Subsequent investigations demonstrated however, that these "ratchets" are organized in much more complicated way as beforehand suggested. Most important role is played by a dimeric protein, formin, located at the +end of a double helical microfilament and oscillating with about 10 Hz rate between a screwed and relaxed state. As a result, the torsion energy of about 1 eV is periodically accumulated and discharged. This energy is equivalent to that released by splitting of about 10 ATP molecules. However, in this process no such splitting is taking place: by a model suggested [5] all what we have here is the exchange of elastic energy between formin and the adjacent microfilaments subunits, with some gain got due to establishment of hydrophobic bonds between the newly inserted actin subunits.

The main common feature of all the protein-machines is a *considerable retardation of relaxation rate of the accumulated energy and enormous restriction (often up to one) in the numbers of freedom degrees onto which the relaxation is taking place* [6-8]. At the same time, the amplitudes of the directed movements performed along the rested freedom degrees reach the characteristic dimensions of the machine itself (which is impossible in the case of man-made industrial machines). These properties are provided by a specific structure of protein macromolecules, which include:

- a relatively soft elastic spring accumulating energy in the form of elastic stresses. The spring consists of non-spiral parts and electrically charged groups;
- rigid "levers" which transmit stresses (α-spirals);
- hinge joint-points, represented by small aminoacids, for example glycine;
- fixation points: hydrogen bonds, S-S- bonds.

The energy accumulated in elastic elements is estimated as $\approx$0,5 eV per 5 x 10^{-9} m (length of the element), while the lifetime of the metastable mechanically stressed states ranges between 10^{-6} and 10^3 s, which in about 15 orders exceeds the rates of self-frequencies of protein molecules (10^{11} – 10^{12} s^{-1})! [8].

Thus, the main ones and tightly interconnected properties of the protein-machines are:

- very much retarded relaxations of accumulated elastic energy, producing quite large (in a macromolecular scale) and essentially vectorized shifts;

- coexistence and practical indivisibility of mechanical and electrostatic components, permitting to define these machines as electro-mechanical.

Let us comment the latter point. The electrically charged groups (mostly negatively charged carbonyls and positively charged aminoacids) contribute to the maintenance of mechanically stressed state of the protein macromolecules. Same role should be played by the charged ligands, in particular ATP, containing three negatively charged phosphate groups. Therefore, binding ATP to proteins should increase electrostatic repulsion onto the latter's surface and hence straighten the protein molecules, supplying them by additional elastic energy. By Pollack [9] estimation the latter's amount is twice as great as that provided by ATP molecule splitting. Taking into consideration, that each next step of practically all the signaling pathways is associated with phosphorylation of proteins, this kind of molecular deformation should play a great, however up to now practically neglected role in cell signaling.

Until now we were dealing with chemo-electro-mechanical transformers in the form of isolated macromolecules (for example enzymes) acting in their immediate vicinity (non-exceeding the radii of molecular interactions). Meanwhile, just discussed example of ATP-protein interactions led some authors to postulate the existence of quite another class of machines, characterized by very much extended ranges of action, including billions of uniformly oriented water molecules. These will be described in the next section.

Low entropy machines.

In his "association-induction" theory Ling [10] points out that an ATP molecule adsorbed by a protein is a powerful electron acceptor, reducing the electron density onto a protein surface within considerable distances. As a consequence, "the liberated carboxyl groups start to absorb K^+ while NH^- and CO^- groups liberated from participating in creating secondary structures become to interact with water, forming a multilayer structure consisting of polarized and uniformly oriented water structures" [10].

This is one among several scenarios of the low entropy water domains formation. As theoretically predicted ([11, 12]; see also this volume), if the density of any ensemble of microscopic particles exceeds a certain threshold, the particles start to oscillate in a coherent way. Such "coherent water domains" (CWD) may extend up to 100 nm and consist of about 10^7 water molecules which oscillate with 10^{11} Hz frequency. Due to a high degree of order, CWD are the areas of very low entropy and are thus able to absorb non-coherent (heat) energy from the surrounding bulk water and transform it into a "high grade" coherent one. Noteworthy, CWD absorb light in the infrared and UV spectral regions. The absorbed energy may reach 12, 06 eV which is only $\approx$ 0,5 eV lower than the energy of water ionization. In other words, CWD possess a substantial amount of free or almost free electrons, which results in the electrical potential difference between CWD and bulk water. In such a way, low entropy areas are real machines, producing electrical energy. Also, they are similar to the above described macromolecular machines (as well as to the photonic machines to be described below) by providing a long life time for exciting electrons. Moreover, these machines are postulated to produce direct mechanical effects by bringing together the molecules oscillating in unison with each other.

What might be a biological role of such machines and, in particular, of a coherent state of water domains? This is as yet non-answered question of a basic importance. Del Giudice et al speculate [11], that such a state is important for providing an extreme temperature stability within the living systems, an avoidance of Joule effects during ions traffic and, most important, a precise meeting of the partner molecules during enzymatic reactions. It is worth mentioning in this respect, that already several decades ago the Russian genetic Josef Rapoport, a discoverer of the chemical mutagens, as discussing why genetic reproduction is so precise, came to the idea that "genetic material" should have a capacity to be in a low entropy state under room temperatures. He emphasized also a high correlation of

Journal of Physics: Conference Series **329** (2011) 012008 doi:10.1088/1742-6596/329/1/012008

mutagenic effects of chemical substances with their dipole moments [13]. (Strange as it may look today, by the urgent requests of a couple of outstanding Russia scientists, including a Nobel prizer Igor Tamm, the entire first edition of Rapoport's book was destroyed, because the author's suggestions "contradicted firmly established physical laws"! Significantly, several years before same arguments were used for rejecting the first B.P. Belousov paper about periodic chemical reaction, so famous in the near future).

A sensitivity of the living matter to microwaves can be also associated with the presence of CWD [12]. Here we are confronted with a rare for biology situation where theoretical predictions are ahead of experimenting; this is a challenge to the lab biologists to make everything possible for exploring the role of low entropy machines, because this can revolutionize the entire image of cell biology.

Macroscopic electrical circuits and their coupling with morphomechanics

With the use of vibrating electrodes techniques it became possible to detect measurable electric fields in the immediate vicinity of the living samples, including single cells [14]. So far as the force lines of electric fields must be closed, they should also pass, in a reverse polarity, through the cells themselves. The first measurements performed on the egg cells of brown algae indicated 2-20 mV potentials between the opposite cell poles and 5 $\mu A/cm^2$ current density throughout the cell. The negative pole corresponded to the site of a rhizoid growth (Fig. 1A). Similar indices and, what is mostly remarkable, similar polarity (the growing/extending part corresponding to the negative pole, while a shrinking/contracting one to the positive one) took place in other samples as well.

Mostly interesting for morphogenesis are electrical circuits established in developing Vertebrate embryos [15]. In these, force lines of electrical fields enter the apical domains of morphogenetically inactive epithelial cells and leave the same cells through their opposite, baso-lateral domains. As a result, the apico

-basal flow of the positive sodium and potassium ions (accompanied by chloride ions) is established, increasing osmotic pressure in subepithelial cavities and promoting thus their swelling and a stretching of

the overlain epithelium. On the other hand, same embryos apically contracted cells acquire a reverse electrical polarity and become the sites where the force lines are leaving embryonic tissues (Fig. 1B). These cells occupy morphogenetically most active contracted regions (blastopore, primitive groove). As a result, we get essentially *macroscopic* electrical circuits, which at the same time precisely coincide with what we call the morphomechanical feedbacks (Fig. 1C). The latter are the positive feedback loops between the extended and contracted areas of an epithelial layer: extension enhances contraction and vice versa [16]. As development proceeds, the corresponding patterns become even more complicated (Fig. 1 D-G).

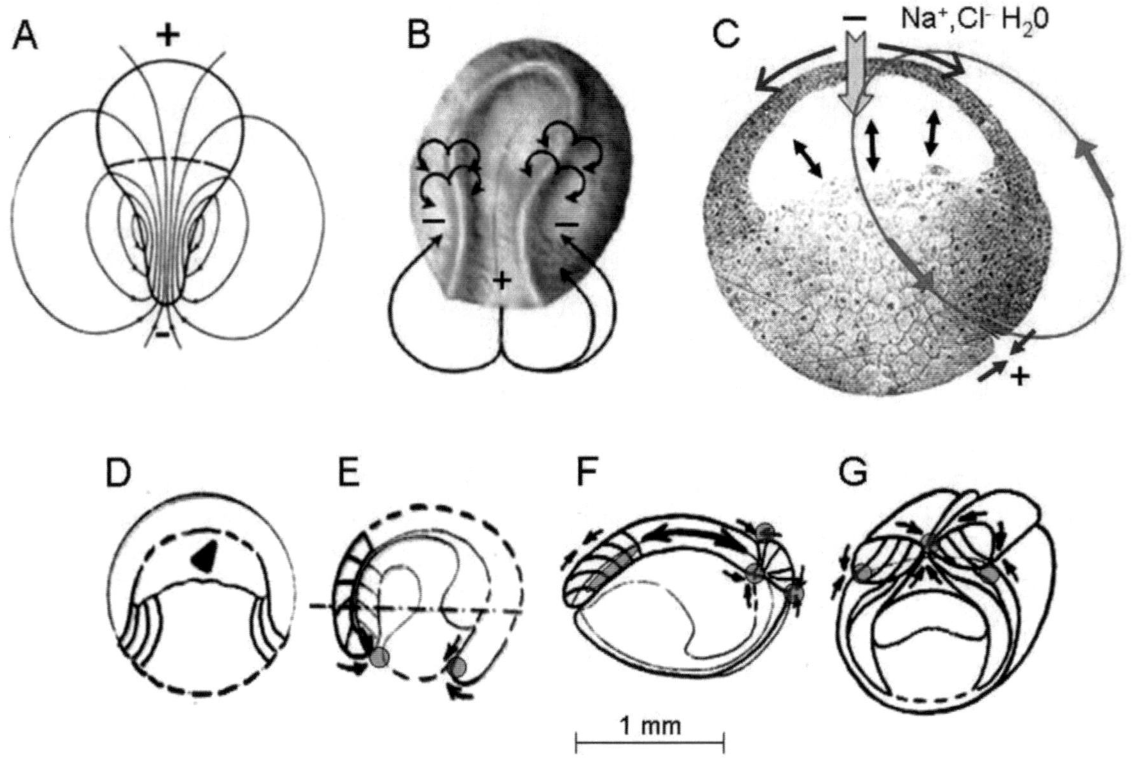

Fig. 1. Electromechanical coupling in eggs and embryos. In this and the next figure electrical vectors correspond to the movements of positive charges. A: outgrowing egg of brown algae, Pelvetia. Negative pole fits the growing rhysoid (below). B: Amphibian embryo at neurula stage. Positively charged are the grooves, negatively charged is the skin ectoderm. C: a mechanoelectric map of a blastula stage of amphibians embryo. Animal ectoderm (above) is negatively charged and stretched due to transport of ions and corresponding osmotic water inflow. To the opposite is the contracted positively charged blastoporal zone. D-G: patterns of tension lines (dense) at the successive developmental stages of a frog embryo (from blastula to neurula). Red areas correspond to the points of tension lines convergence (contracted regions) and presumably to the positive poles of electric fields.

Mechanoelectric machines for pulsatorial growth.
Hydroid polyps, the representatives of lower Invertebrates, form the colonies with quite precise and species-specific branching patterns (Fig. 2A). The branched stems are closed tubes consisting of two cell layers which behave in a roughly similar way, permitting, at the first approximation, to regard the stems as one layered formations. During growth several dozens cells arranged at the stem tip (distal part) for each next minutes are swelled and rotate towards perpendicular orientations, to be later on shrunk and taking oblique positions. The first phase coincides with stem's elongation (extension phase) while the second one with its partial shortening (retraction phase) (Fig. 2B, **A-D**). These periods are called growth pulsations (GP). GPs are characterized by finely regulated temporal-amplitude patterns (Fig. 2C, **A-E**). Extension GP phases are associated with the osmotic water inflow within cell vacuoles while retraction phases with water outflow.

By the model suggested [16], GPs belong to the category of autooscillations with a considerable mechanical component. The main feedback loop is created by the activation of mechanodependent Ca^{2+} channels due to stretching of cell membranes at the extension phase. As a result, the isolated vacuoles (Fig. 2D) are fused into the outside opened channels, leading thus to water outflow and hence to the shift towards retraction phase (Fig. 2E).

Meanwhile, it turned out that GP can be also regulated by external electrical fields, properly oriented in respect to the disto-proximal stem polarity [17]. Namely, if applying to a growing stem a *descending* (positive pole to the distal and negative to the proximal stem pole) 1-3 V/cm electric field (current density 20-150μA/mm^2) GPs were arrested at the retraction phase (Fig. 2F), while in the case of an *ascending* field (negative pole to the distal) the arrest took place at the extension phase (Fig. 2G). If the electrical field was applied perpendicularly to the stem axis, in the case of the negative pole apical location the reactions were the same as in the case of its distal location. All the reactions were completely reversible. The electrically induced arrests were characterized by specific vacuolar patterns: at the extension phase cells were full of small spherical vacuoles, while at the retraction phase large intercellular spaces could be observed.

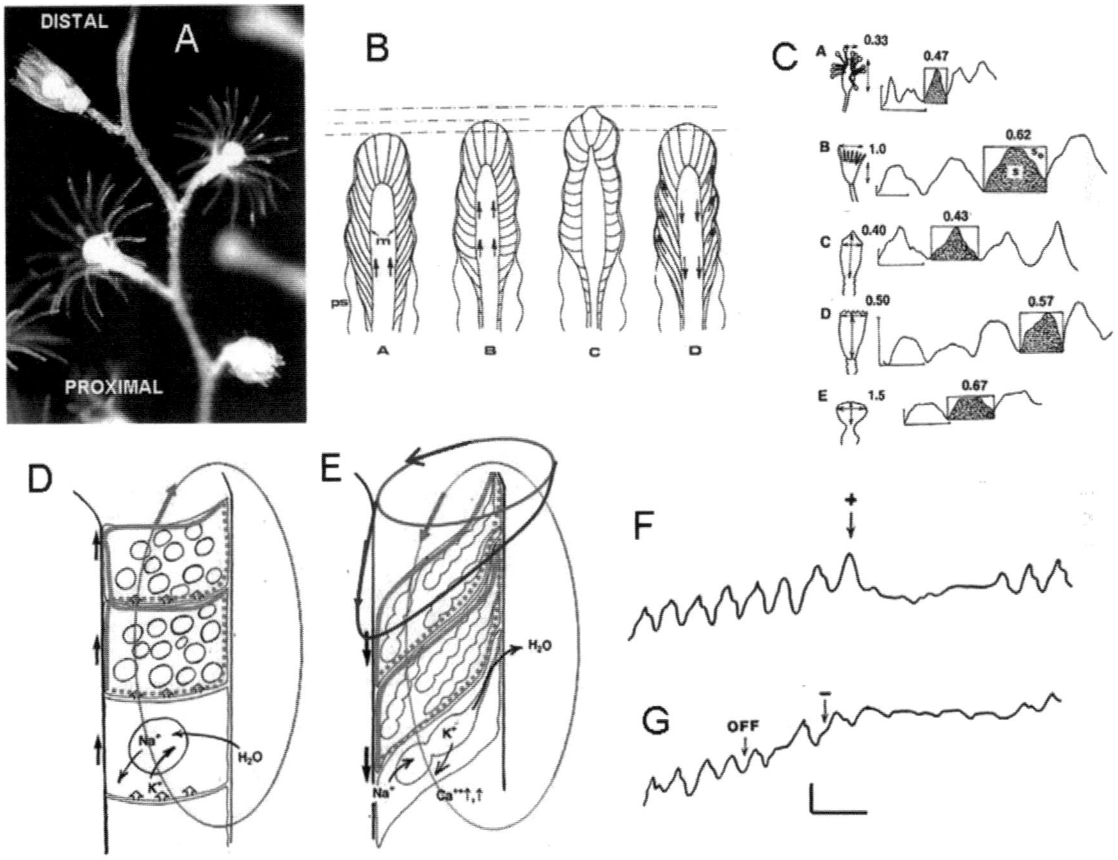

Fig. 2. Growing hydroid polyps as macroscopic dynamo-machines. A: view of a colony. B: Phases of a single GP. **A,D**: retraction, **C,D**: extension. C: Shapes of hydranths and patterns of growth pulsations (GP) in 5 different species. D: vacuolar patterns at the extension phase, E: same at the retraction phase. In both frames red contours display external electric fields which arrest GP at the given phase, pink solid contours outline disto-apical domains of cell membranes assumed to contain electrically non-excited Ca^{2+} channels while pink dotted contours outline proximo-basal membrane domains containing electrically excited channels. Blue contours in E display endogenous currents which according to our model act periodically, triggering the passage from F to G. F: growth arrest at the

retraction phase, positive pole of external field to the distal. G: growth arrest at the extension phase, positive pole to the proximal. Scaling angle under F, G: horizontal line is 10 min, vertical line 10 μm.

In order to explain the polar effects of external electrical fields one can assume that the electrically non-excitable Ca^{2+} channels are predominantly located at the distal and apical areas of cell membranes (Fig. 2 D, E, solid pink contours) while those excited by electrical depolarization in the proximal and basal areas (dotted pink contours). Obviously, under these conditions a descending electrical currents should stimulate, while the ascending one hamper Ca^{2+} ionophoretic inflow. These flows are depicted in Fig. 2 D, E by red contours. As a result, in the first case the cells will be arrested at the retraction, while in the second one at the extension phase, which is just the case. Such an arrangement of Ca^{2+} channels should be regarded as a manifestation of a universal property of so called planar cell polarity playing an important role in cell differentiation [18].

Most interesting would be to apply these suggestions to the normal growth patterns of hydroid polyps, taking into consideration the mechanosensitivity of Ca^{2+} channels, induced by cells periodic swelling. Within such a context, we can regard their growing tips as a kind of biological dynamo-machines, transforming periodic stretching of cell membranes into electric currents mediated by Ca^{2+} flows. These latter are depicted in Fig. 2E by two conjoined blue contours, indicating the currents via non-excitable Ca^{2+} channels of the distal and apical cell membranes. These currents should reach their maximal values just before the retraction GP phase (shown in E) and induce the transition to this phase.

Generation and temporal patterns of photon emission in cells and embryos

Ultraweak photon emission (UPE) of the living objects has been discovered almost 90 years ago and since then its existence has been confirmed in a great number of studies (recent reviews: [19, 20]), although the views upon its biological role and physico-chemical mechanisms of photons generation remain to be quite diverse. Two main different (but not necessarily mutually exclusive) theories upon UPE origin are forwarded. According to the first one, photonic level energy is produced as a result of recombination of free radicals which are the by-products of many chemical reactions, mostly oxidative. By the second view, the photonic level may be reached by summation of smaller portions of energy released during fast enough collective molecular interactions in the dense and/or ordered medium. Slawinsky in his very instructive paper [21] gives several examples of such processes. For example, relaxation of 9 supertwists in DNA molecule liberates 2 eV per 0,01 s only; enzymatic splitting of acetylcholine at the postsynaptic membrane releases up to 1,2 eV even within a smaller time period ($4x10^{-7}$ s). On the other hand, a slow relaxation of excited states, mentioned beforehand as a main feature of the proteins-machines, should also increase the probability of a summation of small portions of energy up to photonic level. Therefore, chemo-mechanical and photon-generating devices should share the same fundamental properties.

Similar estimations make it probable that the photonic level energy is accumulated in cytoskeletal structures and can be released during their physiological or artificial destruction. Thus, by Pokorny [22] estimations, a single microtubule stores a power of about 10^{-14} W/cm, obtained due to a rapid post-polymerization GTP hydrolysis. Taking an average microtubule length of about 10μm, and assuming that no less than 100 microtubules per second are polymerized in a close vicinity, we get a power of about 10^{-15} W, produced in a proper proximity and thus be potentially able to sum up to a photonic level. This amount of energy exceeds in two orders the power of a photon emission, sensed by a standard photomultiplier. Last but not least, the photonic level energy can be accumulated in the above described coherent water domains.

Here we give a brief account of our UPE measurements from cell cultures and embryos, obtained within two last decades in the International Institute of Biophysics (Neuss, Germany). The main aims of the measurements were:

- to check a UPE role in cell signaling;

- to look whether and to what extent UPE is affected by destruction of cytoskeletal elements and cells-substrate contacts;
- to follow UPE dynamics in the developing embryos of a fish, *Misgurnus fossilis.*

In all these studies our interest was focused on the temporal (periodic) patterns of UPE in several Hz range. Correspondingly, we have measured UPE Fourier spectra in terms of either periodograms or spectral densities for $10^0 - 10^2$ s successive time periods. Some of results were also presented as autocorrelation diagrams of Fourier spectra because they give more obvious and detailed image of the signals structure. For more details see [23-25].

1. Photonic component of cell signaling.

FGF (fibroblast growth factor) is a universal ligand for a number of signaling pathways playing quite different roles [3]. In particular, FGF stimulates the entrance of fibroblast cells into a cell cycle, promoting thus their proliferation. We were interested to know, whether FGF-1 addition (10^{-9} g/ml) to a monolayer culture (about 10^5 cells) of mouse fibroblasts (line C3H10T1/2) affects in some way the UPE temporal patterns.

First of all, it turned out that the addition of FGF did not increase the average UPE intensity (Fig. 3A). Moreover, the presence of cell cultures also did not exceed UPE as compared with that of a cell-free medium (Fig. 3, cf B and C), while the latter one was significantly greater than the background photomultiplier level. From this we conclude that the measured signals are related to UPE produced by oxidative reactions within a cell-free medium. Meanwhile, addition of the FGF-induced cell cultures led in few minutes to the appearance of pronounced and gradually evolved regular enough UPE oscillations within $10^1 - 10^2$ s range (Fig. 3C, D).

Interestingly, UPE measurements performed on the monolayer cultures of hyppocamp (mid-brain) cells revealed a similar evolution of periodic patterns, now going on spontaneously, without any intervention from outside (Fig. 4). Such an activity may be attributed to inherent properties of nerve cells.

2. Effects of cytoskeletal drugs and trypsin upon UPE Fourier spectra.

As seen from Fig. 5C, the addition of cytochalasine D (10μg/ml), the drug preventing self-assembly of microfilaments, produced in few minutes a group of spectral density peaks (2s, 13s and 130s periods), two of which were preserved up to 20 min and accompanied by 28s peak. Worth mentioning, most of the periods relate to each other as integer numbers. Much more pronounced periodicity, taking place immediately after the drug addition, can be seen in autocorellogram (Fig. 5B).

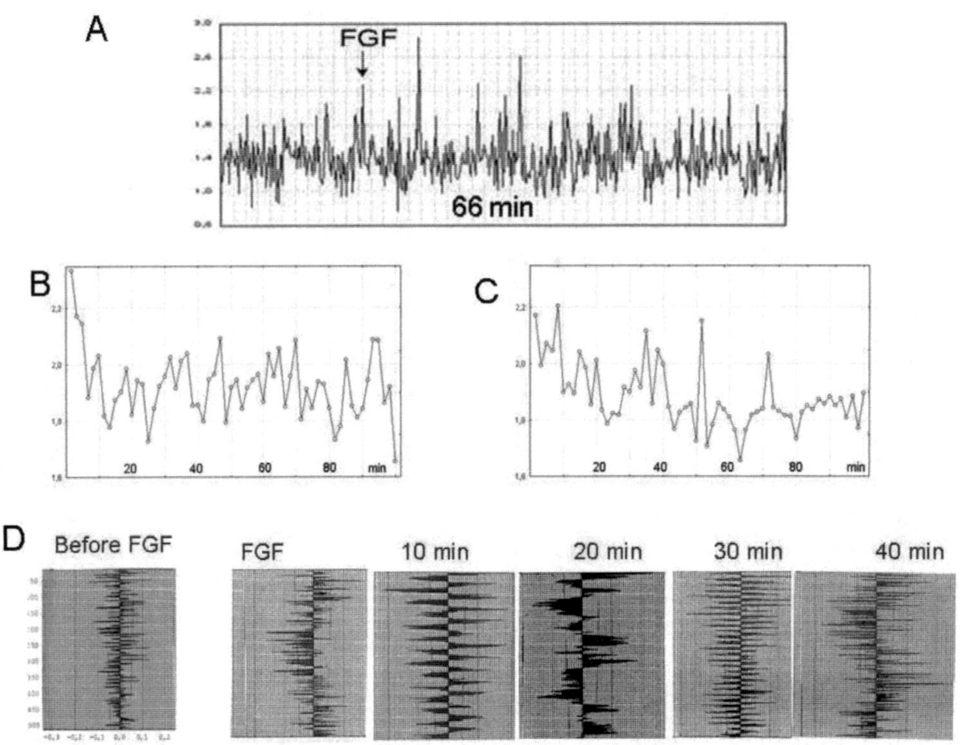

Fig. 3. UPE of a monolayer fibroblast culture under the action of fibroblast growth factor (FGF-1). A: a record. B: a 100-fold aggregated record of a cell-free medium. B: same with fibroblast culture. Note several outstanding peaks, no increase in the overall intensity. In A-C horizontal axis is minutes, vertical axes UPE intensities count/sec. D: autocorellograms of 5 min long successive UPE Fourier spectra, displaying evolution of spectral patterns.

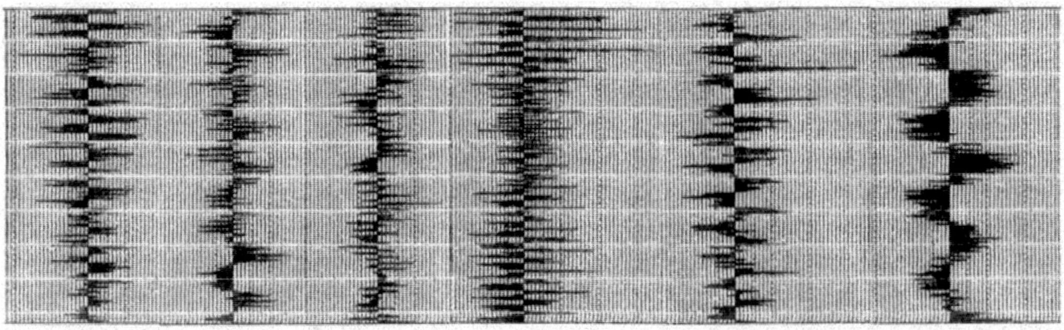

Fig. 4. Successive 5 min long (separated by 10 min intervals) Fourier spectra (in terms of autocorellograms) of spontaneous UPE from monolayer culture of hyppocamp cells.

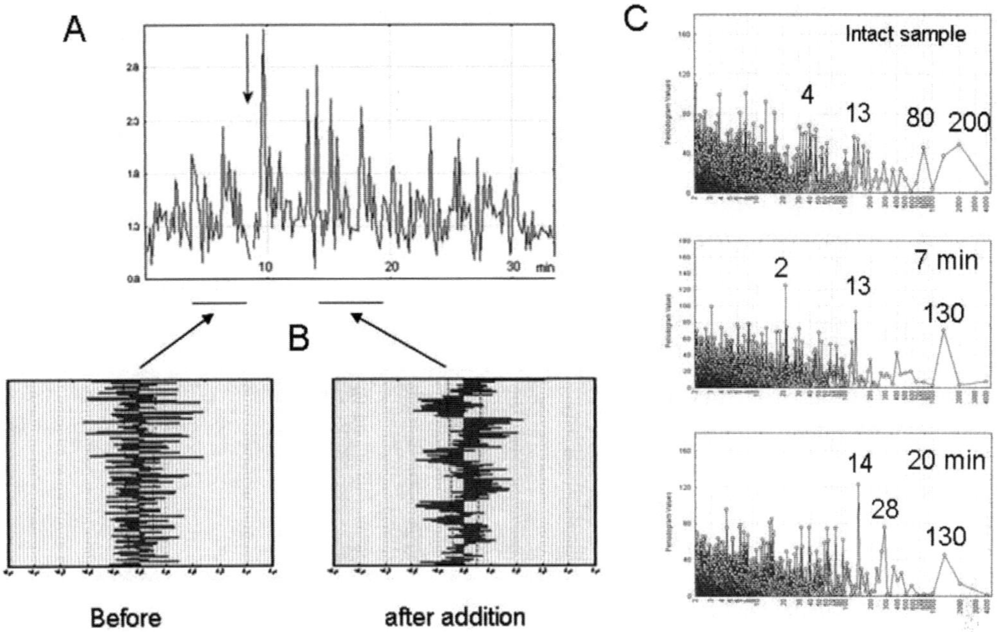

Fig. 5. Fourier patterns of UPE under the action of Cytochalasine D (CD) upon monolayer fibroblast culture. A: UPE record (vertical arrow indicates the moment of CD application) and 5 min long records of Fourier patterns before and after CD application in terms of autocorrelograms. Note increase of spectral coherency soon after CD application. B: evolution of Fourier spectra (periodograms) within 20 min after CD addition, as compared with the same intact sample. Periods (in seconds) of the most outstanding peaks are shown.

Similarly, less than 1 min after adding 10^{-4} M colchycine (agent preventing microtubules assembly) monolayer fibroblast culture responded by appearance of new spectral density peaks which also relate to each other as integer numbers (2s, 10s, 40s and 200s periods). Later on these peaks gradually smoothed (Fig. 6A).

UPE reaction to trypsin/EDTA (0,025%) addition (which led to the loss of cell-substrate contacts) was very dramatic, leading to the formation of several outstanding peaks (some of which also related to each other as integer numbers) already in few minutes after drug administration (Fig. 6B). Later on the peaks gradually smoothed out.

Therefore, the results (6-1) and (6-2) indicate the existence of rather specific and gradually evolving UPE regimes which point, most probably, to the periodical exchange of low intensity photonic pulses between the external medium and living samples. In the cases (6-2) such pulses can be characterized as "alarm signals", since they are associated with destructive processes.

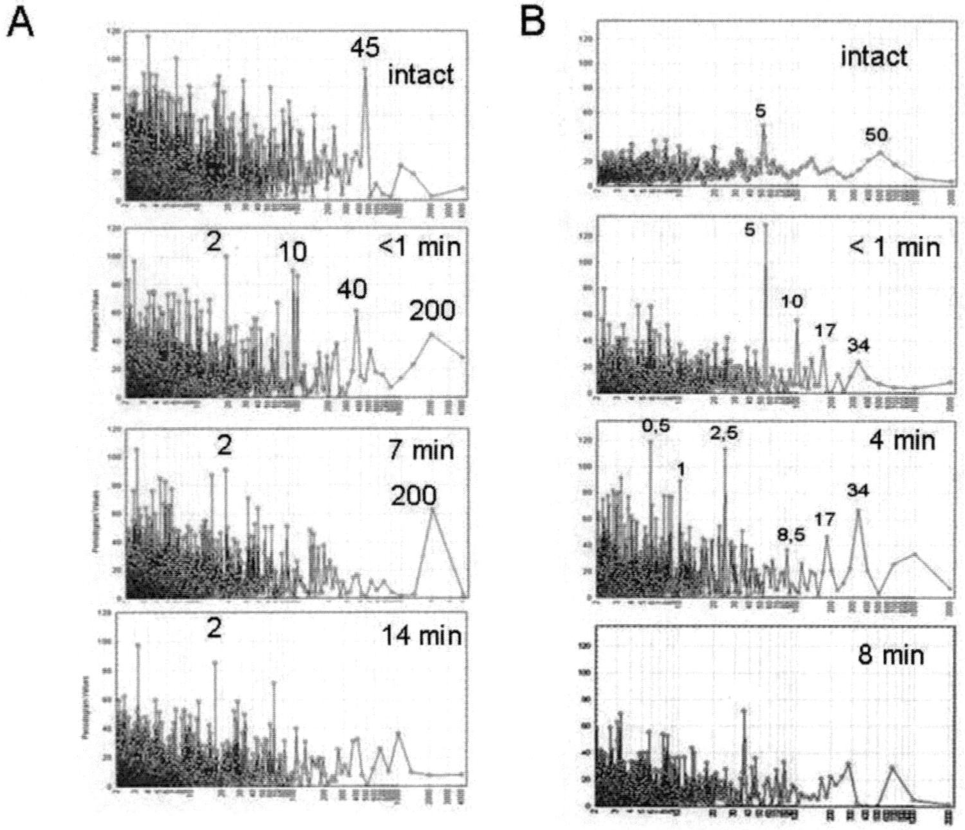

Fig. 6. Evolution of UPE Fourier spectra (in terms of periodograms) after the action of colchycine (A) and trypsine (B) to the monolayer culture of fibroblast cells. Designations as in Fig. 5.

3. Dynamics of UPE Fourier patterns during the development of a fish, *Misgurnus fossilis.*

Fig. 7 displays typical Fourier patterns in terms of spectral densities (middle row) and autocorrelation diagrams (right row) for the batches of eggs and embryos of several successive developmental periods (shown at left). While non-fertilized eggs are characterized by rather chaotic and uncorrelated Fourier spectra, those entering the cleavage period exhibit highly ordered periodic patterns, which later on become again less ordered. However, some other, hardly formalized but nevertheless significant signs of temporal organization can be traced at the advanced developmental stages as well (three lower lines).

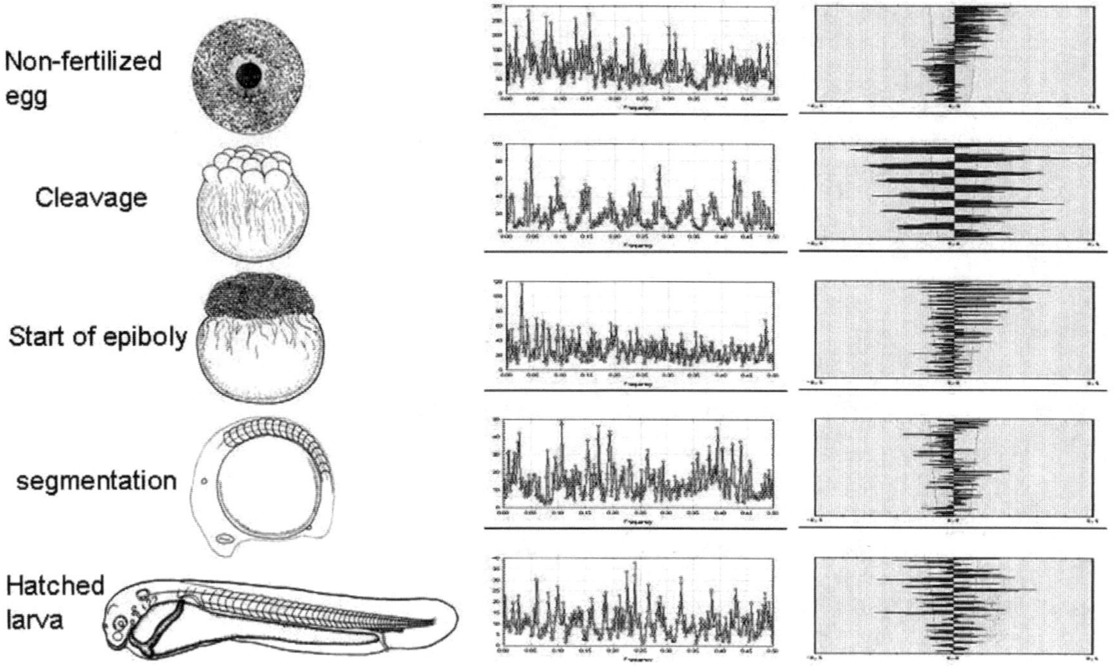

Fig. 7. Evolution of UPE Fourier spectra at the successive stages of Misgurnus fossilis development (left column) in terms of spectral densities (middle column, horizontal axes cover 5 Hz) and the corresponding autocorellograms (right column). Each spectrum was recorded within 2 min.

Discussion

As it was briefly mentioned beforehand, a founder of a morphogenetic field theory, Alexander Gurwitsch, proposed that the dynamic order in the living organisms is supported by a single vectorizing factor, acting in the same way in all the structural levels, from molecular to organismal. Today we can see that the reality is much more complicated and fragmented. What we observe at the molecular level, is a set of devices (machines) accurately collected from specific molecules (mostly proteins) which prevent a chemical energy stored in molecular bonds from chaotic dissipation by transforming it into a high grade (low entropy) mechanical, electrical and photonic energy. In doing this, all the molecular machines (not only electromechanical, but also photon-producing), in spite of all their intrinsic differences, share the common property of retarding the relaxation times and extensively reducing the number of freedom degrees for the liberated energy.

This is however only the first step which is required for extending the dynamic order up to the macroscopic space-temporal dimensions. For fulfilling such a task, that is, for passing from, by rough estimations, 10^{-8}cm/10^{0}s range (typical for supramolecular events) to a macro-morphological realm (10^{1}cm/10^{3}s range) another set of powerful "+" and "-" feedbacks between the molecular/supramolecular machines should come into action, which is to be regulated by macroscopic parameters. Only under these conditions, a set of machines should be transformed into what we call the morphogenetic fields.

At the present time just the first evidences on the existence and nature of such feedbacks are accumulated (e.g. [26, 27]). It becomes widely accepted, that a great repertoire of molecules, supramolecular structures and signaling pathways becomes involved in these feedbacks. Under these conditions, it looks rather ineffective to construct a deterministic picture of such feedbacks, enumerating their details one after other. Instead, following the SOT principles, we should concentrate ourselves in outlining a restricted number of so called order parameters, controlling the behavior of an entire system. It is highly plausible, that mechanical stresses, electrical fields and the factors regulating UPE temporal regimes belong to such parameters.

In addition, the traditional SOT approaches should be enriched by a notion of coherency. Such events as a maintaining of a dynamic order over enormous range of the dimensional and temporal scales and, more specifically, a wide presence of integer frequencies in UPE Fourier spectra unequivocally point to such a coherency. On the other hand, as already mentioned, a high degree of coherency has been described to intracellular water domains. By bringing together these as yet scattered evidences, one may hope to reach much higher level of understanding of the energy transformation in living systems.

Acknowledgements
This study was supported by the Russian Fund for Fundamental Research (RFFI), grant 11-04-01-718.

References

[1] Gurwitsch A G 1944 A theory of biological field (in Russian) Sovetskaya Nauka Moskva.
[2] Krinsky V I Zhabotinsly A M 1981 (in Russian). In: Autowave Processes in Diffusional Systems (M.T.Grechova ed.) Gorky Inst. Appl. Phys. Acad. Sci. USSR 6.
[3] Alberts R et al 2003 Molecular Biology of the Cell Garland Science Taylor and Francis Group
[4] Hanggi P Marchesoni F 2009 Rev. Mod. Phys. 81 387
[5] Shemesh T Otomo T Rosen M K Bershadsky A D and M.M.Kozlov 2005 J. Cell Biol. 170 889.
[6] McClare C W F 1971 J. Theor. Biol. 30 1
[7] Blumenfeld L A 1983 Physics of Bioenergetic Processes. Berlin Springer
[8] Chernavskii D S Chernavskaia N M 1999 Protein-machine. Biological macromolecular constructions. Moscow University Press. (in Russian)
[9] Pollack G H 2001. Cells, Gels and the Engines of Life: A New, Unifying Approach to Cell Function. Seattle WA Ebner & Sons.
[10] Ling G N 2001 Life at the Cell and Below-Cell Level. Pacific Press N Y
[11] Del Giudice E and 6 coauthors 2005 Electromagnetic Biol. and Med. 24 199.
[12] Del Giudice E. and Giuliani L. 2010 Eur. J. Oncology Lab 5 7
[13] Rapoport J A Microgenetics 2010 Moskva (in Russian)
[14] Jaffe L Nuccitelli R 1977 Ann. Rev. Biophys. Bioeng. V. 6: 445-476.
[15] Stern C D 1986 Bioessays 4 180.
[16] Beloussov L V 2008 Phys. Biol. 5 015009
[17] Labas Yu A, Beloussov L V, Kazakova N I, Badenko L A 1987 Ontogenez 18 154 (in Russian).
[18] Wang Y, Nathans J 2007 Development 134 647.
[19] Trushin M V 2004 Rivista di Biologia/Biology Forum 97 399.
[20] Cifra M Fields J Z Farhadi A. 2011 Progress in Biophys. Mol. Biol. 105 223.
[21] Slawinski J 1988 Experientia 44 559.
[22] Pokorny J 2006 In: Herbert Frohlich FRS: A Physicist Ahead of his Time. Univ. of Liverpool 177.
[23] Beloussov L V, Luchinskaia N N 1998 In: Biophotons (Jiin-Ju Chang, J Fisch and F.-A. Popp eds). Kluwer Acad. Publishers. Dordrecht /Boston/ London 121.
[24] Beloussov L V 2002 BioSystems 68 199.
[25] Beloussov L V 2006 In: Biophotonics and Coherent Systems in Biology (L V Beloussov, V L Voeikov and V S Martynyuk eds). Springer. N Y 139.
[26] van der Gucht J and Sykes C 2009 Cold Spring Harb. Perspect Biol. 1 a001909.
[27] Rong Li and G G Gundersen 2008 Nature 9 860.

Journal of Physics: Conference Series **329** (2011) 012009 doi:10.1088/1742-6596/329/1/012009

Ionic Pulses along Cytoskeletal Protophilaments

M V Satarić [1] and B M Satarić [1]

[1] Faculty of Technical Sciences, University of Novi Sad, Serbia Trg D. Obradovića 6

E-mail: bomisat@neobee.net

Abstract. The experimental evidences regarding ionic waves generation and propagation along both microtubules (MT_s) and actin filaments (AF_s) motivated us to develop the physical models that provide the framework for the explanation and analysis of these interesting biophysical phenomena.

In pertaining analysis we partly relied on some experimental as well as numerical data, but also on theoretical estimations enabling us to establish the concept of nonlinear transmission lines which could lead to reasonable clearing up of experimental facts.

We are convinced that these ionic currents actually exist and serve to sustain some important biological cellular mechanisms.

1. Introduction

The cytoskeleton is basic ingredient of all living cells. It is composed of three major types of filamental structures, including tubulin-protein based MT_s, actin-protein based AF_s and intermediate filaments. All of them are organized into networks which are interconnected through different auxiliary proteins [1].

There are at least three well-studied mechanical functions of the cytoskeleton in vivo; providing mechanical strength of the cell, segregating the chromosomes in cell division and active participation in the transport of macromolecules via motor proteins, primarily kinesin, dynein and myosin [2,3,4].

Here we pay the particular attention on the roles of MT_s and AF_s in facilitating intra cellular ionic currents which participate in some fundamental biological processes.

First experimental evidences for ionic waves propagating along AF_s are found out by the group led by H Cantiello [5,6]. These assays had clearly shown that the input ionic signals are being localized and propagate along AF_s in the forms resembling solitonic pulses. The theoretical explanations of these features were done by J Tuszynski et al [7] and later by M Satarić et al [8].

On the other hand, the experiment of measurement of ionic current along a single MT was performed within the same group [9]. Theoretical considerations of ionic currents along MT_s were performed by J Tuszynski et al [10] and in many stages by M Satarić et al [11,12,13,14].

In the following we will first present the outlines of polyelectrolitic concept of cytoskeleton and then briefly consider the main results concerning localized ionic pulses along AF_s and MT_s.

2. Manning-Oosava counterion condensation

Manning showed [15] that a highly charged cylinderlike polymers exert strong attraction on their counterions that a certain fraction condenses onto the polymer forming the so called ionic cloud (IC). Typical representative of this class of biopolymers is DNA. This condensation is triggered when the Manning parameter $M = \frac{l_B}{l_0}$ is greater than unity. Here the Bjerrum length is defined as follows

$$l_B = \frac{e^2}{\varepsilon k_B T} \quad , \tag{1}$$

where e is the electron charge, ε is the dielectric constant of cytosol and $k_B T$ is the thermal energy ($k_B = 1,38 \times 10^{-23} J$).

Taking the physiological temperature $T = 310K$ and $\varepsilon = 80 \times 8,85 \times 10^{-12} F/m$, we get

$$l_B = 0,67nm. \tag{2}$$

On the other hand, l_0 follows from cylindrical polymer's linear charge density q_0:

$$q_0 = \frac{e}{l_0} = \frac{q}{L} \quad , \tag{3}$$

where L is the cylinder's total length and q its total surface charge.

This charge causes counterion condensation around polymer in the cases where the concentration of counterions (n) is small enough that the inequality $\xi_D \gg a$ holds (a is the radius of the cylinder).

Here the Debye length ξ_D is defined as follows

$$\xi_D = (8\pi n l_B)^{-1/2}. \tag{4}$$

Relying on the above conditions we found out that both AF_s and MT_s in physiological conditions could be safely considered as polyelectrolites of Manning's class.

We estimated, on the basis of Manning's theory [15], that the thickness of IC in physiological conditions is for AF and MT respectively

$$\lambda_{AF} = 5nm \quad \text{and} \quad \lambda_{TD} = 2,5nm.$$

On the basis of such concept we established the models where these cytoskeletal cylindrical polymers can be viewed as "coaxial cables" having the features of nonlinear transmission lines. These lines in general could be sequenced in ladders of repeated identical electric elementary units (EEU) which posses specific values of capacity, resistivity and conductivity. We estimated that corresponding inductivities of EEU_s are small enough that can be safely ignored [12].

Between IC of condensed counterions around EEU and repelled rest of bulk ions there exists the depleted layer (with no ions present) with the thickness equal to l_B, Fig. 1. Thus we consider the IC around cylinder as one conductive plate and the depleted layer as the dielectric corroborating the validity of analogy with coaxial cable.

The important point of developed models is the presence of nonlinearity in this concept of "transmission lines".

In both cases, for AF_s and MT_s, the nonlinearity is attributed to capacitance of EEU_s. For AF_s this nonlinearity originates from local pockets between actin proteins arranged helically, while in MT_s the main contribution arises from so called tubulin tails (TT_s) which are very flexible. These TT_s can shrink or stretch and oscillate, thus changing capacity of IC around MT.

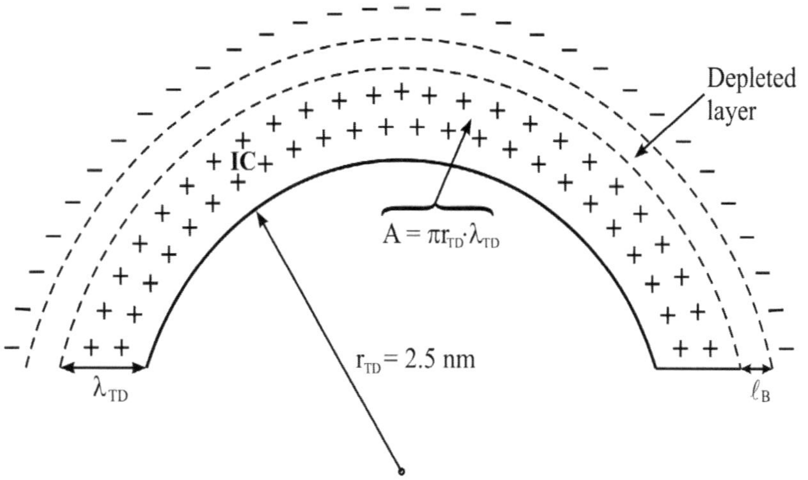

Figure 1. Cytoskeletal cylindrical polymers viewed as "coaxial cables".

3. The case of AF_s

The detailed consideration of AF_s in the context of nonlinear transmission lines is presented in Ref [8].

The equation which describes the time-space voltage of ionic pulse reads (x is the orientation of AF)

$$l^2 \frac{\partial^2 V}{\partial x^2} - R_1 C_0 \frac{\partial V}{\partial t} + 2bR_1 C_0 V \frac{\partial V}{\partial t} = 0. \tag{5}$$

The length $l = 5,4nm$ is the dimension of an EEU, ie. the diameter of one actin protein molecule; C_0 is the leading constant term of EEU capacitance and R_1 is its longitudinal resistance in direction of x-axis; b stands for the parameter of nonlinear term of $EEU's$ capacitance. The solution of Eq (5) reads [8]:

$$V(x,t) = W_1[1 - \exp(\alpha\xi)] \cdot [1 - \frac{W_1}{W_2}\exp(\alpha\varsigma)]^{-1} \tag{6}$$

with following abbreviations

$$W_{1/2} = (2bV_0)^{-1}\left[1 \pm \left(1 + 4\frac{AbV_0}{v_0}\right)^{1/2}\right]$$

$$\alpha = (W_1 - W_2)bV_0v_0; \quad \varsigma = \left(\frac{x}{l} - v_0\frac{t}{R_1C_0}\right). \tag{7}$$

The characteristic wave velocity is

$$v_0 = \frac{l}{R_1C_0} = \frac{5,4 \times 10^{-9}}{2 \times 10^8 \times 2 \times 10^{-16}}\left(\frac{m}{s}\right) = 13\frac{cm}{s}.$$

V_0 is the voltage amplitude defined by the influx of ions directed along AF; At last,

$$A = \left(\frac{dW}{d\xi}\right)_{W=0} < 0.$$

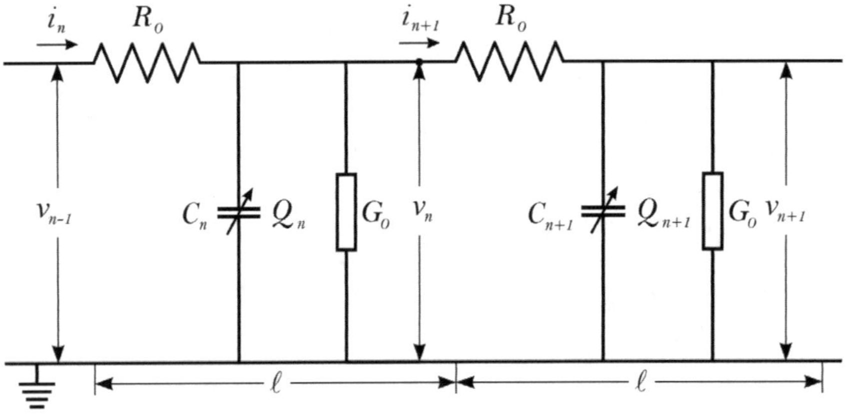

Figure 2. An effective circuit diagram for the n-th EEU with characteristic elements involved in Kirchoff's laws.

The solution Eq (6) is a typical anti-kink wave propagating with the velocity of the order of (cm/s), which has very reasonable order of magnitude.

4. The case of MT_s

Microtubules are hollow cylinders of 13 parallel protofilaments consisting of $\alpha - \beta$ tubulin heterodimers [1].

The experimental evidences regarding ionic currents along MT_s revealed that MT_s exhibit some kind of amplifying roles which resemble the "biotransistors" [9].

Our theoretical modeling of these currents was developed in three stages [11,12,13,14].

The first stage [11,12] is similar to that one applied for AF_s [8] in the way that the roles of nano-pores (NP_s) existing between protofilaments of MT were ignored.

Second version [13] is more complete, where the roles of NP_s are modeled in the context of an approach developed by Siwy et al [16]. This model involves the nonlinear negative incremental resistance of NP_s in parallel with nonlinear capacitance of EEU_s caused by mentioned property of TT_s. The important point of this stage is that the ohmic loss of transmission line should be balanced by the fresh ions injected into IC from NP_s since these act as ionic pumps in suitable voltage regime along MT. As the result, the solution of Kirhoff's equations of nonlinear circuit is the stable bell-shaped soliton voltage pulse which propagates along MT with the velocity of the order of mm/s.

Recently the more advanced version of the same concept was presented [14] in which the NP_s are treated according to within the numerical simulations performed the group led by Tuszynski [17].

We established simple periodic ladder network composed of the lumped sections equal to identical EEU_s (every tubulin dimer of one protofilament is an EEU) as represented in Fig.2

According to Kirchoff's laws from Fig.2 we have

$$\begin{aligned} i_n - i_{n+1} &= \frac{\partial Q_n}{\partial t} + G_0 v_n, \\ v_{n-1} - v_n &= R_0 i_n, \end{aligned} \tag{8}$$

where Q_n represents the charge of EEU capacitor as nonlinear function of voltage v;

$$Q_n = C_0[1 - \Gamma_0\Omega(t - t_0) - b_0 v]v \tag{9}$$

which is deeply elaborated in Ref (14).

Introducing the characteristic impedance Z of an EEU corresponding to characteristic frequency ω

$$Z = \frac{1}{\omega C_0}, \tag{10}$$

and the function $u(x,t)$ unifying the voltage and ionic current along a MT

$$u_n = Z^{1/2} i_n = Z^{-1/2} v_n, \tag{11}$$

and going over to the continuum approximation with respect to space variable x, we get from Eqs (8), in terms of Eqs (9,10),

$$\frac{\partial^3 u}{\partial \xi^3} + 3\left(\frac{ZC_0 s}{T_0} - 2\right)\frac{\partial u}{\partial \tau} + 3ZC_0\Gamma_0\Omega(\xi - \xi_0)\frac{\partial u}{\partial \xi} + 6\frac{Z^{3/2} b_0 C_0 s}{T_0}\left(u\frac{\partial u}{\partial \xi}\right)$$
$$+3(ZG_0 + Z^{-1}R - ZC_0\Gamma_0\Omega)u = 0. \tag{12}$$

Here the characteristic charging (discharging) time of an EEU capacitor C_0 through the resistance R_0 is $T_0 = R_0C_0 = 6,2 \times 10^7\Omega \times 1,92 \times 10^{-16}F = 1,2 \times 10^{-8}s$.

The dimensionless space and time variables are

$$\xi = \frac{x}{l} - \tau; \quad \tau = s\frac{t}{t_0}; \quad s = \frac{v}{v_0}.$$

The characteristic speed is $v_0 = \frac{l}{T_0} = \frac{8\times10^{-9}m}{1,2\times10^{-8}s} = 0,67\frac{m}{s}$; $l = 8 \times 10^{-9}m$ is the length of one tubulin dimer (EEU).

G_0 is the conductance of a NP and Ω the frequency of thermal oscillations of a TT; b_0 is the parameter of nonlinear capacity of a TT.

After a tedious procedure the solution of Eq (11) reads

$$u(\xi,\tau) = \frac{u_0\exp(-2\gamma_0\tau)}{\cos h^2\left\{\left[\frac{\alpha u_0}{4\beta}\exp(-2\gamma_0\tau)\right]^{1/2}\left[\xi - \xi_0(1-\exp(\gamma_0\tau)) + \frac{\alpha u_0}{3\beta}(1-\exp(3\gamma_0\tau))\exp(-2\gamma_0\tau)\right]\right\}}, \tag{13}$$

where the abbreviations have the values

$$\alpha = \frac{2Z^{3/2}b_0C_0s}{T_0\left(\frac{ZC_0s}{T_0}-2\right)}; \quad \beta = \left[3\left(\frac{ZC_0s}{T_0} - 2\right)\right]^{-1}$$
$$\gamma_0 = ZC_0\Gamma_0\Omega\left(\frac{ZC_0s}{T_0} - 2\right)^{-1}. \tag{14}$$

u_0 is the amplitude defined by the initial conditions.

We here mention one of possible numerical solutions of Eq (11) for the specific set of parameters, Fig. 3

$$\alpha = 0,5; \quad \beta = 0,1; \quad \gamma_0 = 0,1; \quad \xi_0 = 0,1.$$

It is apparent that localized pulse-like wave propagates with slightly decreasing amplitude along distances of hundreds EEU_s with constant velocity.

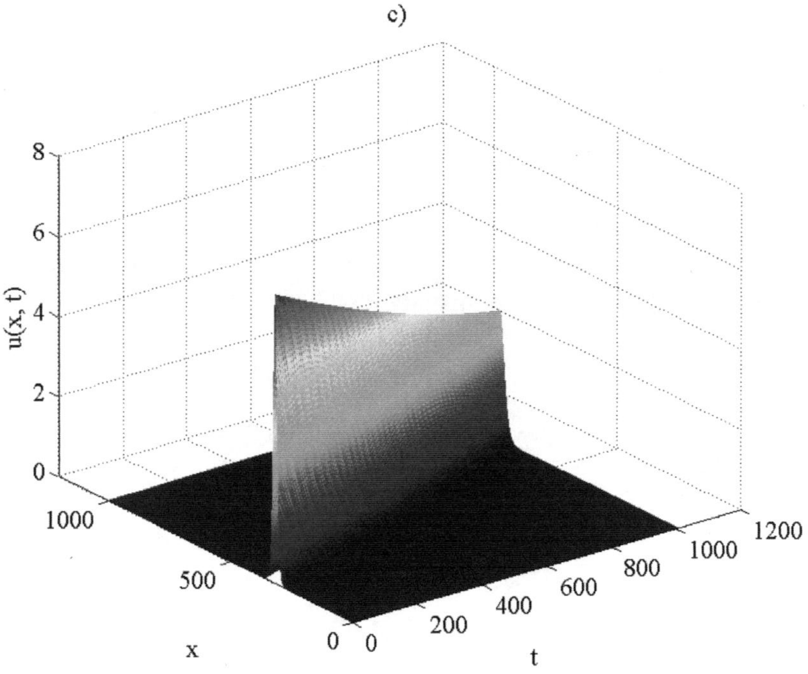

Figure 3. One of possible numerical solutions of Eq (11).

This velocity can be estimated from Fig. 3 as follows:

$$\Delta x = 400l = 400 \times 8nm = 3,2\mu m$$

$$\Delta t = 1000T_0 = 10^3 \times 1,2 \times 10^{-8}s = 1,2 \times 10^{-5}s$$

$$v = \frac{\Delta x}{\Delta t} = 0,26\frac{m}{s}. \qquad (15)$$

The range of this soliton is $3,2\mu m$ which is of the order of cell's diameter.

5. Discussion and conclusion

In this paper we briefly reviewed the recent contributions regarding the role of cytoskeletal filaments (AF_s and MT_s) in ionic propagation within the living cells.

The general feature of all mechanisms is the concept of nonlinear transmission lines. The origin of nonlinear effects is mostly attributed to capacitance of these filaments. The coaxial cable-like geometry of transmission lines relies on their polyelectrolite properties in accordance with Manning's theory.

The order of magnitude of soliton like localized pulses arising within the scope of our theory ranges from mm/s to a several cm/s which are very reasonable values.

Acknowledgements

This research was supported by funds from the Serbian Ministry of Education and Sciences, Grant OI 171009.

References

[1] Dustin P, Microtubules (Springer, Berlin 1984).
[2] Gross S P, Vershinin M and Shubeita G T, 2007, Curr. Biol. **17**, R, 478.
[3] Vale R D, 2003, Cell **112**, 467.
[4] Satarić M V, Budinski-Petković Lj and Lončarević I, 2007, IJMPB **21**, No32, 5387.
[5] Contiello H, Patenande C and Zaner K, 1991, Biophys. J. **59**, 1284.
[6] Lin E C, Cantiello H F, 1993, Biophys. J. **65**, 1371.
[7] Tuszynski J A, Portet S, Dixon J M, Luxford C and Cantiello H F, 2004, Biophys. J. **86**, 1890.
[8] Satarić M V, Bednar N, Satarić B M and Stojanović G, 2009, IJMPB **23**, No22, 4697.
[9] Priel A, Ramos A J, Tuszynski J A and Cantiello H F, 2006, Biophys. J. **90**, 4639.
[10] Pril A and Tuszynski J A, 2008, EPL **83**, 68004.
[11] Satarić M V, Ilić D and Ralević N, 2009, Chin. Phys. Lett **26**, No3, 073101.
[12] Satarić M V, Ilić D, Ralević N and Tuszynski J A, 2009, Eur. Biophys. J. **38**, 637.
[13] Satarić M V, Sekulić D and Živanov M, 2010, Journ. of Comp. Theor. Nanosci. **7**, 2281.
[14] Sekulić D L, Satarić B M, Tuszynski J A and Satarić M V, 2011, Eur. Phys. J. **34**, 49.
[15] O'Shaughnessy B and Yang Q, 2005, Phys. Rev. Lett. **94**, 048302.
[16] Siwy Z S, Powell M R, Petrov A, Kalman E, Trantmann C and Eisenberg R S, 2006, Nano. Lett. **6**, No8, 1729.
[17] Freedman H, Rezania N, Priel A, Carprenter E, Noskov S Y and Tuszynski J A, 2010, Phys. Rev. E **81**, 051912.

Living systems as coherent anharmonic oscillators

M Molski[1]
Theoretical Chemistry Department, Faculty of Chemistry
Adam Mickiewicz University of Poznań
Grunwaldzka 6, 60-780 Poznań, Poland

E-mail: mamolski@amu.edu.pl

Abstract. A model of living systems considered as coherent, time-dependent anharmonic oscillators is presented. It is based on the concept of space-like coherent states minimizing the time-energy uncertainty relation, adapted to the case of biological systems whose growth is described by the Gompertz or West-Brown-Enquist functions. The coherent states of biological growth evolve coherently in space being localized along the classical time trajectory; hence, the growth is predicted to be coherent in space. It is proven that the Gompertz function is a special solution of the space-like Horodecki-Feinberg equation for the time-dependent Morse oscillator in the dissociation state. Its eigenvalue represents the momentum of biological growth, associated with a space-like component whose properties resemble those attributed by vitalists to the life momentum or vital impulse. The physical characteristics of the life energy and momentum and their connection with the concept of zero-point momentum of vacuum are presented.

1. Introduction

In 1968 Herbert Fröhlich [1] discovered the effect of the long-range bio-coherence appearing in living systems treated as coupled oscillators in a heat bath with energy supplied at a constant rate. When the rate exceeds a certain mean one, the oscillators condense into one giant dipole whose sub-elements are spatially interrelated to each other like constituents of the low-temperature macroscopic quantum Bose-Einstein condensates. The Fröhlich model has been confirmed by Molski and Konarski [2] who investigated the biological growth described by the Gompertz function [3]. They found that this function represents the ground coherent state of the Gompertzian growth, which belongs to the class of the non-local coherent states minimizing the time-energy uncertainty relation. Such states evolve coherently in space being localized along the classical time-trajectory; hence the Gompertzian growth has been predicted to be coherent in space [4]. One of the most remarkable achievements presented in [2,4] was derivation of the second-order growth equation which contains a time-dependent Morse potential [5], whose space-dependent form is widely used to describe anharmonic vibrations of diatomic molecules. The model of non-local bio-coherence based on the Gompertz function has been generalized [6] to include the West-Brown-Enquist [7] growth function, which satisfies the second-

[1] To whom any correspondence should be addressed.

order differential equation for the oscillator characterized by the Wei Hua potential [8]. The results obtained reveal that the living systems can be considered as the time-dependent oscillators whose growth is governed by the anharmonic potentials. The main purpose of the present study is to investigate the consequences of the oscillatory model for our understanding of the essence of life and nature of biological growth. In particular, it will be proved that oscillatory model of living systems permits revitalization of the old concept of vitalism completely removed from the area of modern biology, medicine and physics. We shall also be concerned with generalization of the Leggett's classification [9] of the macroscopic quantum phenomena to include a new class of the quasi-quantum phenomena: macroscopic quantization, uncertainty, coherence and non-locality appearing in the biological systems whose growth is characterized by the Gompertz or West-Brown-Enquist functions.

2. The basic concepts of the model

The oscillatory model of the living systems is based on the well-known biological and physical concepts: the biological growth and the generalized quantum theory including non-local (space-like) quantum states and systems. The most important basic concepts of the model are presented below.

2.1. The biological growth

In 1825 Benjamin Gompertz [3] introduced in the framework of actuarial mathematics the sigmoidal (S-shaped) function

$$G(t) = G_0\left\{\left(\frac{b}{a}\right)\left[1-\exp\left(-at\right)\right]\right\} \tag{1}$$

which permits calculating the force of mortality. Nowadays it is employed in biology and medicine to describe the growth of living systems (organ, tissue, organism, population of organisms, bacterial colony, tumour). Here G_0 stands for the initial mass, volume, radius or number of cells or entities, whereas the constants a and b (a, b>0) characterize the dynamics of growth (regression) according to the first-order Gompertzian kinetics

$$\frac{d}{dt}G(t) = b\exp\left(-at\right)G(t) \qquad \frac{d}{dt}G(t)^{\dagger} = -b\exp(-at)G(t)^{\dagger} \tag{2}$$

Here $G(t)^{\dagger}=G(t, b \rightarrow -b)$ defines the Gompertzian system in the regression (decay) phase. One may prove that function (1) is also a solution of the differential equation

$$-\frac{d^2G(t)}{dt^2} + \frac{a^2}{4}\{1-\exp[-a(t-t_e)]\}^2 G(t) = \frac{a^2}{4}G(t) \tag{3}$$

in which the second term represents (with accuracy to the constant) the time-dependent Morse potential

$$V(t) = D_e\left\{1-\exp\left[-a(t-t_e)\right]\right\}^2 \tag{4}$$

Here, D_e is the dissociation energy of the Morse oscillator, a is the so-called range parameter, whereas

$$t_e = \frac{1}{a}\ln\left(\frac{2b}{a}\right) \tag{5}$$

is an equilibrium time at which the potential (4) attains a minimum $V(t=t_e)=0$. One may prove that the growth function (1) and differential equation (3) expressed in the dimensionless variable

$$\tau = \frac{a(t-t_e)}{\sqrt{2}} \tag{6}$$

can be specified in the form

$$G(\tau) = G_\infty \exp\left[-\frac{1}{2}\exp(-\sqrt{2}\tau)\right] \qquad G_\infty = G_0 \exp\left(\frac{b}{a}\right) \tag{7}$$

$$-\frac{1}{2}\frac{d^2G(\tau)}{d\tau^2} + \frac{1}{4}\{1-\exp[-\sqrt{2}\tau]\}^2 G(\tau) = \frac{1}{4}G(\tau) \tag{8}$$

2.2. The extended quantum theory

In 2006 Molski [4] proved that equation (8) is a special case of the more general non-local Horodecki-Feinberg equation [10,11]

$$-\frac{\hbar^2}{2mc^2}\frac{d^2\Psi}{dt^2} + V(t)\Psi = Pc\Psi \tag{9}$$

for a space-like object of mass m moving in the field of the Morse vector potential (4). Here, Ψ is a non-local matter wave associated with the superluminal particle of momentum P, $\hbar=1{,}05457266\cdot 10^{-34}$ J·s is the Planck constant divided by 2π, c is the light velocity. Equation (9) represents the non-relativistic version of the relativistic Feinberg equation for a particle of mass m moving at a velocity v close to infinity and endowed with the relativistic energy and momentum [11]

$$E = \frac{mc^2}{\sqrt{\frac{v^2}{c^2}-1}} \xrightarrow{v\to\infty} 0 \qquad P = \frac{mv}{\sqrt{\frac{v^2}{c^2}-1}} \xrightarrow{v\to\infty} mc \tag{10}$$

In the case of infinite speed, the particle has zero-energy $E=0$ and non-zero momentum $P=mc$ and is called the *transcendental tachyon* [12]. The Horodecki-Feinbereg equation (9) with the potential (4) can be specified in the dimensionless form [4]

$$\left\{-\frac{1}{2}\frac{d^2}{d\tau^2} + \frac{1}{4x_e}\left[1-\exp\left(-\sqrt{2x_e}\tau\right)\right]^2 - \left(v+\frac{1}{2}\right) + x_e\left(v+\frac{1}{2}\right)^2\right\}\Psi_v = 0 \tag{11}$$

in which

$$\tau = \frac{a(t-t_e)}{\sqrt{2x_e}} \qquad x_e = \frac{\hbar a}{c\sqrt{8mD_e}} \tag{12}$$

Here τ is a dimensionless variable whereas x_e is the anharmonic parameter characterizing the Morse oscillator [4]. The eigenvalues of the quantum equation (11)

$$P_v = \frac{\hbar\omega}{c}\left[\left(v+\frac{1}{2}\right) - x_e\left(v+\frac{1}{2}\right)^2\right] \qquad \omega = \frac{a}{c}\sqrt{\frac{2D_e}{m}} \tag{13}$$

represent the quantized momentum, which means that it takes only discrete values according to the quantum number $v=0,1,2\ldots$ whereas the wave function of the ground states ($v=0$) reads [4]

$$\Psi_0 = N_0 \exp\left[-\frac{1}{2x_e}\exp\left(-\sqrt{2x_e}\tau\right)\right]\exp\left[-\frac{1}{\sqrt{2x_e}}(1-x_e)\tau\right] \tag{14}$$

2.3. The second-order micro-macro correspondences

The quantum Horodecki-Feinberg equation (11) and its ground state solution (14) can be transformed to the macroscopic growth equation (8) and its solution (7), upon the substitutions $x_e=1$, $v=0$, which characterize the dissociation ground state of the Morse oscillator [2,4]

$$\left\{-\frac{1}{2}\frac{d^2}{d\tau^2}+\frac{1}{4x_e}\left[1-\exp\left(-\sqrt{2x_e}\tau\right)\right]^2-\left(v+\frac{1}{2}\right)+x_e\left(v+\frac{1}{2}\right)^2\right\}\Psi_v \xrightarrow{x_e=1,v=0}$$
$$\left\{-\frac{1}{2}\frac{d^2}{d\tau^2}+\frac{1}{4}\left[1-\exp\left(-\sqrt{2}\tau\right)\right]^2-\frac{1}{4}\right\}G(\tau)=0 \quad (15)$$

$$N_0\exp\left[-\frac{1}{2x_e}\exp\left(-\sqrt{2x_e}\tau\right)\right]\exp\left[-\frac{1}{\sqrt{2x_e}}(1-x_e)\tau\right]\xrightarrow{x_e=1}$$
$$G_\infty\exp\left[-\frac{1}{2}\exp(-\sqrt{2}\tau)\right] \quad (16)$$

They evidently prove that the macroscopic second-order equation (8) is a special case of the more general quantum non-local Feinberg-Horodecki equation (11), whose ground state eigenfunction (v=0) for $x_e=1$ is reduced to the Gompertz function (7). Equation (8) does not contain the Planck's constant but on the other hand it can be derived from the quantum equation (11). Hence, it is neither classical nor quantum formula. It belongs to the class of quasi-quantum equations whereas the Gompertzian growth is a quasi-quantum phenomenon [6].

2.4. The non-local coherent states of the Morse oscillator

The Horodecki-Feinberg equation (11) can be factorized using the creation-annihilation operators [4]

$$\hat{A}^\dagger\hat{A}\Psi_v=P_v-P_0 \quad \Rightarrow \quad \hat{A}^\dagger\hat{A}\Psi_0=0$$

$$\hat{A}^\dagger=\frac{1}{\sqrt{2}}\left[-\frac{d}{d\tau}+\frac{1-\exp\left(-\sqrt{2x_e}\tau\right)}{\sqrt{2x_e}}-\sqrt{\frac{x_e}{2}}\right] \quad \hat{A}=\frac{1}{\sqrt{2}}\left[\frac{d}{d\tau}+\frac{1-\exp\left(-\sqrt{2x_e}\tau\right)}{\sqrt{2x_e}}-\sqrt{\frac{x_e}{2}}\right] \quad (17)$$

The non-local coherent states of the time-dependent Morse oscillator can be defined as eigenstates of the annihilation operator [4]

$$\hat{A}|\alpha\rangle=\alpha|\alpha\rangle \quad |\alpha\rangle=\exp\left[-\frac{1}{2x_e}\exp\left(-\sqrt{2x_e}\tau\right)\right]\exp\left[-\frac{1}{\sqrt{2x_e}}(1-x_e)\tau\right]\exp[\alpha\sqrt{2}\tau] \quad (18)$$

Such states minimize the time-energy uncertainty relation expressed in the Morse variable T(τ) [4]

$$\Delta T(\tau)^2\Delta E^2=\frac{\hbar^2}{4}\langle\alpha|\exp\left(-\sqrt{2x_e}\tau\right)|\alpha\rangle^2 \quad T(\tau)=\frac{1-\exp\left(-\sqrt{2x_e}\tau\right)}{\sqrt{2x_e}}-\sqrt{\frac{x_e}{2}} \quad (19)$$

hence they represent the non-local minimum-uncertainty coherent state. The ordinary local (time-like) coherent states minimize the position-momentum uncertainty relation and propagate coherently in time on the well-defined space trajectory. The non-local (space-like) coherent states propagate coherently in space along the well-defined time trajectory [4].

2.5. The first-order micro-macro correspondences

The micro-macro correspondences for Gompertzian growth and quantum coherence can be obtained from equations (17) and (18) by introducing $x_e=1$ into the quantum annihilation-creation operators and the ground coherent state $\alpha=0$. Then one gets the macroscopic equations (2) describing biological growth and regression [4]

$$\hat{A}|0\rangle = \frac{1}{\sqrt{2}}\left[\frac{d}{d\tau} + \frac{1-\exp\left(-\sqrt{2x_e}\tau\right)}{\sqrt{2x_e}} - \sqrt{\frac{x_e}{2}}\right]|0\rangle \xrightarrow{x_e=1}$$
$$\frac{1}{\sqrt{2}}\left[\frac{d}{d\tau} + \frac{1-\exp\left(-\sqrt{2}\tau\right)}{\sqrt{2}} - \sqrt{\frac{1}{2}}\right]|0\rangle \xrightarrow{\tau\to t} \frac{dG(t)}{dt} - b\exp(-at)G(t) = 0 \tag{20}$$

$$\langle 0|\hat{A}^{\dagger} = \langle 0|\frac{1}{\sqrt{2}}\left[-\frac{d}{d\tau} + \frac{1-\exp\left(-\sqrt{2x_e}\tau\right)}{\sqrt{2x_e}} - \sqrt{\frac{x_e}{2}}\right] \xrightarrow{x_e=1}$$
$$\langle 0|\frac{1}{\sqrt{2}}\left[-\frac{d}{d\tau} + \frac{1-\exp\left(-\sqrt{2}\tau\right)}{\sqrt{2}} - \sqrt{\frac{1}{2}}\right] \xrightarrow{\tau\to t} \frac{dG(t)^{\dagger}}{dt} + b\exp(-at)G(t)^{\dagger} = 0 \tag{21}$$

Here, |0> and <0| denote the ground state eigenfunctions of the annihilation and creation operators identified with the Gompertzian growth and regression states of the biological system [2]. The results obtained reveal another quasi-quantum effect in biological systems i.e. non-local macro-coherence.

2.6. The macro-uncertainty

The biological growth according to the Gompertz growth function (7) is characterized also by the macroscopic version of the quantum time-energy uncertainty relation to be obtained by substituting $x_e=1$ in (19). Hence, the macroscopic function (7) can be shown to minimize the microscopic time-energy uncertainty relation [4]

$$\Delta T(\tau)^2 \Delta E^2 \geq -\frac{1}{4}\left\langle 0\left|\left[T(\tau),\hat{E}\right]\right|0\right\rangle^2 \qquad \left[T(\tau),\hat{E}\right] = i\hbar\exp(-\sqrt{2}\tau) \tag{22}$$

Here, $T(\tau)$ is the Morse variable for $x_e=1$, <0| and |0> denote the Gompertz functions of regression and growth respectively, whereas the remaining quantities are defined as follows

$$T(\tau) = -\frac{1}{\sqrt{2}}\exp\left(-\sqrt{2}\tau\right) \qquad \hat{E} = i\hbar\frac{d}{d\tau}$$
$$\Delta T^2 = \langle 0|T(\tau)^2|0\rangle - \langle 0|T(\tau)|0\rangle^2 \quad \Delta E^2 = \langle 0|\hat{E}^2|0\rangle - \langle 0|\hat{E}|0\rangle^2 \tag{23}$$
$$\Delta T^2 \Delta E^2 = \frac{\hbar^2}{4}\langle 0|\exp\left(-\sqrt{2}\tau\right)|0\rangle^2$$

This result indicates that the quantum uncertainty formalism can be directly applied to the macroscopic function (1) describing biological growth and regression. It also proves that the Gompertz growth function (7) represents the so-called *intelligent* coherent states [13]; which not only minimize the time-energy uncertainty relation but also maintain this relation in space due to its spatial stability. Such non-local states differ from the ordinary time-like coherent states, which maintain the position-momentum Heisenberg relation in time due to its temporal stability [4]. Consequently, the model presented predicts that formation of the specific growth patterns during biological growth is a non-local and spatially coherent phenomenon. It is a result of the long-range cooperation between the micro-level represented by the singular cell and the macro-level i.e. the bio-system as a whole. This conclusion is consistent with the Fröhlich [1] model of the long-range bio-coherence appearing in living systems treated as coupled oscillators

2.7. Generalization of the model

The model presented can be extended to include the anharmonic Wei Hua oscillator [9] described by the time-dependent potential

$$V(t) = D_e \left\{ \frac{1 - \exp[-c_1(t - t_e)]}{1 + s\exp[-c_1(t - t_e)]} \right\}^2 \tag{24}$$

which is a generalization of the Morse function (4).Then the quantum Horodecki-Feinberg equation (9) for the time-dependent Wei-Hua oscillator

$$\left\{ -\frac{1}{2}\frac{d^2}{d\tau^2} + D\left[\frac{1 - \exp(-\tau)}{1 + s\exp(-\tau)} \right]^2 - P \right\} \Psi_0 = 0 \tag{25}$$

expressed in the dimensionless form with

$$\tau = c_1(t - t_e) \qquad D = \frac{D_e mc^2}{\hbar^2 c_1^2} \qquad P = \frac{P_0 mc^3}{\hbar^2 c_1^2} \tag{26}$$

has the ground state solution [6]

$$\Psi_0 = M[1 - c_0 \exp(-\tau)]^{\frac{1}{c_2}} [c\exp[-\tau]\}^{\frac{c_3}{c_1^2}} \xrightarrow{c3=0 \Rightarrow P=D} M[1 - c_0 \exp(-\tau)]^{\frac{1}{c_2}} \tag{27}$$

In the dissociation state $c_3=D-P=0$, it reduces to the West-Brown-Enquist function [7]

$$m(t) = M\left[1 - c_0 \exp(-c_1 t)\right]^{\frac{1}{c_2}} \tag{28}$$

$$c_0 = 1 - \left(\frac{m_0}{M} \right)^{\frac{1}{4}}, \quad c_1 = -\frac{a_0}{4M^{1/4}}, \quad c_2 = \frac{1}{4}, \quad m(t=0) = m_0, \quad m(t=\infty) = M \tag{29}$$

describing the ontogenic growth. The function (28) has been derived by West et al. [7] on the basis of the first principles including the conservation of metabolic energy, the allometric scaling of metabolic rate, and energetic costs of producing and maintaining biomass. It fits very well the data for a variety of different species from protozoa to mammalians. In (29) m_0 denotes the initial mass of the system, M is the maximum body size reached whereas a_0 is the metabolic parameter. The West-Brown-Enquist function (28) better describes the biological growth than the Gompertz function (1) as the Wei Hua potential (24) for s=*0* reduces to the Morse function (4) being a special case of the former.

3. The quasi-quantum vitalism

The nature of life is commonly attributed to its functional and structural aspects. The contrasting viewpoint called *vitalism* assumes [14-18] that all living systems possess some unique component, which endows them with the special quality – life. This component is said to be the source of life and is commonly associated with power to growth and vital force. Despite many attempts of revitalization [19-22], vitalism is still treated as a metaphysical and non-falsifiable concept completely removed from the area of modern biology, medicine and physics. According to Daniel Denet [23]: *Vitalism - the view that living things contain some special physical but equally mysterious stuff - élan vital - has been relegated to the trash heap of history.* The presented oscillatory model of living systems permits revitalization of the relegated concept of vitalism.

3.1. The momentum of life

The eigenvalues of the Horodecki-Feinberg equation (11) characterize the momentum of the time-dependent Morse oscillator. In the limiting case $v=0$ and $x_e=1$ they are reduced to the momentum

$$P_v = \left(\frac{\hbar\omega}{c}\right)\left[\left(v+\frac{1}{2}\right)-\left(v+\frac{1}{2}\right)^2 x_e\right] \xrightarrow{x_e=1 \quad v=0} \left(\frac{\hbar\omega}{4c}\right)=P_0 \tag{30}$$

of growing Gompertzian systems. The existence of the non-vanishing momentum associated with the biological growth is the first theoretical evidence of the momentum of life (*élan vital*), characterizing all biological systems growing according to the Gompertz function (1). Because this quantity is a purely mechanical concept it is possible to associate with it a mass and energy by making use of the anharmonic parameter (12), which in the Gompertzian systems is equal to one

$$x_e = \frac{\hbar a}{c\sqrt{8mD_e}} = 1 \Rightarrow D_e = \frac{\hbar^2 a^2}{8mc^2} \tag{31}$$

Consequently, we can calculate the dissociation energy D_e appearing in the Morse potential (4). On the other hand, from (30) one can determine the ground-state momentum P0 associated with the growing system. By analogy to the zero-point energy of the harmonic oscillator, equation (30) describes the so-called zero-point momentum P_0 [24, 25] of the anharmonic Morse oscillator. Combining equations (30) with (31) and using the definition of the vibrational frequency of the Morse oscillator [4] appearing in (30)

$$\omega = \frac{a}{c}\sqrt{\frac{2D_e}{m}} \tag{32}$$

one may calculate the approximate momentum associated with the Gompertzian growth

$$P_0 = \frac{D_e}{c} = \frac{\hbar^2 a^2}{8mc^3} \approx mc \tag{33}$$

The approximation (33) is possible as the Horodecki-Feinberg equation (9) describes an object moving almost instantaneously [10]. In such circumstances P_0 is approximately equal to the momentum of a transcendent tachyon (10). Consequently, equation (33) permits estimation of the mass of the particle being a carrier of the biological growth momentum

$$m = \frac{\hbar a}{\sqrt{8}c^2} \tag{34}$$

It depends on the value of a-parameter appearing in the Gompertz function (1). The combination of fundamental constants appearing in (34)

$$\frac{\hbar}{c^2} = 1.17336901 \cdot 10^{-51} \ Kg \cdot s \tag{35}$$

is extremely small, hence the mass (34) associated with momentum of life takes values much lower than the electron neutrino mass measured in MINOS experiment $m_v=7.132 \cdot 10^{-38}$ Kg (0.04 eV). Additionally, the Gompertzian a-parameter is usually expressed in hours, days or years, so the calculation of mass (34) requires application of a suitable factor conversion to seconds which decreases the value (34) by a few orders. For example, the Gompertz function (1) fits the data for the mean weight of a male body in the period (0,18) [year] [26] providing the parameters G_0=7.58(65) [kg], b=0.2005(208) [year^{-1}] and a=0.0638(101) [year^{-1}]. Hence, the mass (34) takes the value $m=8.3927 \cdot 10^{-61}$ [kg].

3.2. The life expectancy and time-energy uncertainty

In the Gompertz model, the life expectancy at birth L_0 can be calculated from the population's survival function S(t) [27]

$$L_0 = \int_0^\infty S(t)dt \qquad S(t) = \left[1 - \frac{b}{a\lambda}\left(e^{-at}-1\right)\right]^{-k} \tag{36}$$

in which λ is approximated by the relation $\lambda=k$. Integration (36) provides the life expectancy formula

$$L_0 = \frac{1}{a\lambda}\ {}_2F_1\left(k,1;k+1;1-\frac{b}{a\lambda}\right) \approx \frac{1}{a} \tag{37}$$

in which $_2F_1$ denotes the Gaussian hypergeometric function [27]. The obtained result indicates that life expectancy is inversely proportional to Gompertzian parameter a, whereas parameter b has no significant influence on L_0 [27]. So, determining the a-parameter by the fitting to the biological growth data one can approximately calculate the life expectancy of a biological system. For instance, for $a=0.0638(101)$ [year^{-1}] determined from the data for male body of mean weight in the period (0,18) [year] [26] one gets

$$L_0 \approx \frac{1}{a} = 15.67 \pm 2.48 \quad \text{[year]} \tag{38}$$

which very well approximates the upper time-limit (18 years) of the data employed in evaluation of the a-parameter. The error $\sigma=\pm 2.48$ [year] propagates from the standard error $\sigma=\pm 0.0101$ [year] of the a-parameter. Taking into account the momentum-mass relationship (33) given in an alternative form

$$\left(\frac{1}{a}\right)(D_e) = \frac{\hbar}{2\sqrt{2}} \quad \text{or} \quad \left(\frac{1}{a}\right)(E) = \frac{\hbar}{2} \tag{39}$$

in which

$$E = \frac{\hbar a}{2} = D_e\sqrt{2} \tag{40}$$

is the life energy defined at the equilibrium time by equations (A2) and (A3). We conclude that (39) has identical form as the minimum time-energy uncertainty relation

$$\Delta t \Delta E = \frac{\hbar}{2} \qquad \Delta t = \frac{1}{a} \qquad \Delta E = \frac{\hbar a}{2} \tag{41}$$

which links the life energy and approximate life expectancy of a growing biological system. The proof of the correctness of relations (41) is presented in Appendix. It is noteworthy that from (32) and (33) one can calculate the frequency of growth

$$\omega = \frac{a}{c}\sqrt{\frac{2D_e}{m}} = a\sqrt{2} \tag{42}$$

which reveals that each growing biological system according to the Gompertz function (1) can be considered as a time-dependent Morse oscillator endowed with a characteristic frequency (42).

3.3. The vital force

In the oscillatory model, the Gompertzian growth of biological systems takes place under the influence of the Morse potential energy (4) including the dissociation term to be calculated from (40)

$$V(t) = \frac{\hbar a}{\sqrt{8}}\left\{1-\exp\left[-a(t-t_e)\right]\right\}^2 \tag{43}$$

Consequently, biological growth according to (1) should be endowed with a growth (vital) force or using the Winsor's notion *power to growth* to be calculated in the same manner as the electric force is calculated in electrodynamics: it is a negative temporal derivative of the vector potential (43)

$$F(t) = -\frac{\hbar a^2}{\sqrt{2}}\left\{1-\exp\left[-a(t-t_e)\right]\right\}\exp\left[-a(t-t_e)\right] \tag{44}$$

In the case of dimensionless potential form appearing in (8) the vital force takes the form

$$F(\tau) = -\frac{1}{\sqrt{2}}\left\{1-\exp[-\sqrt{2}\tau]\right\}\exp[-\sqrt{2}\tau] \tag{45}$$

It is well-known that the Gompertz function (1) describes exponential growth, which then is exponentially retarded and saturated as time continues. According to equation (44), the Gompertzian growth is governed by the force (44) which stimulates growth in the period (t, t_e) and constrains it in the period (t_e,∞) leading to the saturation. The plots of the Gompertz function G(τ), Morse potential V(τ) and associated vital force F(τ) are presented in Fig. 1.

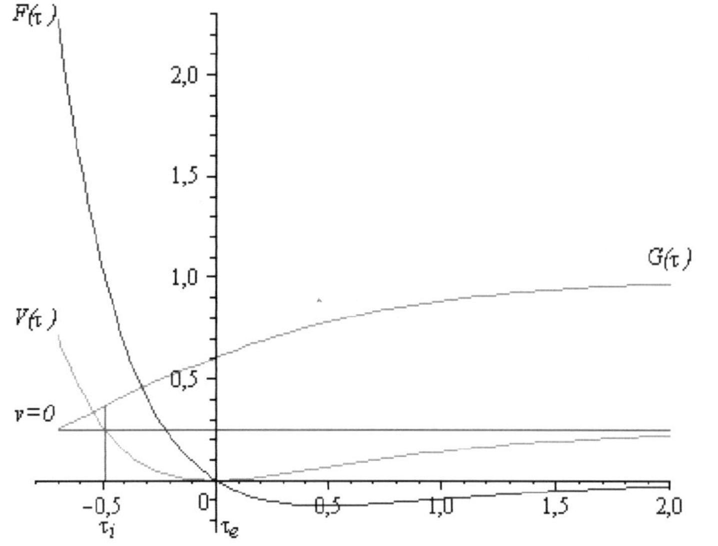

Fig.1 Plots of the dimensionless Gompertz function G(τ), Morse potential V(τ) and vital force F(τ). The straight line represents the momentum of growth P_0 for quantum number v=0; τ_i is inflection point of the Gompertz function; τ_e stands for the equilibrium point at which potential V(τ) attains minimum, whereas force F(τ) passes from accelerating to decelerating stage of growth.

4. Conclusions

The Gompertz function (1) is a ground state eigenfunction of the Horodecki-Feinberg equation for the time-dependent Morse oscillator with the anharmonicity parameter $x_e = 1$. This solution corresponds to the dissociation state, so the growth of biological systems is consistent with the arrow of time (is not oscillating type) and its dynamics is governed by the time-dependent Morse potential. The oscillatory model of growth permits generalization of the Leggett's classification [9] of the macroscopic quantum phenomena to include a new class of the quasi-quantum phenomena: macroscopic quantization, uncertainty, coherence and non-locality appearing in the biological systems whose growth is characterized by the Gompertz or West-Brown-Enquist functions. For the latter case the coherent states of growth and macro time-energy uncertainty relationship have been derived in [6]. The eigenvalue of the Horodecki-Feinberg equation represents the momentum, which, in the Gompertzian systems, can be interpreted as the momentum of growth or vital impulse. It takes a constant value, which does not alter during the growth, in contradistinction to the life energy, which changes according to equation (A2). The quasi-quantum interpretation of vitalism proposed has interesting philosophical consequences. According to the Bergson's theory [15] the only way to account for the origin and functioning of life is to posit a non-material component being a carrier of momentum of life - élan vital, which cannot be reduced to the material (chemical, physical) constitution of biological bodies. The results reported here demonstrate that élan vital can be interpreted in an exactly materialistic manner employing the concept of space-like quantum states and objects in the framework of the Horodecki-Feinberg theory. In particular the carrier of élan vital is a superluminal particle endowed with non-vanishing momentum, which moves in the space-time along the spatial world-line at the velocity close to infinity. Such an infinite-speed particle is called the transcendental tachyon [12]. According to Corben [31,32], there is a possibility of interaction and coupling of such space-like objects (tachyons) with ordinary subluminal time-like objects (bradyons) leading to formation of composite systems built up of both time- and space-like components, which trap each other in a relativistically invariant way. It can be proven that absorption of a tachyon by a bradyon at rest changes the kinematic state of the latter – this effect is called tachyo-kinematic [33,34]. In view of this the Corben's theory predicts that the carrier of momentum of life can couple with somatic matter causing its motion (growth). This property is consistent with the vitalistic concept of anima, which is assumed to be the constituent of the living system endowing them with motion [34]. As it has been proved recently [2], the Gompertz function (1) represents also the ground eigenstate of the annihilation (growth) operator, which minimizes the time-energy uncertainty relation; hence it describes the biological systems that evolve coherently in the space. Consequently, the Gompertz function (1) represents the minimum-uncertainty coherent state of the Gompertzian growth. The uncertainty relation (22) can be given in the approximate form (41) valid for the time-dependent harmonic oscillator. Consequently, the time uncertainty can be expressed by a-parameter appearing in the function (1) and it is possible to connect the microscopic time-energy uncertainty with the macroscopic a-parameter characterizing the Gompertzian growth. Since equation (41) can be specified in the equivalent form

$$\left(\frac{1}{a}\right)m = \frac{\hbar}{\sqrt{8}c^2} \tag{46}$$

there is a direct link between the mass carrier of the life momentum and the approximate life expectancy of a growing Gompertzian system. Because mass m, according to (35), is extremely small, a hypothesis can be put forward that it can be created from vacuum as a virtual particle existing in the period approximately equal to the life expectancy 1/a. This conclusion, although controversial, is in full agreement with the time-energy uncertainty relation permitting the creation of virtual particle-antiparticle pairs from vacuum. Considering the space-like particle, its antiparticle moves in the opposite direction [12] . Virtual particles are real and have measurable effects (e.g. Casimir and Unruh effects, Lamb shift) although they cannot be directly observed. According to the Heisenberg's uncertainty principle "something" can arise from "nothing" provided that the "something" annihilates to the "nothing" after a period of time characterized by the time-energy uncertainty relation. Hence, the "empty" space-time should be considered as [seething] with an infinite number of particle-antiparticle pairs that come into being and then annihilate each other in a suitable time-interval. Virtual particles are manifestations of the fluctuations of vacuum endowed with the so-called zero-point energy. Recent investigation carried out by Feigel [24] has revealed that the vacuum fluctuations contribute to motion of dielectrics in crossed electric and magnetic fields. They induce a flow in the dielectric liquids endowed with velocity 50 [nm/s] predicted theoretically for electric field 105 [V/m] and magnetic 17 [T] [24]. The possibility of creation of the zero-point momentum from vacuum (momentum from nothing [25]) is an important evidence supporting the aforementioned hypothesis on the origin of life momentum. It is worth mentioning that Bergson [15] considered evolution as a tension between virtuality and actuality, which in the framework of the presented quasi-quantum model of vitalism can be taken literally.

Appendix

The quantum Horodecki-Feinberg operator appearing in the wave equation (5) can be derived by quantization of the classical formula [10]

$$\frac{E^2}{2mc^2} + V(t) = Pc \qquad E \to \hat{E} = i\hbar\frac{d}{dt} \tag{A1}$$

in which $\hat{E}$ denotes the energy operator and i is the imaginary constant. Introducing the Morse potential (4) to (A1) and employing relations (33), (42) one can calculate the energy of the growing Gompertzian system

$$E = \frac{\hbar a}{2}\sqrt{\exp[-a(t-t_e)]\{2-\exp[-a(t-t_e)]\}} \tag{A2}$$

Equation (A2) has a quasi-quantum form: on the one hand it contains Planck's constant, while on the other it does not contain quantum operators. One can easily demonstrate that (A2) attains a maximum at the equilibrium point

$$E(t_e) = \frac{\hbar a}{2} = \Delta E \tag{A3}$$

equal to the energy uncertainty in the vicinity of the equilibrium time t_e. To prove this thesis, we expand the Morse potential (4) in the Taylor series in the vicinity of the equilibrium time

$$V(t) = V(t_e) + \left[\frac{dV(t)}{dt}\right]_{t_e}(t-t_e) + \frac{1}{2}\left[\frac{d^2V(t)}{dt^2}\right]_{t_e}(t-t_e)^2 + \ldots.. \tag{A4}$$

The first two terms in A4 are equal to zero, so neglecting the high-order terms one can approximate anharmonic potential (4) by the harmonic one

$$V(t) = \frac{1}{2}mc^2\omega^2(t-t_e)^2 \qquad \left[\frac{d^2V(t)}{dt^2}\right]_{t_e} = 2D_e a^2 \tag{A5}$$

in which ω is the vibrational frequency defined by (42). Consequently, a classical counterpart of the quantal Horodecki-Feinberg equation for the time-dependent harmonic oscillator can be specified in the form

$$\frac{E^2}{2mc^3} + \frac{1}{2}mc\,\omega^2(t-t_e)^2 = P \tag{A6}$$

If we replace in (A6) the time and energy by their uncertainties and assume that they minimize the Heisenberg relation

$$E \to \Delta E \qquad t - t_e \to \Delta t \qquad \Delta E \Delta t = \frac{\hbar}{2} \tag{A7}$$

then, consequently, (A6) takes the form

$$\frac{(\Delta E)^2}{2mc^3} + \frac{1}{2}mc\,\omega^2(\Delta t)^2 = P(\Delta E, \Delta t) \qquad \Delta E = \frac{\hbar}{2\Delta t} \qquad \Delta t = \frac{\hbar}{2\Delta E} \tag{A8}$$

From the mathematical point of view the momentum in (A8) is a function of uncertainties, so assuming that it attains a minimum one can calculate values of uncertainties at assistance (A7). In the case of the time uncertainty, we have

$$\frac{\hbar^2}{8mc^3(\Delta t)^2} + \frac{1}{2}mc\omega^2(\Delta t)^2 = P(\Delta t) \tag{A9}$$

$$-\frac{\hbar^2}{4mc^3(\Delta t)^3} + mc\omega^2\Delta t = \frac{dP(\Delta t)}{d\Delta t} = 0 \Rightarrow \Delta t = \sqrt{\frac{\hbar}{2mc^2\omega}} = \frac{1}{a} \tag{A10}$$

whereas for the energy uncertainty, similar calculations provide

$$\frac{(\Delta E)^2}{2mc^3} + \frac{\hbar^2 mc\omega^2}{8(\Delta E)^2} = P(\Delta E) \tag{A11}$$

$$\frac{\Delta E}{mc^3}-\frac{\hbar^2 mc\omega^2}{4(\Delta E)^3}=\frac{dP(\Delta E)}{d\Delta E}=0 \Rightarrow \Delta E=\sqrt{\frac{\hbar mc^2\omega}{2}}=\frac{\hbar a}{2} \tag{A12}$$

Now it is apparent, that uncertainties (A10) and (A12) satisfy the Heisenberg relation (A7), whereas (A3) represents the energy uncertainty in the vicinity of the equilibrium time t_e.

References

[1] Fröhlich H 1968 *Int. J. Quantum Chem.* **2** 641
[2] Molski M, Konarski J 2003. *Physical Review* E **68** 021916(1-7)
[3] Gompertz B 1825. *Philosophical Transactions of the Royal Society, London* **115** 513
[4] Molski M 2006. *The European Physical* Journal D **40** 411
[5] Morse P M 1929. *Physical Review* **34** 57
[6] Molski M. 2010. *Biosystems* **100** 57
[7] West GB, Brown JH, Enquist BJ 2001. *Nature* **413**: 628
[8] Hua W 1990. *Physical Review* A **42** 2524
[9] Leggett A J 1987. *The Problems of Physics* (Oxford: University Press)
[10] Horodecki R 1988. *Il Nuovo Cimento* B **102** 27
[11] Feinberg G 1967. *Physical Review* **159** 1089
[12] Recami E 1986. *Rivista Nuovo Cimento* **9** 1
[13] Aragone C, Guerri G, Salamo S, Tani JL 1974. *Journal of Physics A: Math. Nucl. Gen.* **7** L149
[14] Reinke J 1904. *Biologisches Zentralblatt* **24** 577
[15] Bergson H 1911. *Creative Evolution* (New York: Macmillan)
[16] Driesch H 1914. *History and Theory of Vitalism* (New York: Macmillan)
[17] White J, Krippner S 1977. *Future Science: Life Energies and the Physics of the Paranormal.* (New York: Garden City)
[18] Wheeler LR 1930. *Vitalism: Its History and Validity* (London: Witherby)
[19] Holmes F 1983. *Science* **221** 146
[20] Hameroff S 1997. *The Journal of Mind-Body Health* **13** 13
[21] Peters D 2001. *Journal of Bodywork and Movement Therapies* **5** 198
[22] Mayo DJ, Gunderson M 2002. Vitalism Revitalized: Vulnerable Populations, Prejudice, and Physician-Assisted Death. The Hastings Center Report **32** 14
[23] Dennet CD 1997. *Kinds of Minds: Toward An Understanding Of Consciousness* (Basic Books).
[24] Feigel A 2004. *Phys. Rev. Lett.* **92** 020404
[25] Tiggenen von B A 2008. *Eur. Phys. J.* D. **47** 261
[26] Haddat S, Restieri C, Krishnan K 2001. *Journal of Toxicology and Environmental Health* A. **64** 453
[27] Missov TI 2010. Analytic expressions for life expectancy in Gamma-Gompertz mortality settings. Working paper 2010-020 of the Max Planck Institute for Demographic Research.
[28] Winsor CP 1932. *Proceedings of the National Academy of Sciences.* **18** 1
[29] Iencinella D, Matteucci G 2004. *Eur. J. Phys.* **25** 249
[30] Konopinski EJ 1978. *Am. J. Phys.* **46** 499
[31] Corben H C 1977. *Lett. Nuovo Cimento* **20** 645
[32] Corben H C 1978. *Lett. Nuovo Cimento* **22** 116
[33] Molski M 1998. *Advances in Imaging and Electron Physics* **101** 144
[34] Molski M 2009. *BioSystems* **96** 136

New Perspectives of Bioelectromagnetics in Biology and in Medicine: DNA Spectra for Diagnostic Purposes

L Giuliani[1,3], E D'Emilia[1,3], M Ledda[2], S Grimaldi[2,3], A Lisi[2,3]
[1] INAIL ex ISPESL, Rome, Italy
[2] CNR – IFT, Rome, Italy
[3] ICEMS, Venice, Italy

Abstract. A new perspective for the use of bioelectromagnetics in biology and in medicine is open. Montagnier and his collaborators highlighted a physical approach to the diagnosis of several diseases, base on detecting the spectra of the DNA of cells, pathogenic agents or tumor cells. The DNA is prepared in an aqueous solution. The method uses the Schuman frequency, or any ELF, to induce the DNA solution to emit electromagnetic signals in the range 300 - 4000 Hz that are producing spectra that result to be typical for each disease. Preliminary tests performed at the facility of Italian CNR – Area Tor Vergata (Rome) – seem to confirm the effectiveness of this diagnostic approach. Further tests have to be performed. The method seems to be related to the same biophysical theory – based on Quantum Electrodynamics – that is the basis of other important effects, now employed to new therapeutic approaches.

I. Background

I.1 – ISPESL-CNR research: the magneto-induced differentiation of stem cells

I.1.1 – The Ion Cyclotron Resonance (ICR)

Before others could notice the many possibilities of application of electromagnetism in biology, a group of ISPESL and CNR researchers (Giuliani L., D'Emilia E., Grimaldi S., Ledda M., Lisi A.), on the basis of preliminary results which correlated the exposure to low frequency magnetic fields with the tissue response and with the processes of cellular maturation [1-3], decided to focus on the use of weak magnetic fields for the differentiation and maturation of stem cells [4]. A breakthrough was made with the Liboff-Zhadin's effect study [5-7], which allowed the creation and implementation of a *magnetic cell differentiator*, patented jointly by CNR and ISPESL [8], and with the partnership with N. Bobkova and M.N. Zhadin [9-10]. It was understood that the greater biological effectiveness was related to the use of ICR, rather than the frequency (50/60 Hz) and this allowed us to refine the protocols of cell exposure, thus achieving the first successes of magneto-induced cell differentiation [11]. A collateral discovery was given by the experimental confirmation of the fact that the ICR is capable of moving currents of ions into the cell, as explained in [12[a]], whence comes the following figure 1, and as can be seen better in the videos available on You Tube [13].

I.1.2 – A possible treatment of heart attack

The collaboration with the Sapienza University of Rome - Department of Experimental Medicine - made it possible to successfully apply these techniques to the differentiation of *primary cardiac cells* in *cardiomyocytes* [12[b]]. The result is a promising approach for the treatment of heart attacks [14]. The research *in vitro* needs to be followed by the preclinical study *in vivo* on nude mice, and then the clinical experimental study, phase 1. A research project for a preclinical study is starting today with the grant of the Italian Ministry of Health. The project involves the collaboration of the mentioned research group ISPESL-CNR, of the Florence University (Cardiolgy and Nuclear Medicine), of the Institute for Cancer Research *B. Ramazzini* of Bentivoglio in collaboration with Luc Montagnier, Emilio Del Giudice, Prigogine Prize 2009 and others. The project involves the removal, by biopsy from consenting patients, of primary cardiac cells at the Careggi Hospital, and replication *in vitro* in [12[b]], with the replacement of the culture technique *cardiosphere*™ (patented by the researchers of La

Journal of Physics: Conference Series **329** (2011) 012011 doi:10.1088/1742-6596/329/1/012011

Sapienza) with the technique of *cultured in microgravity,* thanks to a change in the magnetic differentiator designed and built by the same research group ISPESL-CNR, still to be patented, independently or as an extension of the previous patent.

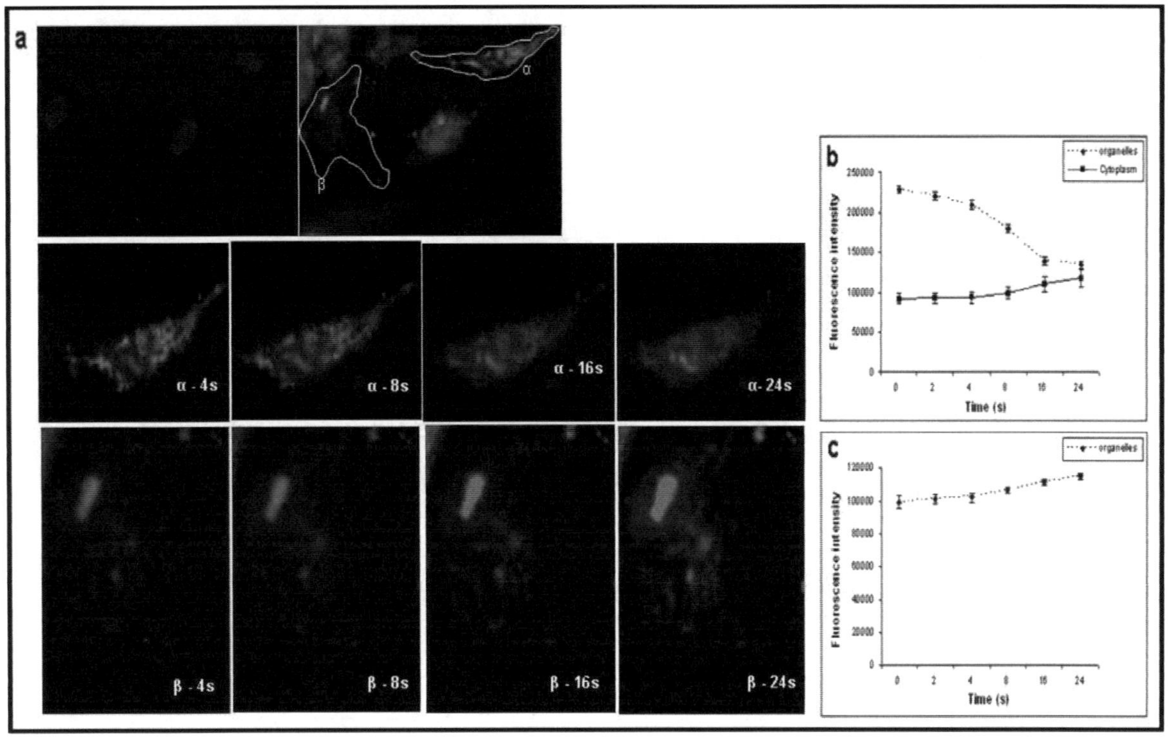

Fig. 1 (from [12]) – Scan every 4 seconds of organelles within a cardiac human cell, incorporating a fluorescent marker of intracellular calcium. The blue areas at the top of the first slide and identify the areas α and β better resolved in subsequent slides.

α: mitochondria that lose calcium during exposure to Ca-ICR.

β: cytoplasmic organelles that incorporate the calcium released from mitochondria.

I.1.3 – ICR for musculoskeletal developing

The technique of cell differentiation and of the magneto-induced cell maturation appears likely to also apply to other tissues, particularly the epithelial [4[a,c]] and muscular one [16]. The latter figure was anticipated already in a study by researchers from ENEA and Padoa University [17], which in fact confirms the already known effect of electromagnetic fields of promoting ostheocitary proliferation and maturation [5[a]], on which is based the work of machinery used in orthopedics for fractures reduction in all the hospitals in the world.

The effect, which can lead to an increased and more rapid development of bone and muscle, could be exploited to design garrisons, such as clothing, bracelets, and collars, both in the adjuvant therapy of development that in the production of sports clothing to be spread both in the professionals' area as in the amateurs' world.

I.2 - ICR for the treatment of cancer

There are many patents for electrical devices using magnetic fields to treat cancer [15]. The total lack of understanding, till now, of the magnetic fields interaction mechanisms with the living matter resulted in the created of machines based only on limited clinical experiences that have not been well validated.

However, recently, V.V. Novikov, Zhadin's colleague who worked on the discovery of the eponym effect, has published, with others, an important work showing the effectiveness of ICR in the Ehrlich cancer treatment in rats [18].

In the current state of knowledge it is not completely surprising. It should be considered that tumor cells are immature cells that have not completed their cycle of differentiation and maturation. So that a method, such as the one for the magneto-induced differentiation, capable of inducing differentiation in stem cells, which are also immature cells, it should not fail to cause a differentiation in cancer cells. The malignant cells could differentiate into benign tumor cells or at least less aggressive. This will lead to an improvement or a containment of the cancer effects, such as the containment of ascites in rats by Novikov *et al.*. On the other hand, the discovery of coherence in living matter opens a promising perspective, any new understanding of carcinogenesis [19].

Other suggestions are coming from the improvements of other applications of ICR [20-21] or of the coupling between the electric polarity of cancer tissues with electromagnetic fields [22-30].

I.3 - Risks related to use of electromagnetic devices

On the other hand we cannot neglect the existence of risks related to the use of electromagnetic devices. Evidences of this kind of risks are not coming from healthy practices but from the daily use of electric and electronic devices.

I.3.1 – Early evidences of risk

Since 50s years, thermal effects of electromagnetic non-ionizing fields were evidenced, due to professional diseases of radar-technicians [31]. In the same period, thermal mechanisms of interaction of MWs with living tissues were fully highlighted. The main mechanism was recognized in the interaction of an electromagnetic wave with the electric dipole of the molecule of water. A living tissue is full of molecules of water, whose rotational motion induces an increase of temperature due to the Joule effect.

Twenty years later many scientists became persuaded of the existence of non thermal effects on living organisms due to radiofrequencies and MWs and it was more and more related with some other properties of waves than amplitude, as the waveform, i.e. the presence or absence of modulation and the frequency of modulating wave [32]. In some cases, the efficacy of electric field alone was considered too while the efficacy of magnetic field alone, with reference to the band of ELF, was early adopted [33]. Despite the knowledge of action mechanisms of ELF were not satisfying; soon epidemiologic data indicated a strong relationship between ELF and arising of cancer, peculiarly of childhood leukemia, while the nexus between RF/MWs and cancer was assumed by a minority of scientists only almost twenty years later [34]. At that time an unifying hypothesis were suggested involving both effects due to high voltage lines and to AM radiofrequencies or TV signals: amplitude modulated radiofrequencies acted with a mechanism overlaying the penetrating efficacy of the carrier way and the direct magnetic interaction of the ELF, due to the amplitude modulation in ELF band [35]. The same hypothesis supported the analysis of the risk related to the use of mobile phones by children [36].

I.3.2 – Brain tumor risk

In the last decade strong epidemiologic evidence was reached of a nexus between ipsilateral head exposure to microwaves from handy and arising of brain tumors and new evidences for the risk related to occupational exposure to RF was achieved [37]. This fact can be related with both thermal [38[a]] and non-thermal effects of RF/MWs [38[b]].

I.3.3 – What do we learn from beneficial about risks?

We recognize that a nested net of signals in ELF/ULF/LF band is acting in vital processes. These signals are usually very weak and are limited in narrow windows of frequency and of amplitude.
From the point of view of survival, this fact is very important, because macro phenomena are filtered in band and in amplitude: in fact only signals strictly tuned with specific frequencies and amplitudes seem to be efficacy or impacting. On the other hand this required exactness makes hard to decode and to reproduce signals useful for diagnostics and therapeutics.

However the Shannon principle on the ratio of signal and noise teaches us that we need keep As Low As Reasonably Achievable all the artificial signals that become noise in the context of the network of our own inner vital signals.

I.4 – Physical diagnostics "selon Montagnier".

A new perspective to diagnostics has been open by the Montagnier developing a physical approach instead the usual biochemical approach to detect bacterial or viral inflammations.

I.4.1 – Spectra of DNA in the LF band

In his speech to the Meeting of Nobel Laureates in Lindau in 2010 [39], Luc Montagnier illustrated the stages of the progressive knowledge of DNA, up to his latest discovery: the existence of molecular complexes in the double helix that are able to act as an antenna working in the band of LF, typically between 300 and 4000 Hz.

In two works of 2009, the Nobel Prize highlighted that the emission spectrum, due to the DNA of bacteria and viruses in inflammatory processes, is typical of the infecting organism and can therefore be considered as a means of diagnosis of infections [40]. Furthermore, in the case of HIV, the response of the physical diagnostic method seems to be more sensitive than the biochemical analysis. The physical method reveals the echo of the virus also in the serum of patients who, following treatment with antiretrovirals, are negative for biochemical analysis [40[b]]. The discovery immediately aroused great controversy, resulting in the skepticism of several scientists, mostly from USA, while the University of Shanghai has dedicated a laboratory to the great French biologist, where he will develop applications of his revolutionary discovery.

However his discovery seems to be based on a strong theoretical basis, whose links with Quantum Electrodynamics has been highlighted in [41].

II – Improving physical diagnostics "selon Montagnier"

II. 1 - A test at Tor Vergata CNR laboratories

During the last visit of Luc Montagnier, on February 16, 2011, to the laboratory of Tor Vergata, where the ISPESL-CNR patented magnetic differentiator is installed, experiments were carried out to compare the spectra of the DNA of Escherichia coli and of human colon carcinoma, detected and recorded in the non-magnetic room at Tor Vergata, with the spectra previously registered by the group of Montagnier in his labs in Île-de-France and in Camerun. Comparative experiment has been successful for Escherichia coli, while DNA of human colon carcinoma cells did not result comparable spectra.

However our preliminary results are showing that the spectrum of human colon carcinoma is quite different from that of serum only or from that of distilled water, the same water used for the preparation of the aqueous solutions of DNA on the same occasion.

II. 2 Materials and methods

The facility of CNR – Tor Vergata, including the shielded room toward magnetic and electromagnetic fields, has been equipped in such a way to replicate the experiments of the Montagnier's group (fig. 2).

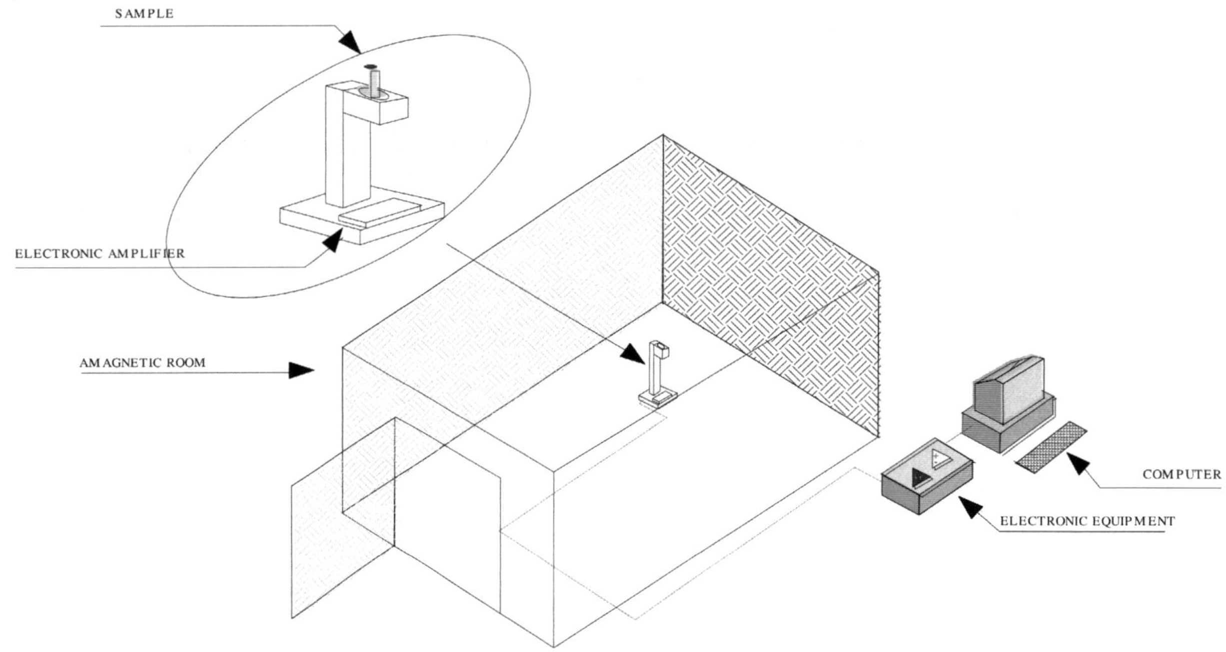

Fig. 2 – The room shielded against magnetic and electromagnetic noise as it was equipped to replicate the experiments of the Montagnier's group.

The detecting coil, connected to the electronic amplifier, is shown in the upper left corner. A vessel with the sample is put in the coil. This equipment was located within the amagnetic room. Electronic devices, including a notebook powered with batteries, were located outside of the room, linked by means of a shielded wire, and then replaced with an optical fiber.

The electronic equipment is able to provide a signal tuned to the Schuman frequency (7.8 Hz). According to the protocol of the experiments of Montagnier and his collaborators, such a frequency, as well any ELF, is able to stimulate the emission of electromagnetic signals from the DNA of bacteria or of cells, diluted in the sample.

Our test has been led first recording the noise, then recording the spectra detected from the sample, initially full of pure water, subsequently filled with an aqueous solution of DNA of cells belonging to a line of human colon carcinoma coming from the bank of the CRN – Area Tor Vergata. Finally the spectrum coming from the PSB serum ehere tumoral cells were cultured has been recorded.

II.3 - Results

The resulting spectra are shown in the following Fig. 3.

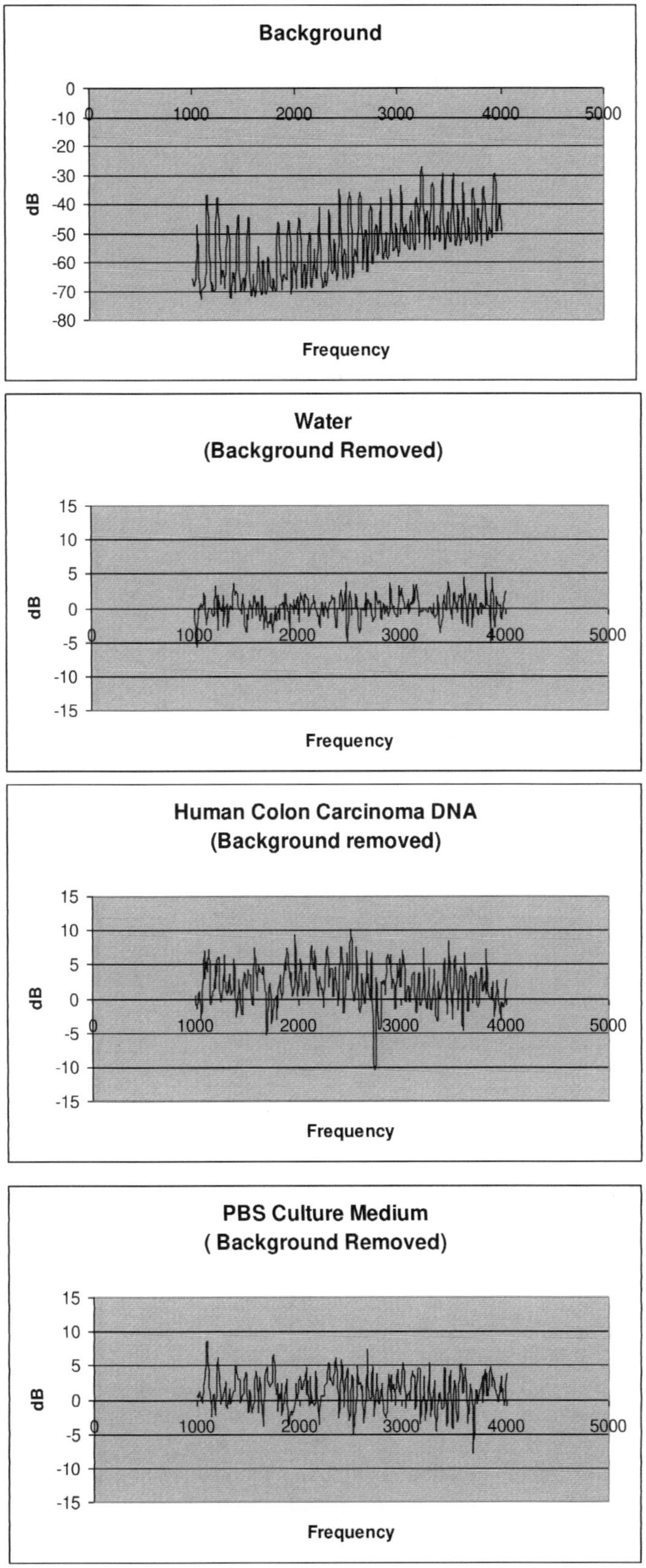

Fig. 3 - Detected spectra. In all the spectra, except the first one, the noise due to the electromagnetic background is removed.

It is impressive that the spectrum of the vessel fill of aquaeous solution of DNA of human colon carcinoma shows an absorption peak in the neighborhood of 2,8 kHz.

However the upper results are preliminary results that have to be replicated and confirmed.

III - Conclusions

Preliminary results are confirming that the physical approach for diagnosis of inflammations or other also severe diseases, of Luc Montagnier and of his collaborators, seems suitable to enable us to early detect the diagnosis, discriminating spectra related to the DNA, of several pathogenic agents, including bacteria and virus.

Furthermore the method can be extended to the case of several degenerative diseases, including some kind of tumors, of which we ignore today the possible viral nature [40a].

Acknowledgements. Many thanks to Dr. Devra Davis for her friendly suggestions.

References

[1] Santoro N, Lisi A, Pozzi D, Pasquali E, Serafino A, **Grimaldi S,** *Effect of extremely low frequency (ELF) magnetic field exposure on morphological and biophysical properties of human lymphoid cell line (Raji)***,** Bioch. Biophys. Acta **1997** Jul 24;1357(3):281-90 **(I)**.

[2] **Lisi A,** Pozzi D, Pasquali E, Rieti S, Girasole M, Cricenti A, Generosi R, Serafino AL, Congiu-Castellano A, Ravagnan G, Giuliani L, Grimaldi S., *Three dimensional (3D) analysis of the morphological changes induced by 50 Hz magnetic field exposure on human lymphoblastoid cells (Raji),*Bioelectromagnetics **2000** Jan;21(1):46-51 **(I)**.

[3] [a] Manni V, Lisi A, Pozzi D, Rieti S, Serafino A, Giuliani L, **Grimaldi S,** *Effects of extremely low frequency (50 Hz) magnetic field on morphological and biochemical properties of human keratinocytes,* Bioelectromagnetics **2002** May;23(4):298-305 **(I)**. [b] Rieti S, Manni V, Lisi A, Giuliani L, Sacco D, D'Emilia E, Cricenti A, Generosi R, Luce M, **Grimaldi S**, *SNOM and AFM microscopy techniques to study the effect of non-ionizing radiation on the morph-ological and biochemical properties of human keratinocytes cell line (HaCaT),* J. Microsc., **2004** Jan; 213(Pt 1):20-8 **(I)**. [c]

[4] [a] Manni V, Lisi A, Rieti S, Serafino A, Ledda M, Giuliani L, Sacco D, D'Emilia E, **Grimaldi S**, *Low electromagnetic field (50 Hz) induces differentiation on primary human oral keratinocytes (HOK)*, Bioelectromagnetics **2004** Feb;25(2):118-26 **(I)**. [b] **Lisi A**, Ciotti MT, Ledda M, Pieri M, Zona C, Mercanti D, Rieti S, Giuliani L, Grimaldi S, *Exposure to 50 Hz electromagnetic radiation promote early maturation and differentiation in newborn rat cerebellar granule neurons***,** J Cell Physiol **2005** 204(2):532 **(I)**. [c] **Lisi A**., Giuliani L., Grimaldi S**,** *ELF non ionizing radiation changes the distribution of the inner chemical functional groups in human epithelial cell (HaCaT) culture***,** Electrom. Biol Med **2006** 25(4):281 **(I).**

[5] [a] Shaberg SJ**, Liboff AR,** Falk MC, *Wire induced osteogenesis in marrow,* J Biomed Material Res **1985** July/Aug, 19(6):673-684 **(USA).** [b] **Liboff AR**, *Geomagnetic Cyclotron Resonance in Living Cells*, J. Biol. Phys. **1985**, 9:99 **(USA)**. [c] **McLeod, B. R. and Liboff, A. R.,** *Cyclotron resonance in cell membranes: the theory of the mechanism*, in Blank, M. J. and Findl, E., eds., *Mechanistic Approaches to Interactions of Electromagnetic Fields with Living Systems,* Plenum, New York, **1987,** pp. 97–108 **(USA)**. [d] Diebert M C, McLeod B R, Smith S D and **AR Liboff,** , Ion resonance electromagnetic field stimulation of fracture healing in rabbits with a fibular ostectomy. *J. Orthop. Res.* **1994,** 12:878 **(USA)**. [e] **Liboff A.R. 1997**, *Electric field ion cyclotron resonance*, Bioelectrom. 18(1):85; [f] **Liboff AR** *The cyclotron resonance hypothesis: experimental evidence and theoretical constraints*, in Ramel C. and B. Norden, eds. *Interact. mechanisms of low-Level* EMF *with living syst.,* Oxford University Press, London, **1991,** p. 130 **(USA)**; [g] **Liboff, AR,** McLeod BR and Smith SD**,** *Resonance transport in membranes,* in Brighton C.T. & S.R. Pollack, eds., *Electrom. in Med. and Biol.* San Francisco Press, San Francisco **1991 (USA).** [h] **Liboff, AR**

The Ion Cyclotron Resonance hypothesis, in Barnes, F.S., Greenebaum, B., eds., Handbook Biol. *Effects of E.M.F.: Bioengin. and Biophys. Aspects of EMF*, 3rd Edition, p. 261, CRC Press, Boca Raton, **2007 (USA).**

[6] [a] Novikov VV and **MN Zhadin**, *Combined action of weak constant and variable low-frequency magnetic fields on ionic currents in aqueous solutions of amino acids*, Biophysics **1944** 39(1):41(**RUS**); [b] **Zhadin MN**, Novikov VV, Barnes F.S. & N. Pergola, *Combined Action of static and alternating magnetic fields on ionic current in acqueous glutamic acid solution*, Bioelectromagnetics **1998**, 19(1):41 **(RUS-USA**).

[7][a] **Del Giudice E**, Fleishmann M, Preparata G, Talpo G, *On the "unreasonable" effects of ELF magnetic fields upon a system of ions,* Bioelectromagnetics **2002,** 27:522 (**GB-I**). [b] **Pazur A**, *Characterisation of weak magnetic field effects in an aqueous glutamic acid solution by nonlinear dielectric spectroscopy and voltammetry,* Biomagn. Res. Technol. **2004** 2:8 **(D);** [c] Comisso N, Del Giudice E, De Ninno A, Fleischmann M, Giuliani L, **Mengoli G**, Merlo F, Talpo G, *Dynamics of the ion cyclotron resonance effect on amino acids adsorbed at the interfaces,* Bioelectromagnetics **2006** Jan., 27(1):16 **(I)**; [d] Alberto D**,** Busso L**,** Crotti G, Gandini M., **Garfagnini R,** Giudici P, Gnesi I, Manta F, Piragino G**,** *Effects of Static and Low-Frequency Alternating Magnetic Fields on the Ionic Electrolytic Currents of Glutamic Acid Aqueous Solutions*, Electromagn Biol Med **2008,** 27(1):25-39 (**I**). [e] **Alberto D,** Busso L, Garfagnini R, Giudici P, Gnesi I, Manta F, Piragino G, Callegaro L, Crotti G,. *Effects of Extremely Low-Frequency Magnetic Fields on L-glutamic Acid Aqueous Solutions at 20, 40, and 60 uT Static Magnetic Fields*, Electromagn Biol Med **2008** 27(3):241-3 **(I).**

[8] **D'Emilia E**, Giuliani L, Grimaldi S, Lisi A, Sacco A, *An electronic device for efficient human stem cells differentiation by physical stimulation,* patent PCT/IB2006/002551 by ISPESL and CNR **2006 (I)**

[9] [a] **Zhadin MN and L Giuliani,** *Some problems in modern bioelectromagnetics,* Electrom. Biol. Med. **2006**, 25(4): 269 **(RUS-I)**. [b] **Zhadin MN**, Barnes F and L Giuliani*, Response to "A Few Remarks on 'Combined Action of DC and ACMagnetic Fields on Ion Motion in Macromolecules'" by Binhi,* Bioelecttromagnetics 2007, 28(6):412 (**RUS-USA-I**). [c] **Del Giudice E and L Giuliani,** *Coherence in water and the kT problem in living matter,* in **Giuliani L and M Soffritti eds.** *Non-thermal effects and mechanisms of interaction between electromagnetic fields and living matter***.** Eu J. Oncology – Library, Fidenza **2010**, vol 5:7-24 **(I)**. [d] **Tigrek S and F Barnes**, *Water structures and effects of electric and magnetic fields,* ibidem, **2010** vol 5:25-50 **(USA).**

[10] [a] **Giuliani L**, Grimaldi S, Lisi A, D'Emilia E, Bobkova NV, Zhadin M.N., *Action of combined magnetic fields on aqueous solution of glutamic acid: the further development of investigations,* Biomagn Res Technol. **2008** Jan.; 6: 1 **(RUS-I).** [b] **Giuliani L**, D'Emilia E, Lisi A,Grimaldi S, Bobkova N.V, Zhadin MN, *Investigating the Icr Effect in a Zhadin's Cell* , Int, J. Biomed. Sc., **2009** 5(2):181-6 **(RUS-I**).

[11] [a] **Lisi A.,** Foletti A., Ledda M., Rosola E**,** Giuliani L, D'Emilia E, Grimaldi S. *et al. Extremely low frequency 7 Hz 100 microT electromagnetic radiation promotes differentiation in the human epithelial cell line HaCaT,* Electrom. Biol Med. **2006**;25(4):269 **(I** [b] **Lisi A,** , Ledda M, De Carlo F, Foletti A, Giuliani L, D'Emilia E, Grimaldi S, *Calcium ICR transfers information to living systems*, Electrom. Biol Med. **2008**;27(3):230 **(I).**

[12] [a] **Lisi A,** Ledda, M, de Carlo F, Pozzi D, Messina E, Gaetani R, Chimenti I, Barile L, Giacomello A, D'Emilia E, Giuliani L, Foletti A, Patti A, Vulcano A, Grimaldi S, ICR *as a tool in regenerative medicine*, Electrom. Biol Med. **2008**;27(2):127 **(I)**; [b] Gaetani R, Ledda M, Barile L, Cimenti I, de Carlo F, Forte F, Ionta V, Giuliani L, D'Emilia E, Frati G, Miraldi F, Pozzi D, Messina E, Grimaldi S, Giacomello A and **A Lisi,** *Differentiation of human adult cardiac stem cells exposed to extremely low-frequency electromagnetic fields,* Cardiovascular Research **2009** June, 82(3): 411 **(I)**.

[13] [a] http://www.youtube.com/watch?v=Gd7u2wBcWRs . [b] http://www.youtube.com/watch?v=0pTj4Y4VGCM .

[14] **Rovelli G.,** *Con i campi elettromagnetici l'autotrapianto è più facile,* Almanacco della Scienza, Quindicinale del CNR **2010** Nov., 15 **(I). Cfr.** pure **Il Sole 24 Ore,** 26 Ottobre **2010**

[15] Brevetti USA (lista incompleta): 3991770, 4106488, 4154246, 4230129, 4303636, 4469103, 4622952, 4662359, 4690130, 4702262,.4767611, 4800899, 5045050, 5183456, 5197940, 5211622, 5437600, 5466576, 5658234

[18] Signori E, de Carlo F, Ledda M, Pierimarchi P, Zonfrillo M, Pozzi D, Giuliani L, D'Emilia E, Scorretti R , Grimaldi S and **A Lisi,** *Effect of Extremely Low Frequency Electromagnetic Field on Mouse Skeletal Muscle Cells Differentiation,* submitted to Bioelectromagnetics 2010 **(I)**.

[19] Gerardi G, **De Ninno A**, Prosdocimi M, Ferrari V, Barbaro F, Mazzariol S, Bernardini D, Talpo G, *Effects of electromagnetic fields of low frequency and low intensity on rat metabolism,* Biomagnetic Reas. Technol. 6(2) **2008, (I).**

[18] **Novikov V.V.,** Novikov G.V. & E.E: Fesenko, *Effect of weak combined static and extremely low-frequency alternating magnetic fields on tumor growth in mice inoculated with the Ehrlich ascites carcinoma***,** Bioelectromag. **2009,** 30(5):343 **(RUS)**

[19] [a] **Del Giudice E,** De Ninno A, Fleischmann M, Milani M, Talpo G, Mengoli G, Vitiello G, *Coherent Quantum Electrodynamics in living matter*, Electromagn. Biol. Med. **2005** 24:199-210 **(I-GB).** [b] **Popp F.A.** 2009 *Cancer Growth and Its Inhibition in Terms of Coherence,* Electrom. Biol. Med. 2009, 28(1):53 **(D),**

[20] Mazzaro R., *Viviamo tutti sulla cresta dell'onda,* New Media P.H., Padova **2010 (I).**

[21] [a] http://www.astelweb.com/index.asp?link=294. [b] http://www.seqex.com. [c] http://www.quec-phisis.com .

[22] [a] **Fricke H and S Morse,** *The Electric Resistance And Capacity Of Blood For Frequencies Between 800 And 4(1/2) Million Cycles,* J Gen Physiol. 1925 Nov 20;9(2):153-67. [b] **Fricke H and S Morse**, The electric capacity of tumors of the breast, J. Cancer Res. **10**, 340-376, 1926 **(USA).**

[23] [a] **Vedruccio C.,** Polarizability of normal and cancerous tissues, a Radiofrequency Nonlinear Resonance Interaction non invasive diagnostic Bioscanner Trimprob detector in **Giuliani L. and M. Soffritti eds.** Non-thermal effects and mechanisms of interaction between electromagnetic fields and living matter**.** Eu J. Oncology – Library, Fidenza **2010**, vol 5:177-185 **(I).**

[24] **Pokorny J**, Vedruccio C, Cifra M, Kucera O, *Cancer physics: Diagnostics based on damped cellular elastoelectrical vibrations in microtubules*, Eu Biophys J, Online First™, 10 March **2011 (CZ-I).**

[25] [a] **Bellerofonte C.** Vedruccio C, Tombolini P, Ruoppolo M, Tubaro A., *Non-invasive detection of prostate cancer by electromagnetic interaction,* Eur. Urology 47: 29 **2005** Jan. **(I)**. [b] **Tubaro A.**, De Nunzio C, Trucchi A, Stoppacciaro A, Miano L., *The Electromagnetic Detection of Prostatic Cancer: Evaluation of Diagnostic Accuracy,* Urology 72(2):340 **2008** Aug. **(I).**

[26] [a] **O. Gokce**, O. Sanli, A. Salmaslioglu, A. Tunaci, C. Ozsoy and F. Ozcan, *Tissue Resonance Interaction Method (TRIMprob) has the potential to be used alongside the recognized tests in the screening protocols for prostate cancer*, Int. J. Urology,;16(6):580-3 **2009** Jun **(TR).**

[27] **Da Pozzo L**, Scattoni V, Mazzoccoli B, Rigatti P, Manferrari F, Martorana G, Pietropaolo F, Belgrano E, Prezioso D, Lotti T, Villari D, Nicita G, *Tissue-resonance interaction method for the noninvasive diagnosis of prostate cancer: analysis of a multicentre clinical evaluation,* BJU Int. **2007** Nov;100(5):1055-9 **(I).** [b] **Di Viccaro D,** Perugia G, Cerulli C, Olivieri V, Bova J, Zanza C, Teodonio S, Liberti M**,** *The accuracy of tissue resonance interaction method probe (Trimprob tm) in non-invasive diagnosis of prostatic cancer. Analysis of the results of 782 patient.* Urologia, 76 Suppl 15:1-3 **2009 (I).**

[28] Gervino G, Autino E, Kolomoets E, Leucci G, **Balma M.,** *Diagnosis of bladder cancer at 465 MHz*. Electromagn Biol Med. **2007**;26(2):119-34 **(I-GR).**

[29] **Carrozza F. 2008** (I), Europ. J. Oncol. **2008** 13(4)

[30] [a] **Vannelli A**, Leo E, Battaglia L, Poiasina E, *Diagnosis of rectal cancer by electromagnetic interactions: preliminary results*, Dis Colon Rectum. **2009** Jan;52(1):162-6 **(I).** [b] **Vannelli A**, Battaglia L, Poiasina E, Leo E, *Diagnosis of rectal cancer by Tissue Resonance Interaction Method*, BMC Gastroenterology **2010**, 10:45 **(I).**

[31] Robinette CD, Silverman C, Jablon S, *Effects upon health of occupational exposure to microwave radiation (radar) 1950-1974*, J. Am. Epid. 1980, 112:39-53 **(USA).**

[32] [a] Taylor, L.S. and A.Y. Cheung (editors). *The Physical Basis of Electromagnetic Interactions with Biological Systems.* Proceedings of a workshop held at the University of Maryland, College Park, Maryland, June 15-17, **1977,** 410 pp. **(USA).** [b] **Stuchly** MA, *Interaction of radiofrequency and microwave radiation with living systems: A review of mechanisms,* Rad. Env. Biophys. 1979, 16(1):1-14 **(USA).** [b] Bawin MS, Gavalas-Medici RJ and **WR Adey,** *Effects of modulated very high frequency fields on specific brain rhythms in cats,* Brain Research 1973, 58(2):365-384 **(USA).** [c] **Blackman C F**; Benane S G; Elder J A; House D E; Lampe J A; Faulk J M, *Induction of calcium-ion efflux from brain tissue by radiofrequency radiation: effect of sample number and modulation frequency on the power-density window.* Bioelectromagnetics **1980**;1(1):35-43 **(USA).**

[33] [a] **Schwan HP and KR Foster**, *RF-field interactions with biological systems: Electrical properties and biophysical mechanisms,* Proceedings of IEEE 1980, 68:104-113 **(USA).** [b] **Ahlbom A.,** *A review of epidemiologic literature of magnetic fields and cancer*, Scand J Work Env Health **1988**, 14:337-343 **(SW).**

[34] [a] **Wertheimer N and E Leeper**, Electrical wiring configurations and childhood cancer, American Journal of Epidemiology **1979**, 109, 273-284 **(USA).** [b] **Feychting M. and Ahlbom A.,** *Magnetic Fields and Cancer in Children residing near Swedish High-Voltage power Lines***,** Am. J. Epidem. **1993** 139(7):467-481 **(SW).** [c] **Goldsmith JR,** *Epidemiological Evidence Relevant to RADAR (MW) Effects,* Env. Helath Perspect. **1997** Dec., 105 Suppl. 6:1580-87 **(ISR).**

[35] **Vignati M and L Giuliani**, *Radiofrequency exposure near high voltage power lines,* Env. Health Perspect. **1997** Dec., 105 Suppl. 6: 1569-73 **(I).**

[36] **Hyland GJ**, *The physics and biology of mobile telephony,* The Lancet 2000, **356:**1833-6 **(USA).**

[37] [a] **Hardell L**, *Wireless phone use and brain tumor risk,* , in **Giuliani L and M Soffritti eds.** *Non-thermal effects and mechanisms of interaction between electromagnetic fields and living matter***.** Eu J. Oncology – Library, Fidenza **2010**, vol 5:363-78 **(SW).** [b] **Szmigielski S.,** *Carcinogenic risks in workers exposed to radiofrequency and microwave radiation,* ibidem, **2010,** vol. 5:357-362.

[38] [a] **Han** Y.-Y, Ghandi OP, De Salles A, Herberman RB, **Davis DL,** *Comparative assessment of models of electromagnetic absorption of the head for children and adults indicates the need for policy changes,* **2010,** vol. 5:301-18. [b]. **Seyhan N**, Guler G, Canseven A, Sirav B, Ozgur E, Tuysuz MZ, *Investigation on blood-brain barrier permeability and collagen synthesis under radiofrequency radiation exposure and SAR simulations of adult and child head,* ibidem, **2010,** vol. 5: 319- 332. [c] **Salford LG**, Nittby H, Brun A, Eberhardt J, Malmgren L, Persson BRR, *Effects of microwave radiation upon the mammalian blood-brain barrier,* ibidem, **2010,** vol. 5: 333-355. [d] **M. Havas M,** Marrongelle J, Pollner B, Kelley E, Rees CRG, Tully L, *Provocation study using heart rate variability shows microwave radiation from 2.4 GHz cordless phone affects autonomic nervous system,* ibidem, **2010,** vol. 5:27-300.

[39] http://montagnier.net/montagnier/index.php/download_file/view/17/.

[40] [a] **Montagnier L**, Aïssa J, Ferris S, Montagnier JL, Lavallée C, *Electromagnetic signals are produced by aqueous nanostructures derived from bacterial DNA sequences,* Interdiscip Sci Comput Life Sci **2009** Jan., 1: 81–90 **(F).** [b] **Montagnier L**, Aïssa J, Lavallée C, Mbamy M, Varon J, Chenal H, *Electromagnetic detection of HIV DNA in the blood of AIDS patients treated by antiretroviral therapy*, Interdiscip Sci Comput Life Sci 1(4):245-53 **2009 (F- CAM).** [c] **Montagnier L**, Aïssa J, Del Giudice E, Lavallée C, Tedesco A, Vitiello G, *DNA waves and water*, IOP Publishng - Journal of Physics: Conference Series 306 **2011 (012007**Dec.)**, (F-I).**

Electromagnetic Resonance in Biological Form: A Role for Fields in Morphogenesis

Alexis M Pietak

Kingston, Ontario, Canada, K7L 4V3

alexis.pietak@gmail.com

Abstract. In morphogenesis, the mechanisms through which homogeneous, symmetric collectives of self-same cells are able to consistently and precisely establish long-range pattern remain an open question of scientific research. This work explores the hypothesis of developing biological structures as dielectric microwave resonators, using plant leaves as a working example. A finite element analysis (FEA) model was designed to determine if suitable resonant modes were physically possible for geometric and electrical parameters similar to those of developing leaf tissue. Using the FEA model, resonant EM modes with patterns of relevance to developing leaf vein modalities were detected. Here I show how the single physical mechanism of EM resonance can self-consistently account for different kinds of key symmetry-breaking operations characteristic of a variety of leaf vascular patterns. On account of the existence of shared geometric signatures in a leaf's vascular pattern and the electric field component of EM resonant modes supported by a leaf-like structure, further theoretical and experimental investigations are warranted. Significantly, this hypothesis is not limited to leaf vascular patterning, but may be applicable to a variety of morphogenetic phenomena in a number of living systems.

1. Introduction

The diverse, intricate, and even mathematically precise patterns of many life forms are fascinating. Plants, for instance, have been called "living crystals" on account of the systematic patterning inherent in the placement of seeds, leaves, and branches [2]. Plant leaves, ubiquitous features of our everyday lives, actually exhibit an incredible wealth of intricate and rather precise patterns in their macroscopic vascular tissue structure (Figure 1). Interestingly, the main features of many leaf vascular patterns show long-range global symmetries and periodicities that can be mapped in relation to sets of nested circles [1] (Figure 2). The features of leaf patterns are therefore consistent with a holistic description in terms of a wave pattern, where each circle (or quasi-circle) is a wave-front [1]. In addition, some plant leaves show strikingly precise angular symmetry in the development of main axes (Figure 1 G-I). On account of these observations involving global features interrelated in systematic ways, a leaf pattern can be understood in terms of a field [1].

What makes biological patterns, such as those found in leaf vasculature, particularly awe inspiring is that they self-assemble from a macroscopically homogeneous and typically quite symmetrical initial

Journal of Physics: Conference Series **329** (2011) 012012 doi:10.1088/1742-6596/329/1/012012

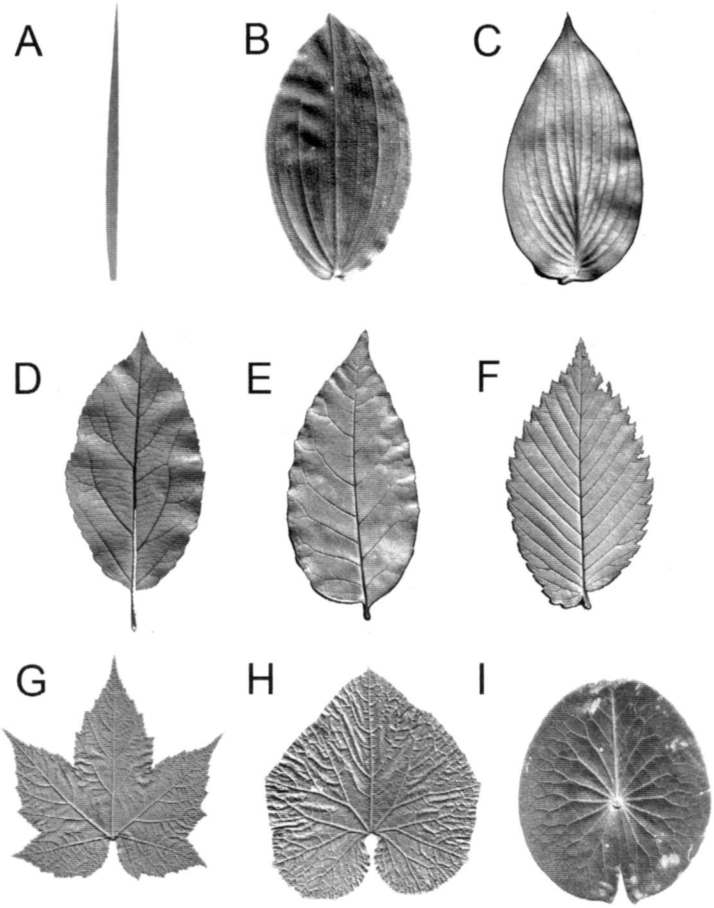

Figure 1: Plant leaves exhibit a wide variety of interesting vascular patterns. The top row shows leaves with a set of veins running roughly parallel to the central midvein, characteristic of pattern class 1 and monocots (panel A, crabgrass; panel B, twisted stalk; and panel C, hosta). The middle row shows leaves with secondary veins running at an angle to a central midvein, characteristic of pattern class 2 and dicots (panel D, apple; panel E, red ash; panel F, American beech). The bottom row shows pattern class 3 leaves with multiple major axes separated by a common angle (panel G, purple bush raspberry; panel H, zucchini; panel I, North American water lily).

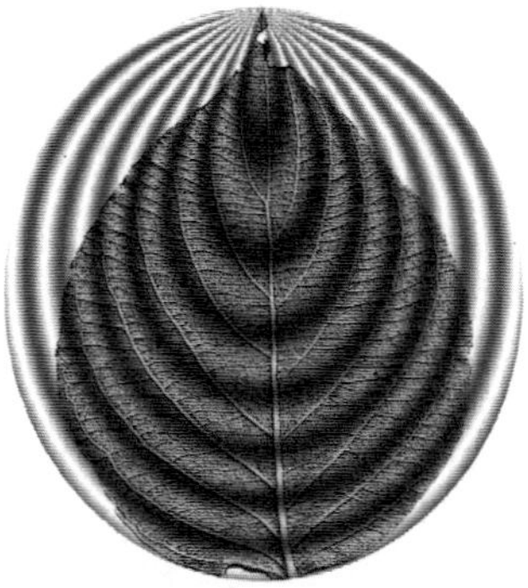

Figure 2: Many patterns in plant vasculature can be described in relation to a set of nested circular or quasi-circular forms with periodic or quasi-periodic spacing reminiscent of wave patterns. Shown here is the leaf of flowering dogwood with a wave pattern superimposed on top of the image [1]. Note how secondary vein spacing is specified by the wave's periodicity and how secondary veins tend to follow the trajectory of the wave fronts.

state where cells in the collective initially appear and behave in nearly identical ways. While today many of the self-assembly processes of morphogenesis can be handled using various complex systems models involving coupled reactions between cells, molecules, and genetic expression [3, 4], large gaps remain in our understanding of even basic morphogenetic phenomena.

The development of long-range patterns in leaf vasculature remains an example of such a gap in understanding. At a cellular level, it's been shown that flow of the plant hormone auxin through cells ultimately leads to their differentiation towards vascular tissue [5-7]. Essentially, while auxin can enter cells, it has trouble leaving without the help of a protein called PIN1 [5]. PIN1 is allocated to the boundary of cells and its positioning leads to the formation of auxin conduits [5]. This set-up is formally called the *auxin canalization mechanism* [6, 8]. PIN1 may be positioned in cells according to the flow of auxin through the cell [9]. This creates a positive feedback loop which results in the formation of dedicated auxin channels. These dedicated auxin channels in the meristematic tissue of the leaf primordium ultimately go on to differentiate into procambrial tissue, which ultimately becomes the vascular tissue system [5]. However, aspects of this hypothesis remain indeterminate because cellular auxin flux centers have still not been discovered, so it is unclear how cells can sense auxin flow to allocate PIN1 expression. Moreover, in almost thirty years of modeling, this interplay between auxin and PIN1 can generate little more than the primary midvein of a leaf and a scattered assembly of secondary veins [9], and certainly not the precise hierarchical organization of secondary and tertiary veins observed in a variety of real leaves [1]. Another mystery of leaf vascular development are experimental and theoretical indications that long-range vascular patterns grow from an array of high concentration auxin nodes forming in the leaf margin [10, 11]. Algorithms producing realistic depictions of leaf vascular patterns have utilized networks of nodes to account for complex vein configurations in leaves [12]. There remains, however, no indication of how auxin nodes form in the precise configurations in space and time that are required to account for leaf vascular patterns.

This present work explores an entirely new physical mechanism for morphogenesis in hypothesizing that fundamental symmetry-breaking operations occurring early in development, such as the appearance of high concentration auxin nodes at the margin of leaf primordia, could occur in response to spatially-patterned EM radiation forming within a developing organ structure. In fact, the field-like characteristics of a biological pattern, where global features show systematic interrelationships (Figure 1 and 2) can be explained if we understand the final pattern in relation to a resonant mode forming inside the developing organ structure. Resonant modes are standing wave patterns forming when a wave is confined within some kind of container and subjected to reflection from internal boundaries [13]. At internal boundaries, incident and reflected waves combine with one another to form a new pattern, the resonant mode, within the bulk material containing the waves. The resonant mode forms only at specific resonant frequencies, and is highly dependent on the geometry of the wave's container and the properties of waves travelling within the bulk material.

On account of their high water content and the polarity of other constituents like amino acids, many biological tissues have moderately high dielectric constants in the microwave regime [14-16]. For instance, human smooth muscle has a relative dielectric constant (ε_r) of 40 to 65 for frequencies of 1 to 10 GHz, decreasing to 7.5-9 at 100 GHz [16]. The ε_r of a 7% saline solution mimicking the ionic strength of plant juices, as well as juices extracted from corn stalks, were measured as 40 at 20 GHz [15]. And plant leaves (at 86 wt% water content) have ε_r of ~35 between 1 and 20 GHz [15]. As a developing plant is typically surrounded by weakly dielectric air (ε_r=1), any radiation produced within the plant (and able to propagate within it) at frequencies high enough to generate wavelengths smaller than the plant organ should reflect from the internal air-plant interface, leading to the formation of an EM resonant mode within the plant. As these resonant modes are standing wave patterns with amplitudes and lines of force varying in space over a long-range, the EM mode could be a basic symmetry-breaking mechanism in the developing organ, defining the location of cellular differentiation, orientation, or expression of biomolecular substances such as auxin. Of course, a strong dielectric loss characteristic is also expected for microwave radiation in watery tissue, so from a classical perspective this hypothesis of biological forms as dielectric resonators should be impossible.

Thus, this model rests on the validity of biological coherence theories as described by Frohlich [17-19] and/or quantum field theorists [20-23].

The concept of coherence in biological systems makes the electromagnetic phenomena needed to create resonant modes within the container of a developing organ at least a theoretical possibility. In 1968, esteemed physicist Herbert Frohlich presented a theory implying the possibility of long-range coherence in thermodynamically open, far from equilibrium, and dissipative biological systems [17-19]. Coherence domains are a key concept that has developed from Frohlich's initial theory [21]. To use the simplest, crudest picture possible, coherence domains are regions of coherently (*i.e.* synchronized) oscillating charged biomolecules like water or cell structures such as cytoskeletal microfilaments. Because these oscillating structures are electrically charged, a coherent EM field is thought to be associated with their coherent vibration.

There are three key features proposed for coherence domains which enable the hypothesis of biological structures as dielectric resonators. First of all, coherence domains are proposed to form fields at frequencies in the microwave to sub-millimeter wave regions of the electromagnetic spectrum [24, 25]. This is important because radiation in this frequency range could form resonant mode patterns with spatial features relevant to those of developing biological features (in the range of 0.1 to 1 mm). Second of all, coherence domains have been hypothesized to be nearly lossless within the coherence domain [21, 26, 27] and mechanisms for EM signal propagation within a coherent biological system have been put forth [28]. This is important, as from a classical standpoint, the dielectric loss tangent and absorption characteristics of biological tissue are too high to sustain or propagate an EM resonant mode in the microwave-terahertz range [14, 15, 29]. Finally, very large coherent regions have been thought possible [30]. The hypothesis of biological structures as microwave resonators requires coherence domains spanning the range of the developing organ (~0.5-10 mm) as again, from a classical standpoint, microwave-terahertz radiation should not be able to travel any significant distances in a biological tissue due to its high loss and absorption characteristics, and would therefore probably not be able to reach boundaries, reflect, and interfere to form a resonant pattern unless the entire region exhibits coherence.

A previous report has examined the basic feasibility of plant leaves as dielectric microwave resonators [31]. This previous report used finite element analysis (FEA) to examine the form of resonant patterns forming on thin, pointed, elliptical disks with different width to length ratios [31]. The quality factors (Q-factors) of EM resonances on the different leaf-shaped models were measured while parameters such as the dielectric constant, conductivity, dielectric loss tangent, and thickness to length ratio of the models were varied. The results found that, with the exception of the dielectric loss tangent, the expected dielectric constants, conductivities, and thickness-to-length ratios of leaf tissue were in ranges capable of supporting EM resonant modes in the microwave frequency range [31]. In addition, as developing leaf primordia often form in a configuration where they are folded along the midvein, this previous report also explored the effect of folding on resonant pattern formation, which was determined to be minimal [31].

This present report builds on previous work as a more systematic assessment of EM resonant mode patterns forming on model geometries simulating possible developing leaf geometries. Here a more comprehensive comparison is made between EM resonant modes and actual leaf patterns. The single physical mechanism of EM resonance is shown capable of accounting for many different kinds of core symmetry-breaking operations observed in different kinds of leaf vascular pattern.

2. Methods

2.1. Finite Element Analysis Modelling

A finite element analysis (FEA) model based on the same reasoning outlined previously [31] was used to identify EM resonant modes on dielectric structures simulating developing leaves. The eigenmode solver of Ansoft HFSS 13.0 software (Ansys, USA) was utilized. The FEA model focused on four geometries: (i) model A, a pointed ellipse with a length three times its width; (ii) model B, a pointed ellipse with a length twice its width; (iii) model C, a pointed ellipse with a length nearly equal to its width; and (iv) a circle with equivalent length and width. Each model had a length of 8 mm (maximum dimension), a thickness of 200 μm, and a width prescribed by the model type. All models used a dielectric constant of 35, a dielectric loss tangent of zero, and bulk conductivity of zero. The model used resistive boundaries at 10 kOhm/mm. Note that finite dielectric loss tangents up to 0.05, bulk conductivities up to 5 S/m, and boundaries with a finite conductivity up to 5 S/m were also used in models and found to make no significant difference in the form of resulting EM modes, affecting only their Q-values. Finite element meshes were created using the automatic meshing routines of the HFSS software with mesh operations set to limit a maximum possible element length of 400 μm. The resulting FEA meshes depended on the model type and the frequency of excitation, with lower frequencies generating larger wavelengths and requiring fewer mesh elements for an accurate solution. Simulations at 20 GHz and above used meshes with more than 10,000 (and a maximum of 40,000) tetrahedral mesh elements. The simplest simulations utilized meshes with only 2232 tetrahedrons (model A), 2206 (model B), 5492 tetrahedrons (model C), and 4685 tetrahedrons (circle). These low order meshes were only used for the lowest frequencies (below 15 GHz). The three main models (model A, B, and C) and their lowest order meshes are shown in Figure 3.

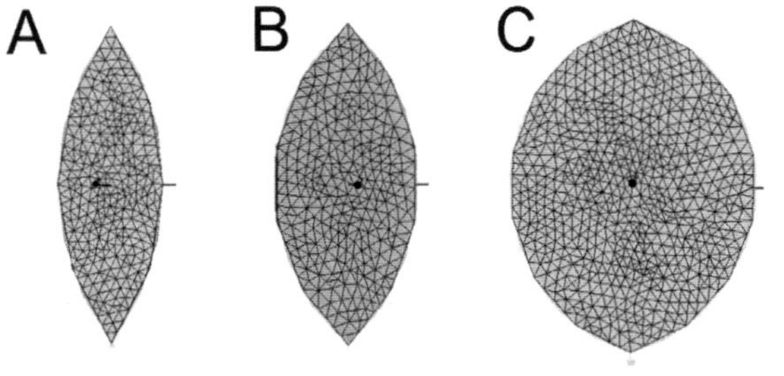

Figure 3: Three basic geometric models and the lowest quality meshes used in finite element analysis. Panel A shows model A, Panel B model B, and Panel C shows model C.

2.2. Leaf Vascular Pattern Analysis

Plant material was collected from 30 different randomly chosen species from the Great Lakes region of Canada (Kingston, Ontario, Canada) with common and Latin names listed in Table 1. Ten leaf samples from each species were digitally scanned (Canoscan Lide 20, Canoscan Inc, USA). Leaf images were measured using image-analysis software (ImageJ 1.37v). The 30 species listed in Table 2 represent a reference library of plant leaf patterns consulted in this work.

In this work, three basic pattern classes were defined for leaf vascular patterns. Pattern class 1 refers to the modality of parallel veins spanning from tip to petiole found in most monocot species, with examples shown in Figure 1 A-C. Pattern class 2 refers to a pattern commonly found in dicot plants characterized by a set of periodically (or quasi-periodically [1]) spaced secondary veins running at an angle off of a primary midvein, with examples shown in Figure 1 D-F. Pattern class 3 refers to a more complex modality characterized by the presence of 3 or more major axis (primary veins) separated by an even angular spacing, with a similar pattern repeated along each axis in the compound leaf (Figure 1 G-I).

Measurements of leaf margin, as defined by the first bifurcation point of secondary veins in pattern class 2 leaves, were made on five samples from each species at the point of maximum leaf width. Averages and standard error of the mean are reported.

Leaf pattern	**Pattern Class**
Crabgrass (*Digitaria sanguinalis*)	1
Hosta (*Hosta capitata*)	1
Twisted Stalk (*Streptopus amplexifolius*)	1
Apple tree (*Pyrus malus*)	2
American red raspberry (*Rubus idaeus*)	2
Common ragweed (*Ambrosia artemisiifolia*)	2
Northern Catalpa (*Catalpa speciosa*)	2
Poplar (*Populus fastigiata*)	2
American Beech (*Fagus grandifolia*)	2
Basswood (*Tilia Americana*)	2
Bitternut hickory (*Carya cordiformis*)	2
Common milkweed (*Asclepias syriaca*)	2
Flowering Dogwood (*Cornus canadensis*)	2
Horsechestnut (*Aesculus hippocastanum*)	2
Marsh milkweed (*Asclepias incarnata*)	2
Slippery elm (*Ulmus fulva*)	2
Staghorn sumac (*Rhus typhina*)	2
White Ash (*Fraxinus americana*)	2
Red Ash (*Fraxinus pensylvanica*)	2
White Elm (*Ulmus americana*)	2
White Birch (*Betula papyrifera*)	2
White Oak (*Quercus alba*)	2
Witch Hazel (*Hamamelis virginiana*)	2
Canada Moonseed (*Menispermum canadense*)	3
N. A. Water Lily (*Nymphaea odorata*)	3
Sugar maple (*Acer saccharum*)	3
Wild Grape (*Vitis riparia*)	3
Zucchini (*Cucurbita pepo*)	3
Common Geranium (*Pelargonium hortorum*)	3
Purple-flowering Raspberry (*Rubus odoratus*)	3

Table 1: List of plant species, Latin names, and pattern class designation for leaf vascular patterns considered in the present work.

3. Results

The electric field magnitudes of the first ten resonant modes forming in the four models are shown in Figure 4. In all EM modes, the electric field was parallel to the boundary of the flat disks, while the magnetic field assumed an angle out of the plane. Numerous EM mode patterns were seen in the four model geometries, all of them iterations upon common pattern modalities. Only the perfect circle

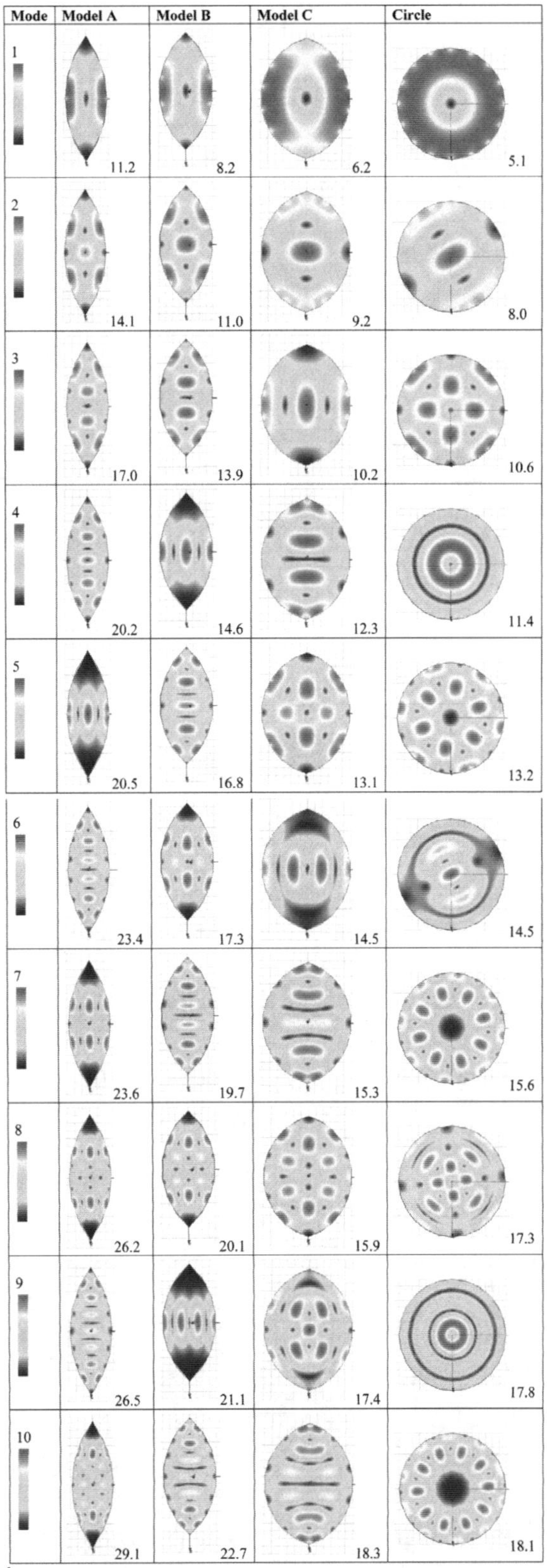

Figure 4: The electric field magnitude of the first ten resonant modes forming in the four geometric models. The bar to the left of each row shows the colour map, with red representing highest field strength and blue the lowest. All models have a length of 8 mm, thickness of 200 μm, and width prescribed by the model. All models used a dielectric constant of 35, conductivity and loss tangent set to zero, and resistive boundary conditions.

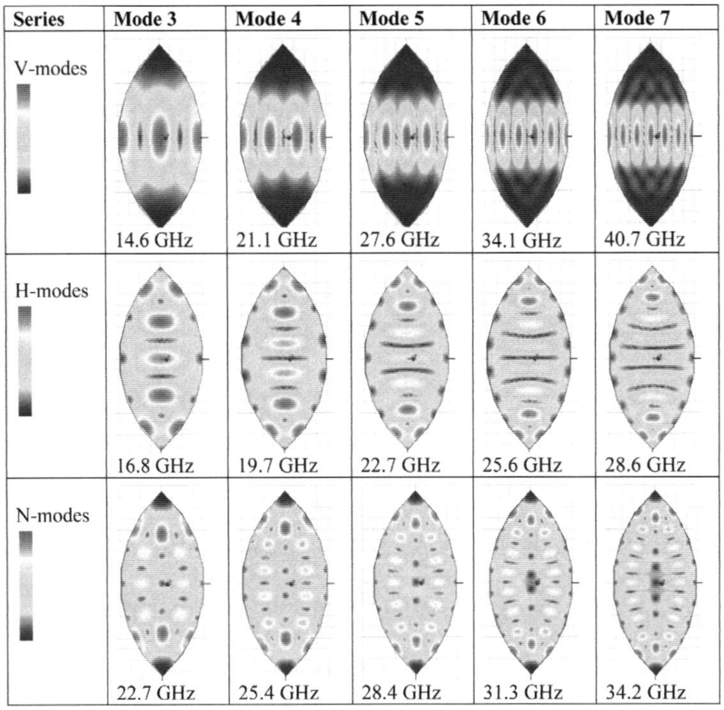

Figure 5: Resonant mode pattern series on model B. Top row shows the electric field strength of the V-mode series. Middle row shows the electric field strength of the H-mode series. Bottom row shows the electric field strength of the N-mode series. Beneath each image the resonant frequency is shown. The bar to the left of each row shows the colour map with red representing highest field strength and blue the lowest.

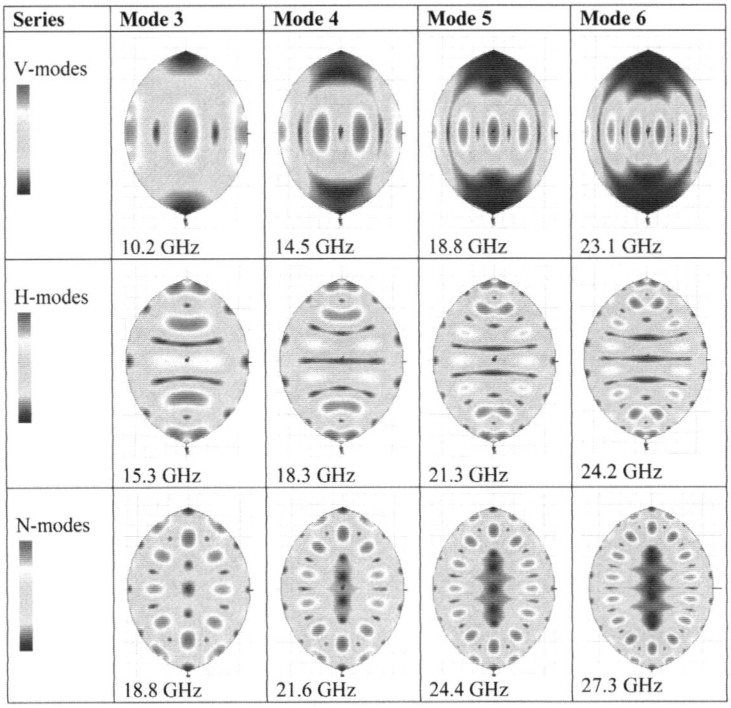

Figure 6: Resonant mode pattern series on model C. Top row shows the electric field strength of the V-mode series. Middle row shows the electric field strength of the H-mode series. Bottom row shows the electric field strength of the N-mode series. Beneath each image the resonant frequency is shown. The bar to the left of each row shows the colour map with red representing highest field strength and blue the lowest.

showed unique pattern modalities, while the three pointed ellipses of model A, B, and C shared similar (but not identical) pattern series (Figure 4). Three of the mode pattern series, referred to here as V-modes ('V' for vertical), H-modes ('H' for horizontal), and N-modes ('N' for node), showed very interesting signatures comparable to real leaf vascular patterns. The progression of these mode pattern series as a function of resonant frequency are shown for model B in Figure 5 and for model C in Figure 6.

The V-mode pattern series was found in all three pointed ellipse models, but not the perfect circle (Figure 4). The V-mode is characterized by equally spaced bands of high electric field strength running parallel to the central longitudinal axis (Figure 5 and 6, top row). As frequency increases, new modes in the V-mode series appear with one additional field maxima (Figure 5 and 6, top row). The V-mode is capable of breaking the symmetry of a homogeneous structure in a manner that is consistent with the vascular patterns observed in monocot leaves of pattern class 1 (Figure 1, top row). A comparison between the 7th V-mode of model B and the leaf of a hosta plant are shown in Figure 7.

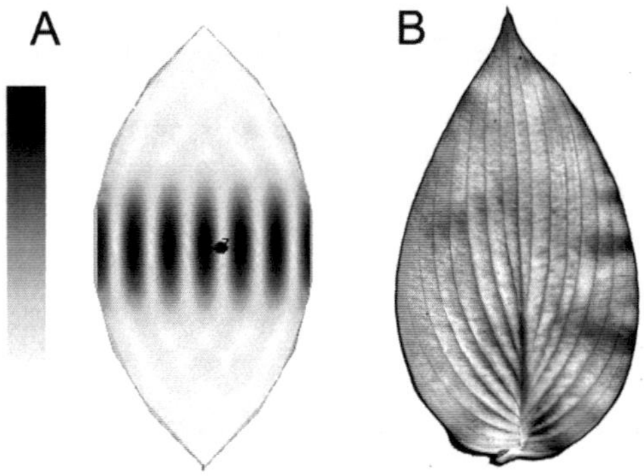

Figure 7: A comparison between the electric field magnitude of the 7th V-mode of model B (40.7 GHz) and the leaf of a pattern series 1 hosta plant (*Hosta capitata*). Scale bar to image left shows electric field strength with black representing high and white low magnitude.

The H-mode pattern series was found in all three pointed ellipse models, but not the perfect circle (Figure 4). The H-mode is characterized by equally spaced bands of high electric field strength running roughly perpendicular to the central longitudinal axis (Figure 5 and 6, middle row and Figure 8). As frequency increases, new modes in the H-mode series appear with one additional field maxima, and at high frequencies field maxima bend into a 'v' shape (Figure 5 and 6, middle row). The H-mode is capable of breaking the symmetry of a homogeneous structure in a manner consistent with the vascular patterns observed in dicot leaves of pattern class 2 (Figure 1, middle row), with an example comparison shown in Figure 8.

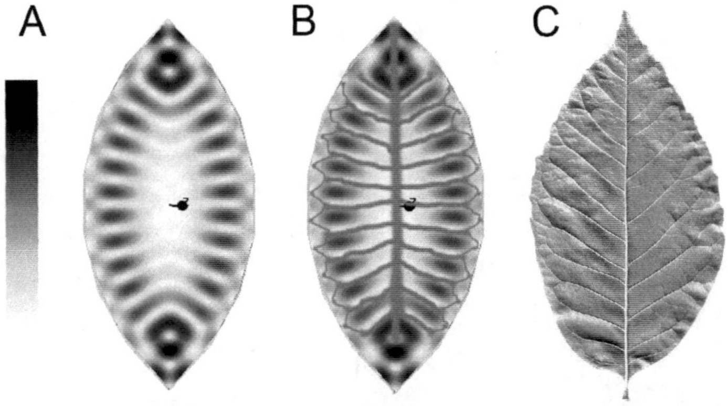

Figure 8: Comparison between the 14th H-mode of model B at 48.4 GHz (panel A), with the vascular structure of a white ash leaf (panel C). The proposed vein development sequence is shown superimposed on the EM mode image in panel B. Scale bar to image left shows electric field strength with black representing high and white low magnitude.

The N-mode pattern series was found in all three pointed ellipse models and in the perfect circle (Figure 4, 5, and 6). The N-mode is characterized by equally spaced nodes of high electric field strength forming around the margin of the structure (Figure 5 and 6, bottom row). As frequency increases, new modes in the N-mode series appear with the addition of two additional field maxima at opposing boundaries (Figure 5 and 6, bottom row). At high frequencies, the H- and N-mode patterns come to appear similar (Figure 6, compare middle and bottom rows). The N-mode is capable of breaking the symmetry of a homogeneous structure in a manner consistent with the vascular patterns observed in dicot leaves of pattern class 2 (Figure 1, middle row). In wider models represented by model C and the perfect circle, the N-mode is capable of selecting multiple axes separated by a fixed angle, which is characteristic of pattern class 3 leaves (Figure 1, bottom row). An example of this capacity is shown in Figure 9, where the 9th N-mode of a circle is shown in comparison to a water lily leaf.

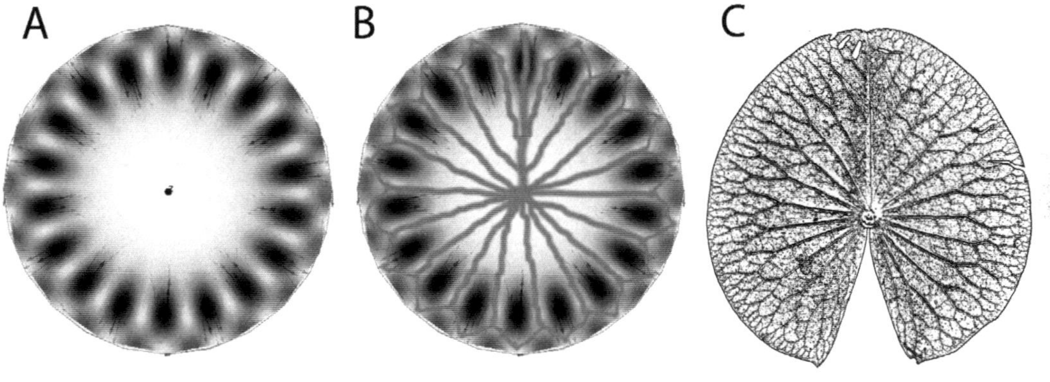

Figure 9: Comparison between the 9th N-mode of a perfect circle (panel A) and the vein structure of water lily leaf (panel C). The proposed vein formation pathway is shown superimposed on the mode image in panel B.

Model A, B, and C all exhibit V-, H-, and N- modes, with only the relative prevalence of particular modes being higher in some models and not others. As a function of frequency, model C showed a higher incidence of V-modes, while models A and B showed higher incidence of H- and N- modes (Figure 4 and Figure 10).

The H- and N- modes appear to influence leaf vascular patterning in a consistent, predictable manner (Figure 11). I hypothesize that the vascular network results from two physical processes. It is hypothesized that first, an EM resonant mode is created which breaks the symmetry of the developing primordium, likely by creating regions of high auxin concentration in relation to electric field strength. Secondly, PIN1 controlled auxin flow from this initial auxin distribution established by the EM resonant mode ultimately leads to differentiation of cells towards vascular tissue. Notably, a central midvein-like feature is not present in any of the resonant mode pattern series shown in Figures 5 and 6. Thus, it is also hypothesized that the midvein of pattern series 2 leaves appears via a mechanism independent of EM-mode formation, perhaps via PIN1-gated auxin flow through the bulk of the leaf and out through the petiole, starting from a homogeneous auxin concentration in the primordium. Comparing plant vascular patterns (Figure 1) and EM resonant modes of the H- and N- series (Figure 5 and 6) implies that secondary veins form in relation to areas of lowest or highest EM field strength. Here it is assumed that veins form in areas of lowest EM field strength. Auxin maxima may form at electric field minima (Figure 11 B). Note that the electric field minima appear in two sets, with fine minima appearing nearest to the boundary and coarse minima (longitudinal bands in H-modes and spots in N-modes) appearing closer to the central region (Figure 11 A and B). It appears as though secondary veins form along the coarse minima, while the positioning of the fine minima influence the location of secondary vein bifurcation points and can thereby account for a connected network in the leaf margin (Figure 11 C). Examples of this hypothesized vein creation mechanism in relation to EM modes and real leaves are shown in Figure 8 and 9.

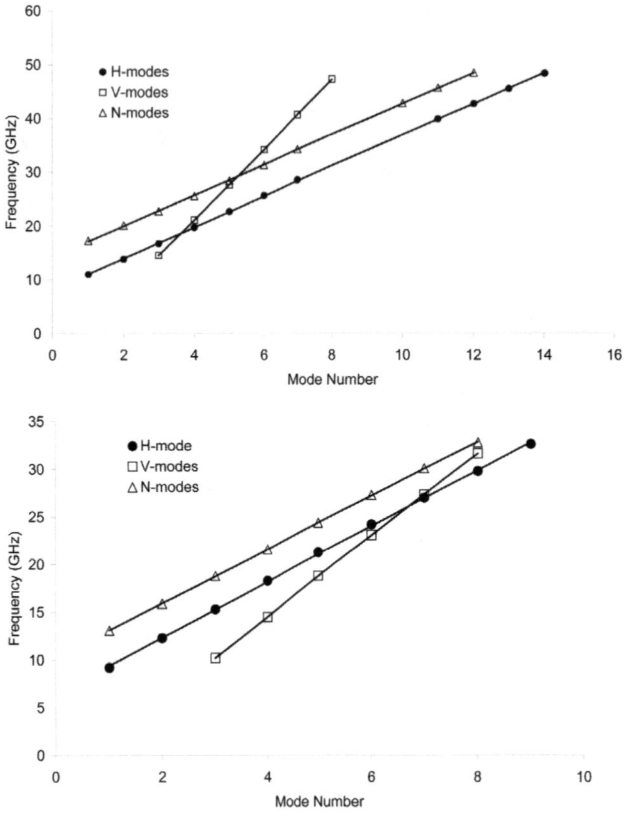

Figure 10: Frequency dependence of V-mode, H-mode, and N-mode pattern series forming on model B (top image) and model C (bottom image). In the wider model C, V-modes are more common at lower frequencies than they are in thinner models B and A. This is indicated by the V-node frequency dependence having a smaller slope in model C than model B.

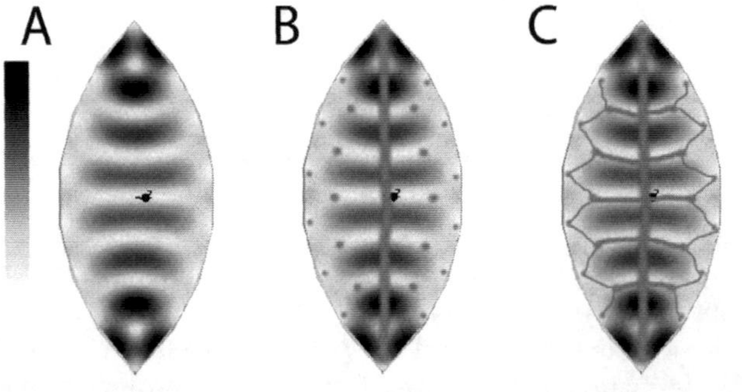

Figure 11: Mechanism of action proposed for H- and N- modes in leaf vascular development. The electric field magnitude of the 6th H-mode of model B is shown in panel A, with field maxima shown in black and minima in white. Panel B shows the primary midvein forming via an alternative mechanism. Auxin maxima are proposed to form in regions of lowest electric field strength (panel B). Connecting regions of high auxin level via a hypothesized auxin canalization mechanism leads to a characteristic leaf vascular pattern (panel C).

Support for this hypothesized relationship between H- and N- modes and leaf vascular network formation exists in the branching characteristics of real leaves (Figure 12). Here the 'margin' is defined in leaves as the distance between the first branch point of a secondary vein and the leaf boundary, and in EM H- and N- modes as the distance between the edge of a course node and the model boundary. Lower frequency H- and N- nodes have a progressively smaller margin size as mode frequency increases and course and fine nodes are pushed towards the outer boundary (compare Figure 12 A-B and E-F, also see Figure 13). This characteristic of H- and N- modes as a function of frequency is consistent with the general behavior of pattern class 2 leaves. As the number of veins (corresponding to mode nodes) gets larger, the margin size tends to decrease dramatically (compare Figure 12 C-D and G-H; also see Figure 13).

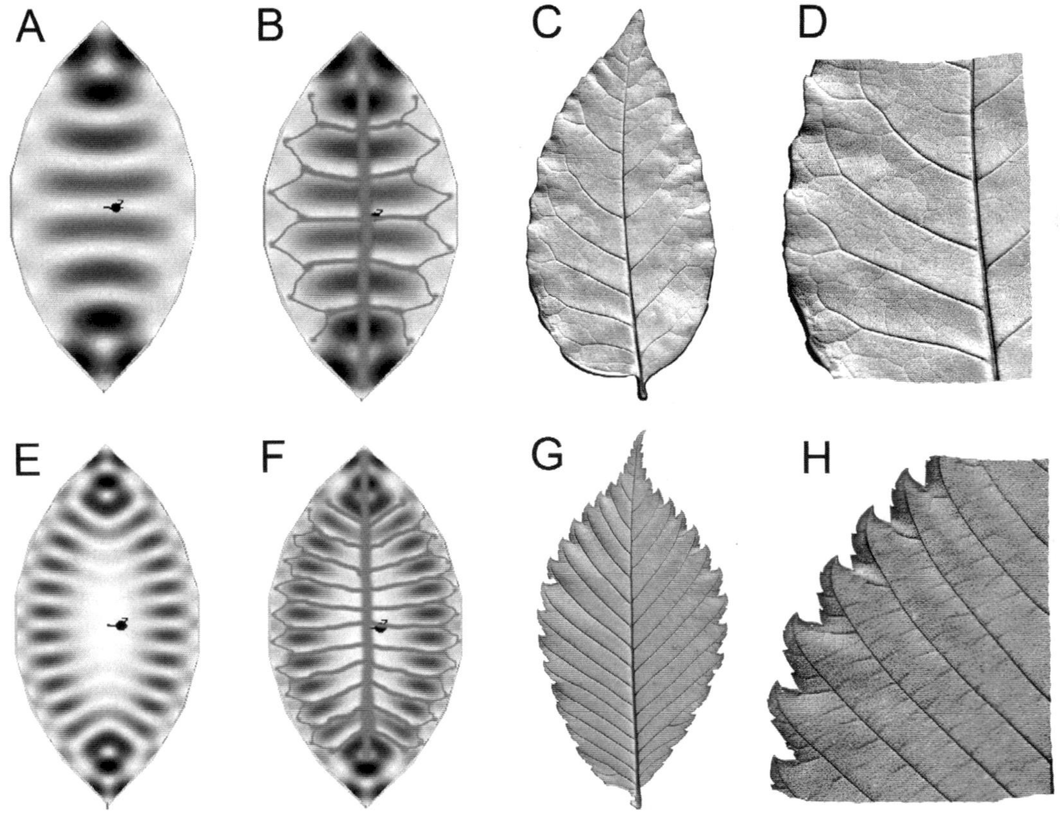

Figure 12: Margin effects in low-frequency (panel A-D) and high frequency (panels E-H) modes and corresponding leaf patterns. Panel A shows the 6th resonant H-mode on model B, while panel B shows the proposed vein creation resulting in large forks due to the positioning of coarse and fine edge nodes. This parallels the margin pattern of an 8 secondary-vein leaf (red ash) shown in panel C with a close up of margin structure in panel D. In contrast, a higher frequency 14th resonant H-mode on model B, with proposed vein creation resulting in smaller forks closer to the boundary due to the positioning of coarse and fine edge nodes (panel F), parallels the 16 secondary-vein leaf (American beech, panel G) and its margin structure (panel H).

Journal of Physics: Conference Series **329** (2011) 012012 doi:10.1088/1742-6596/329/1/012012

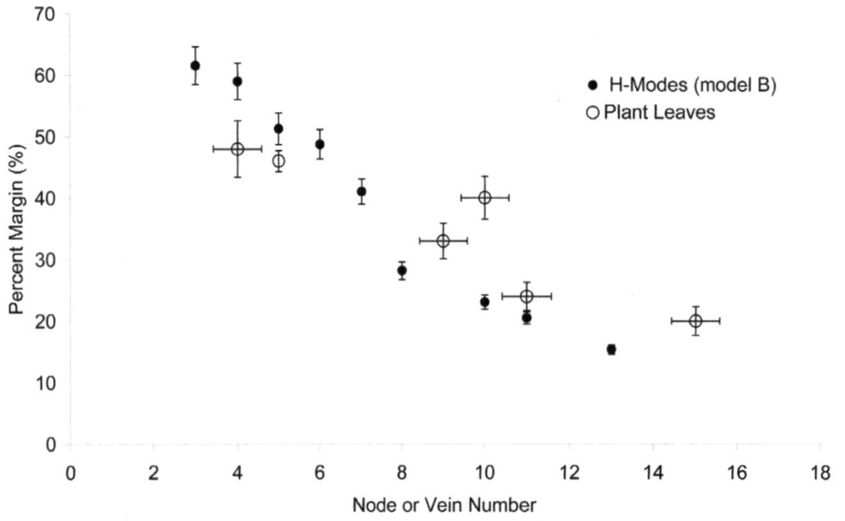

Figure 13: The margin of H-modes decreases with increasing frequency (node number) in a manner paralleling the general behaviour observed in real leaves. The margin is defined as the distance between the boundary and the first bifurcation point of a secondary vein in leaves, or the distance between the boundary and the edge of a course node in resonant H-modes. The percent margin is defined as the percentage distance of the margin distance in relation to the half width of the leaf/mode at its widest point.

4. Discussion

Resonant EM modes forming in a developing biological organ may account for key symmetry breaking events in morphogenesis. Here a FEA model of a rough leaf-primordium-like structure generated three EM resonant mode pattern motifs (the V-, H-, and N- mode series) capable of accounting for basic structural features found in a variety of monocot and dicot leaves. The V- node, consisting of alternating high and low electric field strength bands running parallel to the longitudinal axis of the model, would be capable of symmetry breaking in monocot structures of pattern class 1 leaves. The H- and N- nodes, which appeared similar at higher frequencies, had electric field patterns making them capable of symmetry-breaking in dicot structures of pattern class 2 and 3.

A new hypothesis was also put forth connecting the concept of EM resonant mode activity with more conventional understanding of leaf vascular formation. Here it is proposed that leaf patterns are the consequence of: (i) an the initial symmetry breaking of a homogeneous collective by the electric field of a resonant EM mode, which may influence the initial auxin distribution in the leaf by causing it to form or segregate only in low/high electric field regions, and (ii) the formation of auxin flow paths from this initial auxin distribution via a PIN1 regulated auxin-canalization type mechanism leading to dedicated auxin flow channels that ultimately differentiate into vascular tissue.

Notably, the resonant modes generated on model structures and determined using FEA do not account for all main characteristics of a leaf's vascular pattern. The primary mid-vein of pattern class 2 leaves is not evident in any of the mode series shown in this report. Also, while many dicot leaves have secondary veins with a significant and systematic curvature (Figure 1 and 2 show examples), there is nothing in the horizontal banding patterns of H-modes or the positioning of field nodes/maxima in N-modes, which provides any sign of the curvilinear profile of secondary veins in leaves.

There are numerous deficiencies in the current FEA models which may contribute to inaccuracies such as an inability to account for curvilinear vein profiles. A leaf primordium typically forms as a tissue sheet folded and curved about the meristem into a highly 3-dimensional structure. Also, leaves and leaf primordia are typically more of a teardrop shape with a wider base and pointed tip and are not symmetrical about the horizontal axis like the pointed ellipses modeled here. Therefore, more realistic and complex geometries may need to be modeled to generate resonant modes of greater accuracy and capacity to account for real vein characteristics. In addition, the FEA model considers the leaf tissue to be a homogeneous material, when in fact biological materials at the microscopic scale consist of a variety of parts such as lipid membranes, bound water, and proteins – all of which have significantly different electrical and physical properties. Finally, the FEA model represents a classical solution to Maxwell's equations while the underlying phenomenon of biological coherence rests on a quantum field theory platform. Using a classical model may not account for all features exhibited by the real system, in the same way that modeling a superconductor as an ideal conductor or a superfluid as a zero-viscosity liquid does not yield a complete description of the material and its exhibited phenomena.

5. Conclusions

In conclusion, a finite element analysis model provides electromagnetic mode solutions with standing electric fields sharing key geometric signatures with leaf vascular patterns. The formation of a resonant electromagnetic mode within a developing tissue/organ is therefore capable of explaining how a developing biological system can begin as a homogeneous cell collective and consistently break symmetries in precise ways with global features interrelated in systematic ways. On account of these similarities between this basic model and real leaf patterns, further theoretical and experimental investigations are warranted.

References

1. Pietak, A., *Describing long-range patterns in leaf vasculature with metaphoric fields.* J Theoretical Biology, 2009. **261**: p. 279-289.
2. Jean, R., *Phyllotaxis: The status of the field.* Mathematical biosciences, 1996. **127**: p. 181-206.
3. Koch, A. and H. Meinhardt, *Biological pattern formation: from basic mechanisms to complex structures.* Reviews Mod Phys, 1994. **66**: p. 1481-1494.
4. Meinhardt, H., *Models of biological pattern formation: common mechanisms in plant and animal development.* Int J Dev Biol, 1996. **40**: p. 123-134.
5. Scarpella, E., et al., *Control of leaf vascular patterning by polar auxin transport.* Genes and Development, 2006. **20**: p. 1015-1027.
6. Rolland-Lagan, A. and P. Prusinkiewicz, *Reviewing models of auxin canalization in the context of leaf vein pattern formation in Arabidopsis.* The Plant J, 2005. **44**: p. 854-865.
7. Uggla, C., et al., *Auxin as a positional signal in pattern formation in plants.* PNAS, 1996. **93**: p. 9282-9286.
8. Mitchison, T., *A model for vein formation in higher plants.* Proc Royal Soc London B - Bio Sci, 1980. **207**: p. 79-109.
9. Feugier, F., A. Mochizuki, and Y. Iwasa, *Self-organization of the vascular system in plant leaves: Inter-dependent dynamics of auxin flux and carrier proteins.* J Theor Bio, 2005. **236**: p. 366-375.
10. Rolland-Lagan, A., *Vein patterning in growing leaves: axes and polarities.* Curr Opin Gen Dev, 2008. **18**(4): p. 348-353.
11. Aloni, R., et al., *Gradual shifts in sites of free-auxin production during leaf-primordium development and their role in vascular differentiation and leaf morphogenesis in Arabidopsis.* . Planta, 2003. **216**: p. 841-853.
12. Runions, A., et al., *Modeling and visualization of leaf veination patterns.* ACM Trans Graphics, 2005. **24**: p. 702-711.
13. Staelin, D., A. Morgenthaler, and J. Kong, *Resonators*, in *Electromagnetic Waves*. 1994, Prentice Hall: New Jersey. p. 336-401.
14. Pethig, R. and D. Kell, *The passive electrical properties of biological systems: their significance in physiology, biophysics and biotechnology.* Phys Med Biol, 1987. **32**: p. 933-970.
15. El-Rayes, M. and F. Ulaby, *Microwave dielectric spectrum of vegetation - part I: experimental observations.* IEEE Trans Geosci Remote Sens, 1987. **GE-25**(5): p. 541-549.
16. Gabriel, S. and R. Lau, *The dielectric properties of biological tissue III: Parametric models for the dielectric spectrum of tissues.* Phys Med Biol, 1996. **41**: p. 2271-2293.
17. Frohlich, H., *Long-range coherence and energy storage in biological systems.* International J of Quantum Chemistry, 1968. **2**: p. 641-649.
18. Frohlich, H., *Bose condensation of strongly excited longitudinal electric modes.* Physics Letters A, 1968. **26A**: p. 402-403.
19. Frohlich, H., *Evidence for bose condensation-like excitation of coherent modes in biological systems.* Physics Letters A, 1975. **51A**: p. 21-22.
20. Del Giudice, E., S. Doglia, and M. Milani, *Quantum field theoretical approach to the collective behaviour of biological systems.* Nuclear Physics B, 1985. **B251**: p. 375-400.
21. Del Giudice, E., P. Spinetti, and A. Tedeschi, *Water dynamics at the root of metamorphosis in living organisms.* Water, 2010. **2**: p. 566-586.
22. Preparata, G., *QED Coherence in Matter.* 1995, London: World Scientific. 236
23. Arani, R., et al., *QED coherence and the thermodynamics of water.* Int J of Modern Physics B, 1995. **9**: p. 1813-1841.

24. Cifra, M., et al., *Vibrations of electrically polar structures in biosystems give rise to electromagnetic field: theories and experiments*, in *PIERS Proceedings*. 2009: Moscow, Russia. p. 138-142.
25. Cifra, M., J. Fields, and A. Farhadi, *Electromagnetic cellular interactions.* Prog Biophys Mol Biol, 2011. **105**: p. 223-246.
26. Mesquita, M., A. Vasconcellos, and R. Luzzi, *Near dissipationless coherent excitations in biosystems.* Int J Quant Chem, 1996. **60**: p. 689-697.
27. Mesquita, M., et al., *Large-scale quantum effects in biological systems.* Int J Quant Chem, 2006. **102**: p. 116-1130.
28. Del Giudice, E., S. Doglia, and M. Milani, *Electromagnetic field and spontaneous symmetry breaking in biological matter.* Nuclear Physics B, 1986. **B275**: p. 185-199.
29. Smye, S., et al., *The interaction between terahertz radiation and biological tissue.* Phys Med Biol, 2001. **46**: p. R101-R112.
30. Brizhik, L., et al., *The role of electromagnetic potentials in the evolutionary dynamics of ecosystems.* Ecological Modelling, 2009. **220**: p. 1865-1869.
31. Pietak, A., *Endogenous electromagnetic fields in plant leaves: a new hypothesis for vascular pattern formation.* Electromag Biol Med, 2011. **30**(2): p. 93-107.

Electric field generated by longitudinal axial microtubule vibration modes with high spatial resolution microtubule model

M Cifra[1], D Havelka[2], M A Deriu[3]

[1] Institute of Photonics and Electronics, Academy of Sciences of the Czech Republic, Prague, Czech Republic

[2] Department of Electromagnetic Field, Faculty of Electrical Engineering Czech Technical University in Prague, Prague, Czech Republic

[3] Industrial Bioengineering Group, Department of Mechanics, Politecnico di Torino, Torino, Italy

E-mail: cifra@ufe.cz

Abstract. Microtubules are electrically polar structures fulfilling prerequisites for generation of oscillatory electric field in the kHz to GHz region. Energy supply for excitation of elasto-electrical vibrations in microtubules may be provided from GTP-hydrolysis; motor protein-microtubule interactions; and energy efflux from mitochondria. It recently was determined from anisotropic elastic network modeling of entire microtubules that the frequencies of microtubule longitudinal axial eigenmodes lie in the region of tens of GHz for the physiologically common microtubule lengths. We calculated electric field generated by axial longitudinal vibration modes of microtubule, which model is based on subnanometer precision of charge distribution. Due to elastoelectric nature of the vibrations, the vibration wavelength is million-fold shorter than that of the electromagnetic field in free space and the electric field around the microtubule manifests rich spatial structure with multiple minima. The dielectrophoretic force exerted by electric field on the surrounding molecules will influence the kinetics of reactions via change in the probability of the transport of charge and mass particles. The electric field generated by vibrations of electrically polar cellular structures is expected to play a role in biological self-organization.

1. Introduction

Several experimental works have reported the electrical oscillations of living cells in the frequency region kHz-GHz [1–3]. These works have been motivated by the theoretical analyses indicating that the electrodynamic field is generated by the living cells and that it can have role in organization of intracellular processes [4,5] and interaction between the cells [6,7]. The proposed mechanisms for the organizational role of cellular electrodynamic field lies in the influence of the electrodynamic field force on the probability of the spatial distribution of biomolecules and organelles [4,5,8].

2. Microtubule

The fundamental question is which cellular structures are capable of generation of electrodynamic field. Such structures must fulfil following necessary conditions:

- have sufficient energy supply
- have an ability to convert the input energy to electrodynamic field (either by elastoelectric vibrations or by electronic conduction)
- sufficiently low damping of process which generates the electrodynamic field

Currently, microtubules seem to be the cellular structures which fulfil above mentioned conditions [9]. Microtubules are dynamic structures present in every eukaryotic cell.

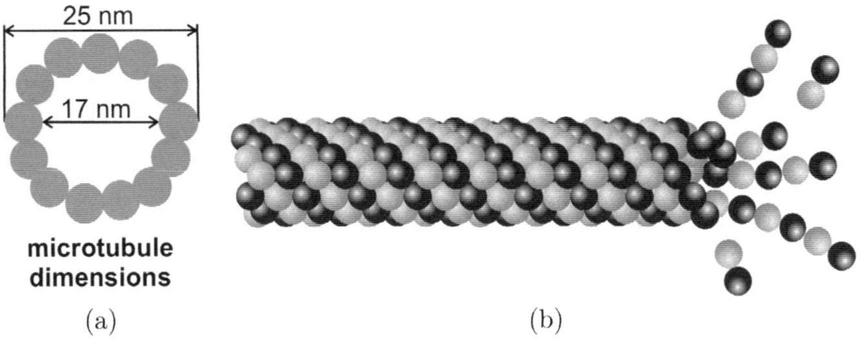

Figure 1. (a) Cross section of microtubule. (b) Detail of chemical plus end of a microtubule undergoing dynamic instability.

Energy may be provided to microtubules from three major sources [1, 8, 9]:

- Energy released from mitochondria as "wasted" energy in the course of citric acid cycle may be the most significant energy source for excitations of MT vibrations.
- Energy released from hydrolysis of guanosine triphosphate (GTP) in processes of dynamic instability of MTs is the energy supply dependent on internal MT process.
- Energy transferred from moving of motor proteins and their interaction with MTs may also contribute to excitation of vibrations in MTs

Once the energy is supplied to microtubule, amplitude of microtubule vibration modes is increased. The problem of the energy spectral distribution (dissipation or condensation) will not be treated here, we refer interested reader to [10, 11]. Here we assume underdamped eigenmode vibrations of the microtubule.

Microtubule is electrically polar structure due to the polarity of its subunits, tubulin heterodimer proteins. Due to the intrinsic electric polarity of microtubules, vibrations of the microtubule lattice will be accompanied by oscillatory electric fields due to the displacement of charges. Vibrations and accompanying oscillatory electric fields are of the same frequency in the first approximation. These type of vibrations can be viewed upon as elastoelectric (electroelastic), acoustoelectric (electroacoustic) [12] or elasto-polarization [13, 14] waves.

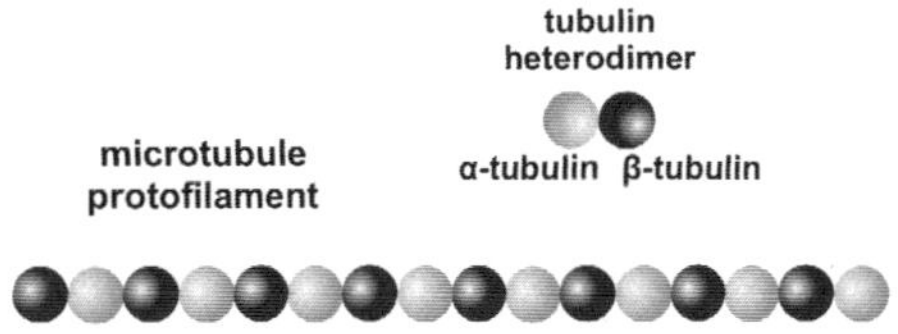

Figure 2. Structural subunits of microtubules: protofilament composed of tubulin heterodimers

Microtubules, treated as tube like orthotropic structures, are able to oscillate from kHz - GHz region [15, 16]. Longitudinal (axial) modes are expected to have the lowest viscous damping due to smallest displacement of surrounding water, which is present *in vivo*, and viscoelastic transition of cellular water [17].

Hypothetically, electric oscillations could be generated by microtubule also in case of charge oscillations based on charge conduction. Special frequency dependent type of ballistic-like electric conduction at specific frequencies in MHz region has been demonstrated in microtubules only very recently [18]. However, it is not described in details yet in any known publication. We will not include possible microtubule charge conduction properties in our model since we describe different frequency region.

In this paper we provide preliminary study of electrodynamic field generated by electrically polar longitudinal axial vibrational modes of microtubules based on the subnanometer resolution of microtubule charge distribution.

3. Materials and Methods

3.1. Subnanometer microtubule charge distribution

Model of microtubule lattice with subnanometer resolution is based on crystallography data of tubulin heterodimers from Nogales et al. [19] (1TUB on RCSB Protein DataBank) and on refined crystallography data of tubulin from Löwe et al. [20] (1JFF on RCSB Protein DataBank). Tubulin heterodimers with subnanometer resolution have been assembled to the 13:3 B-microtubule lattice. Model of microtubule with the length of 1200 nm has been created. This length corresponds to 150 tubulin heterodimers in every microtubule protofilament. Microtubule structure has been refined in order to obtain physiologically realistic lattice. Details of the optimization and refinement procedure of microtubule lattice can be found in [21]. Each atomic group with its coordinates in microtubule has been assigned a charge using Gromos 53a6 force field.

3.2. Dispersion relation

Dispersion relation of stretching (longitudinal axial) mode is based on data from normal mode analysis of coarse grained elastic network model of microtubules published in [21]. The dispersion relation is depicted in Figure 3a.

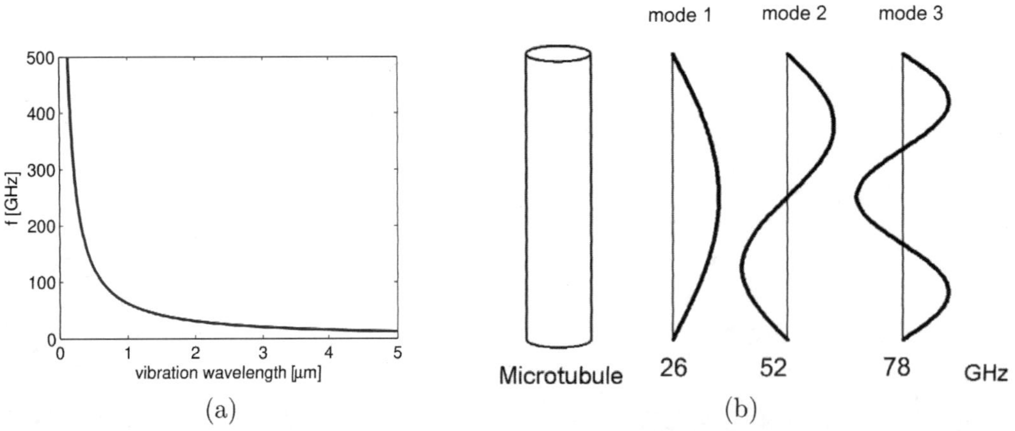

Figure 3. (a) Dispersion relation of longitudinal axial (stretching) vibration modes of microtubules (b) Distribution of vibrational amplitude along the microtubule axis for first three longitudinal axial vibration modes

Journal of Physics: Conference Series **329** (2011) 012013 doi:10.1088/1742-6596/329/1/012013

3.3. Microtubule vibrations and dipole representation

We consider harmonic longitudinal vibrations of microtubules with fixed ends, i.e. vibration amplitude of the first and the last heterodimer is zero. Direction of the vector of the oscillations of the individual charges is parallel to the axis of microtubule. Amplitude of the charge oscillatory motion is same in the cutting plane perpendicular to the microtubule axis and is given by the shape of the longitudinal eigenmode, see e.g. Figure 3b. Electric field generated by the oscillatory motion of every charge is approximated by the electric field of the dipole with the same frequency as microtubule vibrations. Total field is calculated as vector sum of contributions from all dipole fields.

4. Results

Figure 4 and 5 shows microtubule in the coordinate system and depiction of the planes where the electric field has been calculated in following figures.

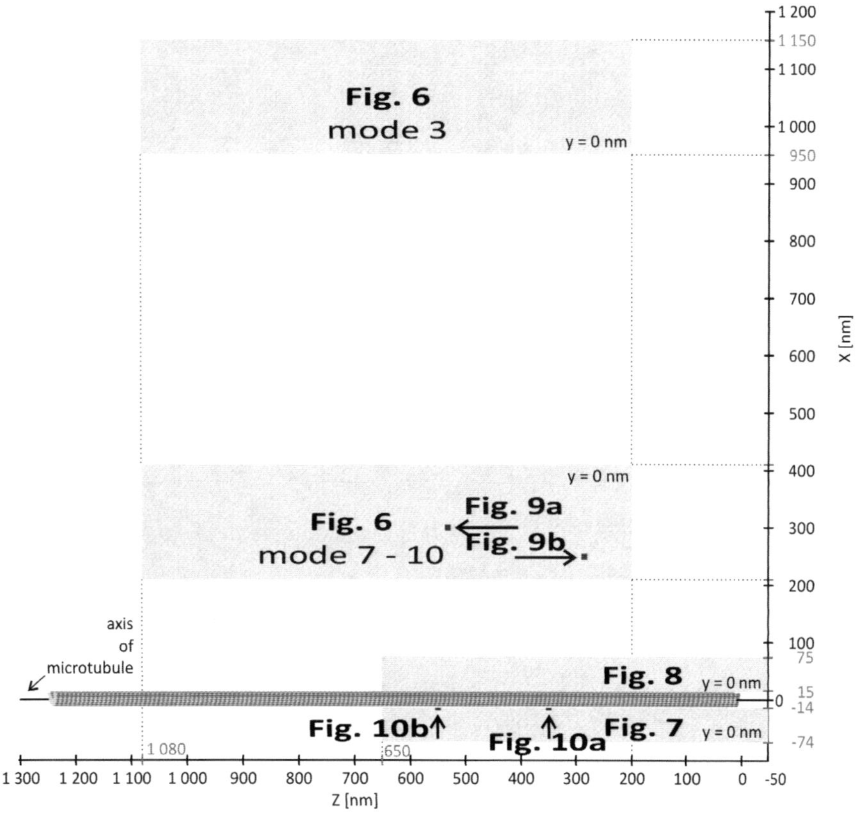

Figure 4. Microtubule in the coordinate system and depiction of the planes where the electric field has been calculated

5. Discussion and conclusion

We calculated electric field generated by collective longitudinal axial vibration modes of microtubule. Every charge involved in the vibrations generates dipole field. Total field is given by the vector sum of fields form all dipoles. We see that electric field of microtubule exhibits

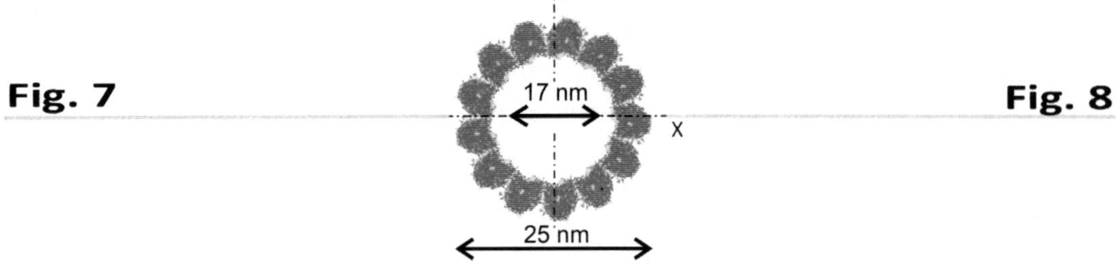

Figure 5. Microtubule in the coordinate system and depiction of angular position of the planes where the electric field has been calculated

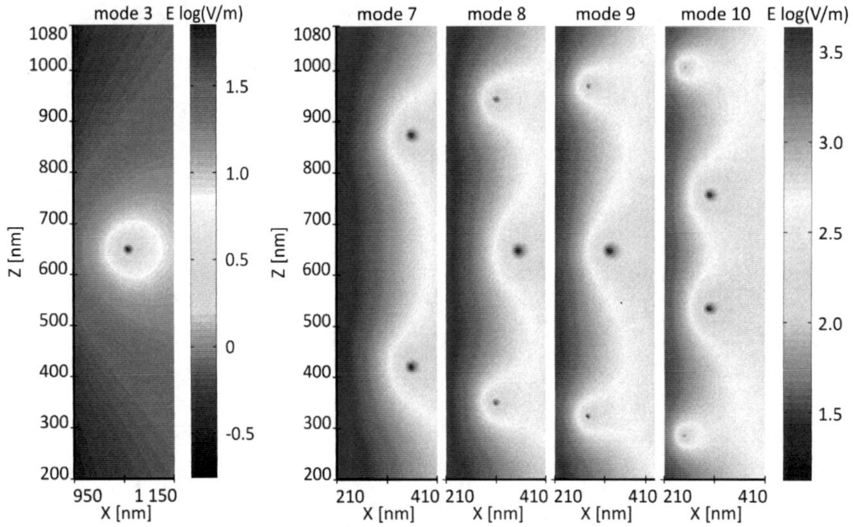

Figure 6. Electric field generated by several microtubule vibrations modes

rich spatial pattern. We can distinguish several levels of the field patterns. First, we can identify local minima of electric field in Figure 6. High resolution calculations in the region of selected field minima generated by the mode 10 are in the Figure 9.

The higher is the mode number, the more local minima occur in the field. These local field minima are caused by the dipole-like fields created by vibrational half-wavelengths as is described in following text. There is one dipole-like field per vibrational half-wavelength created. One can see the dipole-like fields in the Figures 7 and 8. There is high intensity of the electric field at the poles of these dipole-like fields. Neighbour dipole-like fields oscillate with opposite phase and the fields of opposite orientation compensate in certain distance and create local minimum of electric field. Those minima are visible in Figures 6 and 9. Distance of the field minima from the microtubule wall is related to the vibration wavelength. The shorter is the vibrational wavelength (i.e. the higher is the mode number), the closer are located the field minima to the surface of the microtubule wall. It should be emphasized here that this kind of field structure with minima located several tens to several hundreds nanometers from each other is possible

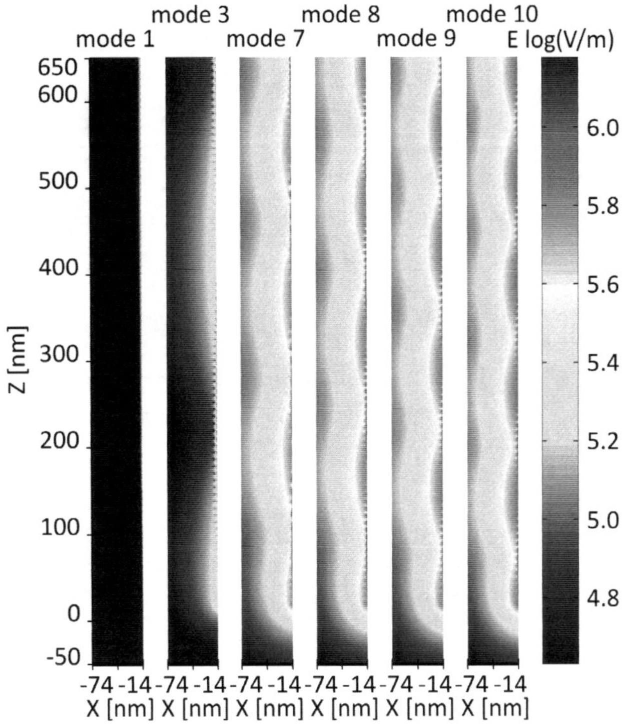

Figure 7. Electric field structure very near to the microtubule surface on the one side of microtubule

due to the elastoelectric nature of the oscillations at this frequency range. Elastic component of the microtubule vibrations propagates approximately with the sound velocity. In the frequency range of longitudinal vibrations of microtubules (tens and hundreds of GHz), wavelength of vibrations will be of the order of hundreds or tens of nanometers.

In the case of multiple microtubules in mutual vicinity (common case within the cell) the generated fields will interfere and create further local maxima and minima of the field. One can easily imagine that the field of the whole microtubule network will have rich three dimensional structure. This kind of field structure seems to be capable of creating field minima and maxima with sufficient spatial resolution (tens of nanometers) to contribute to the organization of biological intracellular processes.

One of the possible mechanisms of action of the cellular field on its internal processes may be dielectrophoresis. Dielectrophoresis [22] is a phenomenon in which nonuniform electric field acts on neutral particle by force via polarization effects. The physical mechanism is following: electric field of an external source acting on a dielectric particle induces separation of charges and exerts forces on them. The forces have opposite directions and in a nonuniform electric field they have also different magnitudes resulting in a non-zero force acting in the direction of the gradient of the intensity of the electric field. Additional force effect on the particle is exerted by the electric field generated by the surrounding polarized medium. The final effect, i.e. motion of the particle depends on the superposed electric fields. The particle is pulled to the region of the highest intensity of the electric field if the polarizability (which depends on permittivity) of the particle is greater than that of the surrounding medium and vice versa in the opposite case.

Journal of Physics: Conference Series **329** (2011) 012013 doi:10.1088/1742-6596/329/1/012013

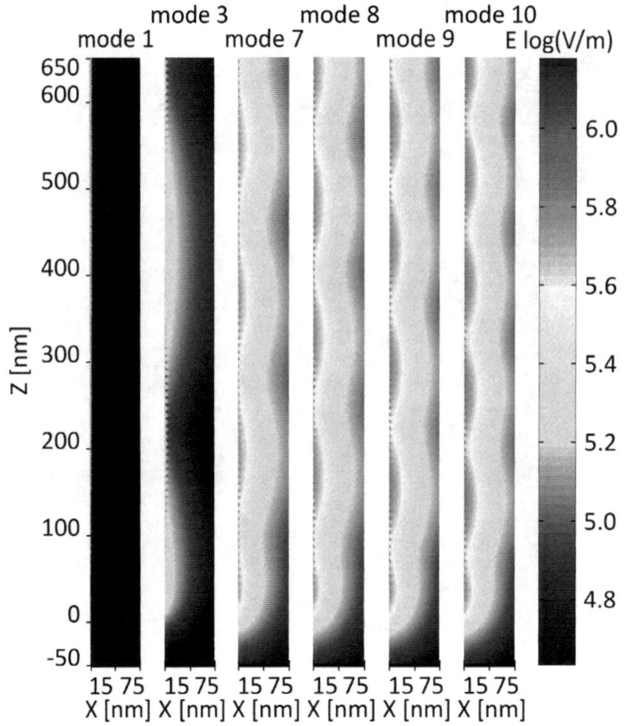

Figure 8. Electric field structure very near to the microtubule surface on the side of microtubule opposite to the that in Figure 7

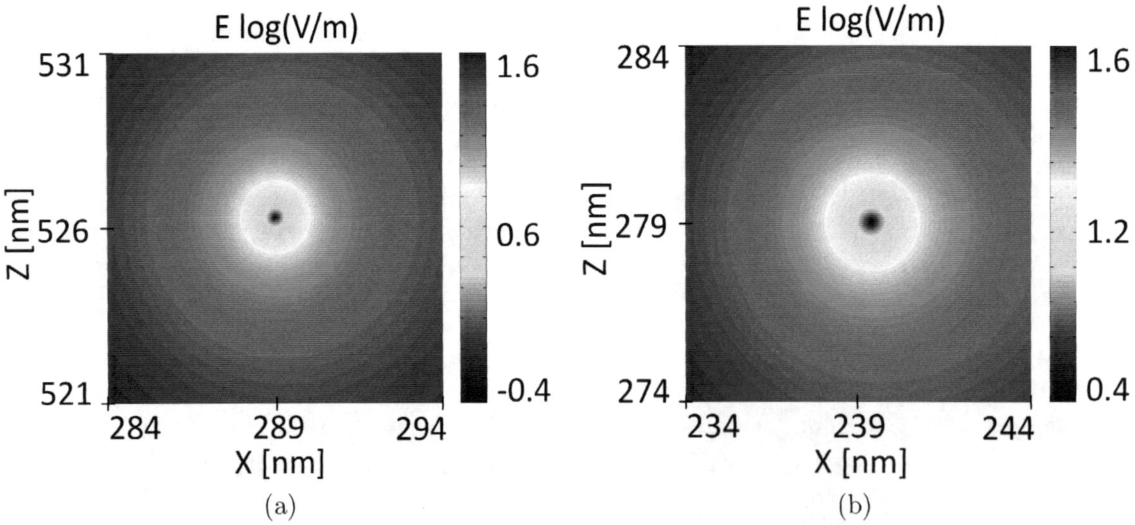

Figure 9. Local electric field minima of mode 10 calculated with 0.1 nm resolution

Underlying physical principle is the same as that of an optical tweezer.

Thus, even neutral particles and molecules can be acted upon by the dielectrophoretic force

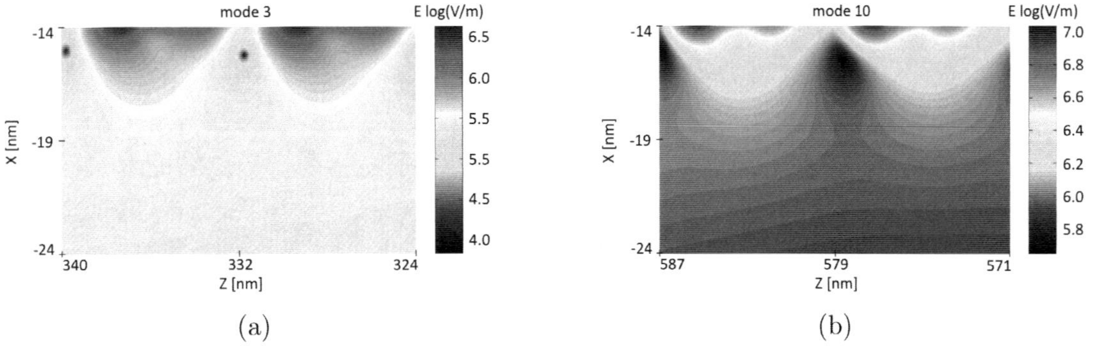

Figure 10. Local electric field minima calculated with 0.1 nm resolution near the microtubule wall

of the non-uniform electric field generated by elastoelectric vibrations of cellular polar structures such as microtubules. The force can influence the probability of the position of the molecules and organelles and influence the chemical kinetics in this manner.

We can conclude that the electric field generated by vibrations of electrically polar cellular structures is expected to play a role in biological self-organization. To what extent needs to be verified by the direct measurement of the electric oscillations of the cells which still remains a technical challenge.

Acknowledgements
This research is supported by the Czech Science Agency, projects P102/10/P454: "Measurement and Analysis of Local Electric Oscillations and Optical Emission of Biomolecular Nanostructures", 102/08/H081: "Non Standard Applications of Physical Fields" and by project P102/11/0649: "Research and measurements of signals generated by nanostructures."

References
[1] Jelínek F, Cifra M, Pokorný J, Hašek J, Vaniš J, Šimša J and Frýdlová I 2009 *Electromagnetic Biology and Medicine* **28** 223 – 232
[2] Pokorný J, Hašek J, Jelínek F, Šaroch J and Palán B 2001 *Electro- and Magnetobiology* **20** 371–396
[3] Jelínek F, Šaroch J, Kučera O, Hašek J, Pokorný J, Jaffrezic-Renault N and Ponsonnet L 2007 *Radioengineering* **16** 36–39
[4] Pokorný J, Hašek J and Jelínek F 2005 *Electromagnetic Biology and Medicine* **24** 185–197
[5] Pokorný J, Jelínek F, Trkal V, Lamprecht I and Hölzel R 1997 *Journal of Biological Physics* **23** 171–179
[6] Pokorný J 1994 *Czechoslovak Journal of Physics* **44** 67–78
[7] Pokorný J and Fiala J 1994 *Neural Network World* **94** 299–313
[8] Cifra M 2009 *Study of electromagnetic oscillations of yeast cells in kHz and GHz region* Ph.D. thesis Czech Technical University in Prague
[9] Cifra M, Pokorný J, Havelka D and Kučera O 2010 *BioSystems* **100** 122 – 131
[10] Pokorný J and Wu T M 1998 *Biophysical Aspects of Coherence and Biological Order* (Academia, Praha, Czech Republic; Springer, Berlin - Heidelberg - New York)
[11] Reimers J R, McKemmish L K, McKenzie R H, Mark A E and Hush N S 2009 *Proceedings of the National Academy of Sciences* **106** 4219–4224 (*Preprint* `http://www.pnas.org/content/106/11/4219.full.pdf+html`) URL `http://www.pnas.org/content/106/11/4219.abstract`
[12] Kharlanov A V 2006 *Acoustoelectric vibrations of cell, in Russian* Akustoelektricheskie kolebaniya kletki Ph.D. thesis Volograd State Technical University
[13] Fröhlich H 1968 *Physical Letters A* **26** 402–403
[14] Fröhlich H 1973 *Collective Phenomena* **1** 101–109

[15] Ghavanloo E, Daneshmand F and Amabili M 2010 *Physica E: Low-dimensional Systems and Nanostructures* **43** 192–198
[16] Wang C, Li C and Adhikari S 2009 *Journal of biomechanics* **42** 1270–1274
[17] Pokorný J, Vedruccio C, Cifra M and Kučera O 2011 *European Biophysics Journal* **40** 747–759 ISSN 0175-7571
[18] Bandyopadhyay A 2011 *9th International Fr¨ohlichs Symposium ELECTRODYNAMIC ACTIVITY OF LIVING CELLS Including Microtubule Coherent Modes and Cancer Cell Physics* p 20
[19] Nogales E, Wolf S G and Downing K 1998 *Nature* **391** 199–203
[20] Löwe J, Li H, Downing K H and Nogales E 2001 *Journal of Molecular Biology* **313** 1045–1057
[21] Deriu M A, Soncini M, Orsi M, Patel M, Essex J W, Montevecchi F M and Redaelli A 2010 *Biophysical Journal* **99** 2190 – 2199 ISSN 0006-3495 URL `http://www.sciencedirect.com/science/article/B94RW-515K8MN-V/2/b04f67bbda5123dade8927db98c5cb93`
[22] Pohl H A 1978 *Dielectrophoresis* (Cambridge Univ. Press, London)

Journal of Physics: Conference Series **329** (2011) 012014 doi:10.1088/1742-6596/329/1/012014

What is more important for radiated power from cells - size or geometry?

D Havelka[1], M Cifra[2], J Vrba[1]

[1]Department of Electromagnetic Field, Faculty of Electrical Engineering CTU in Prague, Prague, Technicka 2, 16627, Czech Republic

[2]Institute of Photonics and Electronics AS CR, v.v.i. Chaberska 57, 182 51 Prague, Czech Republic

E-mail: haveldan@fel.cvut.cz, cifra@ufe.cz, vrba@fel.cvut.cz

Abstract. The cytoskeleton, especially microtubules, is potential source of electrodynamic field of living eukaryote cells. Microtubule network is a very dynamic structure which is composed of highly polar molecules - tubulin heterodimers. Microtubules have their eigenmode vibrations in frequency range from kHz to GHz. We approximated electrical properties of tubulin heterodimer as an elementary electric dipole and described the longitudinal mechanical oscillation of one microtubule by spatial modulation function. Mechanical oscillations of tubulin heterodimers were approximated as a chain of rigid particles and described by the spatial modulation function. An optical branch was used as it is more efficient for generation of electromagnetic field. We also created asymmetrical model of microtubule network of dividing cell and symmetrical model of microtubule network of non-dividing cell. The field around oscillating microtubule was calculated as a vector sum of contributions from all elementary electric dipoles. Finally, we performed calculations with several cell sizes and numbers of microtubules in models and performed comparative analysis.

1. Introduction

Microtubules are cellular structures assumed to be the generators of electrodynamic field [1, 2, 3]. Several experimental works have reported the electrical oscillations of living cells in the frequency region kHz-GHz [4, 5, 6]. Although a theoretical analysis of radiated power from oscillators located in cellular membrane [1], and simple microtubule network [7] has been performed, no works reported parametric analysis of the the dependence of radiated electromagnetic power on the cell size. Yet it is clear that these kind of theoretical predictions are necessary to estimate required dimensions and sensitivity of the sensors for the measurement of cellular electrodynamic activity. In this paper, we provide first results from calculations of radiated electromagnetic power from symmetric radial distribution of microtubule network and the distribution of the microtubules during the cell division for several sizes of a cell.

2. Approximation of microtubule

Microtubules are mostly composed of 13 protofilaments. The basic subunit of one protofilament is tubulin heterodimer. Tubulin heterodimer contains two parts $\alpha-$ and $\beta-$ tubulin dimer and has strong static dipole moment of 337 D (1.12 x 10^{-27}Cm) in direction of microtubule axis (Figure 1b). Microtubule network changes structure during the cell cycle. Mitotic spindle is created during mitosis. This structure of microtubule network ensures accurate distribution of genetic information (Figure 2). Increased intensity of electric field has been experimentally observed around dividing cells [8, 5].

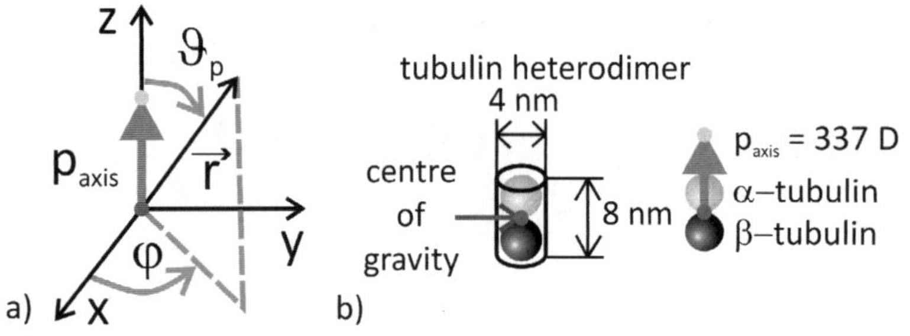

Figure 1. a) Elementary electric dipole in Cartesian coordinate system, b) Approximation of electrical properties of tubulin heterodimer

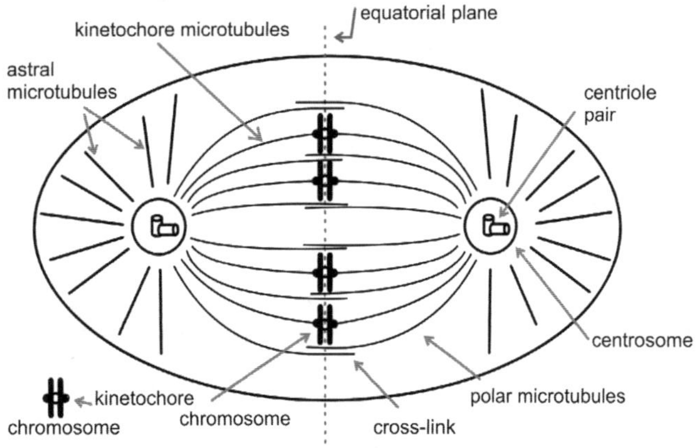

Figure 2. Mitotic spindle

Electric properties of tubulin heterodimer were approximated by an elementary electric dipole. Field of such dipole is described by these three equations 1 - 3.

$$H_\varphi = -\frac{Idl}{4\pi}k^2 sin(\vartheta_p)(\frac{1}{jkr}+\frac{1}{(jkr)^2})e^{-jkr} \tag{1}$$

$$E_r = -\frac{Idl}{4\pi}2Zk^2 cos(\vartheta_p)(\frac{1}{(jkr)^2}+\frac{1}{(jkr)^3})e^{-jkr} \tag{2}$$

$$E_\vartheta = -\frac{Idl}{4\pi}Zk^2 sin(\vartheta_p)(\frac{1}{jkr}+\frac{1}{(jkr)^2}+\frac{1}{(jkr)^3})e^{-jkr} \tag{3}$$

where E is electric field intensity, H is magnetic field intensity, I is equivalent current, dl is length of the dipole, Z is wave impedance, k is propagation constant, ω is angular frequency, p is dipole moment, j is imaginary unit (j^2 = -1) and other symbols are according to Figure 1a.

All equations include:

$$Id\vec{l} = j\omega\hat{p} \tag{4}$$

3. Models of microtubule network

We created two types of models of microtubule network in program Matlab - symmetric and asymmetric model. The symmetric model is model of non-dividing cell (Figure 3). This model has symmetric and uniform distribution of microtubules on sphere [1]. In symmetric model,

[1] Sphere is model of centrosome with diameter 200 nm

There is the same length and number of elementary electric dipoles in each microtubule. Each size of symmetric model of microtubule network has different number of microtubules. The second model of microtubule network is a model of mitotic spindle. The polar and kinetochore microtubules have different length and different numbers of elementary electric dipoles. Astral microtubules have the same length and number of elementary electric dipoles. Each size of asymmetric model of microtubule network has the same number of microtubules - 100 astral microtubules (50 from each centrosome) and 200 polar and kinetochore microtubules (100 from each centrosome). The symmetric and asymmetric models are composed of two types of points - green and red. The red point represents the centre of gravity of one tubulin heterodimer and the green point describes the direction of dipole moment [2] (Figure 1 and 5).

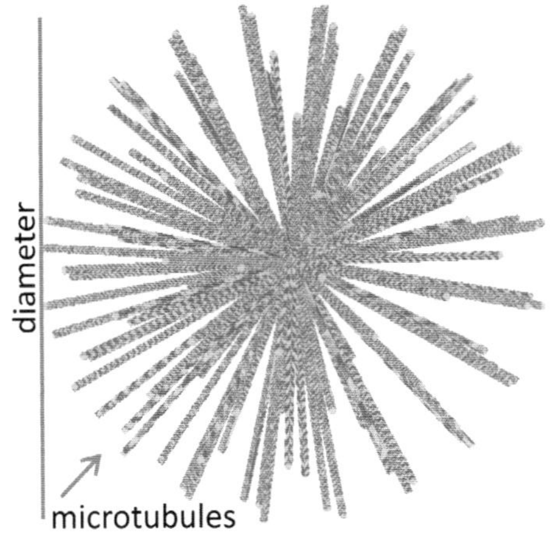

Figure 3. Model of symmetric microtubule network. Each coloured line represents one microtubule.

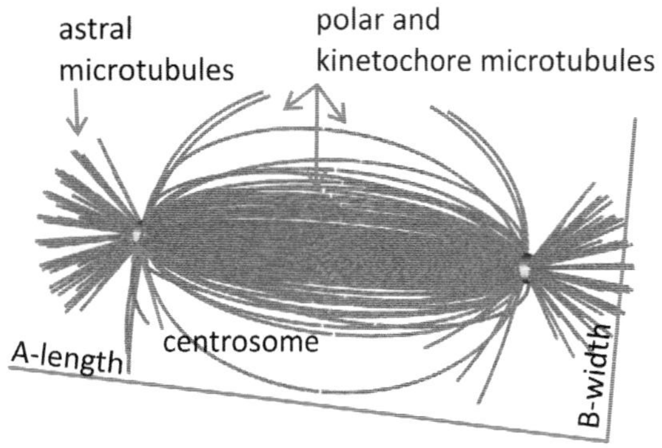

Figure 4. Model of mitotic spindle, asymmetric model.

The dimensions, numbers of microtubules and numbers of elementary electric dipoles in models of four sizes of microtubule networks are in the tables 2 and 1. Comparison of sizes of models are in Figure 6.

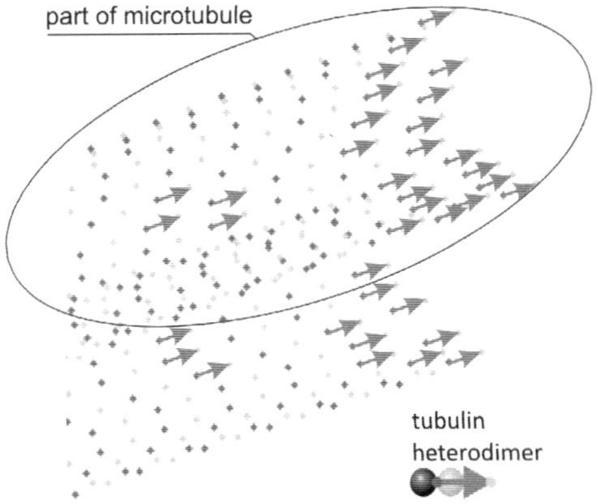

Figure 5. Detail of models of microtubule network. Red points represent centres of gravity of tubulin heterodimers and green points describe the direction of dipole moment of tubulin heterodimers.

[2] the end point of dipole moment of one tubulin heterodimer

Table 1. Parameters of models of symmetric microtubule network. HET - heterodimer, PT - protofilament

model NO.	diameter [μm]	MTs in one model	HETs in one PT	HETs in one model
1)	3	121	175	275 275
2)	5	116	300	452 400
3)	7	114	425	629 850
4)	10	158	613	1 259 102

Table 2. Parameters of models of asymmetric microtubule network. HET - heterodimer, PT - protofilament, K - kinetochore, P - polar

model NO.	A [μm]	B [μm]	K+P MTs in model	HETs in K+P MTs	astral MTs	HETs in 1 astral PT	HETs in 1 astral MT	HETs in model
5)	4.28	1.00	200	448 058	50+50	80	104 000	552 058
6)	6.88	1.67	200	730 808	50+50	134	174 200	905 008
7)	9.48	2.33	200	1 014 078	50+50	188	244 400	1 258 478
8)	17.85	6.69	200	2 162 862	50+50	269	349 700	2 512 562

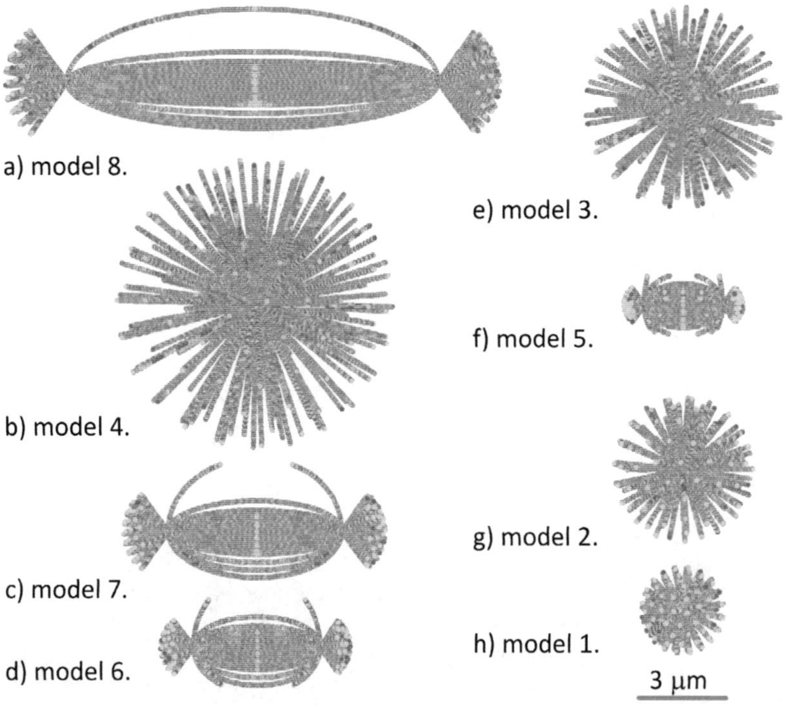

Figure 6. Comparison of sizes of models of microtubule networks sorted by number of elementary electric dipoles in model from a) to h). The models are in scale.

4. Parameters of calculations

We used two different surrounding media: lossy and lossless (physiological solution and lossless medium with permittivity of water). The parameters of surrounding media are in Table 4 (ε_r

Table 3. Parameters of surrounding medium around models of microtubule networks for 8 frequencies. ε_r and σ have subscript 1 for lossy medium and subscript 2 for lossless medium.

f [Hz] (*frequency*)	1 k, 1 M 10 M, 100 M	1 G	10 G	42 G	100 G
$\varepsilon_{r1}[-]$ (relative permittivity)	81	81	71	20	10
$\sigma_1[S/m]$ (conductivity)	1	1.33	17.22	39.67	68.50
$\varepsilon_{r2}[-]$ (relative permittivity)	81	81	71	20	10
$\sigma_2[S/m]$ (conductivity)	0	0	0	0	0

and σ have subscript 1 for lossy medium and subscript 2 for lossless medium). We introduced a longitudinal shift between neighbouring protofilaments 4.92 nm (corresponds to lattice A). Mechanical longitudinal oscillation of tubulin heterodimers in protofilament is represented by oscillations of dipole moments in direction of microtubule axis ($p_{static,axis} = 337D$ static part of one dipole moment). The amplitude of kHz oscillations is estimated: $p_{oscil\_1} = \frac{1}{8}p_{static,axis}$. Pelling [9, 10] measured oscillations of yeast cell wall in range 1 - 10 nm in kHz region, which was attributed to interaction of motor proteins with cytoskeleton. Therefore we used 1 nm amplitude for the microtubule oscillations in kHz region. In frequency region from MHz to GHz we calculated with amplitude of oscillation: $p_{oscil\_2} = \frac{0.1}{8}p_{static,axis}$. This value is assumed to be of the level of thermal motion amplitude of proteins [11].

Oscillations of tubulin heterodimers in one protofilament were approximated by longitudinal oscillations of the chain of rigid two type particles, optical branch ([12]). Optical branch is able to generate and absorb electromagnetic field more effectively than acoustic branch. Movement of particles (tubulin heterodimers) and corresponding oscillations of the dipole moment are described by spatial modulation function:

$$p_{mod} = p_{oscil} \cdot \sin(dis) \tag{5}$$

where p_{mod} is the modulated dipole moment, p_{oscil} is the oscillatory component of the dipole moment and dis is basically distance along the distance in direction of microtubule axis. Modulation function is specific for each mode, minimal mode number is zero, maximal mode number (max) depends on the number of tubulin heterodimers, i.e. length of microtubule. The mode zero represents oscillations of all dipoles in phase and the max mode represents oscillations of neighbouring dipoles in antiphase. Our latest works showed that mode zero is more effective for generation electromagnetic field than higher modes. We assumed that all microtubules oscillate at mode zero in calculations in this paper.

The calculation of power around models was realised as a vector sum of contribution from all subunits (elementary electric dipoles) in one point of field of evaluation (PoFE). We repeated this calculations for all PoFEs. These PoFEs are distributed according to distribution of charges on the spherical surface with the condition of the smallest potential energy. The sphere has centre in the middle of model (Figure 7). In every PoFEs we calculate a radial part of Poynting vector - S and then power is

$$P = \sum_{j}^{np} S_j \cdot A_{PoFE} \tag{6}$$

$$A_{PoFE} = \frac{4\pi r^2}{np} \tag{7}$$

where P is power from one PoFE, A_{PoFE} is surface corresponding to one PoFE, r is radius sphere PoFEs, np number of PoFEs (in our calculations we used 81). We calculate also the electric intensity in planes parallel to axis of Cartesian coordinate system.

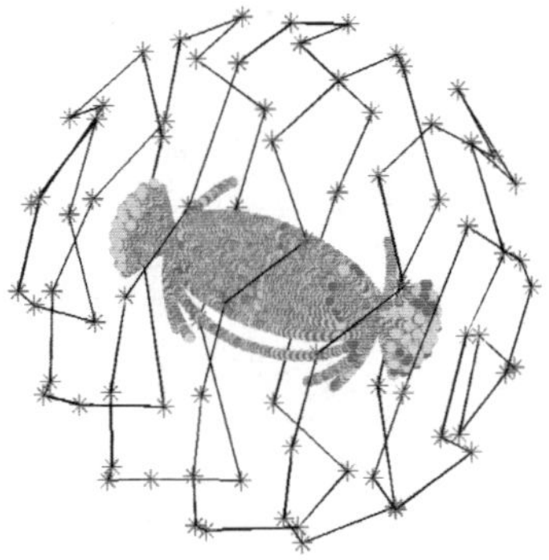

Figure 7. PoFEs - points of field evaluation (red dots) around model of mitotic spindle.

5. Results

We present selected results of calculations of radiated power and electric intensity around eight models of microtubule networks. In calculations we used approximation of electrical properties of tubulin heterodimer by elementary electric dipole (equations: 1 - 3). The protofilaments oscillate in mode 0 which is the most effective mode for generation of electromagnetic field (all dipoles oscillate in phase). The frequencies of oscillations were considered to be from kHz to GHz region. Lossy and lossless medium was considered.

Figures 8 and 9 depict comparison of radiated power from eight models of microtubule networks.

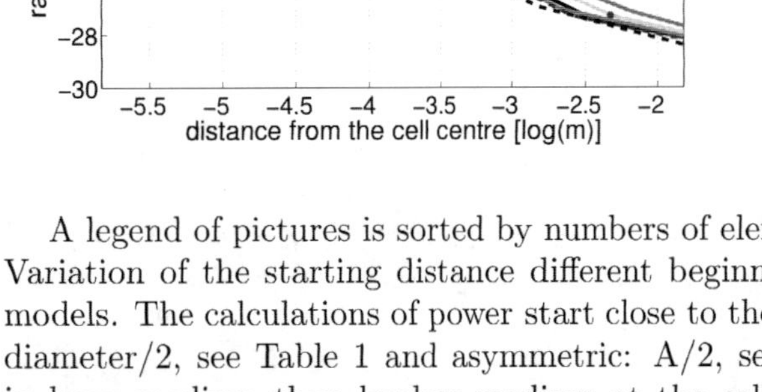

Figure 8. P $[\log_{10}(\mathrm{W})]$ - power generated by models of microtubule networks. Comparison of models. Oscillation frequency is 1 GHz, lossy medium and mode 0.

A legend of pictures is sorted by numbers of elementary electric dipole of models (Figure 6). Variation of the starting distance different beginning of curves is caused by different sizes of models. The calculations of power start close to the border of microtubule network (symmetric: diameter/2, see Table 1 and asymmetric: A/2, see Table 2). The calculated power is higher in lossy medium than lossless medium at the cell wall but decreases faster with distance in radial direction from the model. This effect is caused by phase asymmetry between electric and magnetic field component in lossy medium. The phase shift is $\pi/2+\phi_k$ in lossy medium and $\pi/2$ in lossless medium (ϕ_k is a phase of propagation constant k of electromagnetic field). The phase shift in lossy medium causes that the product of the near field components ($E \sim \frac{1}{(jkr)^3}$ and

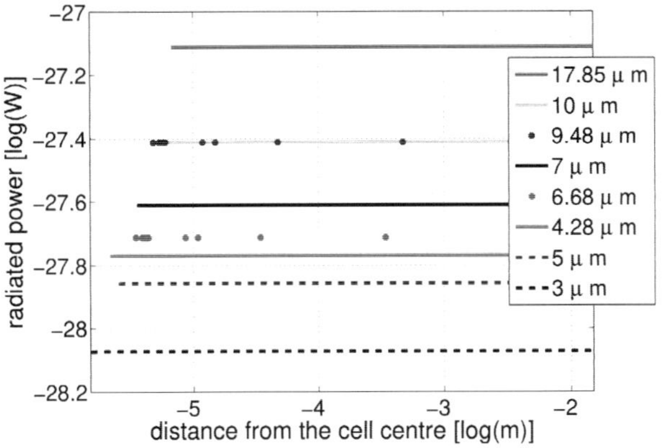

Figure 9. P $[\log_{10}(W)]$ - power generated from models of microtubule networks. Comparison of models. Oscillation frequency is frequency 1 GHz, lossless medium and mode 0.

$H \sim \frac{1}{(jkr)^2}$) is not zero and therefore contributes to the radiation energy in the radial direction. We use only far field components ($E \sim \frac{1}{(jkr)}$ and $H \sim \frac{1}{(jkr)}$) for the calculation of radiated power in lossless medium since the near field components cancel out perfectly. The Figure 9 shows radiated power in dependence on radial distance from the models. The difference between the value of power of models 1 and 8 is one order of magnitude. Model 8 is composed of 10 times more elementary electric dipole than model 1. This dependence is the same for other excitation frequencies. The dependence of excitation frequency on radiated power is depicted in Figure 10. The higher is the frequency, the higher is power in radial direction. Steeper decrease of power is observed in the distance where the field loses its near field character.

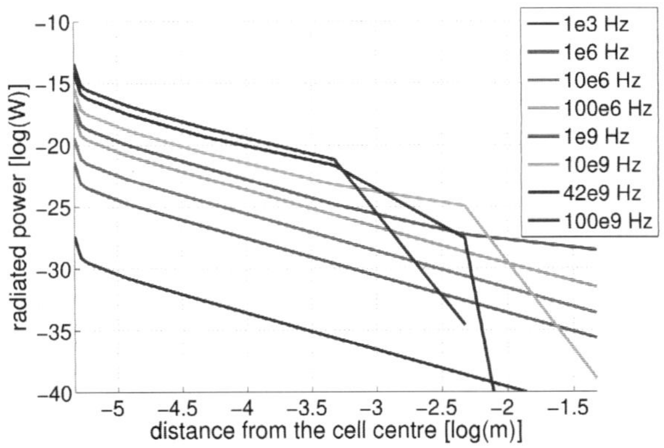

Figure 10. P $[\log_{10}(W)]$ - power generated from model of microtubule network (model 8). Comparison of excitation frequencies. Lossy medium and mode 0.

Figures 12 - 15 depict the electric intensity in logarithmic scale in three orthogonal sections through origin of the Cartesian coordinate system (centre of model) of model 5.

6. Discussion and conclusion

Comparison of calculated radiated power from the eight models gives answer to the question "What is more important for radiated power from cells - size or geometry?" For calculations based on our models, the correct answer is size of the cell. More specifically, the number of elementary electric dipoles (tubulin heterodimers) but the size corresponds to the mass of tubulin heterodimers in the cell. It should be noted here that geometrical and phase symmetry of radiating sources is expected to play a role in the total radiated power. Here we presented model with apparent geometrical symmetry (non dividing cell) and phase symmetry (all dipoles oscillate in phase). However, symmetry is not perfect due to the helicity of microtubules,

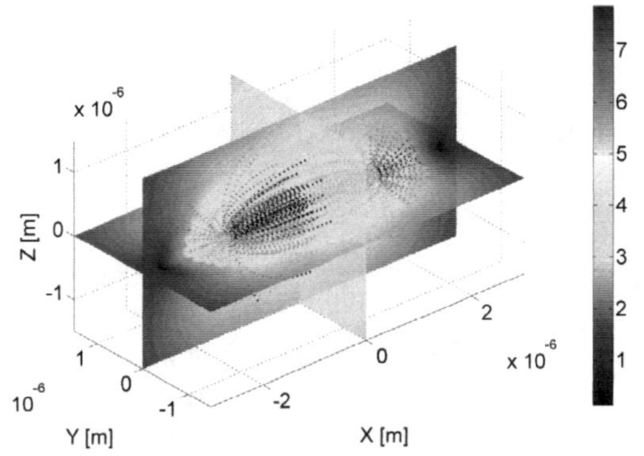

Figure 11. E [$\log_{10}$(V/m)] - electric intensity around model 5. Every tenth elementary electric dipole in protofilament is depicted in this model figure. Black points represent centre of gravity of tubulin heterodimer and green points show direction of dipole moment.

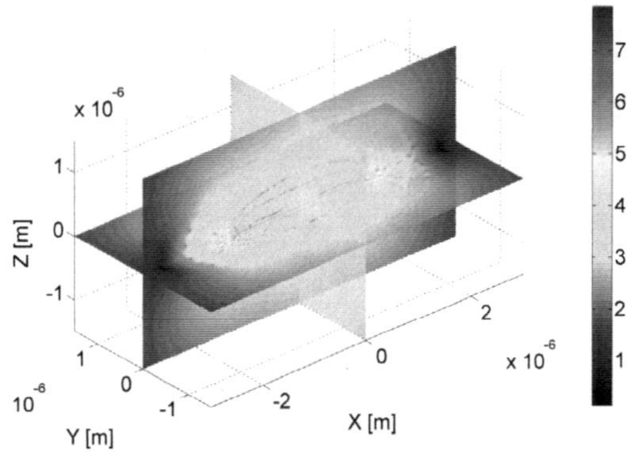

Figure 12. E [$\log_{10}$(V/m)] - electric intensity around model 5.

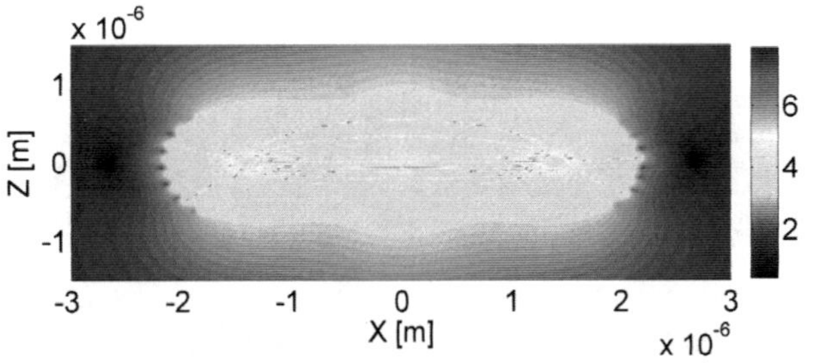

Figure 13. E [$\log_{10}$(V/m)] - electric intensity around model 5 (ZX - plane).

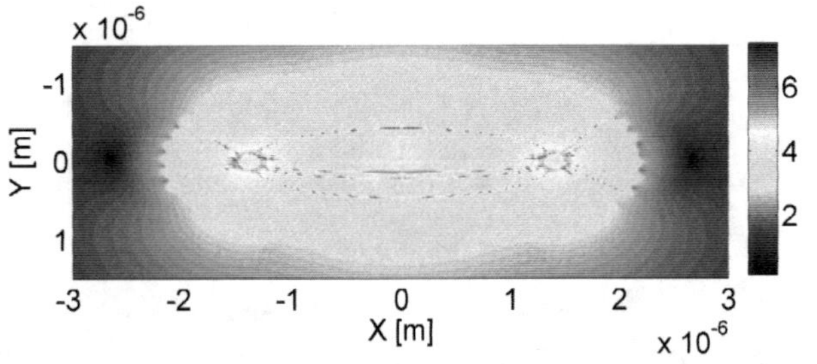

Figure 14. E [$\log_{10}$(V/m)] - electric intensity around model 5 (YX - plane).

therefore the total number of radiators (tubulin dipoles) has greater influence on the radiated power than the geometry of the microtubule network.

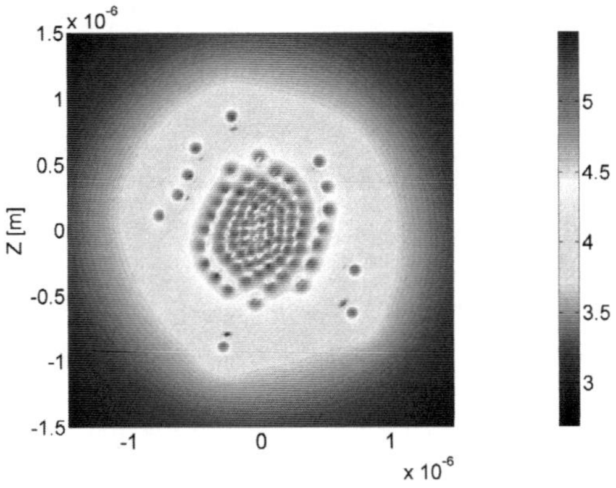

Figure 15. E [$\log_{10}$(V/m)] - electric intensity around model 5 (equatorial plane

Acknowledgements

This research is supported by the Czech Science Foundation, project P102/10/P454: "Measurement and Analysis of Local Electric Oscillations and Optical Emission of Biomolecular Nanostructures", 102/08/H081: "Non Standard Applications of Physical Fields" and by project P102/11/0649: Research and measurements of signals generated by nanostructures. Further it is supported also by research program MSM6840770012: Transdisciplinary Research in the Area of Biomedical Engineering II of the CTU in Prague, sponsored by the Ministry of Education, Youth and Sports of the Czech Republic and by CTU project SGS11/064/OHK4/1T/13.

References

[1] Pokorný J and Wu T M 1998 *Biophysical Aspects of Coherence and Biological Order* (Academia, Praha, Czech Republic; Springer, Berlin - Heidelberg - New York)
[2] Pokorný J 2004 *Bioelectrochemistry* **63** 321–326
[3] Pokorný J 1999 *Bioelectrochemistry and Bioenergetics* **48** 267–271
[4] Jelínek F, Cifra M, Pokorný J, Hašek J, Vaniš J, Šimša J and Frýdlová I 2009 *Electromagnetic Biology and Medicine* **28** 223 – 232
[5] Pokorný J, Hašek J, Jelínek F, Šaroch J and Palán B 2001 *Electro- and Magnetobiology* **20** 371–396
[6] Jelínek F, Šaroch J, Kučera O, Hašek J, Pokorný J, Jaffrezic-Renault N and Ponsonnet L 2007 *Radioengineering* **16** 36–39
[7] Havelka D, Cifra M, Kučera O, Pokorný J and Vrba J 2011 *Journal of Theoretical Biology* **286** 31 – 40 ISSN 0022-5193 URL `http://www.sciencedirect.com/science/article/pii/S0022519311003511`
[8] Pohl H A 1980 *International journal of quantum chemistry: Quantum biology symposium* **7** 411–431
[9] Pelling A E, Sehati S, Gralla E B, Valentine J S and Gimzewski J K 2004 *Science* **305** 1147–1150
[10] Pelling A E, Sehati S, Gralla E B and Gimzewski J K 2005 *Nanomedicine: Nanotechnol. Biol. Med.* **1** 178–183
[11] Chou K C 1988 *Biophysical chemistry* **30** 3–48
[12] Kittel C 1976 *Introduction to Solid State Physics* (John Wiley & Sons)

Charge Transfer in DNA: The Role of Large Polarons

S Zeković[1], S Zdravković[2] and Z Ivić[1]
[1] Vinča Institute of Nuclear Siences, Laboratory for Theoretical and Condensed Matter Physics-020, University of Belgrade, P.O. Box 522, 11001 Belgrade, Serbia.

[2] Vinča Institute of Nuclear Siences, Laboratory for Atomic and Plasma Physics-040, University of Belgrade, P.O. Box 522, 11001 Belgrade, Serbia.

E-mail: zekovic@vinca.rs

Abstract. Polaronic mechanism of charge transfer in DNA is considered on the basis of one-dimensional Su-Schrieffer-Heeger (SSH) model Hamiltonian. It is shown that usually used values of SSH model Hamiltonian parameters, especially transfer integral and electron-phonon coupling constant which are obtained in the framework of ab initio calculations, do not give reliable resolution of the problem of (large) polaron applicability for charge migration in DNA. Because of that, theoretical method is proposed which enables us to estimate explicitly electron-phonon coupling constant on the basis of transfer integral. The value of polaron width obtained for such determined parameters leads to the conclusion that coherent motion of large polaron represents dominant mechanism of charge transfer in DNA.

1. Introduction

Topic of charge transfer in DNA has been in the focus of researchers in various branches of science for a long time. It is the consequence of the fact that the results obtained by the inspection of this problem can be applied in physics, chemistry, biology, medicine, nanotechnology; for instance: damage, mutation and repair of DNA [1], nanoscale sensors for mutations or protein binding [2], nanotechnology aspects of single molecule conduction [3]. At the same time, there exist very conflicting and controversial experimental data concerning the conductivity in DNA. So, DNA was found to behave like metallic wire [4], semiconductor [5] and insulator [6]. The reasons for this are experimental complications in the measurements such as: chemical details of the environment, length and structure of DNA, impurity, temperature, water solution etc.

In order to interpret inconsistent experimental results, a number of charge transfer mechanisms have been proposed. During the past decade polaron motion was actively considered as a possible mechanism of charge transfer in DNA. This possibility was supported by the experimental findings [7, 8]. But, there is not yet clear answer on the question which kind of polarons (large polarons and coherent motion or small polarons and random hopping) is responsible for charge transport.

Our aim in this paper is to give a contribution in clarifying polaronic mechanism of charge migration in DNA.

2. SSH model Hamiltonian

DNA is too large and too complicated system for complete ab initio calculations. For that reason, model investigations have been performed. Here we base our calculations on Su-Schrieffer-Heeger (SSH) model [9].

Tight binding SSH model was firstly introduced to describe the properties of polarons in conducting polymer. Later, its modified form has been applied to DNA by the Conwell group [10, 11, 12].

DNA may be considered as a one dimensional chain composed of the stacked base pairs (A-T and G-C), which enables charge transfer by the $\pi - \pi$ electron interaction between the adjacent nuclei bases. From theoretical point of view, large polaron formation is expected if an extra electron (or hole) is added to one-dimensional chain.

In order to support polaron formation, Hamiltonian must contain the following essential ingredients:

1) electronic system that realizes the charge conduction;
2) electron-phonon coupling to allow the formation of lattice distortion;
3) an elastic restoring force that keeps distortion finite.

SSH Hamiltonian, which fulfils this request, in the usual form is given as:

$$H_{SSH} = \sum_n d_n B_n^+ B_n - t_0 \sum_n \left(B_n^+ B_{n+1} + B_{n+1}^+ B_n\right) + \alpha \sum_n (u_{n+1} - u_n)\left(B_n^+ B_{n+1} + B_{n+1}^+ B_n\right)$$
$$+ \frac{\kappa}{2} \sum_n (u_{n+1} - u_n)^2 + \frac{1}{2M} \sum_n p_n^2 \,. \qquad (1)$$

Here:
$B_n^+ \left(B_n\right)$ is the electron creation (annihilation) operator,
d_n denotes on-site energy,
u_n is the displacement of the n_{th} site from thermal equilibrium position (sites evenly spaced at distance equal to lattice constant $a \approx 3.4\,\mathring{A}$),
κ denotes elastic constant,
M is the mass of the entity at each site,
t_0 denotes transfer integral in thermal equilibrium,
α is electron-phonon coupling constant.

The assumption which is inherent to this Hamiltonian is the linear dependence of transfer integral $t_{n,\,n+1}$ on separation between adjacent sites, namely

$$t_{n,\,n+1} = t_0 - \alpha(u_{n+1} - u_n) \,. \qquad (2)$$

Main and difficult problem in dealing with SSH Hamiltonian is the selection of its parameters, especially the transfer integral t_0 and the electron-phonon coupling constant α. It is very important to emphasize that their precise determination is necessary if we want to know the nature of polaron, character of its motion and mathematical tool for the theoretical description.

We note that the elastic constant κ is determined by the experimental value of speed of sound:

$$\kappa = 0.85 \frac{\mathrm{eV}}{\mathrm{\mathring{A}}^2}. \tag{3}$$

As for the transfer integral t_0, very diverse values may be found in the literature [10,13,14,15]. These values, which mainly depend on the kind of adjacent base pairs, are obtained by ab initio calculations.

According to the same type of calculation, the electron-phonon coupling constant is given as

$$\alpha \approx 2t_0 \, \mathrm{\mathring{A}}^{-1}. \tag{4}$$

Significant remark to such obtained values for t_0 and α is that ab initio calculations are very complicated and with doubtful precision.

3. Theoretical method for explicit estimation of electron-phonon coupling constant

In this paper, the theoretical method which enables one to calculate explicitly the value of α on the basis of the transfer integral t_0 is developed. For the reason of simplicity, in the following we will consider periodic DNA. Such kinds of sequences can be engineered artificially and provide macromolecules that should display fast charge transfer properties. It this case we may use

$$d_n = 0. \tag{5}$$

We begin with the Hamiltonian:

$$H = -\sum_{n,m} t(u_n - u_m) B_n^+ B_m + \frac{1}{2M} \sum_n P_n^2 + \frac{1}{2} \sum_{n,m} \left(V_{n-m} - V_{n-m+u_n-u_m} \right). \tag{6}$$

The next step is to use Fourier transformations

$$t(u_n - u_m) = \frac{1}{\sqrt{N}} \sum_k t_k e^{ika(n-m)+ik(u_n-u_m)}, \tag{7}$$

$$V_{n-m} = \frac{1}{\sqrt{N}} \sum_k V_k e^{ika(n-m)} \tag{8}$$

and

Journal of Physics: Conference Series **329** (2011) 012015 doi:10.1088/1742-6596/329/1/012015

$$V_{n-m+u_n-u_m} = \frac{1}{\sqrt{N}} \sum_k V_k e^{ika(n-m)+ik(u_n-u_m)} . \tag{9}$$

Now, in the approximation of the nearest neighbours and for small displacements u_{n+1} and u_n, the well known SSH Hamiltonian is obtained:

$$H_{SSH} = -t_o \sum_n \left(B_n^+ B_{n+1} + B_{n+1}^+ B_n\right) + \alpha \sum_n \left(u_{n+1} - u_n\right)\left(B_n^+ B_{n+1} + B_{n+1}^+ B_n\right)$$
$$+ \frac{\kappa}{2} \sum_n \left(u_{n+1} - u_n\right)^2 + \frac{1}{2M} \sum_n P_n^2 , \tag{10}$$

where

$$t_o = \frac{1}{\sqrt{N}} \sum_k t_k \cos ka , \tag{11}$$

$$\alpha = \frac{1}{\sqrt{N}} \sum_k k \cdot t_k \sin ka , \tag{12}$$

$$\kappa = \frac{1}{\sqrt{N}} \sum_k k^2 \cdot V_k \cos ka \tag{13}$$

and

$$t_k = \sum_l e^{ikal} t_l ; \quad l = n - m . \tag{14}$$

Taking into account the nature of forces between stacked bases we calculate t_k. Finally, it follows:

$$\alpha = \frac{t_o}{a} . \tag{15}$$

4. Results

Starting point of our analysis is the above tight binding one-dimensional SSH Hamiltonian. Then, Fourrier transformation and quantisation of phonon degrees of freedom are performed. In such a way, we obtain

$$H = -t_0 \sum_n \left(B_n^+ B_{n+1} + B_{n+1}^+ B_n\right) + \frac{1}{\sqrt{N}} \sum_{q,n} F_q e^{iqna} \left(B_n^+ B_{n+1} + B_{n+1}^+ B_n\right)\left(a_q + a_{-q}^+\right)$$
$$+ \sum_q \hbar\omega_q a_q^+ a_q \,. \tag{16}$$

Here:

$$F_q = \alpha \sqrt{\frac{\hbar}{2MN\omega_q}} \left(e^{iqa} - 1\right) \tag{17}$$

and

$$\omega_q = \omega_0 \sin\left|\frac{qa}{2}\right|; \qquad \omega_0 = 2\sqrt{\frac{\kappa}{M}} \,. \tag{18}$$

Polaron solution is used in the form:

$$|\psi\rangle = \sum_n \psi_n B_n^+ |0\rangle \times |\beta\rangle; \qquad |\beta\rangle = \prod_q |\beta_q\rangle; \qquad a_q|\beta_q\rangle = \beta_q|\beta_q\rangle, \tag{19}$$

where ψ_n represents an electron wave function (amplitude) and β_q is obtained by the optimization of the energy $\mathcal{H} = \langle\psi|H|\psi\rangle$.

From the equation of motion for ψ_n

$$i\hbar\dot{\psi}_n = \frac{\partial\mathcal{H}}{\partial\psi_n^*}, \tag{20}$$

in continuum approximation $\psi_n(t) \to \psi(x,t)$, the famous nonlinear Schrödinger equation (NSE) appears in the lowest order:

$$i\hbar\dot{\psi}(x,t) = -2t_0\psi - t_0 a^2 \psi_{xx} - 8E_B|\psi|^2\psi, \tag{21}$$

where indexes x in ψ denote double differentiation over x and the dot represents differentiation over time t. E_B is small-polaron binding energy:

$$E_B = \frac{1}{N}\sum_q \frac{|F_q|^2}{\hbar\omega_q} = \frac{\alpha^2}{2\kappa}. \tag{22}$$

The solution of NSE is well known "soliton-like" solution:

$$\psi(x,t)=\sqrt{\frac{\mu}{2}}\frac{e^{i(kx-\omega t)}}{\cosh\left[\frac{\mu}{a}(x-vt)\right]}. \tag{23}$$

For the purpose of this paper, it is enough to mention that the parameter μ has the meaning of inverse polaron width in unit of a. If we substitute the "soliton-like" solution (23) in NSE (21), it is easy to show that:

$$\mu=\frac{2E_B}{t_o}=\frac{\alpha^2}{\kappa t_o}. \tag{24}$$

Then, the polaron width l_p follows directly:

$$l_p=\frac{1}{\mu}a=\frac{t_o\kappa}{\alpha^2}a. \tag{25}$$

Polaron binding energy can be calculated from the expression:

$$E_{pb}=t_o a^2\int\frac{dx}{a}|\psi_x|^2-4E_B\int\frac{dx}{a}|\psi|^4. \tag{26}$$

Finally,

$$E_{pb}=-\frac{1}{3}\mu^2 t_o=-\frac{1}{3}\frac{\alpha^4}{t_o\kappa^2}. \tag{27}$$

Polaron width and polaron binding energy are completely determined by the parameters of SSH Hamiltonian.

Now, we will compare the results for polaron width (25) and polaron binding energy (27) using ab initio value of α and the value obtained by our theoretical method.

If we take ab initio value $\alpha_{ai}=2t_o\,\mathring{A}^{-1}$, $\kappa=0.85\frac{\text{eV}}{\mathring{A}^2}$ and $t_o\in[0.1-0.3]\text{eV}$ which represent usually used values of t_o in DNA, the obtained results are:

1) $t_o=0.3\text{eV}$, $\alpha_{ai}=0.6\frac{\text{eV}}{\mathring{A}}$; $l_p^{ai}\approx 0.71a$, $E_{pb}(\alpha_{ai})\approx -0.2\text{eV}$,

2) $t_o = 0.2\text{eV}$, $\alpha_{ai} = 0.4\frac{\text{eV}}{\text{Å}}$; $l_p^{ai} \approx 1.06a$, $E_{pb}(\alpha_{ai}) \approx -0.059\text{eV}$,

3) $t_o = 0.1\text{eV}$, $\alpha_{ai} = 0.2\frac{\text{eV}}{\text{Å}}$; $l_p^{ai} \approx 2.12a$, $E_{pb}(\alpha_{ai}) \approx -0.0073\text{eV}$.

As for the theoretical method given above, i.e. $\alpha_{TM} = \frac{t_o}{a}$, the following relations arise:

$$l_p^{TM} \approx 46.3 l_p^{ai}, \qquad E_{pb}(\alpha_{TM}) \approx 4.67 \cdot 10^{-4} E_{pb}(\alpha_{ai}). \tag{28}$$

For instance, if $t_o = 0.3\text{eV}$ and $\kappa = 0.85\frac{\text{eV}}{\text{Å}^2}$ then:

$$l_p^{TM} \approx 33a, \qquad E_{pb}(\alpha_{TM}) \approx -9.34 \cdot 10^{-5}\text{eV}. \tag{29}$$

It is easy to see that when the transfer integral t_o decreases the polaron width l_p^{TM} increases.

5. Discussion and conclusion

In this paper we consider polaron motion as a possible mechanism of charge transfer in DNA. In order to do this we use tight binding one-dimensional SSH model Hamiltonian. The main problem is the selection of its parameters, particularly the transfer integral t_o and electron-phonon coupling constant α. Precise determination of these parameters is very important for adequate description of the nature of polaron and character of its motion.

Ab initio calculations of the above parameters are not reliable enough. From the results obtained for polaron width in this case, it is hard to say that large polaron exists.

In order to understand better polaronic mechanism of charge transfer in DNA, theoretical method is suggested which makes possible to estimate explicitly the value of electron - phonon coupling constant α. On the basis of this value of α we calculate polaron length and polaron binding energy for periodic DNA. From the obtained results for polaron width we conclude that charge transfer in DNA may be achieved by means of coherent motion of large polaron.

Acknowledgements

This research was supported by funds from the Serbian Ministry of Education and Sciences, grant III 45010.

References

[1] Kelley S O and Barton J K, 1999 Science **283** 375.
[2] Treadway C., Hill M G and Barton J K, 2002 Chem. Phys. **281** 409.
[3] Endres R G, Cox D I and Singh R R P, 2004 Rev. Mod. Phys. **76** 195.
[4] Fink H W and Schönenberger C, 1999 Nature (London) **398** 407.

[5] Porath D, Bezryadin A, Vries S and Dekker C, 2000 Nature (London) **403** 635.
[6] Pablo P J, et al., 2001 Phys. Rev. Lett. **79** 3881.
[7] Yoo K H et al., 2001 Phys. Rev. Lett. **87** 198102.
[8] Ly D, Sani L and Shuster G B, 1999 J. Amer. Chem. Soc. **121** 9400.
[9] Su W P, Schrieffer J R and Heeger A J, 1980 Phys. Rev. B **22** 2099.
[10] Conwell E M and Rakhmanova S V, 2000 Proc. Natl. Acad. Sci. USA **97** 4556.
[11] Rakhmanova S V and Conwell E M, 2001 J. Phys. Chem. B **105** 2056.
[12] Conwell E M, Park J-H and Choi H Y, 2005 J. Phys. Chem. B **109** 9760.
[13] Voityuk A A, Siriwong K and Rösh N, 2001 Phys. Chem. Chem. Phys. **3** 5421.
[14] Grozema F C, Siebbeles L D A., Berlin Y A and Ratner M A, 2002 Chem. Phys. Chem. **6** 536.
[15] Park J H, Choi H Y and Conwell E M, 2004 J. Phys. Chem. B **108** 19483.

Journal of Physics: Conference Series **329** (2011) 012016 doi:10.1088/1742-6596/329/1/012016

Endogenous Ion Dynamics in Cell Motility and Tissue Regeneration

N Özkucur[1], S Perike[1], HH Epperlein[1], RHW Funk[1,2]
[1] Department of Anatomy, Medical Theoretical Center, Technical University of Dresden, Germany
[2] CRTD / DFG-Center for Regenerative Therapies Dresden - Cluster of Excellence, Tatzberg 47/49, 01307 Dresden, Germany

E-mail: nurdan.oezkucur@tu-dresden.de

Abstract. Directional cell migration is an essential process, including regeneration of tissues, wound healing, and embryonic development. Cells achieve persistent directional migration by polarizing the spatiotemporal components involved in the morphological polarity. Ion transporter proteins situated at the cell membrane generates small electric fields that can induce directional cell motility. Besides them, externally applied direct current electric fields induce similar kind of responses as cell orientation and directional migration. However, the bioelectric mechanisms that lead to cellular directedness are poorly understood. Therefore, understanding the bioelectric signaling cues can serve as a powerful modality in controlling the cell behaviour, which can contribute additional insights for development and regeneration.

Introduction

Ion flows generated via ion transport proteins located on cellular membranes are well known for maintaining the overall physiological homeostasis such as membrane potential (V_{mem}) and pHi. In combination with these conventional roles, they recently received more and more attention as pluripotent signalling molecules in various biological cascades [1-4]. One rather interesting character of them is their role in directional cell motility, a common and fundamental process in development, tissue regeneration, wound healing, immune response, and tumour metastasis. Here, we focused on investigating the interactions of Na^+,K^+-ATPase (V_{mem}) and of N^+,H^+ exchanger (NHE3, pH_i) with cytoskeletal proteins in persistent directional cell migration [5] and the changes in the intracellular ion contents in tissue regeneration [6].

Methods

Externally applied electric fields were used to induce directional –osteoblast- cell migration *in vitro*. To elucidate the implication of ion transporter proteins in directional cell migration process, we employed various methods ranging from immunocytochemistry to *in vivo* imaging. Specific antibodies were used to visualize the colocalization of ion pumps/exchangers (NaKA and NHE3) with cytoskeletal proteins (vinculin and ß-actin). Cell motility assays combined with the ion transporter-specific inhibitors were used to assess the cellular directivity during migration. For monitoring the changes in intracellular pH and V_{mem}, cells were loaded with ion-specific vital dyes (BCECF-AM and $DiBAC_4(3)$ during migration. Images were taken using an inverted fluorescence microscope.

Results and Discussion

We showed that active Na^+,K^+-ATPase along with vinculin involves in adhesion turnover while N^+,H^+ exchanger activity (pNHE3) associates ß-actin to form de novo protrusions at the leading edge of polarized cells *in vitro* (Fig. 1). Congruent with these, localized activity of NHE3 (pHi monitoring) at the leading edge and, relatively hyperpolarized cell membrane at the front and depolarized ones at the back (V_{mem} monitoring) was observed in polarized cells during directional migration (Fig. 2).

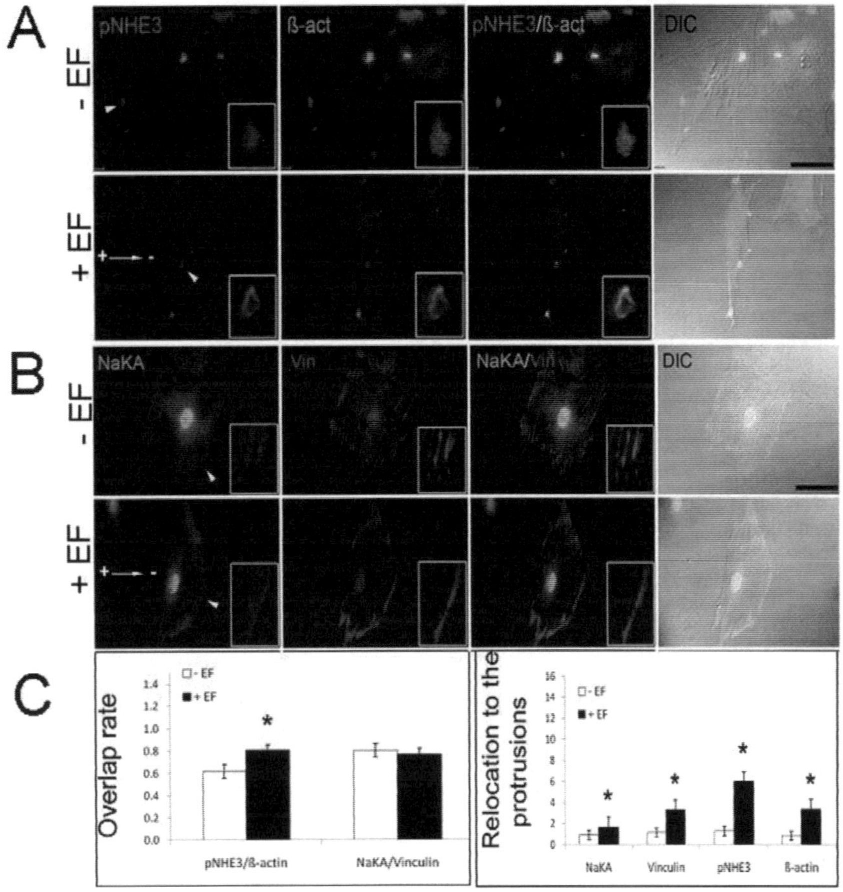

Figure 1. **(A)** Colocalization of Na^+/K^+ATPase (NaKA) and vinculin proteins on the leading edge of osteoblast cells during directional cell migration (+EF) compared to random migration (-EF). **(B)** Colocalization of active NHE3 (pNHE3) and ß-actin proteins on the leading edge of osteoblast cells during directional cell migration (+EF) compared to random migration (-EF). **(C)** Colocalization rate (left) and relocation (right) of the proteins in osteoblast cells during directional cell migration (+EF) compared to random migration (-EF).

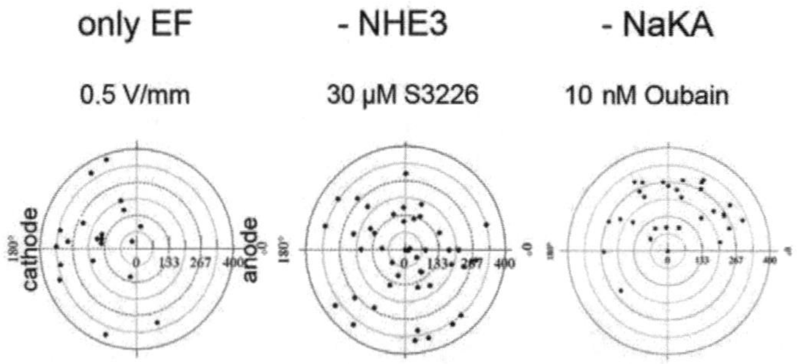

Figure 2. Effect of NHE3- and Na^+/K^+ATPase (NaKA) inhibition on cellular directivity during directional migration. Electric field-induced directional movement (left) of cells was abolished in the presence of inhibitors S3226 (middle) and Oubain (right).

This can be linked to the localized activity of Na^+,K^+-ATPase for maintaining the cellular orientation. Moreover, cellular directedness is disrupted when Na^+,K^+-ATPase or NHE3 is inhibited, confirming that these two proteins are required for perceiving the direction (Fig. 3). Employing ion specific vital dyes, we demonstrated (non-invasively) that intracellular ion contents, Na^+, H^+ and V_{mem}, are dynamically changing during axolotl tail regeneration *in vivo* (Fig. 4).

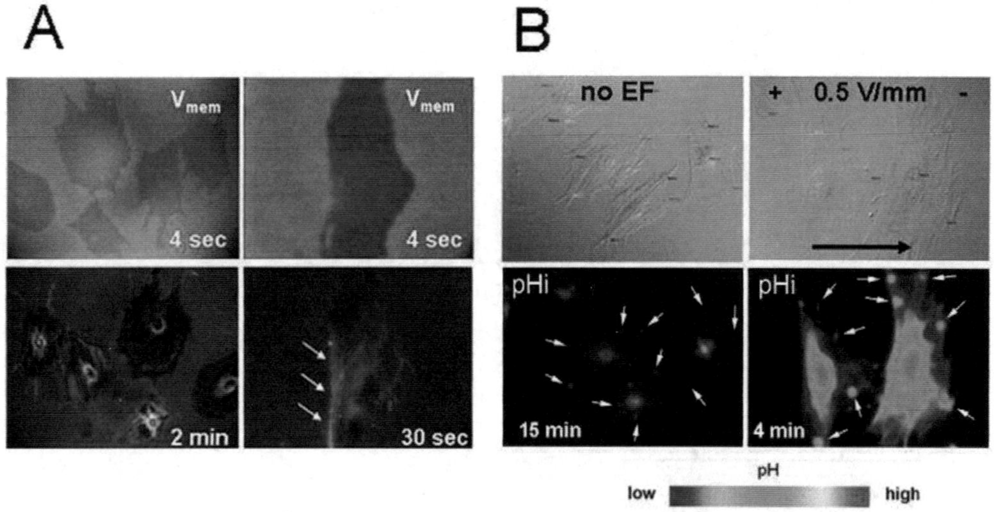

Figure 3. (A) Cell membrane is depolarized at rear end and hyperpolarized at leading edge during directional cell migration. Images taken from cells loaded with V_{mem}-specific dye $DiBAC_4(3)$. **(B)** H^+ outflux at the leading edge of directionally migrating cells. Images taken from cells loaded with pH-specific dyeBCECF-AM.

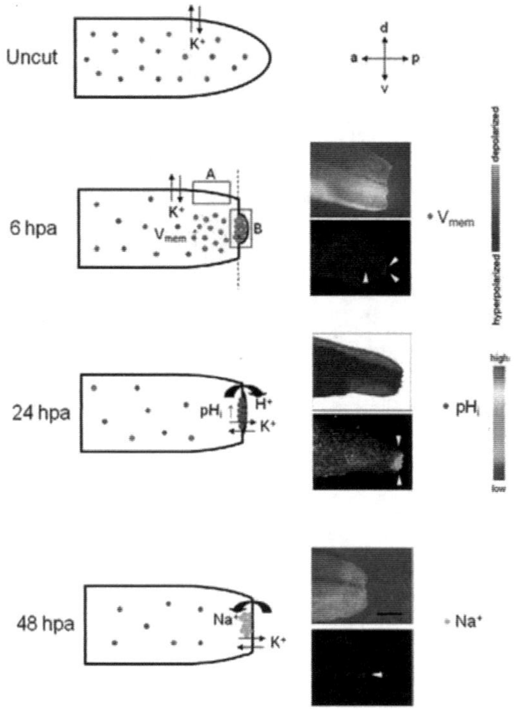

Figure 4. Intracellular ion dynamics during tail regeneration of axolotl larvae. On the left side (column) characteristic changes are schematized for the uncut tail and for different time points after amputation. Right column shows the corresponding images to each time point of interest.

Especially changes of Na^+ and H^+ flows, and V_{mem} are of interest because they are initiated at the very early phase of regeneration process (6 hours post amputation) and last for a few days. None of these changes are observed in uncut tails [6]. Taken together, our data might indicate a critical role for Na^+ and H^+ flows and for V_{mem} during tissue regeneration and their functional roles should be investigated further in more detail. These findings can be important for tissue engineering approaches where manipulation of cell orientation, polarity and migration is in demand.

Acknowledgements

This work was supported by German Research Foundation (DFG-GK NBP, Scholarship to S. Perike) and MeDDrive Grant (Özkucur 60.246) of Technical University of Dresden.

References

[1] Kurashima K, D'Souza S, Szaszi K, Ramjeesingh R, Orlowski J, Grinstein S 1999 The apical Na(+)/H(+) exchanger isoform NHE3 is regulated by the actin cytoskeleton *J Biol Chem* **274** 29843

[2] Rajasekaran SA, Palmer LG, Quan K, Harper JF, Ball WJ, Jr., Bander NH, Peralta Soler A, Rajasekaran AK 2001 Na,K-ATPase beta-subunit is required for epithelial polarization, suppression of invasion, and cell motility *Mol Biol Cell* **12** 279

[3] Denker SP, Barber DL 2002 Cell migration requires both ion translocation and cytoskeletal anchoring by the Na-H exchanger NHE1 *J Cell Biol* **159** 1087

[4] Ozkucur N, Monsees TK, Perike S, Do HQ, Funk RH 2009 Local calcium elevation and cell elongation initiate guided motility in electrically stimulated osteoblast-like cells *PLoS One* **4** e6131

[5] Ozkucur N, Perike S, Sharma P, Funk RH 2011 Persistent directional cell migration requires ion transport proteins as direction sensors and membrane potential differences in order to maintain directedness *BMC Cell Biol* **12** 4
[6] Ozkucur N, Epperlein HH, Funk RH 2010 Ion imaging during axolotl tail regeneration in vivo *Dev Dyn* **239** 2048

Journal of Physics: Conference Series **329** (2011) 012017 doi:10.1088/1742-6596/329/1/012017

Asbestos as 'toxic short-circuit' optic-fibre for UV within the cell-net: — Likely roles and hazards for secret UV and IR metabolism

Robert R Traill

c/- Ondwelle Publications, 29 Arkaringa Crescent, Black Rock (Melbourne), 3193, Victoria, Australia

rrtraill4@dodo·com·au †

Abstract. The most toxic asbestos fibres have widths 250*nm*-10*nm*, and this toxicity is "physical", which could mean either mechanical or optical: Tangling with chromosomes is a •*mechanical* hazard occasionally reported, and fibres <100*nm* wide would probably be most knife-like. Our other concern here is •*optical*: Calculations for fibres ≤300*nm* reveal such a transmission possibility, but only when the amphibole fibres (brown and blue asbestos) are >100*nm* wide — or chrysotile (white asbestos) is >150*nm*. In both cases, UV_A/UV_B -transmission would then predominate. (Chrysotile 150*nm might* be benign — escaping both mechanical and optical!). But what would generate such UV, and why would its transmission be toxic? Thar and Kühl (*J.Theor.Biol.*:2004) explain that the *long* mitochondria on microtubules may be able to act as UV-lasers, (and many observers since Gurwitsch 1923 have reported ultraweak UV emissions escaping from all types of living bio-tissue). That all suggests some universal secret role for UV, apparently related to mitosis. Insertion of fibre "short-circuits" could then cause upsets in mitosis-control, and hence DNA irregularities. Such UV-control could parallel similar lower-powered Infra-Red control-systems (as considered elsewhere for coaxial myelin; or as portrayed by G.Albrecht-Buehler's online animations etc.); and the traditional *short* mitochondria seem better suited for this IR task.

1. Introduction — Are we dealing with Toxic Geometry?

1.1 "Physical damage" rather than chemical

Asbestos fibres are clearly toxic — but why exactly? It seems universally agreed that their effect is *non-chemical*: — In 1978, Stanton and Layard [1] showed persuasively that the pathogenic properties of such fibres *did not correlate with the chemistry* of the fibres, but rather to *their physical dimensions*: — and they went on to list some of the geometrical correlates shown to

† *or*, if obsolete, consult the website: www.ondwelle.com

exist: Thus the stronger causal effects occurred with **diameters of *10-250nm***, and moderate effects with *250–1500nm* thicknesses.[1]

"Non-chemical" implies *"physical"* — but in what way? — It is easy to assume that this means *"mechanical damage"*, and there is evidence that this is sometimes true, as we shall see. But a more interesting possibility is that the cause is *optical*, since crystal-like solids often do feature in our dealing with light, though it may not be immediately obvious just what damage might arise in this way.

(That probably exhausts the list of *feasible primary-cause* possibilities, though we should remain open to the unexpected, and more of the less-likely options are mentioned later. For instance one could imagine acoustic damage, but it is difficult to see that as plausible. So let us concentrate mainly on the two existing suspects.)

1.2 Damage to what?

We are not just talking about membrane-penetration, or blood-letting lesions caused by a needle — the damage in question will affect future growth at that site, and maybe later at secondaries elsewhere. So the damage will almost certainly relate to the DNA, chromosomes, and mitosis. Thus the question narrows to asking *"What* mechanical or optical effects could cause such chromosomal disruption?" — and indeed *"How?"*.

1.3 What do we know about toxic fibres anyway?

First note that, since our concern is with the physics and geometry of the fibres (and not their chemistry), we should extend our study to any other fibres with similar physical properties — notably glass fibres. That should come as no surprise, as such fibres have already been under suspicion for some time. If their toxicity seems less, that could well be due to their geometry being less extreme (and more under design-control) than the asbestos fibres provided by nature.

Thus we may provisionally assume that two fibres with the same geometrical configuration, electro-optical properties, (and perhaps density and stiffness) will be equally toxic-or-benign, regardless of any chemical difference between them. Hence this study will not be confined to asbestos alone.

[1] They also raise some other geometrical issues which we will not pursue here: — (i) the length and/or aspect ratio may also be significant in a less simple way. But if geometry is part of the cause, it is interesting to also see it as part of the effect, and that might be seen as significant. Thus: (ii) "cancer can be viewed as a disease of geometry. Tumor tissue results from growth, which is not patterned appropriately, because it is unable to perceive or execute morphogenetic cues." — quoting Levin.

Table I
(adapted from [2], with permission)

Summary of the various supposedly-homogeneous dielectric fibres which can be seen as an "asbestos-like" toxic hazard — especially if their diameters are very fine (10–250*nm*).
There are two polarized opinions about the relative toxicities of *chrysotile* and the *other (amphibole) asbestos types*, as suggested by the two examples depicted in columns F & G.

	A Names of dielectric fibre-types. w,y,z	B common name	C content-biases, or chemical formulae where: "♣" = $Si_8O_{22}(OH)_2$	D structural pattern	E Refractive Index min – max	F [3] toxic? Merck 1992, *not* 2006!	G toxic? Hester-berg (1987) [4]	soluble?
1	crown "E glass" [v]	(textile)	incl. Ca, Mg, Al, +maybe Na,K,B,F,Ti	amorphous	**1.508 – 1.549**	—		
2	crown glass-wool [v]	(insulation)	incl. Na, Ca, B, Mg, (+maybe Al, K)	amorphous	**1.530 – 1.587**	—		
3	flint glass [v]	lead glass	incl.Pb, (or Ti, Zr)	amorphous	**...** **2.1**	—		**No**
4	(ortho)**chrysotile** = mitaxite	**white asbestos**	$Mg_3Si_2O_5(OH)_4$ [w] (note its uniqueness)	orthorhombic (*serpentine*)[w]	**1.569** – [w] 1.57	slight ?	**6X** query!	**yes**
5	Grunerite [w] = **Amosite** = ferro-	**brown asbestos**	$Fe^{2+}{}_7$ –♣ [w]	monoclinic prismatic [w]	**1.663** – [w] 1.709	**XX**	?	**No**
6	(Magnesio)-riebeckite. = (..)-**crocidolite**	**blue asbestos**	$(Na_2Mg_3Fe^{3+}{}_2)$ –♣ $Na_2Fe^{2+}{}_3Fe^{3+}{}_2$ –♣ [y,w]	monoclinic prismatic [w]	[w] **1.68** – 1.70	**XX** **XX**	**X**	**No**
7	**Tremolite**, (Actinolite) [y] Ferro-actinolite	(rare)	Ca_2Mg_5 –♣ [y,w] $Ca_2(Fe^{2+},Mg)_5$ –♣ [w] $Ca_2Fe^{2+}{}_5$ –♣ [y]	monoclinic prismatic [w]	**1.599** – [w] 1.626	**XX?**	?	**No**
8	Cummingtonite = **Anthophyllite**,	(rare)	1. Mg_7 –♣ [w, y (*some* Fe)]	orthorhombic [y]	**1.598** – [w] 1.685	slight	?	**No**

[v] Tennent [5] [w] webmineral.com [6] [y] Shelley [7]
[z] Mineralogists' usage (1997) [8]

Secondly note that most asbestos types are classed as "amphiboles" with similar physical properties (and similar chemical formulae — See Table I). These amphiboles notably include the notorious "blue asbestos" (crocidolite), and the "brown asbestos" (amosite) — both with a refractive index of about 1·7, and both producing relatively straight and "tidy" fibres.

In contrast, there is "white asbestos" (chrysotile, $n \approx 1{\cdot}57$, often as "untidy" fibres, very slightly soluble) whose toxicity is in dispute, and which is still being used commercially in some countries. This uniqueness raises some interesting questions which we will encounter several times.

2. *What Physical Causes should we consider seriously?*

2.1 Primary causes rather than secondary

It seems sensible to restrict our search to whatever it is that *initiates* the toxic chain-of-events. — After that, the main damage has already been done (even though it may still be curable later on).

Indeed whatever the physical damage from asbestos may be, it will next have biochemical consequences, which are then themselves causes — *secondary* causes (like inflammation) which will probably affect other subsystems further down the causal chain. That constraint helps to clarify the issue — and limit the short-list of likely candidates somewhat:

2.2 Mechanical and Optical taken as the most likely toxic effects

In this context, "physical" is perhaps usually assumed to mean "**mechanical**", as if each fibre were a miniature knife. That has some justification of course, so it deserves our attention.

But physics extends further than that, so we could plausibly blame **direct-optical effects** instead — especially if we view asbestos fibres as *"windows onto a radiation landscape"* (though it is then our duty to identify that radiation-hazard if possible). Such window-like properties would occur if we accept asbestos fibres as either *crystals* or *potential optic fibres* — and their geometry obviously suggests the latter.

2.3 Other conceivable toxic effects from asbestos

Toyokuni [9] lists three physically-toxic possibilities: (i) oxidative stress" from "frustrated macrophages" trying to dispose of the intruding fibres (and perhaps we can class each such fibre as presenting a *physical* "spanner in the biochemical works"); (ii) chromosome tangling (a clear case of the mechanical damage already mentioned); and (iii) adsorption onto the fibre, which could then concentrate any toxic proteins. — We might see the two cases (i) and (iii) as pseudochemical though still perhaps "arguably physical". That does not mean they are necessarily wrong, but they (and other remote possibilities) will not be considered further here. Meanwhile (ii), already on the list, is now discussed further:

3. Mechanical Damage as one Key Cancer-trigger

There is some good evidence that this occurs when *very narrow* fibres (100*nm*–10*nm*) physically tangle with chromosomes. Thus:

•Wang *et al* [10] point to more polyploidy caused by chrysotile which seems inclined to *"sever, puncture, or adhere to and mingle with chromosomes, causing deformities "*

•Ault *et al* [11 (p796, fig5)] : "Chromosome breakage in a mitotic cell containing…fibers. …As the spindle rotated,…[chromosome] arms were dragged through…fibers. …break… and the resulting chromosome fragment…was left behind at anaphase."

•Jaurand, *et al*., (1987), quoted by Voytek [12] *"In rat mesothelial cells, chrysotile… with 84%… having a diameter of less than 0.058 μm, caused chromosomal aberrations… predominantly chromatid breaks… "*

Such entanglement seems less likely for thicker fibres, which would increasingly appear to be like a *perhaps-harmless blank wall* — and therefore less like needles or thorns for the 30*nm* DNA/histone chromosome-threads. Nevertheless a *mobile* "blank wall" could at least be disorganizing.

4. Optical Damage as the other Likely Primary Cause

My own interest in the topic started when I pondered whether asbestos fibres might be transmitting *infra-red* — an idea prompted by the still unproven hypothesis that myelinated nerve fibres do just that (in addition to their acknowledged role of carrying action-potential pulses). [13] Some simple calculations and assessments quickly revealed that IR was absolutely impossible for these much smaller (and *non-coaxial*) asbestos fibres, though UV was a not-too-implausible candidate, especially as it had much more destructive power (in some unspecified way) — but it was far from clear where such UV might come from.

This present revised version of the *Optical Damage* proposal builds on • the now-prolific evidence that living tissue *universally emits occasional ultra-violet photons* — and • a more-specific recognition of the fact that such UV photons have uncharacteristically high quantum energy which is likely to do *DNA-damage* if misapplied.[2]

The perhaps-inevitable conclusions were eventually
(i) that UV-emissions must be serving some (universally?) important metabolic role, but
(ii) such photons needed to be carefully controlled and directed lest they cause damage, and perhaps miss some crucial target as well). So —
(iii) ***If*** asbestos happened to act as fibre-optic "short-circuits", it could seriously disrupt that UV system — causing DNA damage, and hence sometimes initiate cancer.

That leads to the crucial question:

4.1 Can asbestos actually transmit UV? — Experiments needed

Ideally that question calls for systematic laboratory investigation (and let us hope this will occur in due course, despite the obvious expense (♠[3])), but meanwhile we must make do with theoretical calculations:

4.2 Instead: Calculated UV transmissibility under various Conditions

First note that as we go deeper and deeper into the UV range, wavelengths will shorten down towards 200*nm* — and the waves will then cease to be transmitted through any material at all, because such radiation is all absorbed (due to resonance with the outer electron shells of atoms, [14 (p.483)] — so no calculations are needed in that extreme! However this effect does also start to make calculations difficult for *nearby* wavelengths — see [2].

First note that asbestos fibres are *not* those specially-designed optic-cable fibres which have a denser axial core for "steering" any light-beam away from the boundaries. (Nor are they coaxial cables, in the way that myelinated nerves arguably seem to be). Instead any UV/asbestos transmission would have to be by some form of "zigzag" reflection to-and-fro across the fibre — and that can only occur under certain conditions: (i) The refractive index ratio ($\boldsymbol{n}/\boldsymbol{n}_{\mathrm{aq}}$ — *fibre/aqueous-environment*) must be high enough. (ii) The fibre-diameter ***D*** must be large enough for a half-wavelength loop (½λ) to fit across it *and* leave some room for other complications.

Thus if $D \approx \lambda$ it will be a borderline case. With that in mind we can sometimes get a good idea of a fibre's optical capabilities without the need to calculate exactly. For instance, if a certain natural fibre has diameter D=300*nm* (0·3μm), we can confidently guess that it will not be able to transmit visible light, for which λ>400*nm*; *but it might well be able to transmit UV.* So we might then wish to investigate the situation further.

To calculate the transmissibility "should" be a fairly simple task — the above-mentioned use of "$\boldsymbol{n}/\boldsymbol{n}_{\mathrm{aq}}$" within Snell's law should ensure the right conditions for TIR (total internal reflection) so that the light does not escape from the fibre (or slab etc). The trouble is that we need to know

[2] as is shown (obviously) by skin-cancers caused by excess UV exposure from sunlight or solariums.

[3] As in the previous paper [15], I have used "♠" within the text to indicate *points* which ***now seem to be testable-or-retestable experimentally*** (or important enough to put onto the list anyhow!).
Once again such items appear in an Appendix.

Journal of Physics: Conference Series **329** (2011) 012017 doi:10.1088/1742-6596/329/1/012017

actual refractive indices ***n*** (inside and out) — but these ***n*** values change increasingly violently as we approach the absorption wavelength ($\lambda_0 \approx 200nm$) — see Figure 1.

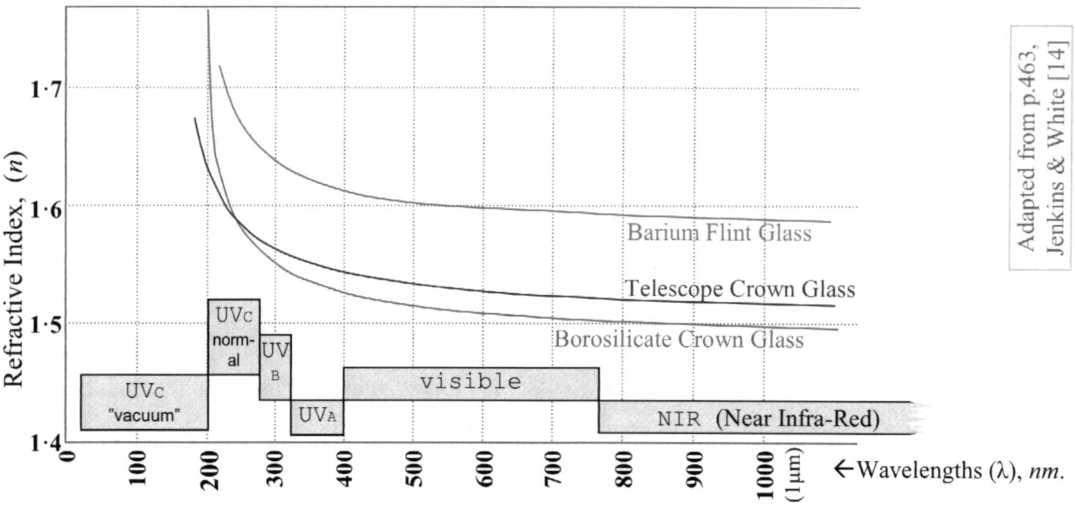

Fig.1 Dispersion curves for three glass-types (asbestos lab-data not available) showing the typical unstable values of *n* within the ultraviolet range, especially $\lambda \approx 200nm$.

Nevertheless, the problem can be solved for this asbestos-case [2] though the procedural details need not detain us here. It may however be helpful to note the main assumptions made:

(i) The validity of Sellmeier's formula.[4] This is used to calculate ***n*** at any relevant wavelength (as in Fig.1), given the value for $\boldsymbol{\lambda_0}$ — and "***A***"[5].
(ii) For want of better knowledge, $\boldsymbol{\lambda_0}$ was taken to be exactly 200*nm* for every relevant medium.

4.2.1 The Raw Results

The calculated figures are set out in Table II (adapted from the earlier report [2]). Thus, e.g., from the "fail" entries, we can see for chrysotile (within either of the two aqueous media) there can be no transmission at all for fibres with diameters of 150*nm* or less. — but a 200*nm* fibre within the "n_{aq}=1.35" environment can just manage to transmit a *maximum wavelength size* of "**325.38***nm*" (which happens to fall within the UVA range of 400–320*nm*).[6]

---

[4] Sellmeier's formula [14 (p469)] states, for the refractive index ***n***:
$\boldsymbol{n^2 = 1 + (A\lambda^2 / (\lambda^2 - \lambda_0^2))}$ — where λ_0 is the resonance-wavelength (at which absorption is theoretically infinite — and so is *n*, as indicated by the denominator "λ^2 - λ_0^2" becoming zero).
"*A*" is the Sellmeier constant, specific to the medium in question.

[5] One can calculate *A*, given some definite value of ***n*** at some *specified* wavelength, (usually using standard sodium light, for which $\boldsymbol{\lambda}$ = 588·995 [14]).

[6] It is helpful to recall these wavelength ranges: ...... *NIR* > [0.78μm=780nm] > *Red* > [610nm] > *Orange* > [590nm] > *Yellow* > [570nm] > *Green* > [500nm] > *Blue* > [450nm] > *Violet* >[400nm] > ***UVA*** > [320nm] > ***UVB*** > [280nm] > ***UVC****(normal)* > [200nm] > *UVC(vacuum)* > [10nm] ......

Clearly that rules out visible light and IR for any such chrysotile fibre of that width, but it does theoretically allow any available *shorter* waves, including UVB (320–280*nm*) — though there is probably a PRACTICAL LIMIT as we shall see shortly.

Note that the rows of the table are spaced out (within practical limits) as a gesture toward making the refractive-index intervals more numerically-uniform and graph-like — as gauged by shading "*1·5, 1·6, 1·7,*" etc. (integer/10 figures in the right-hand *n*-column), and by a green dot half-way between each pair. The *main asbestos-fibre types* are emphasized, which incidentally makes clear much of the distinction between chrysotile (white asbestos) and the other "amphibole" types commonly threatening our health.

External (aqueous) Refractive Index →				n_{aq} (sodium light) = 1.35					n_{aq} (sodium light) = 1.4					
Fibre-Diameters (*nm*) →				300	250	200	150	100	300	250	200	150	100	
Wavelength in vacuo (λ *nm*) →			refract[ve] index	λ	λ	λ	λ	λ	λ	λ	λ	λ	λ	refract[ve] index
"n" data source=[14]		Max?	*n*	≤...	≤...	≤...	≤...	≤...	≤...	≤...	≤...	≤...	≤...	*n*
flint glass		max	*2.1*	**799.94**	675.72	553.74	435.86	**326.58**	779.76	659.14	540.79	426.62	**321.11**	*2.1*
alumotantite			*2*	738.14	624.97	514.17	**407.69**	309.99	716.22	607.00	500.21	**397.80**	304.22	*2*
klebelsberqite			*1.95*	706.44	598.99	493.99	**393.40**	301.67	683.50	580.22	479.43	**383.14**	295.74	*1.95*
Lammerite			*1.9*	674.11	572.53	473.49	**378.95**	293.34	650.04	552.86	458.28	**368.29**	287.23±	*1.9*
parwelite			*1.85*	641.06	545.53	452.62	**364.33**	285±1	615.69	524.85	436.69	**353.23**	*fails*	*1.85*
rouseite			*1.8*	607.16	517.90	431.35	**349.51**	*fails*	580.31	496.07	414.60	**337.92**	*fails*	*1.8*
warikahnite			*1.75*	572.27	489.54	409.60	**334.47**	*fails*	543.70	466.38	**391.92**	**322.34**	*fails*	*1.75*
Amosite	brown	max	*1.709*	542.78	465.63	391.36	321.95	*fails*	512.57	441.22	372.82	309.33	*fails*	*1.709*
Crocidolite	blue	max	*1.7*	536.19	460.30	387.30	319.18	*fails*	505.59	435.58	368.55	306.44	*fails*	*1.7*
Anthophyllite	(rare)	max	*1.685*	525.09	451.33	380.48	314.53	*fails*	493.81	426.10	361.38	301.60	*fails*	*1.685*
Crocidolite	blue	min	*1.68*	521.37	448.32	378.20	312.98	*fails*	489.84	422.91	358.98	299.98	*fails*	*1.68*
Amosite	brown	min	*1.663*	508.57	437.99	370.37	307.67	*fails*	476.20	411.95	350.73	294.44	*fails*	*1.663*
attakolite		min	*1.65*	498.66	430.00	364.33	303.59	*fails*	465.60	403.44	344.34	290.16	*fails*	*1.65*
Tremolite	(rare)	max	*1.626*	480.03	405.02	353.04	295.99	*fails*	445.59	387.44	332.37	*div.oscil*	*fails*	*1.626*
Tremolite	(rare)	min+	*1.6*	459.33	398.42	340.58	287.66±	*fails*	423.21	369.60	319.10	*fails*	*fails*	*1.6*
Anthophyllite	(rare)	min	*1.598*	457.72	397.13	339.61	287.0±	*fails*	421.46	368.20	318.07	*fails*	*fails*	*1.598*
E-CrownGlass	(textile)	max	*1.5874*	449.08	390.22	334.45	286±22	*fails*	412.06	360.74	312.55	*fails*	*fails*	*1.5874*
Chrysotile	white		*1.569*	433.83	378.05	325.38	*fails*	*fails*	395.39	347.54	302.82	*fails*	*fails*	*1.569*
armenite (*n*=...-1.552)		min	*1.55*	417.70	365.22	315.86	*fails*	*fails*	377.61	333.53	292.57	*fails*	*fails*	*1.55*
otherCrownG.	wool-glass	max	*1.5487*	416.58	364.33	315.20	*fails*	*fails*	376.37	332.56	291.86	*fails*	*fails*	*1.5487*
E-CrownGlass	(textile)	min	*1.5302*	400.41	351.52	305.75	*fails*	*fails*	358.40	318.47	*div.oscil*	*fails*	*fails*	*1.5302*
otherCrownG.	wool-glass	min	*1.5076*	380.02	335.42	293.95	*fails*	*fails*	335.46	300.62	*fails*	*fails*	*fails*	*1.5076*
bilinite			*1.5*	372.98	329.89	289.92	*fails*	*fails*	327.47	294.43	*fails*	*fails*	*fails*	*1.5*
gearksutile			*1.45*	323.91	291.69	*fails*								*1.45*

Table II. The longest wavelengths calculated as able to travel through fibres of various types (and 2 different environments)

4.2.2 Table II

4.2.3 The Effect of Substance-as-such

Apart from that, asbestos is here treated just the same as glass *and* the various other minerals (each often chosen simply as a convenient exemplar for its specific *n*-value). This implies the **tentative hypothesis** that, since we are looking for "diseases from geometry+optics" and not from chemistry, the toxic properties are to be gauged entirely from physical variables (mostly represented by the substance's *position within the table*), and not by its identity-as-such. ♠[3]

What saves us from all those other substances is presumably the fact that they either do not readily appear as fibres at all, or their diameters are so *comparatively wide* that any UV within them has *too much freedom*, and thereby loses any dangerous laser-like unanimity-and-focus.[7] Thus consider warikahnite (at n=1.75 on the table): I am not familiar with that mineral, but I presume it never naturally occurs as thin fibres comparable to those discussed here. However if it did, then its 200–150*nm* fibres might well be just as toxic as amosite. Then again, that extreme type of flint-glass (with n=2.1*nm*), if made into fibres 150–100*nm*, could well be particularly toxic; — but probably no-one has ever thought of producing such fibres out of "crystal glass" like that, and indeed such manufacture might be technically impossible or at least prohibitively expensive.[8]

Regarding this particular set of calculated values. It would of course have been more ideal to work with accurate, systematic and comprehensive sets of reliable refractive indices (*in the UV range!*) for the various asbestos types; and maybe we can look forward to that in the future when such data becomes readily available; (♠). Better still would be some direct tests for UV transmissibility within actual fibres (as already suggested ♠). Meanwhile these present calculations might suffice for this early stage of the investigation — a talking point which might inspire more thorough investigations.

4.2.4 Interpreting the Results more generally

Additional shading has converted Table II into two sets of area-graphs to highlight UV-transmissibility. The boundary lines for these shapes merely reflect the figures on the table itself, but they may help our interpretation of it. The right-hand boundaries are just a "joining of the dots" trace of where a particular value does-or-would fall within the table. — Thus the right-most line *joins the extreme figures* (285, 286, 289.92, 291.69*nm* — top-to bottom for the n_{aq}=1.35 case — and 287.23, 290.16, 291.86, 294.45 for the n_{aq}=1.4 case, on the right).[9] As it happens, these figures all approach the UVB/UVC border of λ=280*nm*, without quite getting there.

In principle then, these ≈290*nm* waves should also fit into the wider fibres further to the left, except that there is a PRACTICAL LIMIT as mentioned earlier (when the fibre becomes *too spacious*, and control is lost). There is no strict cutoff in this case, but for reasons explained previously [2 (§3.3.4., p.14)] it is convenient to deem this limit to be when the wavelength has decreased to about 70% of the maximum — or, as in this case, when the available space has increased by (100/70), i.e. ×1·429. That gives us the means to draw the left-hand boundary, seeking cells where correspondingly longer

[7] If so, that apparent laser-like attribute is a partial clue in our search for their source, as we shall see.

[8] However if the manufacture of such bizarre fibres does ever turn out to be feasible, (and diamond would be offer another such example, with n=2.417), then this could make for some interesting experiments. ♠. (Moreover the thin-layer depositing of artificial diamond has already been achieved).

Incidentally there is another way in which bizarre manufacturing could help: One frustration when trying to assess geometrical measurements is that published results comparing width (or length, etc) have always depended on sample fibre-populations with their measurements *defined only statistically* (at best) — thus blurring any sharp effects like cutoff which can occur in optics. *If feasible*, I would encourage the manufacture of asbestos-like glass fibre-sections — all with uniform dimensions and optical properties within each sample. It could then be interesting to replicate many of the existing studies. ♠.

[9] The "dot" (aimed at in this plotting) is deemed to be just below the printed value (if any), and central to it.

wavelengths are only just becoming permitted — namely ($\approx 290nm$)×(10/7) ≈ $414nm$ for the areas in question. That gives us the pale-green area for each of the two cases, (n_{aq} = 1·35 and 1·4) — denoting those fibres deemed to be most toxic if ($\approx 290nm$) radiation (high-energy UVB) were applied to them.

If instead we are interested specifically in UVA at its border with violet (λ=400*nm*, arbitrarily fixed this time), a similar process gives us the right-hand boundary for the pink area; —— and 400×(10/7) = 571.4 gives us the left-hand boundary in the two half-tables. Then again, the intermediate case for the UVA/UVB interface at 320*nm* gives us an intermediate area (plotted in pale-yellow, and barely visible apart from its outline).

Thus any fibre-configuration which falls within any of these areas is (hypothetically) a passive possible-carrier of UV (of one type or another, and some of these will be more toxic than others).

4.2.5 The Effect of Diameter

Perhaps the most important feature is the *absence* of any values at the bottom right of both half-tables. The equations were simply not solvable in those cases. Thus some fibres are just too narrow to carry UV of any type; so if they are toxic at all, it must be for other reasons — and we have already seen their likely role in mechanical damage.

Apart from that however, there is almost nowhere to hide if we insist on using fibres as narrow as those depicted! (Except perhaps in the top left-hand corners, and even that depends on the somewhat arbitrary "PRACTICAL LIMIT" which we might hesitate to trust in life-and-death situations). On the other hand, if the problem is one of interference in some internal metabolic process, the degree of damage may well depend on just what wavelengths are involved metabolically — so exact tuning could turn out to be important, such that some areas of the table could be more important than others after all.

On a more quantitative level, we may recall Stanton and Layard's summary (1978)[1] (already mentioned) that — *the stronger causal effects occurred with **diameters of 10-250nm**, and moderate effects with 250–1500nm thicknesses*. That appears to be **consistent with** Table II if we **(i)** blame mechanical damage exclusively when the fibres are <100*nm* (amphiboles) or <170*nm* (for chrysotile); **(ii)** blame UVB for most of the rest of the <250*nm* cases; and **(iii)** blame *UV in general* (sometimes bypassing the "PRACTICAL LIMIT" putatively offered in wider fibres) for the *"moderate effects with 250–1500nm thicknesses"*

On top of all that, is the complication that there is the frequent "untidiness" of chrysotile fibres. On the one hand, that would tend to upset any tendency to transmit UV; and on the other hand it probably favours mechanical damage (e.g. with fine fragments breaking away from other pieces).

4.2.6 The Effect of the External Refractive Index

For the two cases considered, (n_{aq} = 1·35 and 1·4), there is clearly no great difference (except perhaps for chrysotile) — but it is not clear how much we can validly extrapolate from these cases. Meanwhile it would be helpful here if more research (♠) could be devoted to measuring the optical properties of the various aqueous and other bio-media. (That need for the same data also applies to the mitochondrial discussion raised by Thar & Kühl [16], who used the same two n values as a necessary speculative guess).

Actually this situation *could* turn out to be unexpectedly complicated, due to findings such as those of Preoteasa & Apostol [17] and later work — that local "coherence domains" (CDs) could cause inconstancies in the optical properties of liquids, *especially near boundaries*. As such inconstancies would tend to be dynamic, and if some fibre-optic transmission is only just on the borderline of success (depending on an n-value outside), then that sets the stage for a triode-like situation which can switch the transmission on-and-off — or indeed serve as an amplifier. I had contemplated investigating this possibility further, but in view of the apparent slight n_{aq}-effect shown in Table II, it seems sensible (at least temporarily) just to await events in this area!

5. Politics and Lobbying over White Asbestos (Chrysotile)

Chrysotile ***is*** different, and commercial apologists like Bernstein are quick to point out ways in which it is supposedly benign.[18,19] Table II (if valid) partly supports that claim, but only partly, and rather weakly: To have a comparable UV-transmission, these fibres would apparently need to be some 50*nm* thicker than their amphibole counterparts, perhaps leaving a "harmless-diameter" interval at about 150*nm* before the *mechanical* toxicity begins. (It might be interesting to test that specific point, with stable fibres of that 150*nm* size.♠)

At the same time, that gap (if it exists) could help to *confuse the evidence*, especially if we fail to recognize a likely double-cause phenomenon: In chrysotile's case, the two causes could well be of about equal potency (on different occasions), whereas for amphiboles it is likely that *one cause* (the UV) would largely eclipse any slight "mechanical" effect. It might thus be worthwhile bearing this complication in mind when assessing the politics of the situation — and careful locally-sensitive reviews of such lobbying might be considered, giving due weight to economic realities, and to those participants (on both sides) who are less well funded. (♠).

One point in partial favour of chrysotile is that it is *slightly soluble* (as Bernstein is keen to point out); so it tends to disappear within weeks [18]. That seems desirable, though it might not stop crucial damage which could occur meanwhile. Moreover there is a possibility that, while shrinking, it might briefly expose *new* micro-sites to "short-circuiting"!

6. Mitochondria and their likely UV communication

It is no secret that I support the hypothesis that the human/vertebrate nervous system carries a *double* set of signals: — "[A]" the orthodox Action-potential system — *but also* the yet-to-be-confirmed Infra-Red + RNA system "[R]".

Thar & Kühl's very interesting paper [16] (and the many other minor clues about the separate importance of UV), now raise the possibility of yet another communication system — even if it seems to be much more parochial in its reach. So let us think of it tentatively as the "[U]" system, serving the individual cell and its close environment — apparently via UV, and (at least when necessary) sending a directional UV laser-beam in a (perhaps) highly directional signal-stream or energy-boost.

If that is the case, then it is easy to see that such an arrangement could occasionally be severely upset by anything which misdirected such signal flows — and of course that is consistent with ideas put forward earlier in this paper.

It is perhaps significant to compare long and short mitochondria. For the *long* version, Thar and Kühl [16] argue that the reflection necessary for laser-production could be provided by a particular optical trick of multi-layer pseudo-mirrors (which, here, could only work for UV), whereas Cope [20] had argued in 1973 that the *short*[10] mitochondria were reverberation chambers for IR, using some sort of orthodox reflection (though he did not even mention that such reverberation could generate laser-beams). Meanwhile his account of the supposed reflection did not seem very convincing if applied to visible light (and it certainly would not have worked for UV), nevertheless he may have been right for the case of IR. To sum up then: It may be the case that long mitochondria are "tailor-made" for UV, while short mitochondria are "tailor-made" for IR. That raises the intriguing-but-peripheral question

[10] Long mitochondria were not identified until recently, so Cope would have known only short mitochondria at that time (1973).

of whether these are just the same sort of mitochondrion, reconfigured for the particular task in hand — or whether they would have permanently differentiated beforehand.

In case the [U]-system idea seems a bit far-fetched, it may help to look first at Table III depicting advances in our social communication systems, and then compare it with Table IV which outlines ideas (which seem to correspond) about biological communication. Both tables illustrate the thought that, just because we have already found one or two ways of doing things, that may not be the end of the story — there may be more to come. Of course that is not proof, but it should encourage serious investigation and support on such matters.

Table III Four types of Social Communication

	[M]	[A]	[R]	[U]	#
communication system (society):	"**M**ail": using *paper*	"Telegraph" Paper→Elec.Sig. →Paper (again)	"Telephone"	"Broadband"	S
memory storage (office)	Filing cabinet?	Filing cabinet?	Magnetic tape	DVD, CD	S1
Office-code	Written language	Morse code	Modulated electrical	Digital info-technology	S2
Signal carrier	Post-vans	Wires & cables	Cables & radio	Digital radio	S3
Signal emitter	Letter-writer	+ Morse-key	microphone	Lasers, etc.	S4

Journal of Physics: Conference Series **329** (2011) 012017 doi:10.1088/1742-6596/329/1/012017

Table IV Four types of Biological Communication, (equivalent to Table III?)					
	[M]	[A]	[R] e.g. see [15]	[U] ***speculative!***	#
Biological communication -system: →	Endocrine system (including selective coding)	Chem→ElecSig → Chem at Synapses. Action potentials	RNA→IR/Nerve-myelin →RNA. Local and between nodes of Ranvier	Cell-network of microtubules and mitochondrial UV-lasers(!?) [16]	B
memory storage (biological)	Immune system?	Synaptic configuration	RNA-network of "schèmes"	DNA? centrosome? nucleus?	B1
Bio-code	Protein shapes	?!	IR modulation?	Epigenetics? UV-modulation?	B2
Signal carrier, (biological)	Blood / lymph systems	Nerves, with or without myelin	Myelin (or chitin?)	Microtubules[11] + long mitochondria	B3
Signal emitter	Endocrine gland	Neuron-body; then synaptic bouton	?glia? using *short* mitochondria. [20,21]	*Long* mitochondria in the cell [16,21]	B3

7. Conclusions

There can be no firm inferences at this stage, but the collective evidence does seem to be coming together to suggest a coherent overview. Two areas deserve comment:

7.1 "Secret" metabolism involving UV and/or IR

It is usual to assume that metabolism is just a matter of chemistry. (That view is traditional, and the bias is indeed reflected in the pro-chemistry training of medical researchers). However it is becoming apparent that there are important and sophisticated *physics-based* aspects to metabolism. That can now hardly be doubted regarding IR, thanks to the work of Albrecht-Buehler — e.g. [22].

However our present concern is with UV, and here the evidence is still largely theoretical and circumstantial, as suggested in Tables III and IV — plus investigations such as those of Thar & Kühl [16]. Nevertheless that should suffice to induce us to take the topic seriously — and the corollary that upsetting such systems could have significant consequences.

In comparing UV and IR systems, it is worth bearing in mind (i) that a typical UV photon has about **ten times the energy**, and hence is more likely to do damage if misplaced; and (ii) by the same token it is more suited to any **changes to more stable** configurations such as DNA-complexes, (whereas IR probably confines itself to modifying those comparatively benign changes of day-to-day biochemistry). Meanwhile (iii) the UV photon's wavelength and other linear measurements will be about 1/10 compared to IR — and that seems appropriate if UV's role is **ultra-micro**.

7.2 Some toxicity may entail a signals-upset within UV metabolism

As shown in Table II, asbestos seems theoretically able to act as a conduit for any UV signals which happen to enter the ends of its fibres. The postulate then follows that such conduits would then

---

[11] Microtubules need not be bound by the constraints imposed by Table II. (Table II applies to fibres deemed to be optically uniform internally, and therefore compelled to use the "zigzag" transmission modes which lead to the constraints). In contrast, •coaxial cables, •commercially-designed optic fibres, and (*perhaps?* ♠) •microtubules, can circumvent this constraint. In those circumstances, they could then transmit the usually-preferred "TEM" mode more-or-less as if it were travelling (unconstrained) in free space.

sometimes act as "short circuits" — delivering messages to inappropriate places, and thereby upsetting the normal metabolism. Alternatively they might just degrade the efficiency of normal UV activity — whatever that may be.

7.3 Other toxicity seems due to asbestos's mechanical damage to DNA

It is clear that this can happen with very narrow fibres, (though it is less clear just how common that might be in natural conditions). It has here been assumed that thicker fibres would be less likely to be "knife-like" in this way — but it might be prudent to test that assumption: One perhaps-useful test ♠ would be to manufacture *opaque* fibres (to eliminate any UV-causality), and then do in-vitro studies in the standard way — though no doubt there would be practical difficulties.

7.4 New experimental studies are needed, and perhaps in new directions

Further directed experimental investigations are obviously called for, and some suggestions are offered in the list which follows:

8. List of Suggested Experimental Investigations

Once again, as in the previous paper [15], I offer a list of items, (marked "♠" within the general text) which already seem amenable to experimental testing (or will ultimately need testing despite problems); and I invite those with the relevant resources to investigate them — with or without my collaboration.

§	Foot note	Description of suggested area for experimentation, marked "♠" in the text
4.1 + 4.2.3	(3) —	Direct test for the supposed ability of asbestos fibres to transmit UV (under predicted conditions)
4.2.3	—	Test the hypothesis that fibre toxicity depends *entirely on its geometry and optical properties* — and not (as such) on its chemistry, or its classification into "asbestos" or "glass", etc.
4.2.3	8	*If possible* obtain manufactured fibres of unusual media (*e.g.* with $n>2$ — such as extreme types of flint glass, or diamond), and test such fibres for UV transmissibility
4.2.3	8	*If possible* obtain manufactured fibre-samples in which there is *high uniformity of fibre-properties*, thus eliminating much of the current uncertainty which comes with statistical studies of toxicity.
4.2.3	—	Compile basic tables of refractive indices *within the UV range* — especially for asbestos
4.2.6	—	Compile better basic tables of optical properties (*including the UV range*) for bio-media — especially aqueous solutions and lipids.
5	—	Test the prediction that 150*nm*-wide chrysotile might be *less toxic* than if wider or narrower
5	—	Re-assess the public-health hazard and local politics of chrysotile, given the likelihood that it may have *two different* causal effects (which may have confused the evidence in the past).
6: Table IV	11	More intensive study of the *optical properties of microtubules* (and other features related to mitochondria)
7.3	—	*If possible* get *opaque asbestos-like fibres* (to eliminate UV effects) and then do standard carcinogenic tests via the other causes — especially mechanical damage, and whether it decreases as fibre diameter increases.
—	—	Systematically collect and collate more-exact wavelength details about ultraweak UV emissions

Journal of Physics: Conference Series **329** (2011) 012017 doi:10.1088/1742-6596/329/1/012017

References

[1] Stanton,M.F., and Layard,M. (1978). *The carcinogenicity of fibrous minerals*. National Bureau of Standards Special Publication 506 — from Gaitherburg conference of July 1977.

[2] Traill,R.R. (2010 Nov). "The theoretical case that some asbestos fibres could trigger cancer optically, while others act mechanically". Ondwelle Publications: Melbourne. http://www.ondwelle.com/UV2_trigger.pdf — and also *General Science Journal*.

[3] Merck (1992). *The Merck Manual of Diagnosis and Therapy*. Merck & Co., Inc.: Rahway, N.J.

[4] Hesterberg,T.W., Ririe,D.G., Barrett,J.C., and Nettesheim,P. (1987). "Mechanisms of cytotoxicity of asbestos fibres in rat tracheal epithelial cells in culture". *Toxic. in Vitro*, **1**(2), 59-65.

[5] Tennent,R.M. (ed.)(1971) *Science Data Book*. Oliver & Boyd: Edinburgh.

[6] Webmineral.com (2010 June 21: main access): http://webmineral.com/determin.shtml ; http://webmineral.com/determin/biaxial_minerals.shtml ; http://webmineral.com/determin/isotropic_minerals.shtml

[7] Shelley,D. (1985). *Optical Mineralogy*. Elsevier: New York

[8] Leake,B.E. (chairman) *et al.* (1997). "Nomenclature of amphiboles: …". *The Canadian Mineralogist*, **35**, 219-246. [Their preferred system, which differs from medical usage].

[9] Toyokuni, S. (2009) "Mechanisms of asbestos-induced carcinogenesis" [review]. *Nagoya J. Med. Sci.*, **71**, 1-10.

[10] Wang, N.S., M.C.Jaurand, L.Magne, L.Kheuang, M.C.Pincheon, and J.Bignon (1987). "The interactions between asbestos fibers and metaphase chromosomes of rat pleural mesothelial cells in culture". *Am. J. Pathol.*, **126**(2), 343-349. [http://www.ncbi.nlm.nih.gov/pmc/articles/PMC1899582/pdf/amjpathol00149-0151.pdf]

[11] Ault,J.G., Cole,R.W., Jensen,C.G., Jensen,L.C.W., Bachert,L.A., and Rieder,C.L. (1995) "Behavior of crocidolite asbestos during mitosis in living vertebrate lung epithelial cells" *Cancer Research*. **55**, 792-798.

[12] Voytek, P., M.Anver, T.Thorslund, J.Conley and E.Anderson (1990). "Mechanisms of Asbestos Carcinogenicity". *J.Amer.College of Toxicology*. **9**(5), 541-550.

[13] Traill,R.R. (1988). "The case that mammalian intelligence is based on sub-molecular memory coding and fibre-optic capabilities of myelinated nerve axons". *Speculations in Science and Technology*. **11**(3), 173-181. www.ondwelle.com/OSM10en.pdf

[14] Jenkins,F.A. and White,H.E. (1950). *Fundamentals of Optics*. McGraw-Hill: New York & London.

[15] Traill,R.R. (2011a) "Coherent Infra-Red as logically necessary to explain Piagetian psychology and neuro-microanatomy — Two independent corroborations for Gurwitsch's findings, and the importance of coherent theory". — *In this volume*.

[16] Thar,R., and Kühl,M. (2004). "Propagation of electromagetic radiation in mitochondria?". *J. Theoretical Biology*, **230**(2), 261-270. http://www.mbl.ku.dk/MKuhl/pages/PDF/Thar&Kuhl2004.pdf

[17] Preoteasa,E.A. and Marian V Apostol. (2008) "Collective dynamics of water in the living cell and in bulk liquid. New physical models and biological inferences". Prague conference (July 2008) *proceedings*.

[18] Bernstein,D.M. (2006). *Understanding chrysotile asbestos: A new understanding based on current data*. www.chrysotile.com/data/conferences/Presentation_David_Bernstein.pdf

[19] Bernstein,D.M. (2009). *Review of the differences between chrysotile and amphibole asbestos*. http://www.chrysotile.com/data/

[20] Cope,F.W. (1973). "Electron-phonon (trapped photon) coupling and infrared coaxial transmission line theory of energy transport in mitochondria and nerve". *Bull. Math. Biol.*, **35**, 627-644.

[21] Traill,R.R. (2010 Oct). *Meaningful UV and IR photon-exchange within bio-tissue? — Interdisciplinary evidence, and a new way to view asbestos toxicity*. Ondwelle: Melbourne. www.ondwelle.com/UV1_evidence.pdf — and also *General Science Journal*.

[22] Albrecht-Buehler,G. (2006? — accessed:2010-Sep). *Cell Intelligence*. www.basic.northwestern.edu/g-buehler/summary.htm (and then http://www.basic.northwestern.edu/g-buehler/htmltxt.htm etc. — with animations).

Journal of Physics: Conference Series **329** (2011) 012018 doi:10.1088/1742-6596/329/1/012018

Coherent Infra-Red as logically necessary to explain Piagetian psychology and neuro-microanatomy — Two independent corroborations for Gurwitsch's findings, and the importance of self-consistent theory

Robert R. Traill

c/- Ondwelle Publications, 29 Arkaringa Crescent, Black Rock (Melbourne), 3193, Vic., Australia

rrtraill4@dodo·com·au ‡

Abstract. We can infer mechanisms underlying advanced human intelligence via •*physics, chemistry* and *information-technology*; but also •*epistemology*: analysing *all* knowledge-building processes (based on selection amongst micro-ideas). Hence Piaget offered *"schèmes"* as such items of thought/action, to account for actual human behaviour. In microphysiological terms, basic "schèmes" should have digital properties and a one-dimensional organization. That implied RNA-like molecules (and their "chorus" groups). — However for the necessary fast intermolecular communication, traditional action-potential "spikes" would be much too coarse. The alternative is Infra-Red. It then appears that myelinated nerve fibres seem suitable for an unexpected *second role as coaxial cables for IR!* —— Such *OPTICAL* interpretations also explain several enigmas: •What keeps (myelin-thickness / axon-diameter) ≈ constant? — (a mystery since 1905). •Why PNS myelination is delayed until the axon diameter reaches $1\mu m$. •Why myelin often stops growing at a predictable angle in its wrapping. •Callahan's anomalous failure to detect sensory action-potentials even though his insects were *responding* to invisible IR signals. (Meanwhile *RNA-LIKE CODING* explains ♦inherited behavioural traits, and ♦memory-"recording" as Darwinian!) —— *The important point:* Here IR is independently identified as a necessary solution to logistical problems. In contrast the Gurwitsch tradition first discovered emissions, and then sought explanations. Thus the two approaches corroborate.

Introduction

§1. Mainstream science has not yet really accepted Gurwitsch's 1923 discovery of "ultra-weak radiation" from living tissue, [1,2] despite the increasing evidence for its universal nature. At best, such infra-red and ultraviolet photons are perhaps grudgingly accepted as *"a quaint sideline which need not bother us"* — forgetting the huge importance of those other *quaint sideline*s investigated by Galileo, Oersted, and many others.

Why this neglect of bio-emission? — And can we remedy such negligence?

‡ *or*, if obsolete, consult the website: www.ondwelle.com

Here I report a project which might aid acceptance in two different ways: Firstly, as we shall see in Part B below, it offers independent corroboration — applying Information-Technology to conclude *theoretically* that infra-red (IR) signalling is *necessary to explain advanced intelligence within the brain*; — a conclusion which initially arose in total ignorance of any ultraweak evidence, thus ensuring independence.

The second mode of assistance is less obvious, but more fundamental so I shall deal with it first. (It deals with the theory of knowledge-acquisition — both by any brains we study and by the scientific establishment itself as "Scientific Method". The more physiological issues are deferred until Part B.)

Part A: The finding and acceptance of new knowledge

§2. A shared theme for two different knowledge-domains

This project's chosen topic of *brain-and-intelligence* led naturally to enquiries into "EPISTEMOLOGY" (the nature of *knowledge-generally*, its storage/access, and how it is acquired). But conveniently, such questions can also be readily reassigned from the brain-studies into problems of Knowledge-within-Society (i.e. to Science)[1] — thus encouraging a critical re-examination of doctrines about "Scientific Method", and hence about what helps-or-hinders the acceptance of legitimate innovations.

§3. Popper and some negative impact on research-policy

Such methodological problems often relate to two details in Popper's early formulation of scientific method in 1934 [3] which seems sound otherwise — so his account is a useful place to start the discussion:[2]

One such controversial detail was this: *He overemphasized observation as if it were the only way to "test" a hypothesis (and in a rather uncompromising way)*, a bias which has now too often become a mindless mantra. That climate made things difficult for Gurwitsch in the early days [4,5] — though paradoxically, that is apparently no longer a problem for ultraweak emissions which are surely so well supported observationally by now that they should have been accepted if Popper's principles were indeed the only standard. That anomaly should alert us to look for other criteria which he had overlooked.

The other Popperian detail is that his own rules *do not obey their own testability criteria* (if interpreted as strictly observational in the usual way): His typical disciple will assert these rules, as if they themselves are above any need for testing or rigorous explanation in biological terms. This has relevance to the discussion of "meta-levels" which we will come to later in §5. But now let us look beyond Popper:

§4. Piaget and his added "equilibration" concept

Both "brain" and "society" knowledge-domains were much discussed by the self-proclaimed epistemologist Jean Piaget (1896-1980) of Geneva — who was also a notable biologist and psychologist, well versed in relevant philosophical writers such as Hegel and Kant. He went beyond the philosophical discussion, and considered in some detail just what "in principle" brain-mechanisms and processes would be needed to explain human intelligence and how it develops through childhood. (We shall see that the present project has sought to move further along this path, looking for actual

[1] See below, in the section on "Four Epistemologies" — §10.

[2] Popper himself might well have acquiesced with these two criticisms which do not seem crucial to his own main argument. But by now that is hardly relevant. The problem is that his written-words of 1934 have become sacrosanct within many parts of the scientific community — and its funding bodies.

physiological bases for his "in principle" mechanisms such as his "schème" element as a basic unit of thought/action — leading then to those above-mentioned conclusions about IR).

Another key feature emphasized by Piaget (and largely overlooked by Popper) was that new candidate concepts, whether true or false, had to compete to fit into a perhaps-false pre-existing mind-set — and truth does not necessarily win in such circumstances. There is often a contest in which the pre-existing mind *might sometimes (partly?)* "accommodate" to the new idea, but it is also more likely that the new idea-or-perception will itself be somewhat modified ("assimilated") to fit into the pre-existing mind.

This *"equilibration"* process of adjustment to maximize internal-agreement is clearly fallible, but it does have some positive points. Firstly it has a trial-and-error simplicity about it, allowing it to self-organize in a way similar to other biological activity — in fact it is essentially a Darwinian process, as we shall see. Secondly, as long as its conclusions are *not too rigidly held* (whatever that means), the mind-set remains open to further revision; and indeed competing models may often coexist. Thirdly it seems to work brilliantly for certain relatively-simple tasks such as learning the properties of solid objects[3] where, for instance, the infant has to acquire the surprising concept "that any rotation through 360° brings back the image I started with!!" Fourthly, it probably works statistically as considerably better than random guesswork. Fifthly there is apparently no alternative within simple epistemological systems — and that includes *the start-up phase* of systems (like *parts* of our awareness) which are destined to become more complex and "intelligent".

§5. Intelligence and "meta-levels"

According to both Piaget and Ashby, this intelligence starts to develop when the brain allows part of itself (some of the schèmes) to operate upon other parts instead of interacting directly with channels-to-or-from the environment (Traill 1978b/2006)[5]. This new "meta-level" (ML or M^1L) transcends some of the simple Darwinian activity of the base level (L or M^0L), so creating a sort of local "designer operated" situation. (As they mature, humans then go on to develop a whole cascade of these levels, and that seems to account for human intelligence in comparison to other animals where such hierarchies are of more limited scope. But note that the top level of any such hierarchy will always be unmonitored — with nothing but pseudo-Darwinian "equilibration" to guide it).

§6. The enigmatic power of "equilibration" or "making sense"

It seems then, that when we build up our concepts of the outside world, it is not enough to have observations, not even if this input is impeccably reliable. In the end we also need to be able to make some sort of sense of it when we turn it over in our minds — and this amounts to Piaget's "equilibration", or the "coherence-seeking" considered by some philosopher-epistemologists like Hans Albert [8](1968/1985) and Susan Haack [9](1993).[4]

---

[3] For more complex tasks such as social-interaction, infants need all the help they can get, so some relevant non-trivial schèmes (such as smiling, basic recognition, eye-contact, and various other "intuitions") have evidently evolved in an orthodox Darwinian way *within the species* and are then usually inherited in a way which is easy to explain if these schèmes are indeed RNA-like. The occasional lack of these "prefabricated service-parts" will presumably cause autistic tendencies unless their knowledge-content can instead be learnt by that individual in a belated pseudo-Darwinian way like other less-important learning tasks — or even sometimes by instructive therapy. (Whenever such *complex-basic-skill* schèmes *are* inherited genetically, that comes close to what Kant (1781)[6] mistakenly believed to be true for the easier-to-reconstruct abilities such as our understanding of Solids-and-Space).

[4] Unfortunately the word "coherence" is ambiguous within the present discussions:
It can mean *"self consistency of evidence"* as in a Piagetian equilibration process; but it can also mean

(That then may be why ultraweak bio-radiation has not received the public attention it deserves. No matter how impeccable the observational evidence, the trouble is that hitherto it has *not seemed to make much sense* — and that is where the later parts of this paper might start to help, at least for the case of IR-around-nerves. UV is probably a separate issue relating to the control of mitosis, &/or some *special* brain activity,[5] and if we can ensure that such UV also "makes sense" then no doubt the scientific world will become much more interested in it too.)

§7. How could such a system ever get started?

Pure "sceptical" philosophers like Hume (1748)[12] have long had misgivings about the whole epistemological exercise, because it has seemed to them to be a theoretically impossible task to accumulate knowledge if one starts unaided from nothing![6] The start-up enigma is that any recording-or-interpreting mechanisms will have some knowledge already implicit in their structure, so these cannot be allowed in any genuine start-up system (according to the more-rigorous philosophers). So how could such a system have coped usefully with incoming information? (This dilemma is akin to asking how life can be reproduced unless we start with a living system — and indeed we will see that this is no mere analogy, but a special case of the same thing if we accept that "ability to conduct life" is a form of knowledge).

The "start-up" answer probably lies in the fact that all systems which actually survive "long enough" (whatever that means) will have one thing in common — their *dynamic stability*, and that can be interpreted as "internal coherence" or "some measure of equilibration" or "a sufficiently self-consistent pattern of mutual attractions in real or virtual space". That means that when arbitrary signals about some "observed" environmental system are received by a "start-up brain", that brain will manipulate those signal-traces somehow, and sometimes arrive at some dynamically stable internal configuration using them. That stability, when it occurs, will offer a workable chance of capturing some feature of the "real" setup outside. Well, that may not sound like much, but it does offer a toe-hold which could have allowed the process to get started — and from there (like life) it might well escalate significantly.

That is an argument which may suit biologists. However philosophers are seldom used to thinking in terms of dynamic systems dependent on Darwinian trial-and-error, so they tend not to see coherence as a fully respectable solution to their epistemology-enigma. They do nevertheless sometimes consider it seriously while they continue to look for something more tidy and "logical", or just bemoan the absence of such neatness.

§8. Even then, how is further knowledge-acquisition possible?

In 1968 Popper himself offered a helpful two-page summary of his 1934 book [3] and its objectives (as the preface to its third German edition) — affirming that it was indeed theoretically possible to get

---

"phase unanimity within a beam of radiated waves". However the distinctions should usually be clear from each immediate context, as long as one bears in mind that the ambiguity exists.

[5] Note this interesting difference: In the peripheral nervous system (**PNS**), myelination does not start until the axon-diameter reaches *1μm* — implying a possible association with NIR having wavelengths ≈ *2μm*). However myelination within the *central nervous system* (**CNS**) sometimes starts at much smaller diameters (see images in Peters *et al.* [10]) which suggests a significant presence of UV, though that may just be a byproduct of UV activity unrelated to neural transmissions; — see §8 below, and [11] §5.1.

[6] This *start-up* (ideally a once-only affair) does not refer to any present-day real brain (which clearly does have some inherited abilities) but rather to either (i) hypothetical new start-up situations envisaged legitimately by philosophers, or (ii) whatever it was — countless generations ago — which did actually initiate brain-like activity and the inheritable structures which encoded it. Likewise in the *social* domain the real case does not refer to present-day society, but rather to some start of collective knowledge-and-language (before stone age society evolved).

increasingly closer to the truth, but (as in the previous paragraph) we should not expect perfection. Thus, for instance:

"The answer is not pessimistic ... or 'sceptical' ...: It shows that we can learn from our mistakes". NB: (i) Like Piaget, he invokes trial-&-error (and doesn't that mean "Darwinian"?); but (ii) unlike Piaget, he does not really ask what further mechanisms are required in the learning process. Instead he effectively just takes these for granted (as we all tend to do) — thus overlooking the power of equilibration/coherence.

Interestingly, he then admits to finding out subsequently that his outline (of how epistemology must operate) was already 2500 years old — having been put forward by Xenophanes, whom he then quotes!

§9. Darwinian "Trial-&-Error" or Lamarckian "Instruction"

In 1967, the immunologist Niels Kaj Jerne[13] offered an interesting insight into the knowledge-acquisition process in general — (My *italics* — and quoted previously [14]):

"Looking back into the history of biology, it appears that wherever a phenomenon resembles learning, an *instructive* theory was first proposed to account for the underlying mechanisms. In every case, this was later replaced by a *selective* theory.
Thus ... species [as divinely designed] ... until Darwin...
Resistance of bacteria [purposively! reconfigured] ... until Luria and Delbrück...
Adaptive enzymes [purposively! reconfigured] ...until Monod and his school...
Finally, antibody formation [in **immunology**]..." — [the actual topic of Jerne's paper].

On reflection, that simple selection-of-what-actually-works must surely be true *for any natural epistemological system in its start-up phase* (when it cannot yet have acquired any means whereby it can be "instructed"). Later on, it may have the choice, but "trial-&-error" may still remain preferable for various reasons — such as avoiding the *intellectually-difficult and costly maintenance of designed equipment* (hence the development of "throw-away" modules which *can* sometimes be used in a trial-and-error fashion despite being "high-tech"!)

Yet of course, once we have developed human intelligence (see §5 above), we are free to operate in a Lamarckian way: to create innovative designs, and pass them on for others to use. One minor problem is that we may then fail to recognize the existence or value of Darwinian processes when they occur. Or more seriously, we adults may be so familiar with our own advanced abilities, that we do not notice *any mystery at all* in how we achieve these abilities. Thus (e.g.) if we *"see"* something, then "obviously it must be true" — and usually it is, but not always (as optical-and-other illusions tell us) and then we may just dismiss it all as a weird anomaly. That is where careful attention to Piaget's analyses of childhood mental-development can be useful and instructive — helping to protect us from complacent assumptions about our own "mystical otherworldliness".

§10. The four knowledge-gathering domains

Inspired by Jerne, Popper (1973)[15] identified four main types of epistemological system: (a) The Brain of the individual; (b) Jerne's immunology; (c) Classical Darwinian species-evolution with the knowledge stored within DNA-coding; and (d) Society-as-such including its scientific knowledge, separate from individual brains.

(I have already discussed these four domains at length, so I shall be brief here; but see [16] especially *Table S* on page 31, and [14] Chapter 4).[7]

[7] Respectively http://www.ondwelle.com/OSM02.pdf and http://www.ondwelle.com/BK0_MU6.PDF .

Note firstly that since the epistemological task has been recognized for some 2500 years as "almost impossible", and that only one formal solution to the problem has ever been discovered, it seems likely that all these four domains will be using *that same formal strategy* even if other details are very different. Thus, taking *the* "true-Darwinian" case (c) as our bench-mark (since it is the one most thoroughly understood), we now seem entitled to claim that all four domains will be dependent on Darwinian trial-&-error — at least *at start-up*.

We might also expect uniformity in some of the details which accompany this strategy, such as the need for largish populations (often including replications) of assorted *"candidate-recipés-for-action"* from which to select — more-or-less randomly. Then also note that ultimately such recipés are usually in the form of *linear coding* (like writing) in cases (b), (c), and (d). So what about the brain, case (a)? The text-books favour some arcane configuration of synapses acting alone as in *Figure 1* below — but that does not offer much scope for *organized linear* coding, nor indeed much scope for *stable digitized* code, another apparent requirement for knowledge. So can we suggest something more in line with the (b)-(c)-(d) cases?

It is clear that Piaget's simpler schèmes entail a *linear narrative of action*.[8] These schèmes thus offer a suitable match to the (b)-(c)-(d) cases — (though no doubt the synapses still have their own role to play, perhaps mainly as *analogue* components, e.g. "directing traffic" and/or "volume control", and most crucially providing input and output links to the outside world). Meanwhile though, the new Piagetian interpretation is one key to our discussion of IR emissions in Part B, below — where the schème is considered as being an "RNA-like" molecule.

Differences between the four domains? Certainly some of the other details are very unlike — notably *the time-scale* and *the physical embodiment* for the various linear recipé-types — but these differences need not matter as they do not affect the formal requirements; — [16], Table S again.

§11. Coherence — "internal" and "external"

We can think of encoded-knowledge as a model supposedly representing significant aspects of the world "outside" — though much more accessible than that world, and also offering opportunities for the comparatively-harmless testing of "new ideas" or "mutations" or "sensory perceptions".

In the Pure Darwinian/DNA case, (c), survival certainly depends partly on achieving a model (within the phenotypic body-configuration of the individual) which matches the demands of the environment. Likewise in "(d)", Popper's criterion (that a hypothesis should survive experimental testing within the "real" world) also has some validity. But in neither case is such *external-testing* the whole story:

In "(c)", an embryo whose organs are not properly configured and interrelated *internally* will probably be aborted prematurely long before it is tested in the hostile outside world. That is a failure of internal consistency or "internal coherence" — and it is surely a vitally important type of test. Indeed when life first evolved, this would probably have been *THE most crucial test*, given the temporary absence of competitors or predators. Moreover getting organelles to be mutually beneficial by trial-&-error means would be no small feat — never mind environmental complications (apart from sources of energy and supplies of "arbitrary building materials").

Concerning "(d)": When Popper wrote about "testing" he was clearly just thinking about the external (experimental) type of test. I suggest that this was a mistake, and he should also have allowed about half the tests to be *"thorough tests of internal consistency"* — i.e. tests of internal coherence —

[8] — and his more complex schèmes are *ensembles* of those simpler schemes, linked in a way which is also arguably controlled by "higher level" simple schèmes; [7] Part C.

Journal of Physics: Conference Series **329** (2011) 012018 doi:10.1088/1742-6596/329/1/012018

tests for whether the ideas actually made sense within a dynamic whole.[9] Note that it is important to clarify this point here, because Part B (below) depends heavily on this theoretical tool — given that it is enquiring into mental/neural phenomena which *cannot currently be investigated directly by experiment*. Later the research-possibilities might change somewhat, but it is necessary to establish a plausible *prima facie* case first — and that can only be done via internal coherence, using all the background interdisciplinary understanding we can muster.

However we should recognize the original reason for this twentieth-century bias against theory and internal-coherence arguments. **In the social sciences**, there can be no such thing as a fully-reliable base concept — and even if there were, the outside world would probably soon change thus spoiling that reliability. It is nevertheless still prudent to consider (collectively) such concepts as *"altruism"* and *"the marginal propensity to consume"* to help guide our planning, as long as we remain aware of any vagueness and obsolescence of such ideas. However there are countless examples of overconfident politicians etc. who have taken some social formulae as immutable and infallible, and created havoc thereby. Pol Pot, some current financial "experts", and Robespierre all come to mind.

In contrast, the theory of "what underlies chemistry" *started* like that, with medieval alchemy; but it then progressed through a series of remarkable theoretical developments: • The periodic table; • Chemical formulae with countable atoms; • The Bohr/Rutherford "orbit"-model of the atom; and • The Bohr/Schrödinger "wave-function" model of the atom. The key to such progress largely lies in improvements in identifying what the most basic elements seem to be (and how they interrelate) — even if they end up as the enigmatic mathematically-defined "ψ-wave" of Schrödinger's theory. The key point is that this can be very precisely conceptualized, giving us a very clear model of the hydrogen atom at least — a model we could never match by purely experimental means. And the consequences of this discovery have been immense.

§12. Rethinking Popper's Scientific "Tests"

Popper treated hypotheses as if they were random mutations in a Darwinian environment, though he may not have recognized this explicitly. He did say, on at least one occasion, that it *did not matter where the hypotheses came from* (as long as they were testable), and that looks very much like "random mutation" about to be tested by reality. Of course this has some validity, but it raises at least two points:

Firstly, such extreme sudden-death testing is characteristic of those "start-up" Darwinian situations before the learning-system has any other guide-lines to help it — survival of the supposedly *fit* — but with "fit" considered unrealistically in absolute terms.

In real life, "fitness" may be a more messy concept, for which gradations and statistics may need to be applied. Biologists consider "survival of the fitt*est*" within a competitive world, whereas philosophers like Popper tend to seek *"perfection-or-nothing!"* That may be fine for simple problems, but note that in the above-mentioned history of chemistry, there was a whole sequence of not-quite-right partial truths — and without these, the apparently-right answer would never have been reached.[10] Moreover that does look like Darwinian evolution "(c)" within this new domain "(d)" — and more so than in Popper's prescription for that "(d)".

---

[9] He could also have considered how one might work with conflicts between such evidence — a matter I will not go into here, but see Thagard's book on such matters; [17], or his website.

[10] Intermediate advances can, for instance, help clarify what the basic elements of the problem might be — and that, in turn, might convert an apparently-complex problem into a simple one which would (at last) accord with the philosopher's demands!

Secondly, (as we have seen above in §7), tests of the *internal coherence* of hypothesis-concepts can be just as relevant as Popper's observational testing. Thus in general it seems prudent to aim for an equal balance between the two, but this should probably be varied according to circumstances. Currently there is an oversupply of observations in most fields, and a dearth of coherent explanatory theories[11] — so it seems timely to concentrate on the latter for the time being, especially in the present brain-intelligence project where no adequate direct observations seem possible.

(Moreover if there is any doubt about the boundary between the "internal" and "external", then we might even question whether there is a fundamental difference between plain "coherence" and "agreement with experiment". In a sense they are both coherence-tests, so I sometimes name them "internal coherence" and "external coherence" respectively.)

Anyhow, it seems appropriate to suggest that Popper's meaning of "test" should be extended to include tests for "internal coherence" *whenever appropriate* — but with safeguards to prevent too much reliance on it whenever the basic concepts are vague. Of course, in many fields the basic concepts *are* vague, at least temporarily. That is part of the problem, and that is why it seems so important for progress in psychology to find out what exactly a Piagetian "schème" really is in physical terms — and what its communication system must be. That indeed is the *raison d'être* for the current project, especially Part B below.

Part B: Molecular Memory and its need for IR-linkage

§13. Seeking a physical embodiment for Piaget's abstract "schème"

This project's chosen topic of *brain-and-intelligence* led first to some philosophy-related enquiries into "EPISTEMOLOGY" (the nature of knowledge, its storage/access, and how it is acquired) — and such epistemological issues have been discussed above in *Part A*.

For instance, *Part A* introduced Piaget's concept of **the "schème"** (as the supposed basic unit of thought/action) and how we may provisionally think of each simple schème as linear digitized coding (like writing or computer-tape). By elimination, that schème must apparently be an *RNA-like molecule.* (Whether or not it is actually RNA need not concern us at this stage, though the evidence points strongly that way).

Anyhow, here within Part B we will revisit that argument *from a more biophysical viewpoint*, and explore some logical consequences of that *"RNA-like"* conclusion, especially asking about meaningful communication between such ultra-micro sites and other parts of the nervous system. In broad outline, the argument follows the downward path through this icon-diagram:

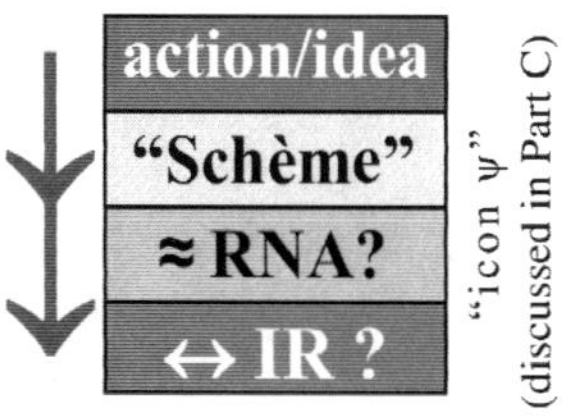

Such theoretical work raises another epistemological issue, also discussed in *Part A*: Thanks to Popper and his disciples, **theoretical treatments** have been regarded as **"not to be trusted *at all* unless corroborated"**; and that qualm definitely has some justification when the underlying concepts are vague, as in the social sciences. In the present case however, the systems being studied are mostly small enough for them to be described using physics and information-technology — disciplines which have largely defeated the vagueness-problem, and therefore do allow us to make significantly reliable theory-based progress. If we can next test such conclusions experimentally, then so much the better

[11] Sir Francis Bacon (1561-1626), said to be the "father of observational science", long ago predicted such pendulum swings between the two biases; (Bacon 1620[18]; also [14] endnote *b*, p.98).

(see the "♠" marks throughout the text)[12] — but many details of such ultramicro ensembles will probably always defy direct observation, so we may sometimes have to content ourselves with the theoretical solutions anyhow!

A sketch-outline of the arguments invoking "molecules+IR"

We can now look at aspects (1)-(4) of molecular-memory theory, partly illustrated by interactions between the layers *i-iv* of the above icon diagram which will appear several times:
(1) Apparent *molecular*-solutions to some enigmas within Psychology, — layers *i-iii*;
(2) Problems of communicating with such sites. IR as the only feasible solution, layers *iii-iv*.
(3) The surprise finding that existing myelin could have a second role as coaxial cable for IR!
(4) Some other surprise findings (including other areas) derived logically from these.

(1) The Memory-code Enigma in Psychology:— the first steps — i.e. the *advanced thought-ability of humans*, including details of its self-organized development?

B(1) – §14. What actual mechanisms could possibly explain thought?

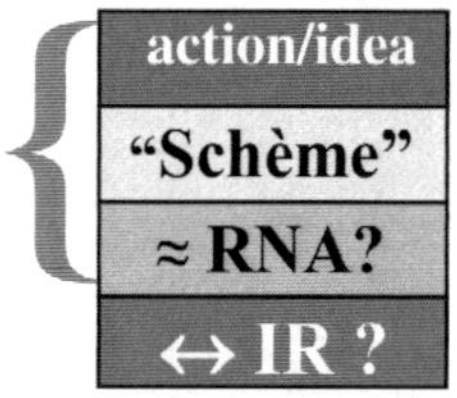

Orthodox physiology focuses on the observable activities of synapses and their electrical "action-potential" signals. This conventional view is that thoughts are *somehow* encoded and marshalled by means of neurons and their synaptic connections — as shown schematically in figures 1*a* and 1*b*, (though of course it has been recognized, at least since the 1960s, that actual neurons are usually much more complicated).

These synaptic arrangements are obviously relevant and they do explain *some* phenomena such as "pattern recognition" abilities — but no-one seems to know how to use such elements to explain *advanced human intelligence* except in vague terms. In particular, if these are the *only* available mechanisms, just how could they store the encoding for a remembered Shakespeare sonnet, or a mathematical theorem, or how to play tennis strategically? — And there are the added insoluble(?) problems of actually *encoding* such thoughts initially, and then separately *retrieving* them at appropriate times; etc. — not to mention the biological need for such systems to *self-organize* during development.

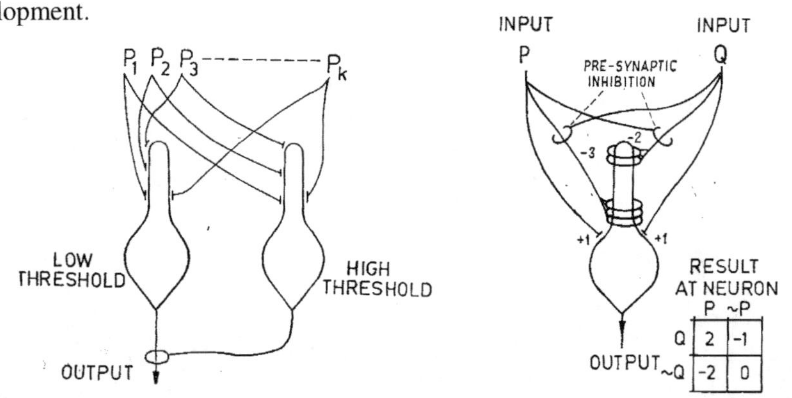

= Figs 3 & 1 — from [19] pp862-3

Fig.1 The text-book view of bio-memory mechanism. These schematic diagrams illustrate the traditional approach. However this is simply not adequate (on its own) for explaining human intelligence. Instead we need to consider a *molecular* arrangement (as suggested in Fig.2).

[12] I have used "♠" in the text to indicate *points* which ***now seem to be testable-or-retestable experimentally***. These items are listed in the Appendix (§33), which summarizes the apparent possibilities at this stage.

Moreover it seems that these are essentially just analogue devices (at least in their more realistic complex form), whereas intelligence surely requires ***digital*** capabilities — so we need to seek some additional memory-mechanism which is digital (AND capable of ***neat sequencing***):

B(1) – §15. **From analogue-neurons to digital "instruction-lists"?**

This is where Piaget's *"schème"* concept can be offered as a semi-solution. — In Piaget's own work, this *schème* (an idea modified from Kant [4]) is an *abstract* entity embodying a mutatable *unit of action*, from which we humans can gradually build increasingly sophisticated concepts; ([5], espec. §C5.2).[13]

One key innovation here was to move our focus away from the mind's concept of *nouns like "chair"* and *onto verbs like "draw"* — suggesting that our initial concept of "chair" is a gradually assembled ensemble of schèmes for •drawing, •outline-tracing, and •sitting, etc. That led to a mass of epistemic-and-psychological theory which we can bypass here; but it seemed that Piaget's "schème" would probably have to be encoded in **linear 1D** form, as strings of **digital** "beads" — all very similar to list-like elementary computer code, or a string of verbal instructions. Anyhow here, in Fig.2, are some diagrams building on this 1D approach.

[13] Those readers who happen to be familiar with "machine code" or other "low level" computer programming, will no doubt recognize an analogy with the *"sub-programming"* technique — with the simplest, lowest level subprogram coding being analogous to the simplest, lowest level Piagetian schème. That comparison is useful, but of course it is important not to take such analogies too far.

Journal of Physics: Conference Series **329** (2011) 012018 doi:10.1088/1742-6596/329/1/012018

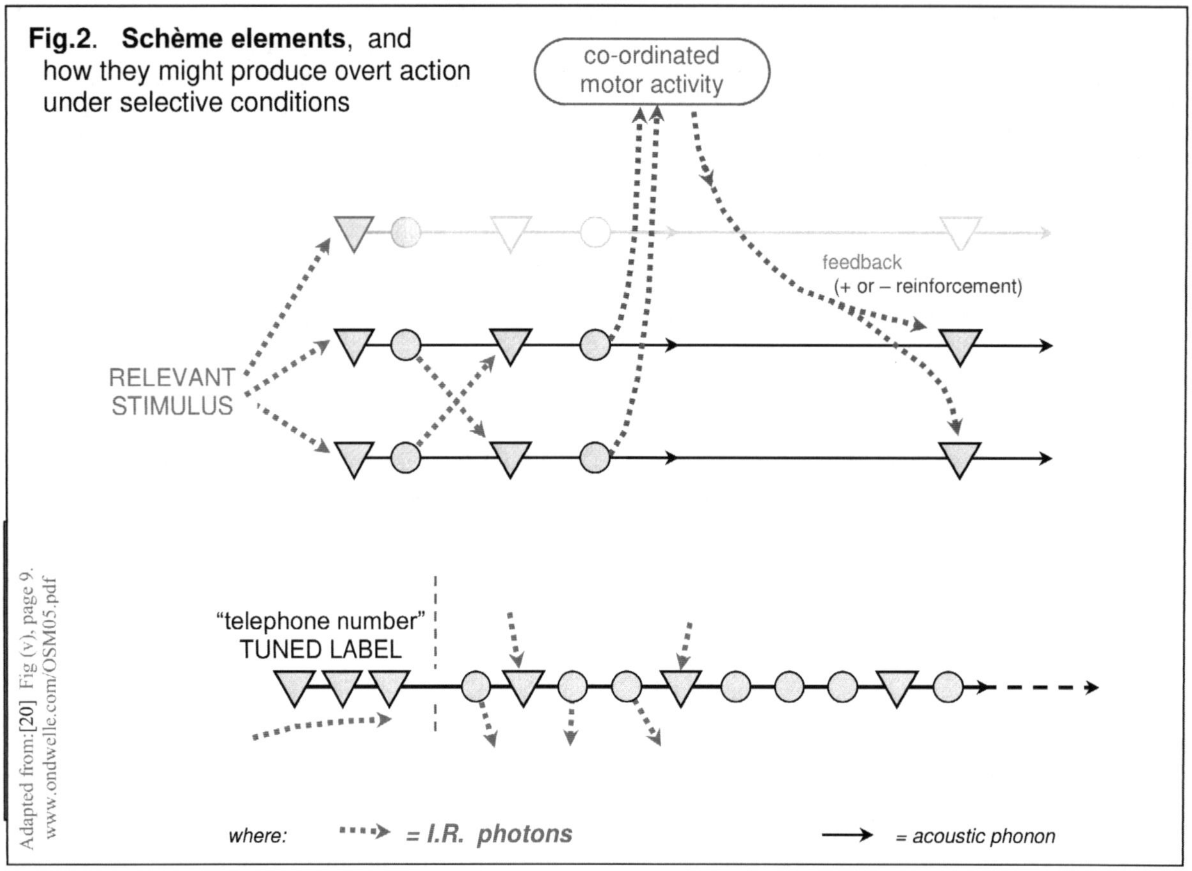

Fig.2. Schème elements, and how they might produce overt action under selective conditions

B(1) – §16. **But what could these abstract "schèmes" be physically?**

There seem to be only four or five plausible candidates for such linear-digital structures: •DNA, •RNA, •PNA, •Protein; — (and in principle such a service could also theoretically be provided by *un*branched rows of dendrite-free neurons within relevant local parts of the brain — but there seems to be no evidence that such rows exist, and they would surely be grossly inefficient and digitally-unreliable if they did). So, *by elimination* it seems that the basic elements of advanced thought must be RNA-like, thus constituting the (still hypothetical) "[R]" system — operating in parallel with the traditional "[A]" system of neurons and synapses which probably offers essential analogue-adjustment and support.

Hence **Hypothesis 1**: *That important components of human memory-coding are embodied in linear macromolecules such as RNA-or-DNA.*

(— Difficult to test except by its implications)[14]

[14] There is no clear direct lab-evidence for this memory-code method, (so *in*direct instead?♠) but it does have a sound *logical* basis — and its *consequences* are encouraging: One unforeseen ♦**bonus** is that it allows us to bypass the very awkward question of how such memories could be "written down" — i.e. "where is the 'tape-recorder'?" In fact (as we saw in Part A) the likely answer lies in **Darwinian trial-and-error** amongst the vast populations which must be available (thus avoiding the Lamarckian "tape-recorder" altogether). Another ♦**bonus** arose externally when Mattick [21-24] made it clear that only 3% of human RNA was needed for its traditional role — leaving 97% available for other tasks (such as memory-coding perhaps?!♠).

For our present purpose (as mentioned above), it probably matters little which type of molecule is involved — so we can be content just to say *"RNA-like"*. But whichever macro-molecule it is, we run into a complication which turns out to be instructive:

action/idea
"Schème"
≈ RNA?
↔ IR ?

(2) Fast-"talk" needed between memory-sites? — IR or what?

B(2) – §17. **Action-Potential Pulses as far too crude for Digital Sites:**
Ordinary molecular bioprocesses involve energy jumps of about 0.5 to 1.3 eV — with a wave-period of 8-to-3 *femtoseconds* (10^{-15} sec.) [15] — clearly in the near infrared range (NIR), and potentially available for local "broadband" signalling if organized correctly. So it seems likely that any molecular memory-coding will come to use this NIR medium, for local communication at least. And if there is also "neuro-talk" via the well-known action-potential spikes, then so be it! There is no law here against bilingualism!

However we should be clear about the contrast: Several *femtoseconds* for a NIR wave-cycle, against the *millisecond* action-potentials. Thus we see a mismatch-factor of more than 10^{11} (or perhaps only 10^9 to be conservative)[16] — like contrasting *"today's internet-electronics"* against "a morse-key of the 1850s".

(Of course we all know that the millisecond "morse-key signals" do exist in nerve-fibres, whereas the evidence for *NIR neural-signalling* is only indirect (♠) despite its vastly greater theoretical capacity. — Yet surely the nervous-system needs all the "bilingual" help it can get from whatever mechanisms are bio-feasible, as long as they can work together. Thus, for instance, it is quite likely that one major role for the action-potential system "**(1)**", is to offer an "analogue" service which fine-tunes and marshals the activities of an NIR "[R]" system, and thus refines the rapid digital activities at RNA-like sites.)

In any case, it is difficult to see how the comparatively clumsy action-potentials could offer a sufficiently fast and precise broadband network. And any molecular memory-system would surely need this broadband precision, since only then would its digital RNA-like sites be able to talk *efficiently* among themselves.

Hence **Hypothesis 2**: *That if bio-molecular memory-coding exists, its chief channel for fast communication will be via NIR signals.*

In other words, we are considering *photons as potential message-carriers* for advanced intelligence activity — normally replacing the action-potentials altogether whenever we are dealing with "conversation" which is amongst molecular sites exclusively. Indeed there seems to be no other plausible possibility for *fast* communication,[17] (though chemical diffusion does offer a slow alternative "postal" service). Anyhow **Figure 2** shows such IR activity (colour-dots) added to the bare 1D molecules (the "strings-of-beads" in grey/black).

---

[15] With corresponding wavelengths of 2.5 to 1 μm, assuming travel at the velocity of light in vacuo. (See [25: "Clue B"]: now online as www.ondwelle.com/OSM10en.pdf).
[16] E.g. to allow for many cycles within each photon.
[17] What Evidence can there be here if direct observation is not feasible? Consider this:
- BY ELIMINATION: The IR solution seems to be *the only explanation* which fits with all the constraints.
- BY ITS USEFULNESS: It also leads to unexpected solutions to enigmas in other fields — discussed below in sub-part "**(4)**" — §23-§27.

(3) Can we identify Paths for such IR communication?

B(3) – §18. Direct Short-range through Watery Media

This is possible, though limited to about 20μm before it is absorbed (depending on the actual frequency). That should suffice for many intermolecular tasks and for adjacent cells. However fatty media such as myelin offer much better prospects:

B(3) – §19. Long-range through Myelin (acting as an optic fibre)?

This too seems theoretically possible, since myelin is effectively *the dielectric within a coaxial cable* (bounded inside and outside by conductors, like any TV coaxial though much smaller as befits the much shorter wavelengths involved). §22 shows pictorially how this might work — but first:

Digression to explain Transmission-Modes within Coaxial Cables

B(3) – §20. The TEM mode of wave-transmission

This is the "straight ahead" transmission-mode for light in free-space — with both **T**ransverse vectors for both the **E**lectric vectors and the **M**agnetic vectors.
But it can also occur between **two separate conductors**:

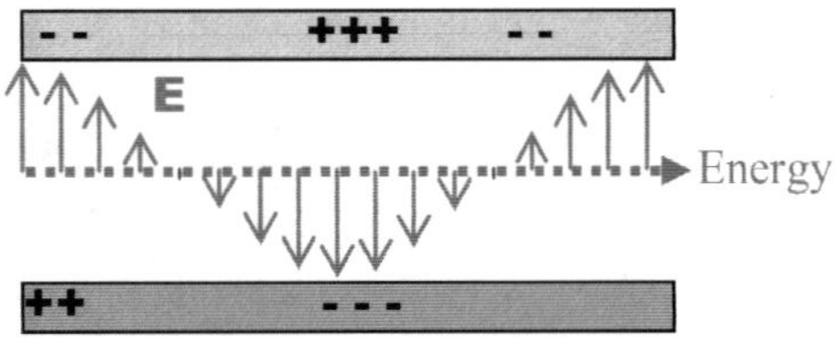

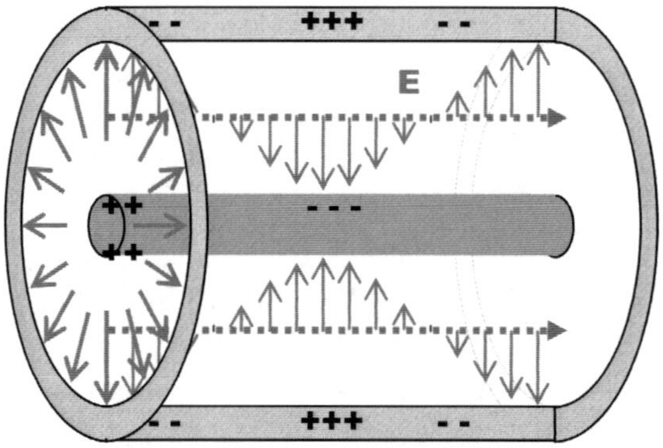

Fig.3. TEM waves travel through the dielectric (e.g. myelin) between *two different conductors* in each case

This "straight-ahead" TEM mode is what engineers usually strive for — but biology has different priorities and constraints, which might allow (or even favour[18]) "zigzag" modes, as depicted in Figure 4:—

[18] Optical dispersion amongst the different frequency components within the brain might usefully (i) help allocate signal-types into various dedicated paths — &/or (ii) transform signals in a repeatable way such as to suit (by trial-&-error?) stringent pre-set requirements of locality X (but no others) — a form of tuned reception *if* it actually works. (Traill 1978b[7], §C6.7).

B(3) – §21. **Zigzag modes (e.g. "$TE_{1,0}$")**
— a frame for external standing-waves

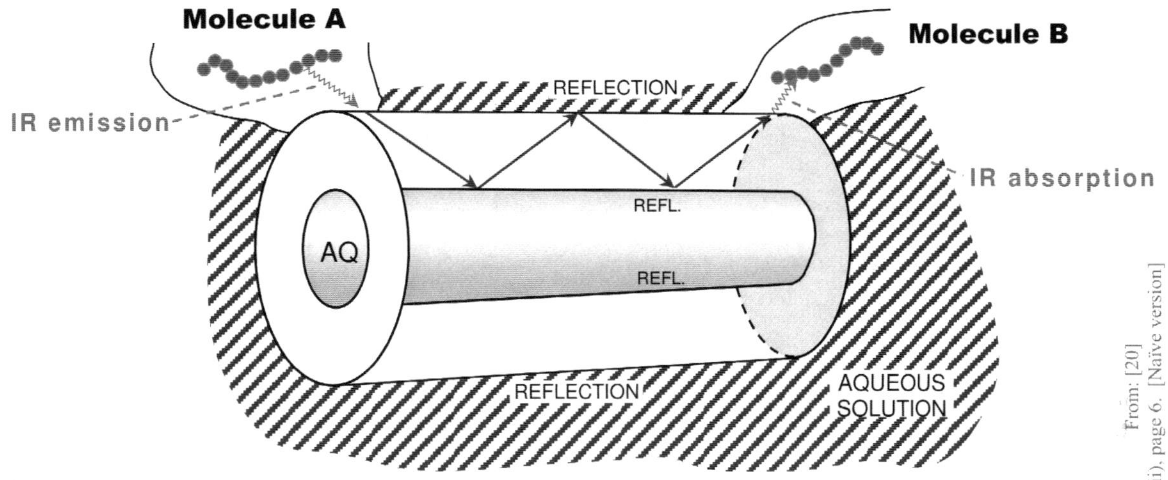

Fig.4a.
Naive depiction of "zigzag" transmission (Geometric optics)

From: [20]
Fig (ii), page 6. [Naïve version]
www.ondwelle.com/OSM05.pdf

Physical optics:

The first non-*TEM* propagation mode in a co-axial cable — the $H_{1,0}$ or $TE_{1,0}$ mode:

"H" because the only <u>axial</u> component is **magnetic**;
"TE" since the **E**lectric fields are entirely **T**ransverse;
"1" = single angular period ; (1 sine-cycle in 360°)
"0" = no radial overtones [26]

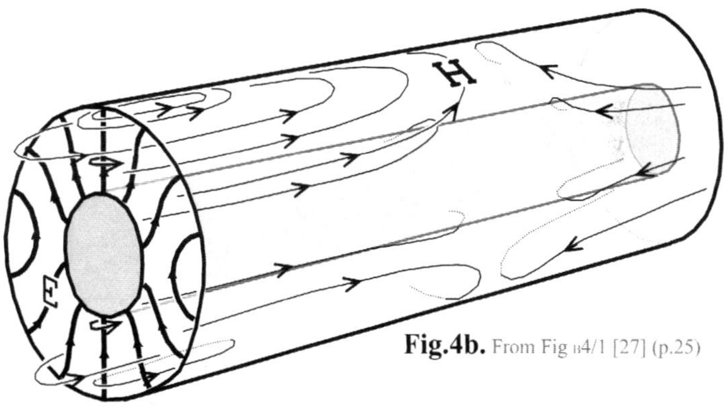

Fig.4b. From Fig II4/1 [27] (p.25)

One relevant point here is that non-TEM waves — *with a sideways component* — may (i) occur weakly in the *initially-bare* watery axon, despite significant absorption by water; (ii) readily leak out and (intermittently?) set up a standing-wave pattern around the nerve-fibre. (iii) This would tend to create alternate concentric energy-filled-zones, and energy-free-zones or "moats". (iv) As we shall see, myelin (a *good* dielectric) may then be likely to grow within any such energized zone, until reaching the low-energy "moat" — as a new coaxial.

End of the Digression

Journal of Physics: Conference Series **329** (2011) 012018 doi:10.1088/1742-6596/329/1/012018

B(3) – §22. **Myelin geometry — its likely "transmit+broadcast" ability**

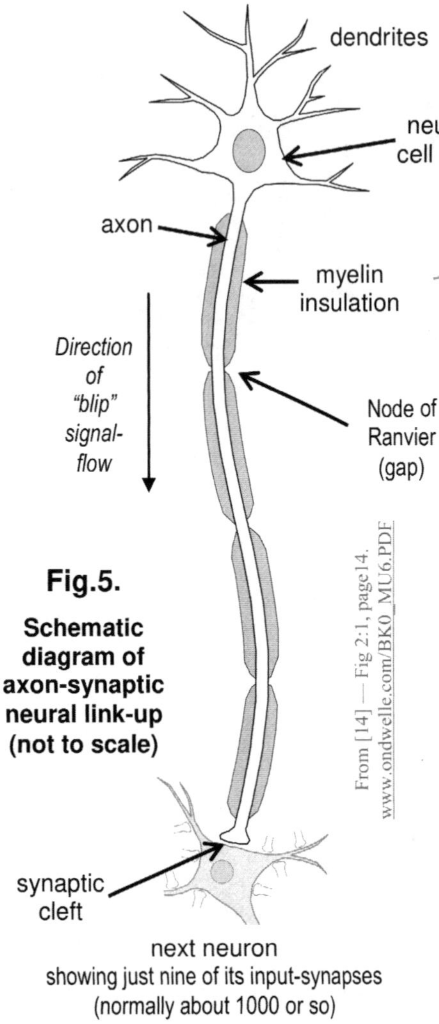

Fig.5.

Schematic diagram of axon-synaptic neural link-up (not to scale)

Fig.6: Node of Ranvier.

(simplified redrawing from photo in Peters *et al.* [10] page 249.)

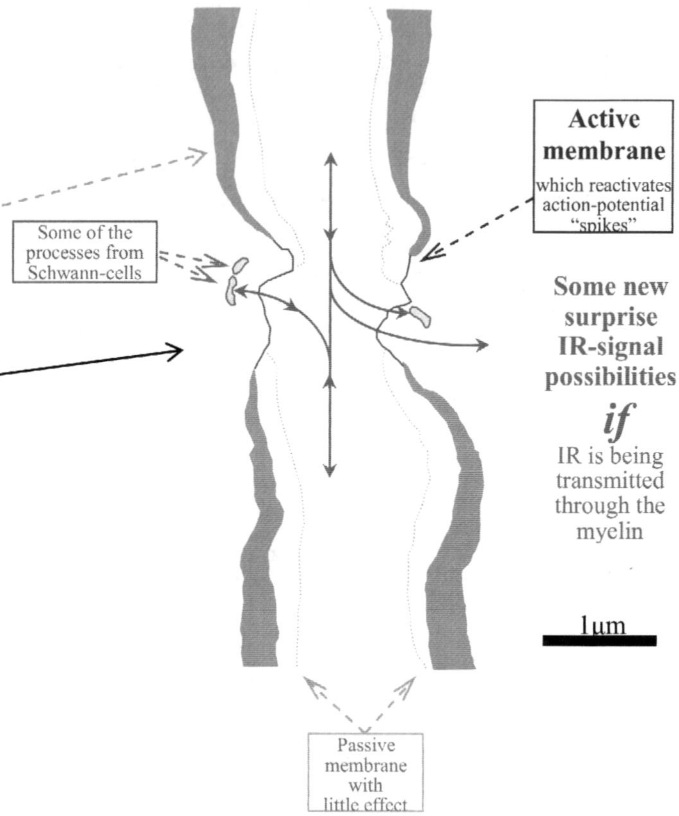

Note that orthodox action-potential signal travels in one direction only (downwards in Fig.5) and it does not stop at the nodes-of-Ranvier (despite their "refuel" importance) — so its role could be compared to *an express train on a one-way system between "cities" (cell-bodies).*

In contrast, it seems likely that IR signals (if they exist) would be more like a *"two-way stop-on-demand service between "small villages" (nodes-of-Ranvier)* — and also between adjacent sites, whether near *the node "station"* or simply near each other. See red arrows in Fig.6. In any case, it would probably be worthwhile now to study the geometry of such sites more closely, bearing in mind their possible optical and "radio aerial" properties.♠

(4) Neural-IR (if true) offers Surprise-Solutions to Problems

Here is a list of five serendipitous theoretical solutions to problems in other areas, arising from the above theory. Perhaps we might also add the two "bonus points" ("♦") listed within footnote 14.

Neuro-anatomy ↑↓ IR wave-patterns

"icon N" (discussed in Part C)

1a. B(4):§23. Myelin Geometry

How thick will the myelin be? (see Fig.7, below). As Donaldson & Hoke[28] noted in 1905:

Area of grey annulus (myelin) ≈ Area of central circle (axoplasm)

And this implied *myelin-thickness* (**m**) / *inner-radius* (**r**) ≈ •414

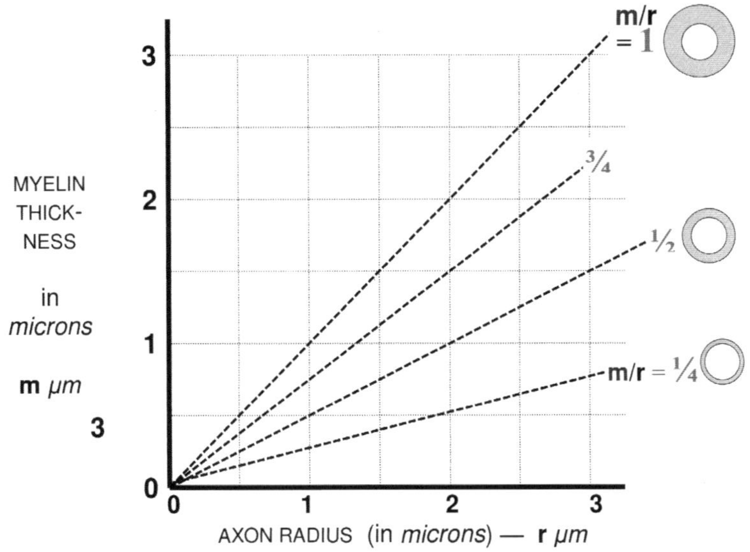

Fig.7. Plotting the geometry of axon+myelin cross-sections.

The *moat hypothesis* plus the subsequent reasoning together predict that the ***m/r*** ratio will mostly take on a small "quantized" repertoire of values.

Each such value is represented by a straight line through the origin — here shown as dashed lines, using fictitious values.

From [14]; Fig 7:7, page 54. www.ondwelle.com/BK0_MU6.PDF

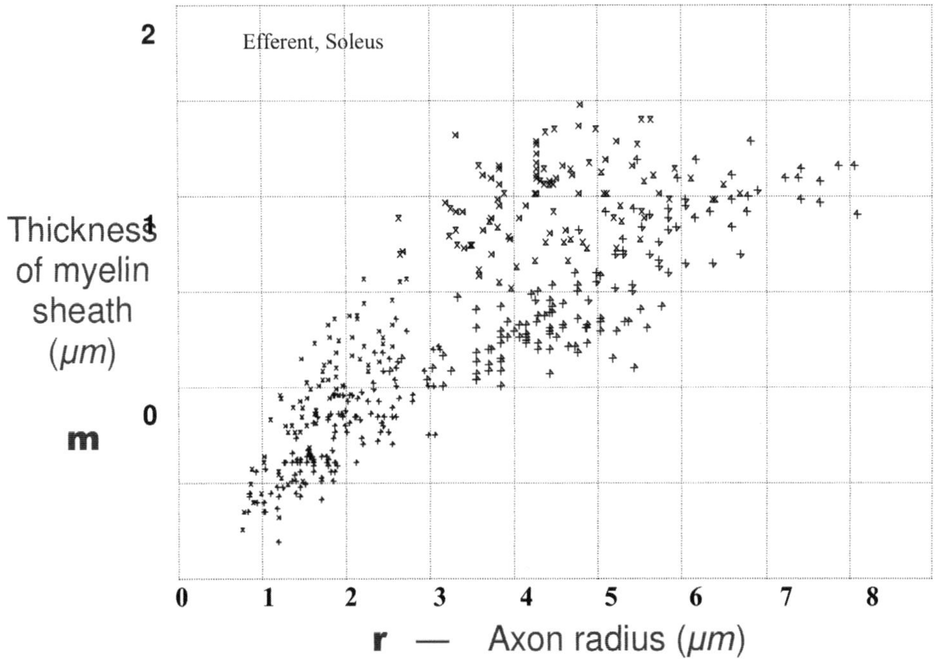

Fig.8
One set of actual *m/r* values

Data kindly provided by Boyd & Kalu

Note the faint trend toward a straight line below-then-amongst the scatter (left-to-right), favouring

m/r = 2 / 6½

i.e. m/r = •308

Note the high resolution. There were also two other such plots, Boyd & Kalu,[29] but more are needed ♠

Adapted (by axis-swop) from [14], Fig 6:3, page 42.

Journal of Physics: Conference Series **329** (2011) 012018 doi:10.1088/1742-6596/329/1/012018

So now, applying IR theory, we come to:

1b. B(4):§24. Mechanism? —Plausible Myelin control via IR

Hypothesis: That myelin growth tends to be encouraged by IR standing-waves around an axon — and tends to stop when the wave-field intensity vanishes (or falls below some threshold).

Fig.9a. POSTULATED TRANSVERSE STANDING-WAVE ACROSS AXON producing → "MOATS" AT NODES

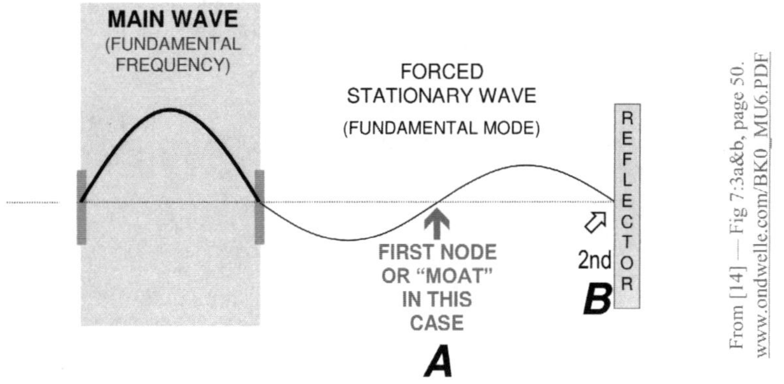

ALSO Fig. 9b.:

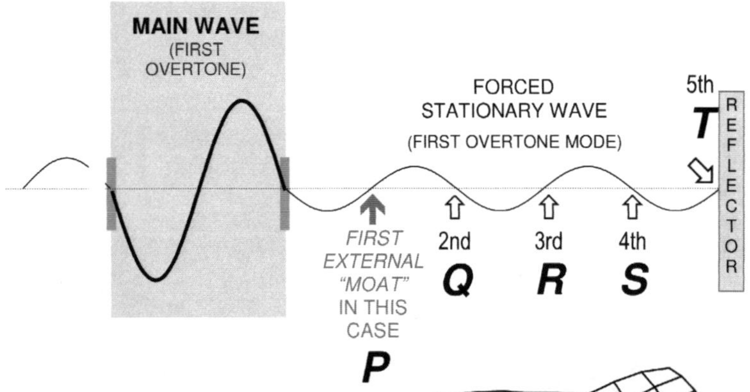

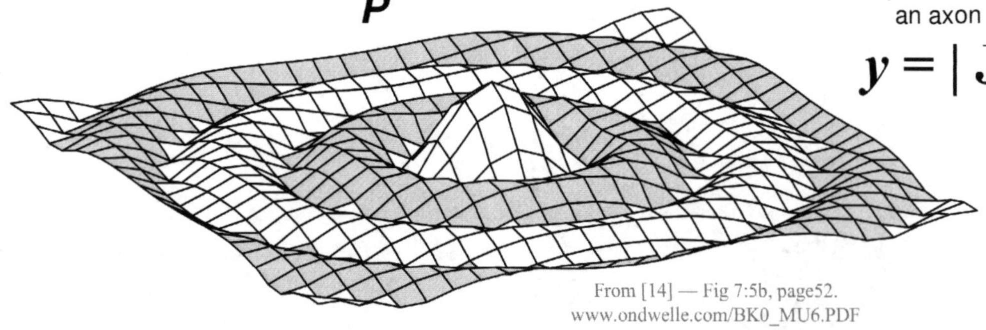

Fig.10.
Transverse standing wave pattern likely around an axon (or a wire)

$$y = |\, J_0(\mathrm{k}r)\, |$$

$$\approx |\, \cos(\mathrm{k}r)\, |$$

2. B(4):§25. Why PNS myelination is delayed until axon diameter ≈1μm

The likely reason for that constraint on the peripheral-nerve development should now be obvious from the above diagrams. The 1μm would correspond roughly (given the circular geometry) to the half-wavelength which forms the basis for the loop at the centre if the concentric pattern.♠

For the central nervous system (CNS), much smaller early-myelination diameters are observed, and that could be taken as a sign of consistent UV presence, though it is not suggested that UV could actually be transmitted effectively along myelin paths. (The mere "reliable" presence of UV for whatever reason, would suffice — and presumably these conditions are not to be found in the PNS).♠ [11] §5.1, and see footnote 5 above.

3. B(4):§26. **Why Myelin often stops its "wrap-around" near where it starts?**

— The "Quadrant" enigma of Peters[30] and Webster [31]; — [11]

— Explicable if we consider the myelin as filling up the annular "virtual" space around the axon (due to IR standing-waves). Thus:

Fig.11. How myelin actually wraps around an axon, but also showing the "virtual" barrier which would (hypothetically♠) constrain this process

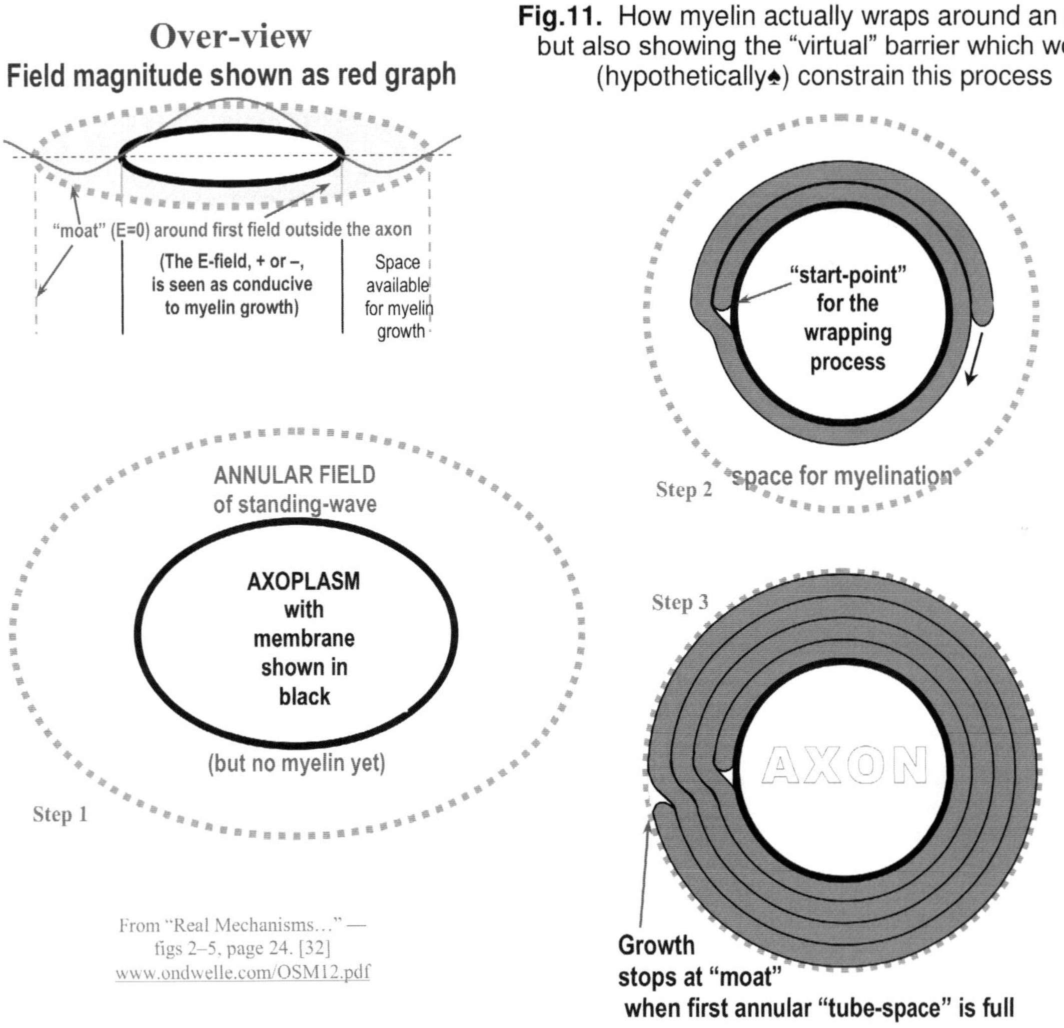

From "Real Mechanisms..." — figs 2–5, page 24. [32]
www.ondwelle.com/OSM12.pdf

4. B(4):§27. **The missing Action-potential when insects "see" IR signals**

Callahan [33] showed that his insects were *responding to* IR — but he could not detect any relevant action-potentials!

The plausible solution is that the IR is received directly into the insect nervous-system, without the need for inefficient "translation" (of IR→Action-potential→?IR). [34,35] ♠

5. B(4):§28. **The surprising efficiency of the optic nerve**

An apparently inadequate channel capacity (for action-potential signals, [36]) might actually have much more "broadband ability" if seen as having fibre-optic properties. ♠

Part C: Piaget + IR + Gurwitsch = Corroboration & Coherence?

§29. Scientific corroboration as "Social Equilibration" or "Coherence"

Paul Thagard [17] gives several historical examples of complex scientific problems — and how they were eventually solved with the aid of a search for *coherence* amongst the often-prolific evidence. In particular he details the coherence-patterns eventually offered within the revolutionary ideas of *Darwinian theory* (Ch.6, p131+), *Continental drift* (Ch.7, p157+), *Gravity* (Ch.8, p191+), and the simpler case of *Oxygen-versus-Phlogiston* (§3.2, p39+).

Such coherence-amongst-our-ideas[19] is all very well in retrospect, but it may be *far from obvious* while one is actually engaged in the original investigation — or only *"apparently obvious"* regarding some small *part of the problem* at any particular time as the project progresses.

Meanwhile there is the danger that such parochial conclusions will be inaccurate or misleading, so our goal should be to get well beyond the parochial into some sort of large "jigsaw-puzzle" where all the relevant evidence will somehow "fit in" eventually. — There is apparently no royal road to achieving that systematically, but we can-and-do find *apparently-relevant pieces* or *sub-assemblies* and try to fit them into that jigsaw-puzzle. (And isn't that just *social* epistemology at work, seeking *equilibration* amongst candidate "schème-analogues"?)

So now, regarding weak bio-emissions and related phenomena largely overlooked by mainstream science: Perhaps it is time to take stock of whatever semi-complete bio-physical "sub-assemblies" are now at our disposal, and try to fit them together into a bigger collective system — a new attempt at a bigger multidimensional jigsaw-puzzle with much more overall coherence which would offer a more persuasive case *IF* the internal coherence does indeed exist. In any case it seems important that we investigate the situation:

§30. Ultraweak IR and UV observations: — One(?) "sub-assembly" of ideas?

As mentioned in §3, these experimental findings (Gurwitsch, 1923…[1,2], etc.) were often simply not believed in those early days. Improved methods and many further investigations have nominally changed that — notably with contributions from Fröhlich, and F.A.Popp, and with publications such as *"Electromagnetic Bio-Information"* [37]. But as this body of work has not yet been taken very seriously by mainstream science, it may be fair to identify it (as it stands) as a *"sub-assembly"* which might well contribute

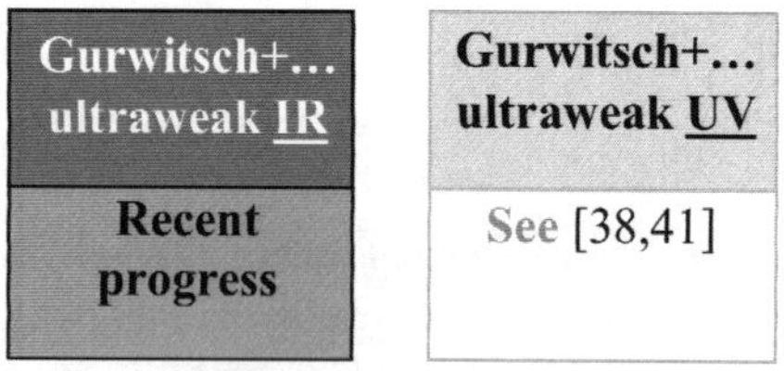

Two "sub-assemblies" from the direct Gurwitsch tradition

[19] At the same time, the actual dynamic system under study must necessarily have enough *physical* coherence to hold it together — at least long enough for us to study it (as discussed above in §7). One trouble is that we may initially have no notion about what causes such physical coherence in the various cases.

to some future general enhancement of our ideas about cell activity.

Here I take the liberty (at least provisionally) of dividing this sub-assembly into two — distinguishing between IR and UV emissions. For one thing, the above discussion of the nervous system had identified IR as a likely participant, but it had very little to say about UV — indeed the gross optical peculiarities of those different bands are notably different. Secondly UV photons would be much more dangerous if misplaced (with about ten times the energy which would also entail special logistical complications)[38,41]. So graphically, I offer two iconic "tiles" to represent these two *"sub-assemblies"*, but on this occasion I shall use only the IR case, (though still referring to its icon as "**G**" as if it were the only one!).

§31. Two "sub-assemblies" arising from Psychology and Neurophysiology

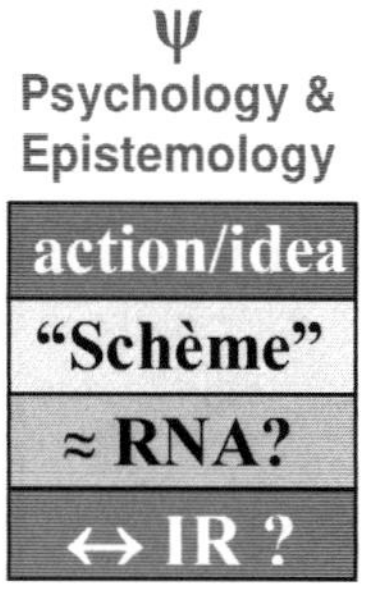

Part B· provided the discussion of Piaget's theory (§13–§15) and then its apparent *molecular* (§16), and *IR* (§17–§22) requirements. That progression was symbolized by the four-layered icon which is now labelled as "ψ": Of course each layer could be offered as a separate minor sub-assembly, but it is only when they are brought together that they offer significant collective coherence as an idea.

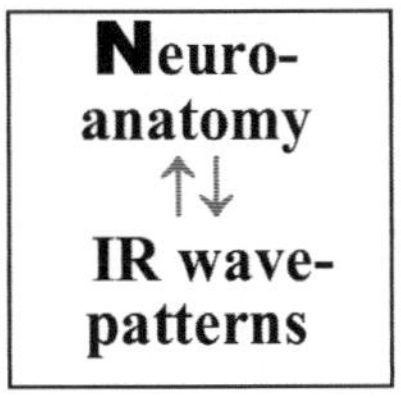

The next advance, "B(4)" (§23-§28) was to seize the windfall opportunity of further coherence which provided unexpected solutions to problems in other areas. — That takes us to the yellow "Neuroanatomy" icon, which can now be referred to as "**N**". This sub-assembly offers a degree of parochial coherence on its own (e.g. suggesting that growth might be controlled by IR *even if that IR had no other role* such as in "ψ") — but clearly it makes a more compelling-and-coherent case if the two sub-assemblies N and ψ" are envisaged as collaborating via IR.

§32. Assembling the (incomplete) jigsaw-puzzle

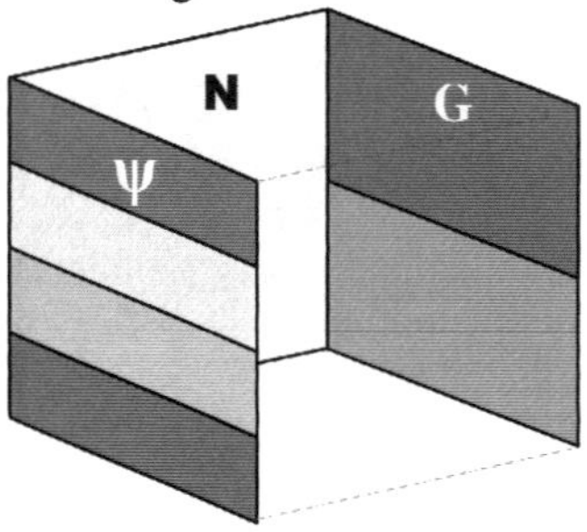

How the three sub-assemblies might fit together — but leaving room for more evidence

This "three walled cube" is not necessarily the best fit, but it offers a useful talking-point. **G** approximates "pure experiment" while ψ approximates "pure theory" — in which case we may seek to find suitable links between them, and to some extent **N** seems to fill this bridging role. Further evidence would clearly be welcome, and closer study of such configurations might help us choose future areas of enquiry — e.g. how to extend-and-integrate the accompanying theorizing about ultra-violet (UV), [41].

Meanwhile we might hope that such enhanced coherence will focus serious attention on such enquiries. That could lead to significant advances.

References

[1] Gurwitsch,A. (1923). Die Natur des spezifischen Erregers der Zeilteilung. *Arch. f. entw. Mech.*, **100**: 11.

[2] Gurwitsch,A. (1924). Les problèmes de la mitose et les rayons mitogénétiques. *Bull. d. Histol. Appl.,* **1**: 486.

[3] Popper, K.R. (1934/1982). *Logik der Forschung* – (1959) *The Logic of Scientific Discovery.* Hutchinson: London.

[4] Traill, R.R. (2010a). *Meaningful UV and IR photon-exchange within bio-tissue? — Interdisciplinary evidence, and a new way to view asbestos toxicity.* Ondwelle: Melbourne. www.ondwelle.com/UV1_evidence.pdf

[5] Langmuir, I.. (1953). *Pathological Science*: Colloquium at the Knolls Research Laboratory, December 18, 1953. http://www.cs.princeton.edu/~ken/Langmuir/langmuir.htm

[6] Kant, I. (1781"A" / 1787"B" / — 1929/2007). *Critique of Pure Reason.* [Transl. by Norman Kemp Smith] Palgrave Macmillan. [Various versions also available online].

[7] Traill, R.R. (1978b/2006). *Molecular explanation for intelligence…,* Thesis, Brunel University: http://hdl.handle.net/2438/729 (2 files) — or Ondwelle Publications: www.ondwelle.com (7 files).

[8] Albert, Hans (1968/1985). *Traktat über kritischen Rationalismus / Treatise on Critical Reason.* Princeton University Press: NJ. [espec. Ch.1]

[9] Haack, Susan (1993). *Evidence and inquiry: towards reconstruction in epistemology.* Blackwell: Oxford.

[10] Peters, A., S.L.Palay, & H.deF.Webster (1991, 3rd ed.). *The fine structure of the nervous system: neurons and their supporting cells.* Oxford University Press. ISBN 019 506571-9

[11] Traill, R.R. (2005a). *Strange regularities in the geometry of myelin nerve-insulation — A possible single cause.* Ondwelle: Melbourne. http://www.ondwelle.com/OSM01.pdf

[12] Hume, D. (1748) An Inquiry Concerning Human Understanding — (1962) in A.Flew (Ed.) *On Human Nature and the Understanding.* Collier Macmillan: London.

[13] Jerne, H.K. (1966) "Antibody formation and immunological memory" In Gaito, J. (1966a) *Macromolecules and Behavior.* Meredith/Appleton-Century-Crofts: New York; Ch.9, pp151-157.

[14] Traill, R.R. (1999). Mind and Micro-Mechanism: a hunt for the missing theory. Ondwelle: Melbourne. http://www.ondwelle.com ISBN: 0-9577737-0-6

[15] Popper, K.R. (1975/1994) "The rationality of scientific revolutions"; (i) in Rom Harré (ed.) (1975) *Problems of Scientific Revolution. Scientific Progress and Obstacles to Progress in the Sciences, The Herbert Spencer Lectures 1973.* Clarendon Press: Oxford. — Also in (ii) M.A.Notturno (ed.) (1994) *The Myth of the Framework: In defence of science and rationality.* Routledge: London.

[16] Traill, R.R. (2008a/2005b). *Thinking by Molecule, Synapse, or both? — From Piaget's Schema, to the Selecting/Editing of ncRNA.* Ondwelle: Melbourne. www.ondwelle.com/OSM02.pdf [French version: www.ondwelle.com/FrSM02.pdf] — Also in *Gen.Sci.J.* www.wbabin.net/physics/traill2.pdf

[17] Thagard, P. (1992/1993). *Conceptual Revolutions.* Princeton University Press: Princeton, NJ.

[18] Bacon, F. (1620/1960). *The New Organon.* Bobbs-Merrill: Indianapolis.

[19] Traill, R.R. (1970). "Boolean operators: A critique of their conceptual usefulness in modelling adaptable neural nets"; in J.Rose (ed.). *Progress of Cybernetics.* **2**[of 3], Gordon & Breach: London. pp861-871. — ISBN 0677 143203.

[20] Traill, R.R. (1976b/2007). *The gulf between behavioural psychology and fundamental physiology: a systematic attempt to bridge the gap.* Monograph 15, Institute of Cybernetics, Brunel University. (www.ondwelle.com/OSM05.pdf)

[21] Mattick, J.S. (2001). "Noncoding RNAs: the architects of eukaryotic complexity". *EMBO Reports* **2**(11), 986-991. http://emboreports.npgjournals.com/cgi/content/full/2/11/986

[22] Mattick, J.S. (2003). "Challenging the dogma: The hidden layer of non-protein-coding RNAs on complex organisms" *Bioessays.* **25**, 930-939. www.imb-jena.de/jcb/journal_club/mattick2003.pdf

[23] Mattick, J.S. (2004-Oct). "The hidden genetic program of complex organisms" *Scientific American.* **291**(4), 30-37. www.sciam.com/article.cfm?articleID=00045BB6-5D49-1150-902F83414B7F4945

[24] Mattick, J.S., & M.J.Gagen (2001). "The Evolution of controlled multitask gene networks: The role of introns and other noncoding RNAs in the development of complex organisms". *Mol. Biol. Evol.* **18**(9), 1611-1630. [Review] http://mbe.oupjournals.org/cgi/content/full/18/9/1611

[25] Traill, R.R. (1988). "The case that mammalian intelligence is based on sub-molecular memory coding and fibre-optic capabilities of myelinated nerve axons". *Speculations in Science and Technology.* **11**(3), 173-181. (www.ondwelle.com/OSM10en.pdf).

[26] Skilling, H.H. (1948). *Fundamentals of Electric Waves.* Wiley.

[27] Traill, R.R. (1977/2006). Toward a Theoretical Explanation of electro-chemical interaction in memory-use: the postulated role of infra-red *components and molecular evolution.* Monograph 24, Institute of Cybernetics, Brunel University. : [www.wbabin.net/physics/traill8.pdf *or* www.ondwelle.com/MolecMemIR.pdf]

[28] Donaldson, H.H., Hoke, G.W., (1905). "The areas of the axis cylinder and medullary sheath as seen in cross sections of the spinal nerves of vertebrates". *J.Comp.Neurol.* **15**, 1–16.

[29] Boyd, I.A., and K.U.Kalu (1973). "The relation between axon size and number of lamellae in the myelin sheath for afferent fibres in groups I, II, and III in the cat". *J. Physiol., London*, **232**, 31P-33P.

[30] Peters, A. (1964). "Further observations on the structure of myelin sheaths in the central nervous system". *J.Cell Biol.*, **20**, 281-296.

[31] Webster, H.deF. (1971). "The geometry of peripheral myelin sheaths during their formation and growth in rat sciatic nerves". *J.Cell Biol.*, **48**, 348-367.

[32] Traill, R.R. (2009). *Real Mechanisms for Natural-Thought? The History of an in-depth Analysis.* www.ondwelle.com/OSM12.pdf

[33] Callahan,P.S. (1977) "Comments on Mark Diesendorf's critique of my review paper" *International J. Insect Morphol. and Embryol.* **6**(2), 111-122.

[34] Traill, R.R. (2005c). *How Popperian positivism killed a good-but-poorly-presented theory — Insect Communication by Infrared.* Ondwelle: Melbourne. www.ondwelle.com/OSM03.pdf — Also in *Gen.Sci.J.* www.wbabin.net/physics/traill7

[35] Traill, R.R. (2008b). *Critique of the 1977 debate on infra-red 'olfaction' in insects — (Diesendorf vs. P.S.Callahan).* Ondwelle: Melbourne. [http://www.ondwelle.com/OSM09.pdf — Conference of the Australian Entomological Society, at Orange, NSW, Australia: 30 September 2008]

[36] Jonas, J.B., Ulrike Schneider, and G.O.H.Nauman (1992). "Count and density of human retinal photoreceptors". *Graefe's Archive for clinical and experimental Opthalmology,* **230**, 505-510.

[37] Popp,F.A., *et al.* (Editors) (1979). *Electromagnetic Bio-Information.* Urban & Schwarzenberg: München.

[38] Traill, R.R. (2010b). *The theoretical case that some asbestos fibres could trigger cancer optically, while others act mechanically.* Ondwelle Publications: Melbourne. http://www.ondwelle.com/UV2_trigger.pdf — also in *General Science Journal.*

[39] Traill, R.R. (2000). *Physics and Philosophy of the Mind.* Ondwelle: Melbourne. http://www.ondwelle.com ISBN: 0-9577737-1-4

[40] Hydén, H. (1967). "Biochemical changes accompanying learning". In G.C.Quarton, T.Melnachuk, and F.O.Schmitt (eds), *The Neurosciences*, pp.765-771, 913, 914. Rockefeller University Press.

[41] Traill, R.R. (2011b). "Asbestos as 'toxic short-circuit' optic-fibre for UV within the cell-net: — Likely roles and hazards for secret UV and IR metabolism". — *In the present volume.*

Journal of Physics: Conference Series **329** (2011) 012018 doi:10.1088/1742-6596/329/1/012018

Appendix: List of Suggested Experimental Investigations

§33. These items, (marked "♠" within the general text) seem worth considering for some sort of experimental testing; and at first sight each such test-type seems feasible, though maybe expensive. I accordingly invite those with the relevant resources to investigate them — with or without my collaboration, though I would appreciate information about any such developments.

§	foot note	Description of suggested area for experimentation, marked "♠" in the text
16	14	Lab-tests for *RNA-like* memory-coding? *E.g.* Re-do Hydén's work, but with RNA now seen as a likely coding-medium *in its own right* — mostly *independent* of any synaptic-coding enigmas. Traill (2000[39], pp.45, 53), Hydén (1967[40] and his other works).
16	14	Of Mattick's 97% genetic ncRNA.s:— Do some serve as raw Piagetian "schèmes"?
17	—	Can NIR neural-signalling actually travel optically through myelin? And if so, can we use its principles to design test-probes for further investigation?
22	—	Re-study the Nodes of Ranvier, checking for optical (radio-aerial) significance
23	—	Replicate and extend Boyd & Kalu's scattergrams ([29]; plus Fig 8 within §23 (above).)
25	—	Suppose we impose unnatural wavelengths at myelination-sites, will that cause myelination to start at some unusual "critical" diameter in the growing axon? And if so, is that new diameter ≈ half the imposed wavelength?
25	—	Is there actually any connection between "early" CNS myelination and any presence of UV?
26	*Fig 11*	Devise false virtual-"moats" to see whether that affects the eventual myelin properties
27	—	Are IR signals received *directly* into the insect nervous-system *non*-chemically? And is chitin (instead of myelin) important as a dielectric conduit for IR in insects?
28	—	Can the optic nerve transmit photon-signals? — If so, what wavelengths? And might this have "broadband" implications regarding that nerve's ability?
—	—	Systematically collect and collate more-exact wavelength details about ultra-weak IR emissions

Increased dielectric constant in the water treated by extremely low frequency electromagnetic field and its possible biological implication

Xun Shen
Institute of Biophysics, Chinese Academy of Sciences, Beijing 100101, China

E-mail: shenxun@sun5.ibp.ac.cn

Abstract. Water is the most abundant compound on the surface of the Earth, and can be considered to be the most important molecule in living systems. Water plays a variety of cellular functions, being the solvent of most biological molecules, a substrate and product of enzymatic catalysis, an important component of macromolecules, and more. Because of importance of water in life, many physical and chemical treatments were invented to improve the quality of drinking water. Among them, the treatment with electromagnetic field is a well-known, but much debatable physical method. Although electromagnetic field has been utilized for treating water for 80 years, many reports on beneficial biological effect of electromagnetic field-treated water were either anecdotal or less convincing. To explore if there is any physical base for understanding possible biological effects of electromagnetic field-treated water, dielectric relaxation spectra of deionized water treated with an extremely low frequency electromagnetic (ELFEM) field were measured and compared with that of untreated water.
It was surprisingly found that the dielectric constant of the ELFEM field-treated water was 3.7% higher than the control over the frequency range of 1-10 GHz, which indicates a higher molecular polarization occurs in the ELFEM field-treated water. Electrostatic and thermodynamic analysis shows that proteins or other biomacromolecules would have more reduced free energy when they are hydrated in high dielectric constant water. Since free energy is of crucial importance for stability of proteins, protein folding and its conformational change, as well as catalytic activity of enzymes, the free energy reduction of the biomacromolecules hydrated with higher dielectric constant water may be responsible for many possible biological effects of electromagnetic field treated water.

1. Introduction

Water is the most abundant compound on the surface of the Earth and it is the principal constituent of all living organisms. The human body is about 65 per cent water by weight, some tissue such as brain and lung being composed of nearly 80 per cent water. For all practical purposes, water can be considered as the only solvent in the body. Water exists in the mammalian body as two compartments, each with several subdivisions. They are intracellular water compartment, which makes up approximately 60-65% of body water, and extracellular water, which makes up the other 35-40% of body water. Intracellular water compartment is found inside the bilayered plasma membrane, and is the matrix in which cellular organelles are suspended, and chemical reactions take place[1]. The extracellular water compartment is found in the space that surrounds the cells of a given tissue and

Journal of Physics: Conference Series **329** (2011) 012019 doi:10.1088/1742-6596/329/1/012019

provides the immediate microenvironment that allows for movement of ions, proteins and nutrients across the cell barrier. Besides these, water molecules even become part of protein structure, and are called as buried water molecules, which have no contact with bulk solvent. The buried water plays a role in structural stabilization of proteins. A cryogenic X-ray crystallographic study demonstrated that large-scale networks of hydrogen bonds covering the entire surface of proteins were quite flexible to accommodate to the large-scale conformational changes of proteins and seemed to have great influences on the dynamics and function of proteins [2].

Water property is of crucial importance because, very often, it determines the details of biological activity. For example, it was shown that a minimum amount of water (of the order of 3%) is necessary for the activity of proteins and that this amount of water corresponds to the percolation of hydrogen bonds in a first hydration layer [3]. More generally, it is a common statement that liquid water is a necessary condition for life. Because of importance of water in life, many physical and chemical treatment were invented to improve the quality of drinking water. Among them, the treatment with electromagnetic field is a well-known, but much debatable physical method since 1930s. The earlier devices mostly employed permanent magnets, but many now use alternating magnetic or electrostatic fields. In this paper, we are not going to deal with many confused reports, such as that Magnetic Treatment can reduce the effects of hard water, as an alternative to water softening [4], but mainly focus on whether the electromagnetic field-treated water is biological effectiveness. Ismail and Shoukry reported recently that the snails after 12 weeks exposure to magnetic water were slightly greater, had a much significantly increased number of eggs and egg-masses than the control group. They also observed that the hatchability rate of treated eggs in magnetic water for 5 days was higher than that in control group. However, a significantly reduced survival rate of newly hatched snails was observed in magnetic water for another 7 days after hatching [5]. D. Zhao and his coworkers reported that increased activity of antioxidant enzymes (including catalase, superoxide dismutase (SOD) and glutathione peroxidase) and reduced lipid peroxidation in the mice fed with magnetic field-treated water for one month, and reduced blood lipid level in rats fed with magnetic field-treated water for 5 weeks were observed in comparison with control groups [6]. It is a pity that the authors did not specify their used magnetic field and how was the water treated. Y.Liu and his coworkers investigated effects of the water, which was treated with an extremely low frequency electromagnetic field (abbreviated as ELFEM field) same as that used in this investigation, on leukocyte adhesion and microcirculation of mesentery, blood viscosity and erythrocyte deformation, potential and peristalsis of intestines and stomach, fatigue endurance capacity and activity of natural killer cell in rats. They found that drinking of the ELFEM field-treated water resulted in obviously reduced adhesion and migration of leukocytes, less aggregation of platelets, increased secretion of IL-1 and IL-6, reduced blood viscosity, improved gastrointestinal peristalsis and contractility, and higher activity of natural killer cell (splenocyte) in rats, as well as longer swimming time was found in mice fed with the ELFEM field-treated water [7-10]. Above mentioned biological effect of electromagnetic field-treated water on animals seems unbelievable. In order to understand such biological effectiveness of electromagnetic field-treated water, on one hand, some better-controlled animal experiments or clinical trials should be done to see if the biological, especially the health-beneficial, effectiveness of electromagnetic field-treated water could be confirmed, on the other hand, some solid physical or chemical bases should be established to explain possible biological effectiveness of the electromagnetic field-treated water. For this purpose, we examined if dielectric property of water, which may influence structure and function of hydrated proteins, has any significant change after treatment with electromagnetic field.

2. Experimantal Procedure

2.1. Electromagnetic field and treatment of deionized water

The extremely low frequency electromagnetic field (ELFEM field) used for treating water was generated in an appratus provided by Guangdong Junfeng BFS Industry Co.Ltd. The principal setup in the water treatment is shown as Fig.1. Deionized water 6 times flowed through the alterate ELFEM field perpendicularly in a non-ferrous pipe. Each time, the water remain in ELFEM field for about 1 second. The magnetic flux density of the ELFEM field $B = 2^{1/2}B_0\mathrm{Sin}(2\pi ft)$, in which B_0=0.15 Tesla, and f = 50 Hz.

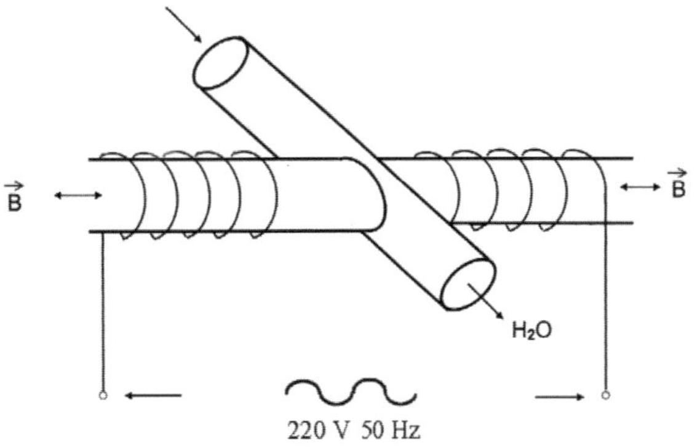

Fig.1. Diagram of the extremely low frequency electromagnetic field used for water treatment.

2.2. **Measurement of Dielectric Relaxation spectra**

All measurements were carried out in State Key Laboratory of New Ceramics and Fine Processing, Tsinghua University (Beijing) using a parameter network analyzer (Agilent/HP 8720ES-S, Agilent, Santa Clara, CA) with an open-end coaxial Agilent/HP 85070E Dielectric Probe (19 mm diameter) immersed in a glass beaker (total volume of 25 mL) containing a sample water. The beaker was held at 23 ± 1°C by a temperature-controlled circulator. The whole measuring system was placed in an air-conditioned room maintained at 23 ± 1 °C. The probe calibration was performed by three separate runs: open-circuited to the air, shortcircuited with mercury and in contact with pure water. Dielectric relaxation (DR) spectra were recorded for the frequency range of 50 MHz–20 GHz (200 frequency points in log-scale). The DR spectra of the ELFEM field-treated deionized water, ε_{em}^* (ν), and untreated deionized water, ε_c^*(ν), were each measured 6 times for obtaining average values and statistical analysis. Here ν is the relaxation frequency.

2.3. **Statistics**

Every 6 data on each frequency point in the DR spectra of ELFEM field-treated water and the every 6 data on each frequency point in the DR spectra of untreated water were subjected to Student's t-test. $p<0.01$ was used as a criteria to adjudge if the dielectric constant averaged over 6 data at each frequency point of the DR spectra from ELFEM field-treated water is significantly different from the corresponding dielectric constant from untreated water.

3. Results

The dielectric relaxation spectrum of ELFEM field-treated deionized water and the control that pass through the apparatus without switching on power are measured, and shown in Fig.2. It clearly shows that the dielectric constants of the ELFEM field-treated water at almost every frequency point are significantly higher than those obtained for untreated water. In average, the dielectric constant of the ELFEM field-treated water is 3.7% higher than the control over the frequency range of 1-10 GHz.

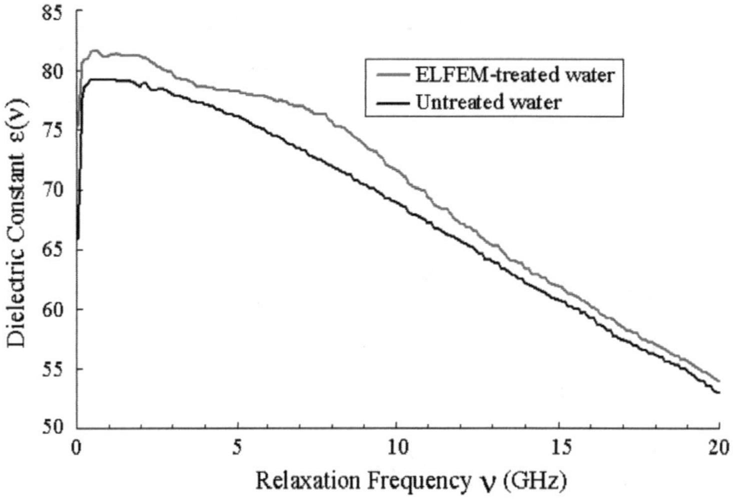

Fig.2. Dielectric relaxation spectra of deionized water with or without treatment of an extremely low frequency electromagnetic field.

The conductivity spectra of the ELFEM field-treated water and untreated water, $\sigma_{EM}(\nu)$ and $\sigma_c(\nu)$ can also be calculated from the imaginary part of the Dielectric relaxation spectra, $\varepsilon''(\nu)$, by following equation [11]:

$$\sigma(\nu) = 2\pi\nu\varepsilon_o\varepsilon''(\nu) \qquad (1)$$

where ε_o is the free space permittivity , and has the value of 8.84 x 10^{-12} *F/m*. The calculated conductivity spectra of ELFEM field-treated water and untreated water are shown in Fig.3. It can be seen that ELFEM field-treatment did not cause any alteration in the conductivity of deionized water.

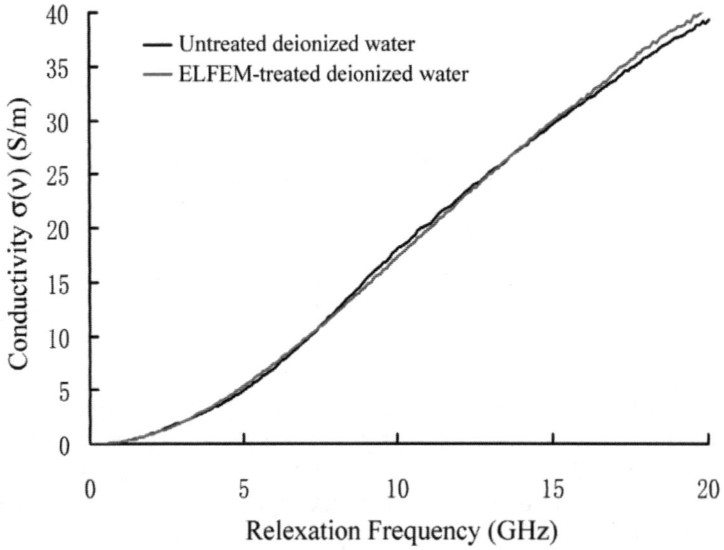

Fig.3. Conductivity spectra of deionized water with or without treatment of an extremely low frequency electromagnetic field.

4. Discussion

In the present investigation, the dielectric relaxation spectrum of deionized water was found significantly increased after the treatment by an extremely low frequency electromagnetic field. However, as expected, the ELFEM field-treatment does not cause any alteration in the conductivity of deionized water, since ELFEM field-treatment can not change either composition or concentration of remaining ions in deionized water. The latter result may even serve as an eviednce that the observed change of dielectric relaxation spectrum in deionized water is hardly due to any systematic error in experimental measurements.

Water is the substance with individual molecular dipole moments, μ_i. Dielectric relaxation spectroscopy probes the response of the total dipole moment, $M(t) = \Sigma\mu_i(t)$, to a time-dependent external electromagnetic field. This inherent ability to monitor the cooperative motion of a molecular ensemble distinguishes dielectric relaxation from methods like NMR or Raman spectroscopy which essentially yield information on the motions of individual molecules. Kirkwood firstly proposed the following famous equation to calculate dielectric constant of water [12]:

$$\frac{(2\varepsilon+1)(\varepsilon-1)}{9\varepsilon}=\frac{4\pi N_0 d}{3M}(\alpha+\frac{\mu^2 g}{3kT}) \quad (2)$$

where ε is dielectric constant of water, N_0 is Avogadro's number; d is the density of water; M is molecular weight of water; k is the Boltzmann constant; α is the molecular polarizability; μ is the molecular dipole moment; g is the Kirkwood correlation factor that accounts for nonrandom orientation of neighboring molecules, and T is the temperature of water. Based on this equation, it can be known that the increased dielectric constant of the ELFEM field-treated water means that ELFEM field-treated water has higher molecular polarization α. Since molecular polarization determines the higher order structural of water, orientation of H_2O dipoles and the angle of H-bonds in bulk water might be altered.

Among the unsolved problems of physics, that of the properties of liquid water is one of the more important, not only because of its fundamental aspects but also because of its impact on life. The importance of electrostatic effects as the primary correlation between structure and function of biological molecules has been pointed out [13, 14] and demonstrated [15] repeatedly. It is now quite clear that electrostatic effects play a major role in enzyme catalysis [16,17], electron transfer [18,19], proton transport [20], ion channels [21,22], ligand binding [23], macromolecular assembly [24], and signal transduction [25]. It has been well established that all bio-macromolecules including polypeptides, polynucleotides, and polysaccharides, are stiff molecular chains with restricted folding capacities due to inclusion of rigid ring structures or double amide bonds in the backbone sequence. This creates ''irreducible spatial charge separation'' between positive and negative partial charges, causing elevated electrostatic energy. As a conceptual diagram, Fig.4a shows that binding of water molecule to protein results in increased distance between amino group and carboxyl group. As Fig.4b shows, in vacuum, the work required to separate a charge pair (NH^+ and CO^-) on a biomacromolecule is [26]

$$W_V = \frac{q_1 q_2(\frac{1}{s_0}-\frac{1}{r_{12}})}{4\pi\varepsilon_v}, \quad r_{12}=s_0+d \quad (3)$$

In water, the work required is

$$W_w = \frac{q_1 q_2(\frac{1}{s_0}-\frac{1}{r_{12}})}{4\pi\varepsilon_w} = W_V\frac{\varepsilon_v}{\varepsilon_w} \quad (4)$$

Where q_1 and q_2 are charge on NH^+ and CO^- group respectively, s_0 and r_{12} are distance between NH^+ and CO^- group in vacuum and in water respectively. ε_v and ε_w are dielectric constant of water in vacuum and in water respectively. The work requried to separate a charge pair (NH^+ and CO^-) in dielectric water is reduced as $\varepsilon_v/\varepsilon_w$ of the work required in vacuum. Thus, the work required in ELFEM field-treated water would be further reduced by a factor of $\varepsilon_{em}^*(\nu)/\varepsilon_c^*(\nu)$ in comparison with the untreated water, which is about 1.037. (See section "Results"). The reduction of the work require to separate charge on NH^+ and CO^- groups means that the free energy of the hydrated protein is reduced in ELFEM field-treated water if the treated water could enters the cells, replaces the previously bound water, and keeps its dielectric property in full or certain degree.

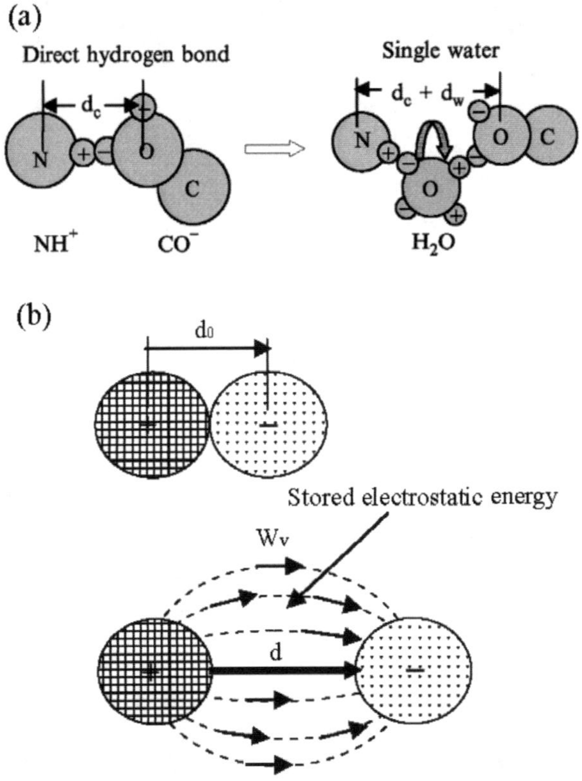

Fig.4. (a) The conceptual diagram showing formation of single water bridges between NH^+ and CO^- groups in hydrated proteins and (b) the Work is required to separate a charge pair. The irreducible separation of partial charges on the backbone of macromolecular polymers such as proteins creates an electrostatic field that stores this energy.

In fact, the conclusion "free energy of macromolecules in the water of high dielectric constant would be reduced" can be easily deduced from thermodynamics. The free energy $\Delta G = \Delta H - T\Delta S$ of water binding is the sum of negative electrostatic enthalpy ΔH partially compensated by the positive contributions from negative entropic change ΔS caused by the increased order of dielectrically aligned water molecules. It has been also well established that free energy is of crucial importance for stability of proteins, protein folding, conformational change of proteins, and catalytic activity of enzymes. The free energy function of protein has been a critical parameter for computational prediction of protein structure and function [27, 28].

Finally, it must be pointed out that all above discussion is based on an assumption that the electromagnetic field-treated water could still keep its high dielectric property in full or certain degree when it enters human body and living cells. The assumption seems to be reasonable, but solid evidence is still lack. Further study in this line would be necessary.

Acknowledgments

Author sincerely thanks Dr.Huanlin Gou (State Key Laboratory of New Ceramics and Fine Processing, Tsinghua University) for performing measurement of dielectric relaxation spectra of water and. Guangdong Junfeng BFS Industry Co.Ltd. for providing the apparatus generating extremely low frequency electromagnetic field.

Reference

[1] W.Kapit,R.Macey,E.Meisami The Physiology Coloring Book 2nd ed p.1-7. Addison/Wesley/Longman, Inc.San Francisco, 2000.

[2] Nakasako M. Large-scale networks of hydration water molecules around proteins investigated by cryogenic X-ray crystallography. 2001 *Cell Mol Biol* (Noisy-le-grand). **47** 767-90.

[3] Careri G, Giansanti A, Rupley JA. Proton percolation on hydrated lysozyme powders.1986 *Proc Natl Acad Sci U S A*. **83** 6810–6814.

[4] "Hardness in Drinking Water". New Hampshire Department of Environmental Services. October 2008. p.2, Retrieved 2009-10-25.

[5] Ismail NM, Shoukry NM. Laboratory studies on the effect of magnetic technologies on some biological parameters of Biomphalaria alexandrina in Egypt. 2007 *J Egypt Soc Parasitol*. **7** 275-86.

[6] Zhao Dayuan, Lu Xialing, Zhu Huiming, et al**.** Experimental study on biologic effect of magnetic field treated water in mice, 1998 *Chin J Phys Ther* , **21** 278-281.

[7] Liu Yuying, Zhao Xiumei, Liu Fengying, et al. Effects of frequency spectrum water on leukocyte adhesion and microcirculation of mesentery in rats. 2002 *Chinese Journal of Practical Medicne* **2** 489-91. (in Chinese)

[8] Liu Yuying,Han Yue, Zheng Sirong, et al. Effects of Frequency Spectrum Water on Blood Viscosity and Erythrocyte Deformation in Rats. 2002 *Chinese Journal of Microcirculation.* **12** 17-18. (in Chinese)

[9] Liu Yuying, Zhao Xiumei,Liu Fengying, et al. Effects of Frequency Spectrum Water on IL-1, IL-6 and Leukocyte s Behavior in Rats. 2003 *Chinese Journal of Microcirculation.* **13** 23-24. (in Chinese)

[10] Yuying Liu, Xiumei Zhao, Fengying Liu, et al. Effect of spectrum water on potential and intestines and stomach in rats. 2004 *Chinese Journal of Clinical Rehabilitation*, **8** 479. (in Chinese)

11. Bishop C. The Relationship Between Loss, Conductivity, and Dielectric Constant, in "Advanced Engineering Electromagnetics" (Ed. By C.A.Balanis), 2001, p1-12.

[12] Kirkwood, J.G. The dielectric polarization of polar liquids, 1939 *J.Chem. Phys.*, **7** 911-919.

[13] L. Pusztai;., J.C. Soetens, P.A. Bopp, The static dielectric constant and molecular geometries in ambient water studied by reverse Monte Carlo simulations, 2003 *Physica A*, **323** 42-50.

[14] Perutz MF. Electrostatic effects in proteins. 1978 *Science*, **201** 1187.

[15] Nakamura H. Roles of electrostatic interaction in proteins. 1996 *Q Rev Biophys.* **29** 1–90.

[16] Warshel A. (1981) Electrostatic basis of structure-function correlation in proteins. 1981 *Accts Chem Res* **14** 284–290.

[17] Warshel A. Computer modeling of chemical reactions in enzymes and solutions. New York: John Wiley & Sons; 1991.

[18] Hilvert D. Critical analysis of antibody catalysis. 2000 Annu Rev Biochem. **69** 751–793.
[19] Parson WW, Chu ZT, Warshel A. Electrostatic control of charge separation in bacterial photosynthesis. 1990 *Biochim Biophys Acta.* **1017** 251.
[20] Gunner M, Nichols A, Honig B. Electrostatic potential in rhodopseudomonas viridis reaction centers: implications for the driving force and directionality of electron transfer. 1996 *J Phys Chem.* **100** 4277–4291.
[21] Okamura MY, Feher G. Proton transfer in reaction centers from photosynthetic bacteria. 1992 *Annu Rev Biochem.* **61** 861–896.
[22] Åqvist J, Warshel A. Energetics of ion permeation through membrane channels: solvation of na1 by gramicidin a. 1989 *Biophys J.* 56:171.
[23] Åqvist J, Luzhkov V. Ion permeation mechanism of the potassium channel.2000 *Nature.* **404** 881.
[24] Simonson T, Archontis G, Karplus M. Continuum treatment of long-range interactions in free energy calculations. application to protein-ligand binding. 1997 *J Phys Chem B* **101** 8349–8362.
[25] Muegge I, Schweins T, Warshel A. Electrostatic contributions to protein-protein binding affinities: application to rap/raf interaction. 1998 *Proteins.* **30** 407–423.
[26] Glennon TM, Villa` J, Warshel A. How does gap catalyze the gtpase reaction of ras? A computer simulation study. 2000 *Biochemistry.* **39** 9641–9651.
[27] Gary D. Fullerton and Ivan L. Cameron Water Compartments in Cells, 2007 *Methods in Enzymology*, **428** 1-28.
[28] Gao, Z. G., and K. A. Jacobson. Allosterism in membrane receptors. 2006 *Drug Discov. Today.* **11** 191–202.
[29] Karplus, M., and J. Kuriyan. Molecular dynamics and protein function. 2005 *Proc. Natl. Acad. Sci. USA.* **102** 6679–6685.

Journal of Physics: Conference Series **329** (2011) 012020 doi:10.1088/1742-6596/329/1/012020

Using multifractal analysis of ultra-weak photon emission from germinating wheat seedlings to differentiate between two grades of intoxication with potassium dichromate

Felix Scholkmann[1], Michal Cifra[2], Thiago Alexandre Moraes[3] and Cristiano de Mello Gallep[3]

[1] Biomedical Optics Research Laboratory (BORL), Division of Neonatology, University Hospital Zurich, 8091 Zurich, Switzerland

[2] Institute of Photonics and Electronics, Academy of Sciences CR, 18251 Prague 8, Czech Republic

[3] Applied Photonics Lab – LaFA, DTT/CESET, State University of Campinas, 13484-370, Limeira, Sao Paulo, Brazil

Felix.Scholkmann@usz.ch

Abstract. The aim of the present study was to test whether the multifractal properties of ultra-weak photon emission (UPE) from germinating wheat seedlings (*Triticum aestivum*) change when the seedlings are treated with different concentrations of the toxin potassium dichromate (PD). To this end, UPE was measured (50 seedlings in one Petri dish, duration: approx. 16.6–28 h) from samples of three groups: (i) control (group C, $N = 9$), (ii) treated with 25 ppm of PD (group G_{25}, $N = 32$), and (iii) treated with 150 ppm of PD (group G_{150}, $N = 23$). For the multifractal analysis, the following steps where performed: (i) each UPE time series was trimmed to a final length of 1000 min; (ii) each UPE time series was filtered, linear detrended and normalized; (iii) the multifractal spectrum ($f(\alpha)$) was calculated for every UPE time series using the backward multifractal detrended moving average (MFDMA) method; (iv) each multifractal spectrum was characterized by calculating the mode (α_{mode}) of the spectrum and the degree of multifractality ($\Delta\alpha$); (v) for every UPE time series its mean, skewness and kurtosis were also calculated; finally (vi) all obtained parameters where analyzed to determine their ability to differentiate between the three groups. This was based on Fisher's discriminant ratio (FDR), which was calculated for each parameter combination. Additionally, a non-parametric test was used to test whether the parameter values are significantly different or not. The analysis showed that when comparing all the three groups, FDR had the highest values for the multifractal parameters (α_{mode}, $\Delta\alpha$). Furthermore, the differences in these parameters between the groups were statistically significant ($p < 0.05$). The classical parameters (mean, skewness and kurtosis) had lower FDR values than the multifractal parameters in all cases and showed no significant difference between the groups (except for the skewness between group C and G_{150}). In conclusion, multifractal analysis enables changes in UPE time series to be detected even when they are hidden for normal linear signal analysis methods. The analysis of changes in the multifractal properties might be a basis to design a classification system enabling the intoxication of cell cultures to be quantified based on UPE measurements.

1. Introduction

Ultra-weak light (~ 10^2 photons/s cm^2 [51]) in the wavelength range of approx. 200–800 nm is spontaneously emitted from the surface of every living organism. This ultra-weak photon emission (UPE) is generally considered to be ultra-weak chemiluminescence caused by the formation of energetically excited products (P^*) from the reactants K and M (which could be atoms or molecules), followed by a relaxation of P^* to a deeper energy level, accompanied by the emission of photons: $K + M \rightarrow P^* \rightarrow P + h\nu$ [1]. The excited states are considered to be mainly caused by interactions with reactive oxygen species (ROS) and/or reactive nitrogen species (RNS) that are produced under physiological states (e.g. by mitochondria in the respiratory chain [35], by phagocytic cells during the "respiratory burst" [14] or by ionizing radiation from internal and external sources [17, 18, 20, 19]) and pathophysiological states (e.g. during postischemic reperfusion [3]). The interaction of ROS/RNS with biomolecules leads to exothermic reactions where photons are emitted, e.g. during the interaction of ROS/RNS with lipids [4, 42] or proteins [43]. On the other hand, electronically excited states and their relaxation are caused by mechanisms other than those based on ROS/RNS. For example, Tuszyński and Dixon [46] showed that the activity of the proton pump in the mitochondrial wall could be a source of UPE as a consequence of the excitation and relaxation of cytochrome enzymes. Additionally, the reaction of triplet oxygen ($^3O_2$) with bilirubin under aerobic conditions and in an alkaline solutions exhibits UPE (with a peak at ~ 670 nm) [52]. Furthermore, non-enzymatic amino-carbonyl reactions between reducing sugars and amino acids (Maillard reactions) cause UPE (with two broad peaks at ~ 500 nm and ~ 695 nm) [53]. These reactions take place in biological systems as a step in the formation of advanced glycation end-products (AGEs) [32]. Thus, while ROS/RNS reactions seem to be the primary source of UPE, other physiochemical processes also appear to be involved.

The physiological significance of UPE is an ongoing debate. While on the one hand it is proven that UPE is correlated with physiological and pathophysiological states (e.g. temperature [10, 38], cell cycle state [27, 31, 5], "respiratory burst" during immune reactions [54], cell density [41], normal vs. cancer tissue [36], physical exercise [28], chronobiological phase [9, 8, 7, 49, 48, 23, 11], growth dynamics of germinating seedlings [24, 25, 47]), the role of UPE as an active factor regulating biological processes is still under investigation on the other hand [50, 6].

Previous experiments revealed that intensity and dynamics of UPE from germinating wheat seedlings are altered by intoxication with potassium dichromate (PD) ($K_2Cr_2O_7$) of different concentrations. PD is a oxidative stress causing [40, 44], carcinogenic [39] and genotoxic [12] agent that can readily cross cellular membranes. It was found that (i) UPE increased with very mild PD intoxication by 1.5 ppm [33] and 15 ppm [33, 2] compared to the UPE intensity from the control group. Also (ii) the UPE dynamics showed enhanced oscillations with a 6 h period length under PD intoxication of 1.5 ppm [2].

The aim of the present study was to further investigate the effects of the treatment of germinating wheat seedling with differing PD concentrations on UPE dynamics. In particular, we investigated whether it is possible to differentiate between three different groups of wheat seedlings (G_{25}: treated with 25 ppm of PD; G_{150}: treated with 150 ppm of PD; C: control) based on the analysis of the UPE dynamics. To this end, we used multifractal time series analysis to investigate the UPE dynamics. To the best of our knowledge, the use of multifractal analysis to analyze UPE signals is reported here for the first time.

2. Material and methods

2.1. Sample preparation and UPE measurements

Three groups of wheat seedling (*Triticum aestivum*) were prepared in Petri dishes, containing 50 wheat grains in 10 mL solution each: (i) control (group C, $N = 9$), (ii) treated with 25 ppm of PD (group G_{25}, $N = 32$), and (iii) treated with 150 ppm of PD (group G_{150}, $N = 23$). After the samples had been kept in a dark chamber with controlled temperature (20 °C) for 48 h, UPE was measured using a photomultiplier tube (Hamamatsu PMT H7360, spectral response range: 300-650 nm, peak sensitivity wavelength: 375 nm, dark noise: 15 cps) and a counting board (Hamamatsu M8784). Every measurement was started at the same time (9:30 am) and for a time span in the range of 16.6–28 h. The UPE values were acquired in 10 s time slots.

2.2. Multifractal detrended moving average (MFDMA) method

The rationale behind the multifractal approach is that a single scaling exponent (the fractal dimension d_f) is often not enough to describe the whole complex dynamics of a given non-stationary time series. To overcome this problem, the framework of multifractal analysis was developed based on the multifractal formalism introduced by Frisch and Parisi [16], Halsey et al. [22] and Mandelbrot [30]. Multifractal analysis allows the multiply scaling laws to be quantified by calculating the multifractal spectrum ($f(\alpha)$) which is a representation of the multifractal scaling of the time series. $f(\alpha)$ can be calculated by different approaches, e.g. the multifractal box-counting (MF-BOX) method [13], the wavelet transform modulus maxima (WTMM) method [37], the multifractal detrended fluctuation analysis (MFDFA) [26], the wavelet leader (WL) based multifractal analysis [29], the partition function and structure function method [15], or the multifractal detrended moving average (MFDMA) method [21]. Since it was shown that MFDMA is superior to MFDFA in quantifying the multifractality [21], MFDMA was used in the present study. A detailed mathematical description of the MFDMA method can be found in the paper of Gu and Zhou [21].

2.3. Data analysis

For the data analysis, the following steps where performed: (i) each UPE time series was trimmed to a final length of 1000 min (starting 100 s after the beginning of the measurement); (ii) each UPE time series was filtered (by applying a Savitzky-Golay filter (span: 3 samples) to remove the noise), linear detrended and normalized by scaling to the range [0, 1]; (iii) the multifractal spectrum was calculated for every UPE time series; (iv) each multifractal spectrum was characterized by calculating the mode (α_{mode}) of the spectrum and the degree of multifractality ($\Delta\alpha = \alpha_{max}-\alpha_{min}$); (v) for every UPE time series, its mean, skewness and kurtosis were also calculated; finally (vi) all obtained parameters where analyzed for their capability to separate the three groups by calculating the Fisher's discriminant ratio (FDR) for each parameter combination. FDR is a measure of the group discriminability (the higher the FDR value the higher the discriminability power). FDR is defined as [45]

$$FDR = \frac{(m_1 - m_2)}{(\sigma_1^2 - \sigma_2^2)} \tag{1}$$

where m_1 and m_2 are the mean values of the two distributions and σ_1^2 and σ_2^2 the respective variances. In addition to FDR, a non-parametric test was used (Wilcoxon test) to test whether the parameter values are significantly different or not.

For the calculation of the multifractal spectrum, the following values for the lower bound of the segment size n (n_{min}), upper bound of the segment size n (n_{max}), the length of n (N), the multifractal order (q) were used according to the suggestion of [21]: $n_{min} = 10$; n_{max} = 10% of the length of the time series used; $N = 30$, $q = [-4, 4]$. Additionally, we used a backward detrending window in MFDMA since it was shown that it was superior to the usage of a centered or forward detrending window [21].

3. Results

The multifractal analysis of the UPE signals revealed that the characteristics of the multifractal spectra were different for each group (see Fig 1-2, Tab 1). When comparing all the three groups, FDR had the highest values for the multifractal parameters (α_{mode}, $\Delta\alpha$); also, the differences of these parameters between the groups where statistically significant ($p < 0.05$) (see Fig 2, Tab 1). The classical parameters (mean, skewness and kurtosis) had lower FDR values than the multifractal parameters in all cases and showed no significant difference between the groups – except for the skewness between group C and G_{150} (see Fig 2, Tab 1).

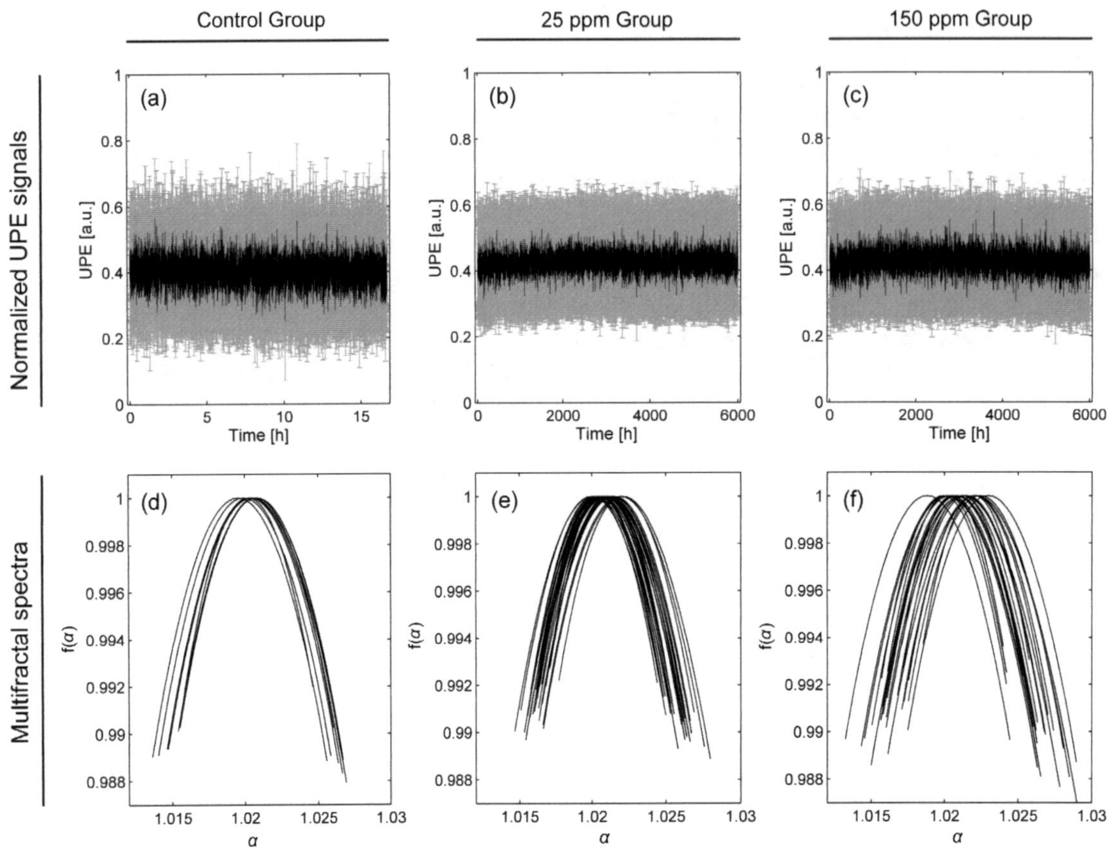

Figure 1. Results of the multifractal analysis. Subfigures (a)–(c): detrended and normalized UPE signals of the three groups (black lines: mean, grey: standard deviation), (d)–(f) corresponding multifractal spectra. It is clear that the distribution of the multifractal spectra is different for the 150 ppm and 25 ppm group compared to the control group.

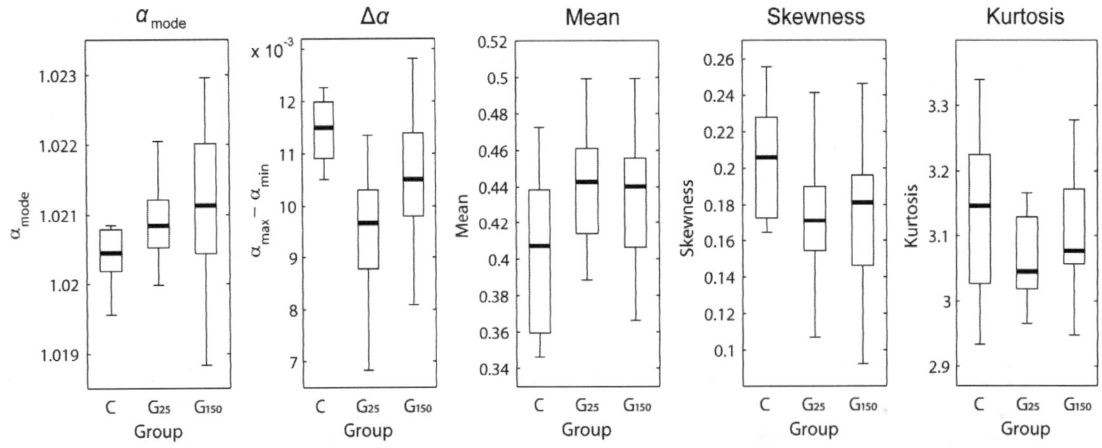

Figure 2. Boxplots of the determined values of multifractal parameters (α_{mode}, $\Delta\alpha$) and classical parameters (mean, skewness and kurtosis).

Table 1. Results of the statistical analysis. The asterisks refer to the significance levels: significant (*, $p < 0.05$), highly significant (***, $p < 0.001$).

Parameter	C vs. G_{25}		C vs. G_{150}		G_{25} vs. G_{150}	
	FDR	*p*	*FDR*	*p*	*FDR*	*p*
α_{mode}	0.4585	0.0326 *	0.5099	0.0244 *	0.0408	0.3106
$\Delta\alpha$	0.2573	0.0715	1.2262	0.0006339 ***	0.1425	0.0267 *
Mean	0.2135	0.1021	0.1648	0.0643	0.0000896	0.5965
Skewness	0.3599	0.0539	0.3682	0.0343 *	0.0000551	0.8505
Kurtosis	0.0463	0.5025	0.0001349	0.2237	0.0138	0.3106

4. Discussion, conclusion and outlook

The results of the data analysis show that the multifractal parameters α_{mode}, $\Delta\alpha$ had the largest FDR values, indicating that they can be used to differentiate between the UPE time series originating from the three sample groups of germinating wheat seedlings. The differences observed in the multifractal parameters indicate that the UPE time series show different non-linear characteristics of their dynamical behavior with respect to the state of the germinating seedlings (not stressed [C], mildly intoxicated [G_{25}], and strongly intoxicated [G_{150}]).

The interpretation of the multifractal parameters α_{mode}, $\Delta\alpha$ is as follows: the local Hölder exponent α_{mode} is related to the degree of irregularity of the analyzed signal (the lower α_{mode}, the more irregular the signal), $\Delta\alpha$ reflects the strength of multifractality (the higher $\Delta\alpha$, the more structure has the signal) [21, 34]. Based on these facts and based on the results obtained for the multifractal parameters (see Fig 2), we conclude that the more stressed the germinating wheat seedlings are, the more irregular and less structured the UPE time series become. Interestingly, the strength of multifractality $\Delta\alpha$ decreases more for the intoxication with 25 ppm PD than with 150 ppm.

In conclusion, multifractal analysis enables the detection of changes in UPE time series that are hidden for normal linear signal analysis methods. The non-linear features of a UPE time series seem to be related to the states of the bio-system. The analysis of changes in the multifractal properties might be a basis to design a classification system enabling the intoxication of cell cultures to be quantified based on UPE measurements. Further investigations also have to address the questions of how different parameters of (i) the UPE data (i.e. the length of the time series), (ii) the preprocessing process (i.e. type of applied filter, span of the filter, type of detrending method, type of normalization), and (iii) the determination of the multifractal parameters (i.e. type of method, type of characterization of the multifractal spectra) influence the results obtained, and how they could be optimized to improve the classification of the UPE time series based on multifractal analysis.

Acknowledgements

The authors thank John and Rachel Folkes for proofreading the manuscript, as well as Dr. Martin Wolf and Reto Kofmehl for stimulating discussions and for important comments on drafts of the manuscript. Michal Cifra acknowledges financial support from the Czech Science Foundation GA CR (grant no. P102/10/P454). LaFA is supported by FAPESP and CNPq.

References

[1] Barenboim G M, Domanskii A N and Turoverov K K 1969 *Luminescence of biopolymers and cells* (New York, London: Plenum Press)

[2] Bertonha E, Moraes T A, Conforti E and Gallep M C 2010 In: *Latin America Optics and Photonics Conference*: Optical Society of America p MB08

[3] Bertuglia S and Giusti A 2003 *Am. J. Physiol. Heart Circ. Physiol.* **285** H1064-H71

[4] Boveris A, Cadins E and Chance B 1980 *Photobiochem. Photobiophys.* **1** 175-85

[5] Chwirot W B 1992 *Recent Advances in Biophoton Research and its Applications,* ed F A Popp, *et al.* (Singapore: World Scientific) pp 259-85
[6] Cifra M, Fields J Z and Farhadi A 2011 *Prog. Biophys. Mol. Biol.* **105** 223-46
[7] Cifra M, Van Wijk E, Koch H, Bosman S and Van Wijk R 2007 In: *2007 17th International Conference Radioelektronika, Vols 1 and 2,* pp 434-7
[8] Cifra M, Van Wijk E, Koch H, Bosman S and Van Wijk R 2007 In: *Measurement 2007: 6th International Conference on Measurement, Proceedings,* pp 421-4
[9] Cifra M, Van Wijk E, Koch H, Bosman S and Van Wijk R 2007 *Radioengineering* **16** 15-9
[10] Cifra M, van Wijk E P A and van Wijk R 2008 In: *PIERS 2008 Cambridge, Proceedings,* pp 907-11
[11] Cohen S and Popp F A 1997 *Skin Res. Technol.* **3** 177-80
[12] Dana Devi K, Rozati R, Saleha Banu B, Jamil K and Grover P 2001 *Food Chem. Toxicol.* **39** 859-65
[13] Feder J 1988 *Fractals* (New York: Pelnum Press)
[14] Forman H J and Torres M 2002 *Am. J. Respir. Crit. Care Med.* **166** S4-S8
[15] Frisch U 1999 *Turbulence: The Legacy of A. N. Kolmogorov*: Cambridge University Press)
[16] Frisch U and Parisi G 1985 ed R Benzi, *et al.* In: *Proceedings of the International Summer School of Physics "Enrico Fermi"*, North-Holland, Amsterdam p 84
[17] Goraczko W 2000 *Med. Hypoth.* **54** 461-8
[18] Goraczko W and Slawinski J 2004 *Nonlinear. Bio. Tox. Med.* **2** 245-58
[19] Goraczko W and Slawinski J 2008 *J. Lumin.* **128** 1155-61
[20] Goraczko W, Slawinski J and Staninski K 2008 *J. Radioanal. Nucl. Chem.* **277** 613-8
[21] Gu G F and Zhou W X 2010 *Phys. Rev. E* **82** 011136
[22] Halsey T C, Jensen M H, Kadanoff L P, Procaccia I and Shraiman B I 1986 *Phys. Rev. A* **33** 1141-51
[23] Jung H H, Yang J M, Woo W M, Choi C, Yang J S and Soh K S 2005 *J. Photochem. Photobiol. B: Biol.* **78** 149-54
[24] Kai S, Mitani T and Fujikawa M 1993 *Jpn. J. Appl. Phys.* **32** L417-L9
[25] Kai S, Ohya T, Moriya K and Fujimoto T 1995 *Jpn. J. Appl. Phys.* **34** 6530-8
[26] Kantelhardt J W, Zschiegner S A, Koscielny-Bunde E, Havlin S, Bunde A and Stanley H E 2002 *Physica A* **316** 87-114
[27] Konev S V, Lyskova T I and Nisenbaum G D 1966 *Biophysics* **11** 410-3
[28] Laager F, Park S H, Yang J M, Song W and Soh K S 2008 *Eur. J. Appl. Physiol.* **102** 463-9
[29] Lashermes B, Jaffard S and Abry P 2005 In: *2005 IEEE International Conference on Acoustics, Speech, and Signal Processing,* pp 161-4
[30] Mandelbrot B B 1999 *Multifractals and 1/f noise* (Berlin, Heidelberg, New York: Springer Verlag)
[31] Mei W P 1992 *Recent Advances in Biophoton Research and its Applications,* ed F A Popp, *et al.* (Singapore: World Scientific) pp 243-58
[32] Monnier V M 2003 *Arch. Biochem. Biophys.* **419** 1-15
[33] Moraes T A, Garofalo R, Ramos S, Martins L and Gallep C D 2009 In: *IMOC 2009, International Microwave and Optoelectronics Conference,* pp 221-3
[34] Munoz-Diosdado A, Jiménez-Flores J R and Méndes-Cruz A R 2010 In: *5th Cairo International Biomedical Engineering Conference,* Cairo, Egypt
[35] Murphy M P 2009 *Biochem. J.* **417** 1-13
[36] Musumeci F, Triglia A and Grasso F 1992 *Recent Advances in Biophoton Research,* ed F A Popp, *et al.* (Singapore: World Scientific) pp 307-26
[37] Muzy J F, Bacry E and Arneodo A 1993 *Phys. Rev. E* **47** 875-84
[38] Nakamura K and Hiramatsu M 2005 *J. Photochem. Photobiol. B: Biol.* **80** 156-60
[39] Norseth T 1986 *Br. J. Ind. Med.* **43** 649-51
[40] Pedraza-Chaverri J, Yam-Canul P, Chirino Y I, Sanchez-Gonzalez D J, Martinez-Martinez C M,

Cruz C and Medina-Campos O N 2008 *J. Food. Chem. Toxicol.* **46** 619-27
[41] Popp F A, Li K H, Mei W P, Galle M and Neurohr R 1988 *Experientia* **44** 576-85
[42] Prasad A and Pospisil P 2011 *PLoS One* **6** e22345
[43] Slawinski J 1988 *Experientia* **44** 559-71
[44] Soudani N, Amara I B, Troudi A, Bouaziz H, Boudawara T and Zeghal N 2011 *Toxicol. Ind. Health* (in press)
[45] Theodoridis S, Pikrakis A, Koutroumbas K and Cavouras D 2009 *Introduction to pattern recognition - A MATLAB approach* (Oxford: Accademic Press)
[46] Tuszynski J A and Dixon J M 2001 *Phys. Rev. E* **64** 051915
[47] Usa M, Kobayashi M, Scott R Q, Maeda T, Hiratsuka R and Inaba H 1989 *Protoplasma* **149** 64-6
[48] Van Wijk E P and Van Wijk R V 2005 *Forsch. Komplementärmed. Klass. Naturheilkd.* **12** 96-106
[49] Van Wijk E P A, Van Wijk R and Cifra M 2007 In: *Biophotonics 2007: Optics in Life Science,* pp J6331
[50] Van Wijk R 2001 *J. Sci. Expl.* **15** 183-97
[51] Van Wijk R V and Van Wijk E P 2005 *Forsch. Komplementärmed. Klass. Naturheilkd.* **12** 77-83
[52] Watanabe H, Usa M, Kobayashi M, Agatsuma S and Inaba H 1992 *J. Biolum. Chemilum.* **7** 1-11
[53] Wondrak G, Pier T and Tressl R 1995 *J. Biolum. Chemilum.* **10** 277-84
[54] Wymann M P, von Tscharner V, Deranleau D A and Baggiolini M 1987 *Anal. Biochem.* **165** 371-8

Journal of Physics: Conference Series **329** (2011) 012021 doi:10.1088/1742-6596/329/1/012021

Photon Counts Statistics in Leukocyte Cell Dynamics

Eduard van Wijk[1,2,*], Jan van der Greef[1,3], Roeland van Wijk[1,2]

[1] Netherlands Metabolomics Centre/ Sino Dutch Center for Preventive and Personalized Medicine, Leiden University, Leiden, The Netherlands

[2] Meluna Research, Amersfoort, The Netherlands

[3] TNO Netherlands Organization for Applied Scientific Research, Zeist, The Netherlands

eduard.vanwijk@sinodutchcentre.nl

Abstract. In the present experiment ultra-weak photon emission/ chemiluminescence from isolated neutrophils was recorded. It is associated with the production of reactive oxygen species (ROS) in the "respiratory burst" process which can be activated by PMA (Phorbol 12-Myristate 13-Acetate). Commonly, the reaction is demonstrated utilizing the enhancer luminol. However, with the use of highly sensitive photomultiplier equipment it is also recorded without enhancer. In that case, it can be hypothesized that photon count statistics may assist in understanding the underlying metabolic activity and cooperation of these cells. To study this hypothesis leukocytes were stimulated with PMA and increased photon signals were recorded in the quasi stable period utilizing Fano factor analysis at different window sizes. The Fano factor is defined by the variance over the mean of the number of photon within the observation time. The analysis demonstrated that the Fano factor of true signal and not of the surrogate signals obtained by random shuffling increases when the window size increased. It is concluded that photon count statistics, in particular Fano factor analysis, provides information regarding leukocyte interactions. It opens the perspective to utilize this analytical procedure in (in vivo) inflammation research. However, this needs further validation.

1. Introduction

Infections cause currently more than 20 percent of all human death in the world. Like all other animal organisms, the human survives this world with hostile and rapidly evolving pathogens by utilizing mechanisms to resist infection by pathogens. These defences are of two kinds: innate immune responses and adaptive immune responses. During the first critical hours and days of exposure to a *new* pathogen, humans rely on the innate immune system. The second, the adaptive immune system, which remembers previous encounters with specific pathogens, is too slow after a first exposure to a new pathogen. It can take a week before specific clones of B and T cells are activated and expanded, whereas a single bacterium with a doubling time of one hour can produce already a 20 million progeny infection in a single day. The innate immune system relies on the recognition of particular types of molecules that are common to many pathogens but are absent in the host. These pathogen-associated immune-stimulants stimulate two types of innate immune responses – inflammatory responses and phagocytosis by cells such as neutrophils and macrophages. The recognition of the microbial invader is usually quickly followed by its engulfment by a phagocytic cell. In vertebrates, macrophages reside in the tissues throughout the body and are especially abundant in areas where infections are likely to arise, including the lungs and gut. They are also present in large numbers in connective tissues, the liver and spleen. They patrol the tissues of the body and are among the first cells to encounter invading microbes. The second major family of phagocytic cells, the neutrophils, are not present in tissue of healthy subjects but are abundant in blood. These cells are rapidly recruited to tissue sites of infection [1].

Once the pathogen has been phagocytised, the macrophage or neutrophil kills the pathogen utilizing different strategies. The phagosome is acidified and fuses with lysosomes, which contain lysozyme and acid hydrolases that can degrade bacterial cell walls and proteins. In addition, the phagocyte assemble a special electron transport system complex (NADPH-oxidase) on the phagosomal membrane that catalyzes the production of a series of highly toxic oxygen-derived compounds, including superoxide anion (O^-_2), whereas subsequent reactions lead to the formation of other toxic agents including hydrogen peroxide (H_20_2), hypochlorous acid (HOCl) and possibly hydroxylradical (OH·) and singlet oxygen ($^10_2$). The production of these toxic compounds is accompanied by a transient increase in oxygen consumption by the cells, called the respiratory burst. The respiratory burst [2] is intrinsically associated with low-level biochemiluminescence or ultra-weak photon emission. The production of ultra-weak photon emission by phagocytosing (polymorphonuclear) leukocytes correlates with their metabolic activation [3].

A fascinating aspect of the neutrophil is the rhythmicity in the process of metabolic activation. In fact, almost every organism examined displays some form of rhythmic activity. Moreover, within a single organism almost every major organ system can, under appropriate circumstances, generate sustained oscillations. The periods of oscillation can range from a fraction of a second to several hours. At cell level, the neutrophilic leukocyte is a well-studied example of this type of regulation, in particular the amplitude and frequency modulation of intracellular NADPH concentrations. With respect to the cellular spatial-temporal organization, it has been observed, that NAD(P)H oscillations are in-phase with oxidative oscillations and that superoxide production follows NAD(P)H amplitude and frequency changes. It has been argued that associated kinases and phosphatases play a role in NADPH oxidase phosphorylation and in decoding these fluctuations into series of superoxide bursts [4] and periodic cytolytic events [5].

Although these studies are limited to the amplitude and frequency modulation of cellular metabolic oscillations, there is also some indication for the existence of intercellular communication. Thus, neutrophils appear to detect time-varying chemical fields displaying this same period [6]. Other research data [2] suggest a role of the neutrophil's ultra-weak radiation in stimulating, in a non-chemical, non-electrical way, the metabolic activity of other neutrophils which are spatially (but not optically) separated. These intercellular communication aspects suggest that sustained oscillations in ultra-weak photon emission may be even displayed within populations of leukocytes cells.

Collective regulation has attracted the attention of researchers from the 1970's and onwards for its association with the statistical properties of the ultra-weak photon emission. However, it is difficult to develop reliable procedures for the use of ultra-weak photon emission in the detection of cooperative phenomena in cell populations. Oscillations in intensity have been demonstrated in a few cases. A distinct type of oscillations in ultra-weak luminescence with a period of about 1 h was for instance observed in cell populations using Nitellopsis obtusa cells [7]. In another study, long-period oscillations with a period of a multiplex of 3 min have been measured in the photon emission of a culture of dark-adapted Acetabularia cells utilizing Fourier analysis [8].

In the present study, we document on the possibility to apply novel photon count distribution analytical methods in cell population studies, in particular for neutrophil cell populatons and their respiratory burst associated photon emission. We have focussed on the detection of ROS related single photon emission without enhancers. This is a novel development since almost all studies record photon emission that has produced by chemical substituents like luminol or lucigenin which react upon reactive oxygen species (ROS) with the production of photons. Studies on the detection of photon emission without enhancers are very limited, mostly due to the sensitivity of the detection equipment. However, just this latter is of increasing importance for obtaining information from time resolution studies on the photon signal directly associated with metabolic processes. The use of high-sensitivity, low-noise photomultiplier equipment in combination with novel photon count statistics permit documentation of photon signals within the concept that photons in a biological system are part of an integrated organization of biochemical reactions. Such studies have indicated that photon count distribution tends to a super-Poisson distribution which implies bunching of emitted photons indicating clustering of excitation processes [9-11]. The opposite phenomenon, anti-bunching of photons indicated by a sub-Poisson distribution of photon counts, may be initiated under conditions of excitation feedback. It has been reported in early laser studies but has not reported for living systems [12].

Super, normal or the sub-Poisson nature of a photo count distribution is easily ascertained by the value of its Fano factor [13]. The Fano factor is the ratio of the variance of the number of photons detected in many repeated measurements with a counting time window of duration T to the mean number of photons in the same set of measurements. It is written as F(T) = Variance [Ni(T)] / mean [Ni(T)] where Ni(T) is the number of events in the ith time window of length T. Within this context, its value is greater than one for super-Poisson, is equal to one for Poisson distributions and is less than one for sub-Poisson distributions. The Fano factor may be dependent on window duration and estimation of Fano factor values for increasing window sizes then results in a Fano factor time curve. In some biological systems, the F(T) shows an increase because rare clusters of high and low photon counts are more apt to be found as more and more data are collected. Such clustering has been considered characteristic of a fractal process [14]. Fano factor time curve was utilised before to study the statistical properties of action potentials [15-18], ultra-weak photon emission data from micro-organisms and human subjects [11, 19-21].

We considered it important to utilize Fano factor analysis in the study of populations of cells when photon emission is a rare event. Thus, when populations of thousands or millions of cells show a photon emission of less than hundred photons per second, photon emission from individual cells is considered very rare. The emission from an entire population of non-interacting cells is expected to result in Poisson distribution of photon counts. Any deviation towards super-Poisson suggests cell interaction; the signal contains information within the sequence of photon counts events about the population as collective phenomenon.

The presented experiments document the results of F(T) analysis of ultra-low level photon emission measured from isolated neutrophil leukocytes, after stimulation of the "respiratory burst" process by Phorbol 12-Myristate 13-Acetate (PMA). The chemiluminescence emitted during this reaction is recorded with the use of highly sensitive photomultiplier equipment without chemical enhancer. Photon count statistics and F(T) analysis are reported and suggest that neutrophils as a population, display an integrated metabolic organization.

2. Material and Methods

2.1. Neutrophils isolation

Neutrophils were isolated from freshly drawn whole blood from pigs that was collected and heparinised at the moment the pigs were aborted. At four occasions, a quantity of 500 millilitre (ml) was collected, divided in three portions and stored at 4°C for use at three consecutive days after collection meaning that we had twelve experimental days. During each of the twelve experimental days, neutrophils were isolated in three independent sessions. In each session, 50 ml blood was utilized. In principle, isolation was performed following the method described earlier by Shen et al (1994) [2]. In short, neutrophils were isolated from 50 ml heparinized whole blood after dextran (Sigma-Aldrich, St. Louis, MO) sedimentation of erythrocytes for 45 minutes. After isolation, neutrophils were washed by centrifugation and resuspension in Hanks' Balanced Salt Solutions (HBSS) (Sigma-Aldrich St. Louis, MO) in three cycles. Cells were finally resuspended in fresh HBSS and counted before use. Neutrophil viability, assessed by the ability of the cells to exclude trypan blue (Sigma-Aldrich St. Louis, MO), was always more than 90%.

2.2. Experimental protocol

The reported experiments on photon emission were carried out with neutrophil cells incubated at 25^{0}C. Before initiating photon emission, 10 ml of neutrophil suspension in HBSS (approximately $1.5x10^7$ cells) was placed in a Petri dish of 5 cm diameter and equilibrated at 25^{0}C for 15 min in the dark chamber of the photomultiplier equipment. After this temperature equilibration, 10μg phorbol myristate acetate (PMA) (Sigma-Aldrich St. Louis, MO) was added to the Petri dish to induce the respiratory burst. There was a delay of 7 sec between adding the PMA and the first moment of measurement. This was due to the fact that PMA was added to the cells, closing the measuring chamber and starting the measurement. The photon emission was recorded during 2400 s. In thirty six experiments, photon emission was measured without using enhancer substances. In twelve experiments luminol was present at a concentration of 1μg/ml during the incubation and photon emission measurement. This resulted in a total of thirty six curves without and 12 curves with luminol for data analysis (see below).

2.3. Photon emission measurements

Photon emission was recorded with a vertically placed, cooled (−25°C) photomultiplier tube (H-R550, Hamamatsu Photonics Co. Ltd, Hamamatsu, Japan), that was placed in a vertical position on the top of the incubation chamber. The opening window of the photomultiplier was 46 mm. The distance between the sample and photomultiplier tube was 10 cm. The temperature of the sample (Petri dish) holder was controlled at 25^{0}C by a Peltier-element. Recording was performed with consecutive 50 millisecond (ms) periods. Technical aspects and data acquisition were computer controlled.

2.4. Data analysis

The photon emission signal was screened for artefacts. The sources of artefacts are primarily cosmic ray-induced photoelectron clusters or micro discharges occurring in the PMT's. These artefacts are clearly detectable. For correction, values larger than 15 times the standard deviation in the entire data row were replaced by mean values. Because the removal of artefacts influences the standard deviation of the overall data set this procedure was repeated until no artefacts were found. Two type of data analysis were made. First, we looked at the intensities in time to estimate the time course of photon emission produced by the cells during their respiratory burst. Subsequently, we estimated the period that the induced signal was quasi-stable. The term quasi-stable period has been used because the overall signal in this period shows fluctuations without an overall upward or downward trend. For this purpose, time fragments were selected which on one hand have no significant upward or downward trend and, on the other hand are sufficiently long to the estimate the Fano factor dependency on window times up to 50 s (see below). This resulted in the selection of 3-4 non-overlapping fragments of 5 min for each emission curve. It ultimately resulted in a set of 120 data rows for the Fano factor analysis without enhancer and 40 data rows with luminol.

2.5. Fano factor analysis

The value of F(T) represents the Poisson nature of a photo count distribution [14]. Its value is greater than one for super-Poisson distributions, is equal to one for Poisson distributions and is

less than one for sub-Poisson distributions. Since background, due to electronic noise also demonstrates a deviation from a Poisson value of unity, one has the option to correct for background assuming that the sources of photon counts originating from signal and background are independent for measurements of all varieties of time windows. In the case of a low signal compared to noise, however, such correction decreases the reliability of the estimations. On the other hand, such correction is not necessary when one is only interested in the relationship between Fano factor value and time window used for analyzing the photon data. In the present analysis, we calculated this relationship both for background signals and neutrophil population signals (which also included the background).
Whether the Fano factor time curve truly reflects long-range correlations of events was tested by constructing surrogate data sets based on the same data but after randomization of these data. These surrogate data sets were obtained 10 times of shuffling randomly the values in the primary observed signal.

2.6. Statistics
Statistica 6.0 was used for basic statistical analyses of photon count data.

3. Results

3.1. Kinetics of PMA-induced photon emission
The PMA-induced photon emission during the respiratory burst of the leukocyte cell population is illustrated in Figure 1 by a representative example.

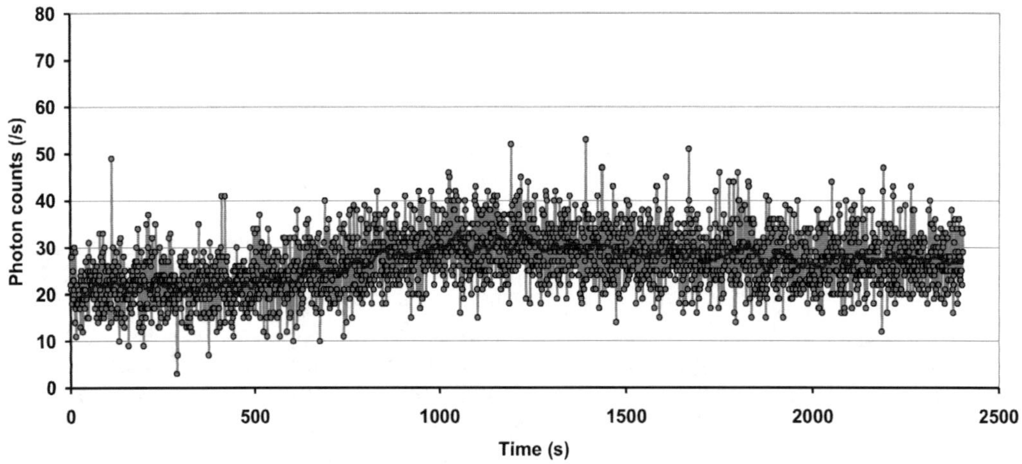

Figure 1. Spontaneous photon emission of leukocytes during respiratory burst after stimulation with PMA (without enhancer).

The data in this representative curve show initially a value of about 20 counts per sec (cps), which is followed by an increase in photon emission starting 500 sec after the addition of PMA. In the initial period we have not detected any difference between the photon signal of cells and background. The initial (0-500 sec) values represent background from electronic noise; it is similar for backgrounds measured with shutter open and shutter closed. In the experimental period 36 backgrounds were determined; the average background values (±std) was 18.3 ± 0.9 cps. The increase starting after 500 sec was highly typical and continuing up to 1000 sec. From 1000 sec on photon emission only slowly decreased and allowed the selection of consecutive 300 s time periods with quasi-stable photon emission in the period of 1000-2400s for photon count statistical analysis and F(T) estimations.

These kinetics suggest that after PMA addition there is a 500 sec initial lag in ROS production , followed by an increase and then a period of relatively constant production. To be certain about the initial lag in ROS production, similar incubations were performed in which the enhancer luminol was present. A representative example of the kinetics of photon emission in the presence of luminol is presented in Figure 2. Data suggest that a similar lag period is observed when the enhancer luminol was used, suggesting that the initiation of ROS production requires such period of time.

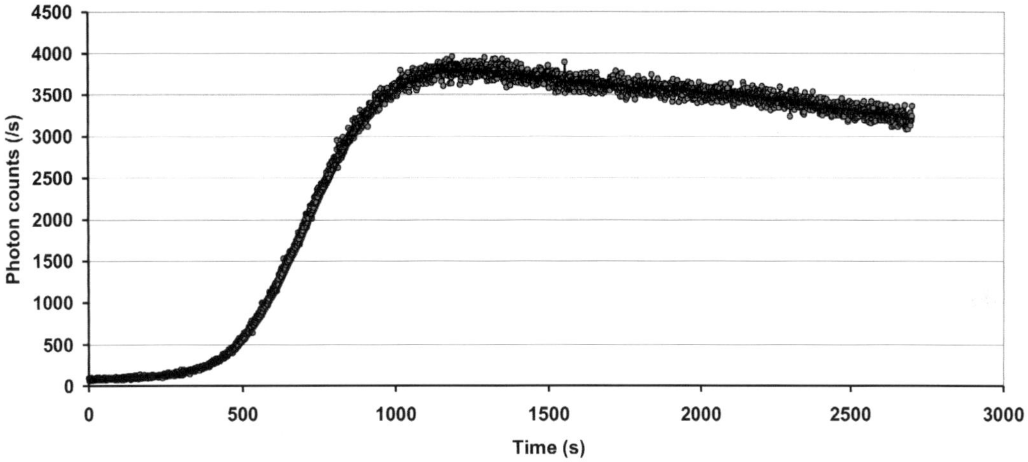

Figure 2. Luminol enhanced photon emission of leukocytes during respiratory burst after stimulation with PMA. The black curve represents moving average.

The presented example with luminol, demonstrates that under these conditions the actual photon signal (corrected for background) in the quasi stable period is about 300 times higher as compared with the emission without luminol.

3.2. F(T) of PMA-induced photon emission

Increased photon signals recorded in the quasi stable period were used to calculate the behaviour of the Fano factor F(T), regarding different window sizes. The window sizes ranged between 50 ms and 50 sec measurement period. The longer periods include combinations of 50 ms measurement periods. The analysis is limited to the 50 sec maximum because reliability strongly decreases with less than 6 windows. This procedure was executed both for leukocyte data and background. Moreover, to demonstrate the presence of sequence related phenomena, the Fano factor of true signal was compared with surrogate signals obtained by random shuffling the original data.

First, the F(T) in background data was estimated utilizing thirty six background datasets. The Fano factor data at different windows were calculated for each individual dataset and then these individual F(T) curves were averaged. The mean F(T) curve is shown in Figure 3. The data show that the Fano factor is about 1.25 and it is not increasing with larger window sizes. It has the same flat curve and average value as permutated background data. The super Poisson nature is a common feature for photomultiplier tubes and has been ascribed previously [20].

Similar analyses were made for cell populations of leukocytes utilizing 120 experimental data sets. It was demonstrated that the Fano factor increased definitively with window time.

Journal of Physics: Conference Series **329** (2011) 012021 doi:10.1088/1742-6596/329/1/012021

The F(T) curve representing the average of the 120 curves is presented in Figure 4. In contrast, the F(T) in permutated series is not increasing. The increase in the F(T) curve of real signals occurs because rare clusters of high and low photon counts are more apt to be found as more and more data are collected; a clustering which has been considered characteristic of a fractal process.

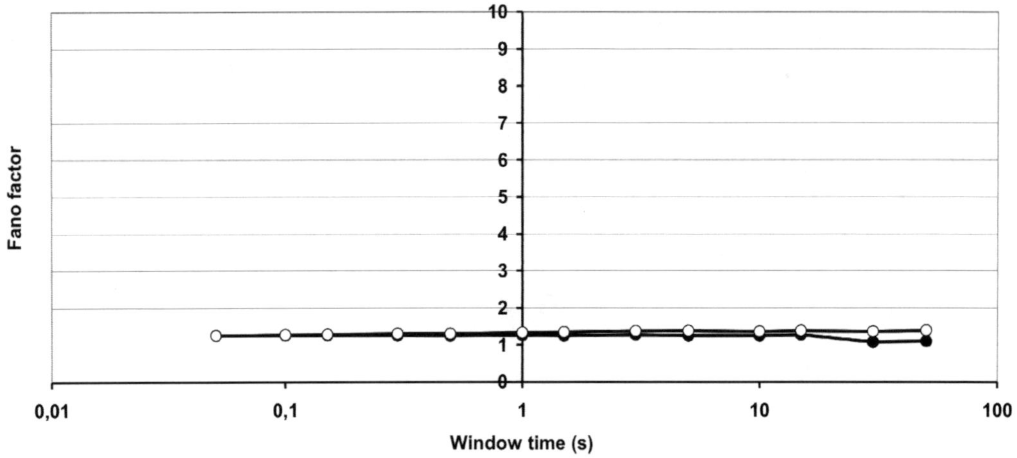

Figure 3. Fano factor curve F(T) averaged over 36 background measurements (open symbols represent average over background curves; closed symbols represent average over background curves after shuffling the data).

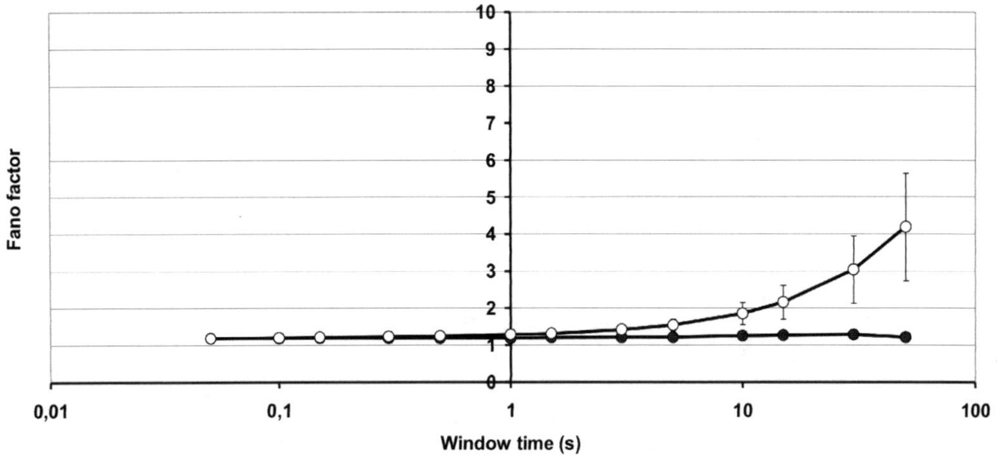

Figure 4. Represents the average F(T) over the 120 data sets of leukocyte cell populations (open symbols represent average over measured data; closed symbols represent average over shuffled data).

It must be emphasized however, that these curves were derived from data including the actually photons from PMA-stimulated photon emission and electronic background. Within the entire set of data, subsets are present with high and with low photon signal compared to electronic counts. It can be hypothesized that high and low photon data sets are significantly different with respect to their F(T) curve. In Figure 5, a comparison is made between F(T) of all neutrophil data sets with and without shuffling and the F(T) curves of the 30 data sets with highest and 30 data sets with lowest intensity signal. The data confirm the effect of signal

Journal of Physics: Conference Series **329** (2011) 012021 doi:10.1088/1742-6596/329/1/012021

intensity on the shape of F(T) curve. The higher the intensity, the higher the increase of the F(T) curve, demonstrating the influence of background values with its flat F(T) curve (Figure 3).

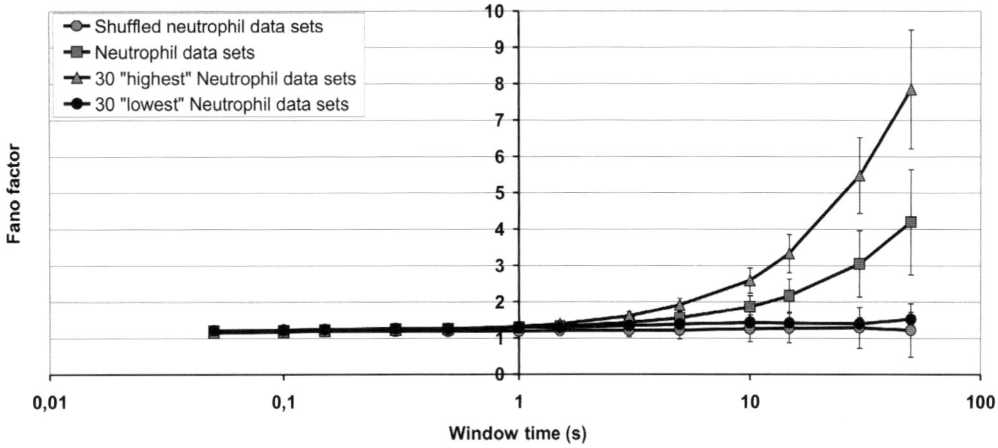

Figure 5. F(T) curve of all leukocyte data sets (with and without shuffling), and F(T) curves of data sets from 30 highest and 30 lowest photon emission intensity.

3.3. F(T) curves of low-intensity artificial light

For comparison the F(T) curve of artificial light of similar intensity as high levels of PMA-induced photon signals was estimated. For this purpose, light emission from a light-emitting diode (LED) (Conrad Electronic, Type TLLG 5400, green) was measured and analyzed following the same procedure. With the use of grey filters (Jenaer Glaswerk Schott & Gen, Mainz, Germany) the intensity was reduced to 30 cps (including background) without compromising the stability of the signal. A total of 50 data sets as collected. The average F(T) curve, presented in Figure 6, illustrates that the Fano factor over the entire range of windows was about 1.25, similar to background. The Fano factor of the surrogate, shuffled data sets has the same value.

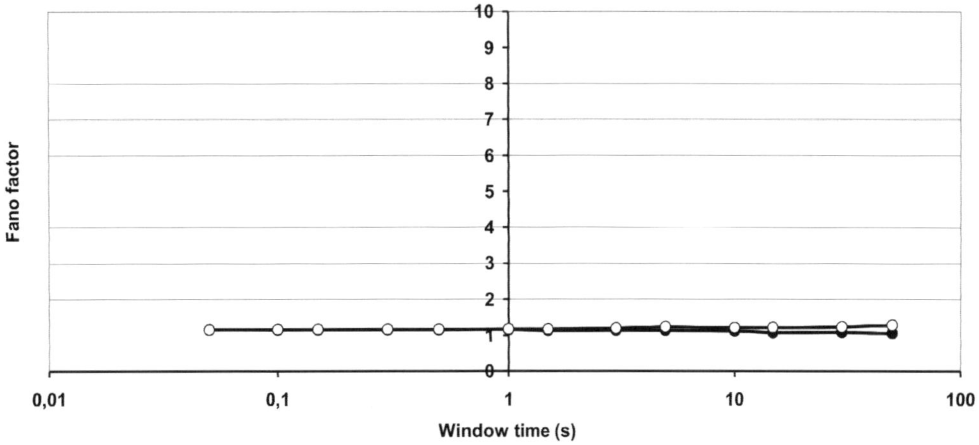

Figure 6. Fano factor curve F(T) averaged over 50 LED data sets (open symbols represent average over LED curves; closed symbols represent average over LED curves after shuffling the data).

These data are another argument that the photon emission signal of the neutrophil cell population, in contrast to artificial light or background, contains some decipherable information in the form of an increasingly long-range correlation with longer window times.

3.4. F(T) curves of PMA-induced luminol dependent photon emission

As indicated before, a series of 40 data sets on PMA-induced photon emission in the presence of luminol have been collected. A representative curve of this collection is shown in Figure 7. We intended to make the same comparison between F(T) curves of luminol-dependent photon emission data sets with high and low intensities.

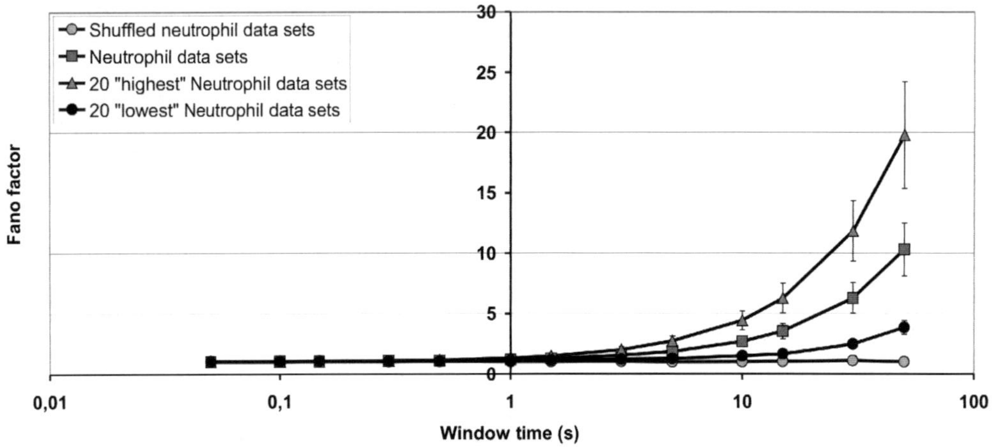

Figure 7. F(T) curve of all neutrophil data sets measured in the presence of luminol (with and without shuffling), and F(T) curves of data sets from 20 highest and 20 lowest photon emission intensity.

Figure 7 illustrates F(T) curve of all luminol measurements (N=40), and F(T) curves of measurements from 20 highest and 20 lowest intensity data sets. Data demonstrate that the slope of F(T) curve also depends on intensity of the signal: the higher the intensity the higher the increase with window size. In shuffled data this increase totally disappeared.

4. Discussion

Production of ROS in health maintenance is of major relevance with regard to the capacity of phagocytic cells, which are a major producer of superoxide $O_2^{\cdot-}$ in response to invading microorganisms. In phagocytic cells, a burst of $O_2^{\cdot-}$ important for killing microorganisms is produced by a multicomponent cell membrane-bound NAD(P) upon receiving specific membrane signals.

The utilization of spontaneous ultra-weak photon emission to register ROS production in cells and tissues has been extensively described. For an overview the reader is referred to extensive classic and recent reviews [22,23]. The detection of photon energy from tissues or cells is a rare event. A spontaneous photon emission of approximately 10 photons / s was registered. It must be emphasized that the actual number of photons emitted needs a variety of factors to be included. Among them are the spectral sensitivity of the photomultiplier, the instrumental characteristics like the size and shape of the sample holder and photomultiplier. Other factors relate to the emitting biological material, namely its homogeneity within the Petri dish and the number of cells. The latter factors influence the absorption by cells other than emitter cells. Apparently, the actual number of photons registered represents a percentage of photons leaving the cell. In tissues, composed of multilayers of cells one can imagine a large fraction of emitted photons may be absorbed. In case of a cell monolayer as in the present experimental conditions with neutrophils, this absorption is relatively small. The spectral sensitivity is estimated at approximately 20%. We have checked microscopically the spatial distribution of cells as well as the image of photon emission from the neutrophil cultures using a highly sensitive CCD camera. It was demonstrated that the cells tend to cluster and adhere to each other, covering the Petri dish surface for about 35%. From the distance between Petri dish and the opening of the photomultiplier we estimated that 1.2% of the photons reach the tube. The entire set of these conditional circumstances led to the conclusion that leukocytes emit approximately 1 photons per cell during a time period of 20 min to 1 h. In this respect, the emission indeed is a rare event. Therefore, the deviation from a Poisson distribution is a good indicator for a form of communication between processes involved in cells producing photon emission.

The fluctuation in observed values of rare events is represented by a Poisson distribution. Fano factor [14], is a good indicator of the deviation of observed values from a Poisson distribution; in case this ratio is higher than one, this is called super-Poisson. This ratio has been previously utilized for characterizing spontaneous photon emission of bacteria [11], Dictyoselium [19] and human hands [20,21]. In the previous studies it was also described that the Fano factor was dependent on time windows. In these studies, like in the present ones, it was checked that such increase completely disappeared in surrogate series obtained by randomly shuffling or permutation of the 50 ms photon data. The common explanation for the increase in variance is that in longer window lengths the more rare clustering of photon signals is revealed.

The data can be interpreted as argument for the occurrence of fluctuations of some sort of multi-frequency nature. Studies on periodic changes in ultra weak photon emission are hardly performed, except in a few cases. Highly regular fluctuations have been described for soybean roots. In those experiments, photon emission as well as the surface potential have been chosen because both can be measured without damaging the specimen [25]. In the time course, photon emission intensity changes almost synchronously with variation of the surface potential. Other example, with less regular periodic oscillations in ultra-weak luminescence are the cell populations of Nitellopsis obtusa cells with a major period of about 1 h [7], and dark-adapted Acetabularia cell populations with a period of a multiplex of 3 min [8].

In most cases, however, the fluctuations in photon emission are highly irregular and questions have been raised about its random or otherwise chaotic (for instance fractal) nature. In the present experiments, the F(T) curve showing increased Fano factor with increasing window size points to the latter explanation, i.e., fractality. This means that cells interact. This interaction however, needs further clarification.

The most attractive model for cell interaction and intercellular dynamics has been the glycolytic pathway of yeast cells in culture. It appeared that under certain aerobic and anaerobic growth conditions, the concentrations of metabolites of the glycolytic pathway oscillate [25,26]. For the oscillations to be observable, they had to be synchronous in the billions of cells in the cuvette. Indeed, mixing of cell populations that oscillated 180 degrees out of phase, caused a transient disappearance of the oscillations, evidencing some sort of active synchronization. It also implies that the dynamic phenomenon of the oscillation propagates within and between cells. These results were in favor of a concept that metabolic processes are able, in principle, to oscillate in a collective manner within intact individual cells as well as within cell populations.

In summary, the data led to the conclusion that Fano factor analysis offers a perspective to estimate frequency modulation of intercellular metabolite interactions. The encoding of this signal information is of interest for research aimed at defining states of (chronic) inflammation, like in rheumatoid arthritis. However, this needs further validation in animal models or humans.

Acknowledgements
This work was supported by the Samueli Institute of Information Biology and the Rockefeller-Samueli Center for Research in Mind-Body Energy, the Netherlands Metabolomics Centre and The Sino-Dutch Centre for Preventive and Personalized Medicine.

References
[1] Alberts B, Johnson A, Lewis J, Raff M, Roberts K, Walter P 2002 *MolecularBbiology of the Cell* (New York: Garland Science) pp 1423-62
[2] Shen X, Mei W, Xu X 1994 Activation of neutrophils by a chemically separated but optically coupled neutrophil population undergoing respiratory burst *Experientia* 50 963-8
[3] Halstensen A, Haneberg B, Glette J, Sandberg S, Solberg CO 1986 Factors important for the measurement of chemiluminescence production by polymorphonuclear leukocytes *J. Immunol. Methods* 88 121-8
[4] Kindzelskii AL, Zhou MJ, Haugland RP, Boxer LA, Petty HR 1998 Oscillatory pericellular proteolysis and oxidant deposition during neutrophil locomotion *Biophys. J.* 74 90-7
[5] Kindzelskii AL, Petty HR 1999 Early membrane rupture events during neutrophil-mediated antibody-dependent tumor cell cytolysis *J. Immunol.* 162 3188-92
[6] Albrecht E, Petty HR 1998 Cellular memory: neutrophil orientation reverses during temporally decreasing chemoattractant concentrations *Proc. Natl. Acad. Sci. USA* 95 5039-44
[7] Śpiewla E 1994 Dependence of amplitude oscillatory changes in potential and electrical resistance of Nitellopsis obtusa cell membranes on the stage of cell development *Acta. Soc. Bot. Pol.* 63187-189.
[8] Van Wijk R, Van Wijk EPA 2003 Oscillations in ultraweak photon emission of *Acetabularia acetabulum* (L.) *Indian J. Exp. Biol.* 41 411-8.
[9] Cohen S, Popp FA 1998 Whole body counting of biophotons and its relation to biological rhythms *Biophotons* ed JJ Chang, J Fisch, FA Popp (Dordrecht: Kluwer Academic Publishers) pp 183–192
[10] Kobayashi M, Devaraj B, Inaba H 1998 Observation of super-Poisson statistics of bacterial (Photobacterium phosphoreum) bioluminescence during the early stage of cell proliferation *Phys. Rev. E* 57 2129–2133
[11] Bajpai RP 2004 Biophoton emission in a squeezed state from a sample of Parmelia tinctorum *Phys. Lett. A* 322 131–136.
[12] Teich MC, Saleh BEA 1988 Photon bunching and antibunching *Progress in Optics, Vol. 26* ed E Wolf (Amsterdam: Elsevier Science) pp. 1–104
[13] Fano U 1947 Ionization yield of radiations. II. The fluctuations of the number of ions, *Phys. Rev*. 72 26–29
[14] Bassingthwaighte JB, Liebovitch LS, West BJ 1994 *Fractal Physiology* (New York: Oxford University Press).
[15] Teich MC 1989 Fractal character of the auditory neural spike train *IEEE Trans Biomed Eng* 36 150–160.
[16] Teich MC 1992 Fractal neuronal firing patterns *Single Neuron Computation* ed T McKenna, J Davis, SR Zormetser (Boston: Academic Press) pp 589–625.
[17] Lewis CD, Gebber GL, Larsen PD, Barman SM 2001 Long-term correlations in the spike trains of medullary sympathetic neurons *J. Neurophysiol.* 85 1614–1622.
[18] Gebber GL, Zhong S, Lewis C, Barman SM 1999 Differential patterns of spinal sympathetic outflow involving a 10-Hz rhythm *J. Neurophysiol.* 89 833–840.
[19] Kobayashi M, Inaba H 2000 Photon statistics and correlation analysis of ultraweak light originating from living organisms for extraction of biological information *Appl. Opt.* 39

Journal of Physics: Conference Series **329** (2011) 012021 doi:10.1088/1742-6596/329/1/012021

183–192.

[20] Van Wijk EPA, Ackerman J, Van Wijk R 2005 Effect of meditation on ultraweak photon emission from hands and forehead *Forsch. Komplmentarmed. Klass. Naturheilkd. (Research in Complimentary and Classical Natural Medicine)* 12 96–106.

[21] Van Wijk EPA, Van Wijk R, Bajpai RP, Van der Greef J 2010 Statistical analysis of the spontaneously emitted photon signals from palm and dorsal sides of both hands in human subjects *J. Photochem. Photobiol. B* 99 133–143.

[22] Boveris A, Cadenas E, Chance B 1981 Ultraweak chemiluminescence: a sensitive assay for oxidative radical reactions *Fed. Process.* 40 195–198.

[23] Van Wijk R, Van Wijk EPA, Wiegant FA, Ives J 2008 Free radicals and low-level photon emission in human pathogenesis: state of the art. *Indian J. Exp Biol.* 46 273-309

[24] Usa M, Kobayashi M, Scott RQ, Maeda T, Hiratsuka R, Inaba H 1989 Simultaneous measurement of biophoton emission and biosurface electric potential from germinating soybean (glycine max) *Protoplasma* 149 64-66.

[25] Hess B, Boiteux A 1968 Mechanism of glycolytic oscillation in yeast. I. Aerobic and anaerobic growth conditions for obtaining glycolytic oscillation *Hoppe Seylers Z. Physiol. Chem.* 349 1567-74.

[26] Boiteux A, Hess B, Plesser T 1977 Oscillatory phenomena in biological systems *FEBS Lett.* 15 1-4

Journal of Physics: Conference Series **329** (2011) 012022 doi:10.1088/1742-6596/329/1/012022

Primary cilium - antenna-like structure on the surface of most mammalian cell types

J Dvorak[1], V Sitorova[2], D Hadzi Nikolov[2], J Mokry[3], I Richter[4], L Kasaova[1], S Filip[1], A Ryska[2], J Petera[1]

[1]Department of Oncology and Radiotherapy, Charles University Medical School and University Hospital, Hradec Kralove, Czech Republic
[2]Department of Pathology, Charles University Medical School and University Hospital, Hradec Kralove, Czech Republic
[3]Department of Histology and Embryology, Charles University Medical School, Hradec Kralove, Czech Republic
[4]Department of Oncology, Regional Hospital, Liberec, Czech Republic

E-mail: dvorakj@fnhk.cz

Abstract. The primary cilium is a sensory solitary non-motile microtubule-based organelle protruding in the quiescent phase of the cell cycle from the surface of the majority of human cells, including embryonic cells, stem cells and stromal cells of malignant tumors. The presence of a primary cilium on the surface of a cell is transient, limited to the quiescent $G_1(G_0)$ phase and the beginning of the S phase of the cell cycle. The primary cilium is formed from the mother centriole. Primary cilia are key coordinators of signaling pathways during development and tissue homeostasis and, when deffective, they are a major cause of human diseases and developmental disorders, now commonly referred to as ciliopathies. Most cancer cells do not possess a primary cilium. The loss of the primary cilium is a regular feature of neoplastic transformation in the majority of solid tumors. The primary cilium could serve as a tumor suppressor organelle. The aim of this paper was to provide a review of the current knowledge of the primary cilium.

Introduction

The primary cilium was discovered in the nineteenth century. Over one hundred and forty years passed between the discovery of this organelle and a deep appreciation of its important functions. It is a solitary sensory non-motile microtubule-based organelle protruding in the quiescent phase of the cell cycle from the surface of the majority of human cells, including embryonic cells, stem cells and stromal cells of malignant tumours, with the exception, for example, hepatocytes and haematopoetic cells. The primary cilium acts as a cellular antenna that senses extracellular signals from the environment and translates them into specific intracellular signalling cascades [1]. The primary cilium consists of an axoneme with nine peripheral microtubule doublets arranged around a central core lacking two central microtubules: the 9 + 0 pattern (contrary to motile cilia with the 9 + 2 pattern), surrounded by the ciliary membrane, which is a specialized domain extension of the cell membrane. The axoneme extends from the basal body, which is derived from the mother centriole of the centrosome [2]. The cilia do not contain ribosomes. Proteins are not synthesized within the cilium. Therefore all proteins required to build and maintain the cilium, as well as the signalling proteins required for cilia functioning, are synthesized in the cell body and transported into and out of the cilium [3]. This bidirectional microtubule-based transport is known as intraflagellar transport [4].

The term "primary cilium" was first applied by Sorokin for airway epithelial cells, which start out with a single cilium, thus called "primary", and then, during differentiation, form large numbers of additional motile cilia, which are called "secondary" cilia [5, 6]. Most cells of the body have just a single cilium and, by analogy with the single initial cilium of the airway cells, these single cilia are now generally called primary cilia. The primary cilium is essential for the orientation of the cytoskeleton in the cell and the orientation of the cell in the tissue. The main difference between primary and secondary cilia is in centriologenesis. Basal bodies/centrioles of secondary cilia are not formed by the duplication of pre-existing centrioles (as primary cilia), but are formed on the surface of spherical precursor structures called deuterosomes, which originate *de novo* in the cytoplasm from fibrogranular material.

The lengths of primary cilia are more variable than the length of motile cilia. It was observed that primary cilia in the rat liver are approximately twice as long in the large bile ducts (7.35±1.32 μm in length) than in the small bile ducts (3.58±1.12 μm in length) [7]. Primary cilia elongate under unfavourable conditions, such as during hypoxia or starvation.

Relationship between the primary cilium cycle and the cell cycle

Primary cilia exhibit dynamic patterns of assembly and disassembly as they progress through the cell cycle [Scheme 1]. Centrosomes, which are composed of two centrioles arranged perpendicularly and surrounded by pericentriolar material, organize the cytoplasmic microtubule network during interphase, whereas during mitosis they form the bipolar spindle. In post-mitotic cells, the centrosome migrates to the cell surface, where the mother centriole differentiates into the basal body. This structure nucleates microtubules to form a primary cilium. The presence of a primary cilium on the cell surface is transient, limited to the quiescent $G_1(G_0)$ phase, as well as to the beginning of the S phase of the cell cycle. Disassembly of the primary cilium must precede mitosis. In asynchronous populations, typically 10% to 25% of cells have observable cilia. As cultured cells become quiescent, an increasing proportion of the population is ciliated, attaining > 90% in some cell types [8]. In the G_2 phase the microtubules of the primary cilium are resorbed and the centrosome migrates to the centre of the cell where it organizes the mitotic spindle. The term "centriole/basal body" is sometimes compared to the term "evening star/morning star": the same organelle, when organizing the mitotic spindle in the centre of the cell, is called a centriole, whereas in the quiescent phase of the cell cycle, when anchored to the plasma membrane, forming the primary cilium, it is called a basal body [6]. Thus, it can never be in both places at the same time.

Journal of Physics: Conference Series **329** (2011) 012022 doi:10.1088/1742-6596/329/1/012022

Scheme 1
The primary cilium cycle and the cell cycle

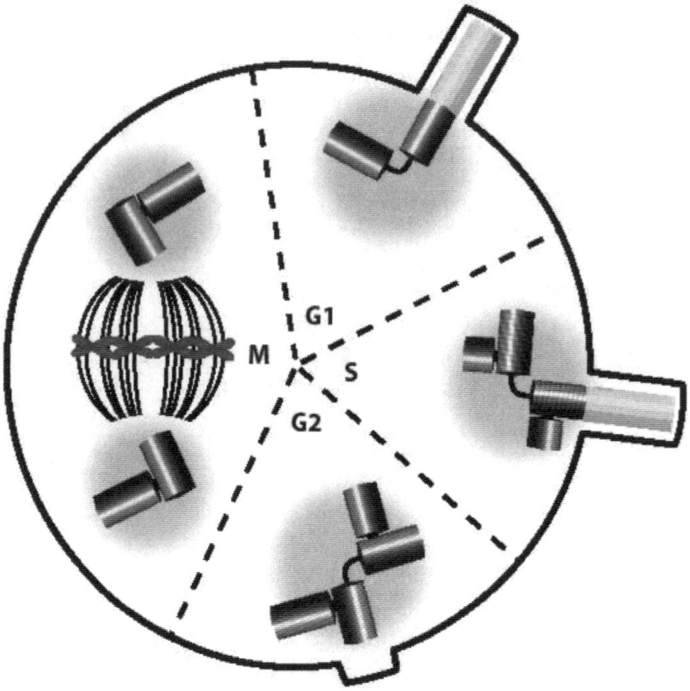

For a long time the primary cilium was considered to be a nonfunctional vestige. The function of the centriole has remained quite controversial for many decades. This controversy arose from the emphasis of the mitotic role of cenrioles and marginalization of their role in the formation of the sensory primary cilium. There is ample experimental evidence and many observations to show that mitosis can proceed without centrioles in many cases [9, 10]; however, the primary cilium cannot exist without centrioles. With the new appreciation of the ubiquitous importance of the primary cilium in physiology, the hypothesis that centrioles primarily exist for the purpose of driving ciliogenesis seems to be quite acceptable [9]. Centrioles may be more important for the formation of the primary cilium than for mitotic spindle development. In an adult organism, the majority of cells exist in a quiescent state. Therefore, for a non-dividing cell, it is more important to respond to transient environmental signals sensed through the primary cilium than to prepare for mitosis. Thus, centrioles are only shortly "borrowed" to act as a mitotic organizing centre during mitosis in order to optimize cell division [11] [Scheme 2].

Scheme 2
Example of a cell cycle with a duration of 24 hours
In non-cancer cells, the centrosome is located asymmetricaly during interphase (22 hours), close to the plasma membrane, forming the sensory primary cilium, whereas during mitosis (2 hours) it forms the mitotic spindle.
In cancer cells, the centrosome is located close to nucleus during interphase (it can probably overcome the restrictions of the cell cycle connected to the primary cilium cycle), and during mitosis it forms the mitotic spindle.
Note: actual times for each phase depend on the tissue, the microenvironment and the physiological state.

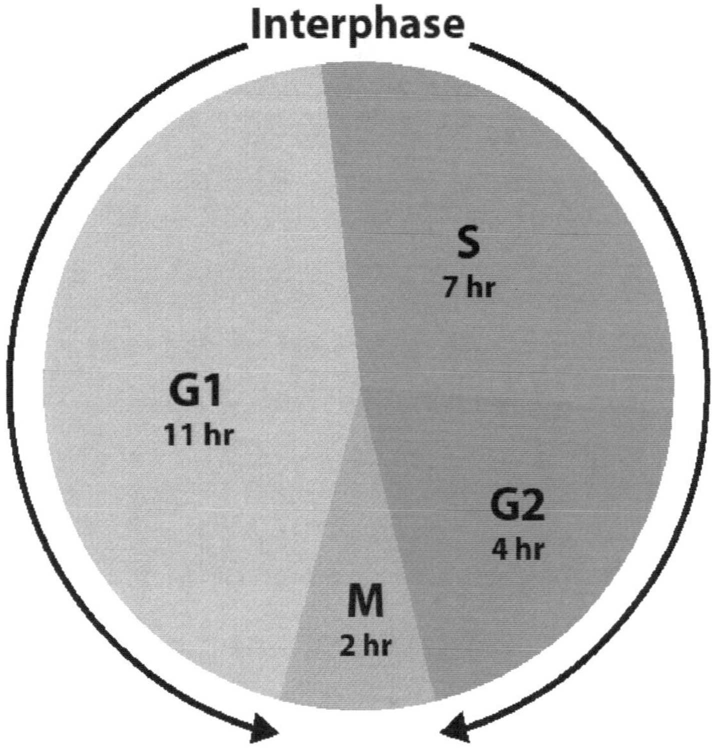

Asynchronous growth and asymmetric signalling of primary cilia on the surface of sister cells according to the age of their centrioles/basal bodies

Centriole duplication in the $G_1(G_0)/S$ phase results in two centrosomes, one with an older mother centriole and one with a new mother centriole; they are segregated during mitosis. It was observed that sister cells resulting from a mitotic division grow primary cilia asynchronously and that the sister cell receiving the older mother centriole usually builds its primary cilium first. The primary cilium on the sister cell with the older mother centriole is transiently longer and contains more receptors. These observations suggest that the segregation of differently aged mother centrioles, an asymmetry inherent in every animal cell division, can influence the ability of sister cells to respond to environmental signals, potentially altering the behaviour or fate of one or both sister cells. Asynchronous primary cilium growth can lead to differential responses of sister cells to an identical environmental signal. In the case of a signal that is transient (short period) or weak, the cell that bears the primary cilium first can respond to that signal, whereas its sister cell cannot [12].

Ciliary rootlet

The ciliary rootlet originates from the proximal end of the basal body, anchoring the cilium, and extends proximally towards the cell nucleus. Prominent rootlets are associated with sensory cilia of photoreceptor cells (sensors of electromagnetic field). In a photoreceptor, the rootlet appears as a very thick, striated filament that traverses the entire cell body all the way to the synaptic terminal [13]. The diameter of the rootlets of photoreceptor cells is approximately 80-100 nm, thus is the largest cytoskeletal structure in any mammalian cell. The rootlet is composed of homopolymeric rootletin protofilaments bundled into filaments. Rootletin (220 kD protein) has a globular head domain and a tail domain consisting of extended coiled-coil structures.

Ciliopathies

Primary cilia are key coordinators of signalling pathways during cell development and in tissue homeostasis. When defective, primary cilia are a major cause of human diseases and developmental disorders, which are now commonly referred to as ciliopathies.

Most cancer cells do not possess primary cilia, although there are some exceptions, such as tumours depending on the Hedgehog pathway, for example, basal cell carcinoma and medulloblastoma. The loss of primary cilia is a regular feature of neoplastic transformation in the majority of solid tumours [Scheme 2]. The primary cilium could serve as a tumour suppressor organelle [14].

Conclusion

The primary cilium is a microtubule-based structure protruding from the cell surface, functioning as a sensor for extracellular environmental signále that regulate cellular differentiation and division [11]. The presence of a primary cilium on the surface of the cell is transient, limited to the quiescent phase of the cell cycle. Defects in cilia cause a broad spectrum of human diseases known as ciliopathies [15]. The loss of primary cilia is a regular feature of neoplastic transformation in the majority of solid tumours. Most cancer cells do not possess primary cilia.

Acknowledgement: Supported by research projects MZO 00179906 and SVV 53251.

References

[1] Molla-Herman A, Ghossoub R, Blisnick T, *et al.* 2010 The ciliary pocket: an endocytic membrane domain at the base of primary and motile cilia. *J. Cell Sci.* **123** 1785–1795
[2] Satir P, Pedersen L B and Christensen S T 2010 The primary cilium at a glance. *J. Cell Sci.* **123** 499–503
[3] Green J A and Mykytyn K 2010 Neuronal ciliary signaling in homeostasis and disease. *Cell Mol. Life Sci.* **67** 3287–3297
[4] Rosenbaum J L and Witman G B 2002 Intraflagellar transport. *Nat. Rev. Mol. Cell Biol.* **3** 813–825
[5] Sorokin S P 1968 Reconstructions of centriole formation and ciliogenesis in mammalian lungs. *J. Cell Sci.* **3** (2): 207–230
[6] Marshall W F and Nonaka S 2006 Cilia: tuning in to the cell's antenna. *Curr. Biol.* **16** R604–614
[7] Masyuk A I, Masyuk T V and LaRusso N F 2008 Cholangiocyte primary cilia in liver health and disease. *Dev. Dyn.* **237** 2007–2012
[8] Plotnikova O V, Golemis E A and Pugacheva E N 2008 Cell cycle-dependent ciliogenesis and cancer. *Cancer Res.* **68** 2058–2061
[9] Marshall W F 2008 Basal bodies platforms for building cilia. *Curr. Top. Dev. Biol.* **85** 1–22
[10] Uetake Y, Loncarek J, Nordberg J J, *et al.* 2007 Cell cycle progression and de novo centriole assembly after centrosomal removal in untransformed human cells. *J. Cell. Biol.* **176** 173–182
[11] Plotnikova O V, Pugacheva E N and Golemis E A 2009 Primary cilia and cell cycle *Primary cilia; Methods in cell biology* (vol 94) ed in Sloboda R: (Burlington, Massachusetts, USA: Elsevier) pp 137-160
[12] Anderson C T, Stearns T 2009 Centriole age underlies asynchronous primary cilium growth in mammalian cells. *Curr. Biol.* **19** 1498–1502
[13] Spira A W, Milman G E 1979 The structure and distribution of the cross-striated fibril and associated membranes in guinea pig photoreceptors. *Am. J. Anat.* **155** 319–337
[14] Seeley E S, Nachury M V 2009 Constructing and deconstructing roles for the primary cilium in tissue architecture and cancer. *Primary cilia; Methods in cell biology* (vol 94) ed in Sloboda

R: (Burlington, Massachusetts, USA: Elsevier) pp 300-313
[15] Marshall W F 2008 The cell biological basis of ciliary disease. *J. Cell Biol.* **180** 17–21

May the variable magnetic field and pulse red light induce synergy effects in respiratory burst of neutrophils in vitro?

H Nawrocka – Bogusz[1], **F Jaroszyk**[2]

[1, 2] The Department of Biophysics, University of Medical Science, 10 Fredry Str., 61-701 Poznań, Poland

E-mail: hnawrocka@ump.edu.pl

Abstract. We investigated the effect of the red light (R) (630 nm), magnetic field (MF) and magnetic field combined with the red light (MF+R) upon reactive oxygen species (ROS) production by neutrophils in vitro. The object of the research was hydrogen peroxide (H_2O_2) formation during neutrophils respiratory burst or within steady-state. Blood from healthy volunteers was used for the purpose of the study. Flow cytometry method, using transformation of DCFH-DA (2'7'-dichlorofluorescin diacetate) to the fluorescent DCF (2'7'-dichlorofluorescin), was used for estimation of hydrogen peroxide production. The variable magnetic field of ELF range of the mean induction equals 26.7(μT), the red light at the energy density of 1.17(J/cm^2) and their combination were applied for 30 minutes each. The fundamental frequency of pulses was 180÷ 195 Hz. A statistically significant decrease of H_2O_2 production by neutrophils was observed. The level of the decrease was in the range of 10-30% and was dependent on the kind of applied physical factors and whether neutrophils were stimulated or not. The observation showed that the variable magnetic field combined with red light do not induce the synergy effect.

1.Introduction

Neutrophils are highly specialized white blood cells. Their basic function is executing the phagocytosis and successive killing of phagocytized microorganisms by secretion of destructive molecules into the phagosome and by the respiratory burst [1, 2, 3]. The neutrophil oxidative burst can be activated by a number of different soluble and particulate stimuli, including chemoattractants, certain cytokines, phorbol esters, calcium ionophores, different lectins, and opsonized as well as various unopsonized microorganisms [1, 3]. The NADPH oxidase enzymatic complex assembles on the plasma (or phagosome) membrane of granulocytes, macrophages and B cells upon stimulation [4, 5]. The neutrophil NADPH oxidase consists of the catalytic subunit gp91^{phox}, which is located at the phagosome membrane together with p22^{phox}. The assembly of gp91^{phox} and p22^{phox} with the cytosolic subunits p40^{phox}, p47^{phox}, p67^{phox} and small GTP-ase Rac2 results in rapid production of superoxide [1,4, 6, 7]. The phagocyte NADPH oxidase produces superoxide anion ($O^{2\cdot-}$) which is a precursor to other ROS by the electrogenic process of moving electrons across the cell membrane [8, 5, 3]. Most of $O^{2\cdot-}$ formed dismutates to H_2O_2, which is converted to HOCl by myeloperoxidase [1, 2, 3, 7]. Each NADPH molecule donates two electrons, which cross the membrane through a redox pathway contained within gp91^{phox} and reduce two O_2 molecules to $O^{2\cdot-}$. In detail gp91^{phox}, also called Nox2, contains an electron transport chain that comprises NADPH, FAD and two heme groups that pass electrons sequentially to O_2 at an external binding site, to produce $O^{2\cdot-}$ [5]. Electron flux depolarizes the membrane, opening proton channels when the proton flux compensates charge and prevents large changes in both cytoplasmic and phagosomal pH. This charge translocation must be compensated to prevent self-inhibition of NADPH by extreme membrane depolarization [5, 6, 8]. After about a minute of stimulation, membrane depolarization reaches a steady state, even though superoxide is still being produced, indicating the movement of a compensatory charge into the external media. This is attributed to the delayed release of protons via an associated proton channel, conductance of which is voltage gated [6]. When proton channels are inhibited, pH drops to levels that inhibit NADPH oxidase [5, 9].

Protein kinase C (PKC) isoforms are implicated in the oxidative burst as the potential key activators of NADPH oxidase and they are required for phosphorylation of p47phox, resulting in the assembly and activation of phagocytic NADPH oxidase [4]. PKC is also important for gating of the proton channel. Phosphorylation of proton channels enhances the opening of proton channels [9]. PKC-δ is required for full assembly of NADPH oxidase and activation of the respiratory burst. Neutrophils also express PKC-α and -β, which may be involved in adhesion, degranulation and phagocytosis. The conventional PKC (α, β, γ) contains two regulatory domains: C1 and the calcium-binding C2 domain. The PKC-δ has lack of the C2 domain and is therefore calcium-insensitive [4]. Activity of the NADPH complex also requires the involvement of at least two G proteins. These are proteins that act as molecular switches, alternating between binding GDP in the inactive state and GTP in the active state [6].

Binding of respiratory burst activators such as: LTB_4 (leukotriene B4), PAF (platelet activating factor) or FMLP (N-formyl-L-methionyl-L-leucyl-L-phenylalanine) to receptors on the neutrophils plasma membrane releases intracellular calcium ($[Ca^{2+}]_i$) via the phospholipase-inositol triphosphate pathway [10, 11]. This depletion of $[Ca^{2+}]_i$ in turn leads to Ca^{2+} influx via store operated channels resulting in elevated Ca^{2+} levels that are thought to be implicated in several Ca^{2+} dependant pro-inflammatory responses. NADPH oxidase inhibition will also abrogate membrane depolarisation and augments Ca^{2+} entry into the cell. Depolarization alters the electrochemical gradient across the neutrophil plasma membrane and limits Ca^{2+} entry into the cell [10]. Both intra- and extracellular Ca^{2+} Are required for full activation of the respiratory burst of human neutrophils, and the Ca^{2+} influx from extracellular space plays an important role either in generation of reactive oxygen metabolites or in activation of protein kinase C [11].
We have investigated whether the red light combined with variable magnetic field can cause double or higher effect (synergy effect) on hydrogen peroxide formation by neutrophils in vitro. According to the definition, the synergy effect means that the combination of the factors can cause a greater reaction than the sum of the individual effects of each physical factor when used separately.

2.Material and methods

2.1.Material. The blood from healthy volunteers was used for the purpose of the study. As the anticoagulant was the lithium heparin.

2.2.Reagents. DCFH-DA (2'7'-dichlorofluorescin diacetate; Sigma-Aldrich, USA) was used for showing the presence of hydrogen peroxide [12, 13, 14,15, 16]. The 0.0609 g of the substance was dissolved in 5 ml of 96% ethanol. The solution was kept in - 20°C. Before every measurement DCFH-DA solution was diluted in PBS (phosphates buffered physiological salt solution) in proportion 1 µl for 10µl PBS. PMA (phorbol 12-myristate, 13-acetate; Sigma-Aldrich, USA) was used to produce submaximal stimulation of the respiratory burst [3, 12, 16]. The PMA solution was prepared by dissolution 1 mg of PMA in 1 ml of 96% ethanol. The solution was kept in - 20°C. For the red blood cells lysis the special solution was prepared [13]. The lysis solution was made by dissolution 8.26 g NH_4Cl, 1.0 g K_2CO_3 and 0.037 g four-sodium salt EDTA in 1l of distilled water. pH of the solution was 7.2÷7.4. The lysis solution was kept in 4°C.

2.3.Method. For one measurement the blood from one person was divided into 4 test-tube containing 50 µl of blood each. 10µl of DCFH-DA diluted solution was added to each test-tube. Two of the examined samples were incubated in the presence of the red light (R) of wavelength 630 nm or variable magnetic field (MF) or combination of those factors (MF+R) for 30 minutes each in the room temperature in dark. The rest of test-tubes as a control was kept also in the same environment conditions. After the physical factor application 2 µl of PMA solution was added to the one of control test-tubes and one which was influenced of R/MF/MF+R. After incubation at room temperature for 15 minute in the dark the lysis solution was added. 700µl of this solution was added to each of test-tube. After 15 minute in the dark at room temperature the flow cytometry measurement was made [12].

2.4. Viofor JPS. Generator of the red light and variable magnetic field was the device Viofor JPS. The mean induction of the variable magnetic field of ELF range equaled 26.7(µT]), the red light energy density was 1.17(J/cm^2). The distance of 1cm was kept between the source of physical factors and test-tubes to avoid raise of the sample temperature. The level of the magnetic induction or/and the density of light application was growing at 12 seconds interval starting from 0.30(J/cm^2)/4.45(µT) to the selected, then the intensity drops to the initial value, the process is repeated cyclically. The fundamental frequency of pulses was 180 ÷ 195 Hz. The form of impulses is close to the peak shaped. The pulses were administered in the form of packets of pulses (12.5 ÷ 29 Hz), groups of packages (2.8 ÷ 7.6 Hz) and series (0.08 ÷ 0.3 Hz).

2.5. Flow cytometry. The induced respiratory burst was assessed by the intracellular oxidative transformation of DCHF-DA to the fluorescent DCF ($\lambda_{excitation}$=498nm, $\lambda_{emission}$=522nm) via the flow cytometer. The DCF fluorescence is linearly dependent on the quantity of the respiratory burst [12, 16, 17]. Cytofluorymeter FACScan Becton Dickinson equipped with 15 mW argon laser was used for DCF fluorescence measurement [12, 17].

3.Results

We used the following marks of blood samples:
1-contol blood sample unstimulated with PMA, without the influence of MF/R/MF+R
2-neutrophils activated by PMA (control sample for „4")
3-a sample of blood unstimulated with PMA, under the influence of MF/R/MF+R
4-neutrophils activated by PMA, under the influence of MF/R/MF+R
(4-3)-respiratory burst in samples influenced by MF/R/MF+R
(2-1)-respiratory burst in native samples (not exposed to MF/R/MF+R)
We compared samples "1" with "3", and samples "2" with "4" by using Wilcoxon test. We also calculated difference between respiratory burst in native samples and respiratory burst in samples influenced by MF/R/MF+R (Figure 2).
We present our data on graphs (Figure 1, 2), where Y axis presents DCF fluorescence in arbitrary units for each person (case).

A)

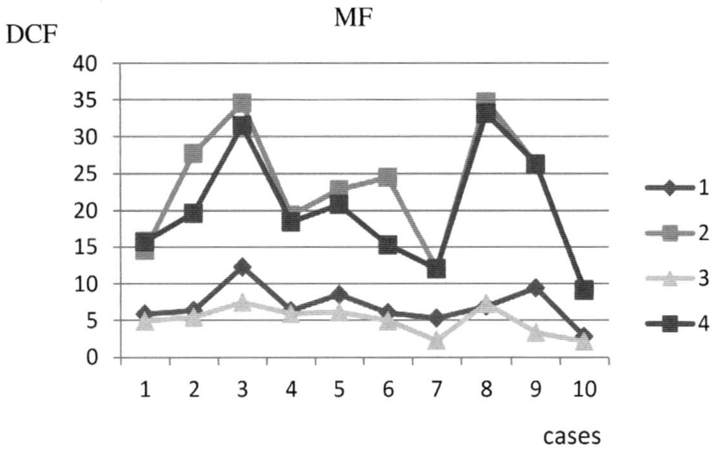

B)

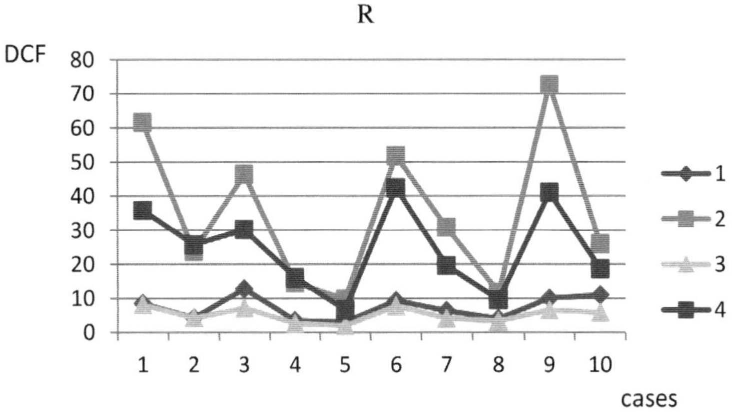

C)

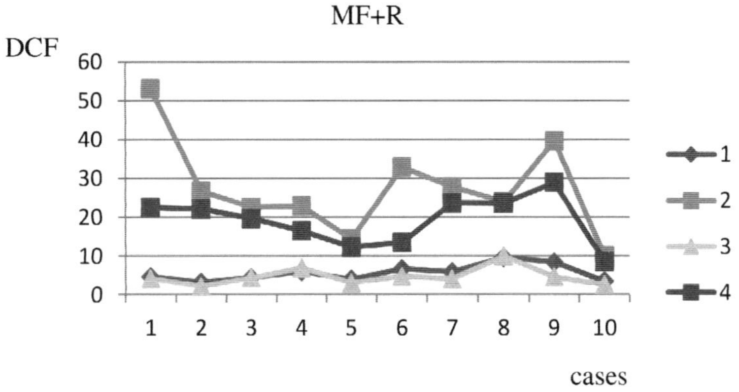

Figure 1. DCF fluorescence after red light (A), variable magnetic field (B), magnetic field combined with red light (C) application.

The magnetic field (MF), the red light (R) and the magnetic field combined with the red light (MF+R) caused statistically significant changes of hydrogen peroxide production by unstimulated (blue and yellow line on each graph) and PMA-stimulated neutrophils (red and green line on each graph) (Figure 1, 2, Table 1).

Each of the physical factors applied separately and its combination caused reduction ($p \leq 0.007$) of DCF fluorescence of PMA unstimulated neutrophils average by 30% (Table 1). MF resulted 10.6% DCF fluorescence decrease of stimulated neutrophils ($p=0.028$). The red light caused the largest reduction (29.5%) of hydrogen peroxide formation by stimulated neutrophils ($p=0.01$). MF+R caused 30.3% ($p=0.005$) decrease of DCF fluorescence of those neutrophils.

Table 1. Percentage of DCF fluorescence decrease and "p" value calculated according to the Wilcoxon test.

	1&3	**2&4**	**(2-1)&(4-3)**
MF	28.6 % p=0.007	10.6 % p=0.028	—
R	28.8 % p=0.005	29.5 % p=0.01	29.7 % p=0.037
MF+R	17.8 % p=0.0468	30.3 % p=0.005	33.6 % p=0.005

The production of hydrogen peroxide was statistically significant (Figure 2, Table 1) between the neutrophils under the influence of red light or magnetic fields combined with red light (4-3) and the neutrophils without their influence (2-1).

A)

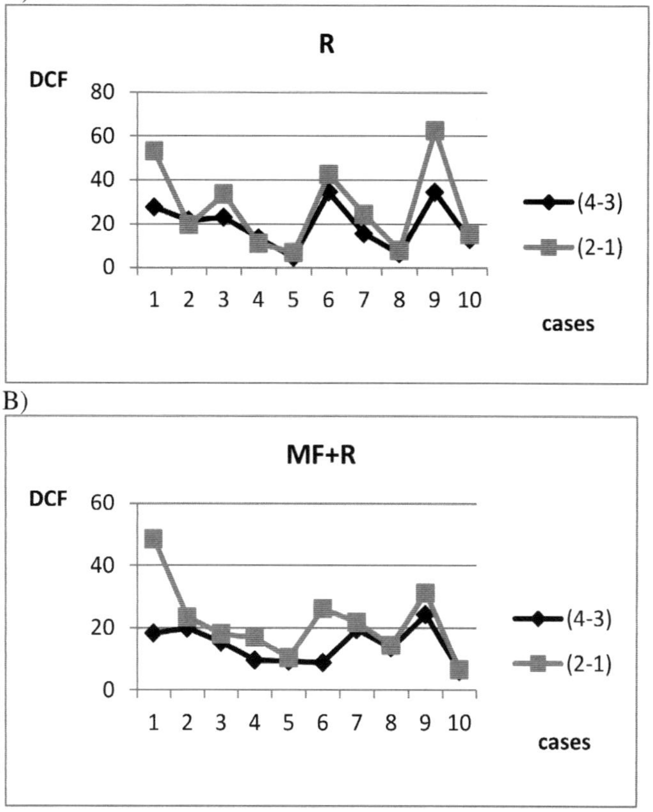

Figure 2. Quantity of respiratory burst after the red light irradiation (A), and the magnetic field combined with red light application (B).

The red light caused the biggest changes of respiratory burst of stimulated and unstimulated neutrophils. Magnetic field made the lowest decrease.
We didn't observe the synergy effect after application of combined factors as variable magnetic field and red light (Table 1).

4.Discussion

The absorption of light by molecules causes electronic excitation. When molecules, such as chromophores, are excited an improvement in kinetics chemical reactions is observed. However, the mechanism of light absorption by chromophores does not explain many of the experimental observations. Light induces a wave-like alternating electric field in a medium that is able to interact with polar structure and produce dipole transitions. This electrical component of electromagnetic energy has an effect on dipolar molecules such as enzymes, ionic pumps, nuclear material and nucleotide molecules [18].

There is evidence indicating that phototherapy has a significant influence on neutrophils [19]. For example: Fujimaki at al. found that Ga-Al-As laser [wavelength: 830 nm, doses: 9.5 and 19 (J/cm^2)] radiation significantly reduced the production of reactive oxygen species by human neutrophils in vitro [19, 20]. Lopes-Martins at co-authors evaluated the effect of the In-Ga-Al-P laser [(wavelength: 650 nm, doses: 3, 7.5 and 15 (J/cm^2)] in vivo on inflammatory cell migration in acute mouse pleurisy. They observed the largest migration reduction for neutrophils after application of 7.5 (J/cm^2) [19,21]. Other studies showed that Ga-Al-As laser [wavelength: 685 nm, dose: 2.5 (J/cm^2)]

irradiation decreased number of neutrophils, the levels of neutrophils anti-apoptotic factors, the IL-1β concentration, and lung myeloperoxidase activity after lipopolysaccharide-induced lung inflammation in rats [19, 22].

According to Panagopoulos at all an oscillating, external electric or magnetic field will exert an oscillating force on every free ion on both sides of the plasma membrane, as well as on the ions within channel proteins, while they pass through them. This external oscillating force will evoke in every ion, a coherent forced-vibration. This vibration of electric charge is able to irregularly gate electrosensitive channels on the plasma membrane and thus cause disruption of the electrochemical balance and function of the cell. The oscillating ions will then represent a periodical displacement of electric charge, able to exert forces on every fixed charge of the membrane, such as the charges on the voltage sensors of voltage-gated channels. This way, the oscillating ions could be able to disrupt the membrane's electrochemical balance by gating such channels [23]. The authors concluded on the base of calculations that the low frequency fields are the most bioactive ones and pulsed electromagnetic fields (PEMF) can be more biologically active than the continuous ones [23]. Balcavage at all stated also that pulsed electromagnetic fields can modify cation flow across biological membranes and alter cell metabolism. PEMF has the potential of regulating the flow through cation channels, changing the steady state concentrations of cellular cations and thus the metabolic processes dependent on cation concentrations [24].

Magnetic field can also change cell proliferation, activation of several enzymes, increase of certain protein concentrations [25]. MF can influence the process of ion binding with proteins [26]. There are reports about the influence of magnetic field on neutrophils.

Nody at all presented the results of AC (60 Hz) and DC magnetic fields influence on respiratory burst in neutrophils. MF was in the range of 0.1 to 2 mT. They observed that both 60 Hz and DC magnetic field enhanced the production of DCF fluorescence during the respiratory burst [27].

Sheiko at co-authors evaluated the effects of various physical factors of electromagnetic nature on the activities of neutrophils coming from patients' blood with breast cancer. They found that alternating magnetic field (10 mT induction, 1.5-3-6 Hz frequency, total exposure 15 min), low-intensity laser or photodiode radiation (both: λ=637nm, 7.5 mW/cm^2, duration of exposure up to 5 min), and their combinations induced mobilization of blood neutrophils function [28].

Varani at all has investigated the effect of the magnetic field of the intensity varying from 0.2 to 3.5 mT on human neutrophil A_3 receptors. In human neutrophils A_3 adenosine receptors exert their anti-inflammatory properties by inhibiting degranulation, chemotaxis and superoxide anion production. Authors observed a significant increase of adenosine A_3 density when the magnetic intensity was in the range 1–3.5 mT, not lower [29]. According to the same author in another paper PEMFs evokes an upregulation of the A_{2A} adenosine receptors and alters the response of this receptor subtype in human neutrophils. PEMF treatment causes a reduction of the superoxide anion production as a result of upregulation of the A_{2A} receptors located on the neutrophil surface [30].

Summarizing, the variable magnetic field or the red light affects the cell differently, e.g. the light-induced electric field is able to change the membrane potentials in the mitochondrial membrane by transferring charges from the outer side to the inner side or directly by the polarization of membrane dipoles. Irradiation with light at 633nm increases proton gradient and causes changes in mitochondrial optical properties [31]. A varying magnetic field can cause the normalization of cell's membrane potential or increases the activity of enzymes. Therefore, supposed targets of red light or/and variable magnetic field influence taking into account the mechanism of respiratory burst could be: calcium ions flux/influx (Ca^{2+}), NADPH oxidase complex, protein kinase C (PKC) isoforms, proton channel, the system of antioxidant enzymes: SOD (superoxide dismutase), CAT (catalase) and GSH-Px (glutathione peroxidase). SOD is a metal protein containing zinc and copper, CAT contains hem and GCH-Px selenocysteine. Cieślar at co-authors observed reduced activity of SOD in human blood in vitro study after exposure to the variable magnetic field generated by the device Viofor JPS. The authors concluded that the variable magnetic field indirectly inhibits the influence of reactive oxygen species in living organisms [32].

Our observation did not confirm the synergy effect when the magnetic field and the red light were used together. Probably, the path of the respiratory burst activation prevents evoking of the synergy effect by the variable magnetic field combined with red light.

It is well known that the red light as well as variable magnetic field is used singly as physiotherapy methods [12]. Phototherapy or the variable magnetic field could be applied in autoimmune and inflammatory diseases that are mediated in some part by neutrophils e.g. gout, rheumatoid arthritis, and also in a variety of airway diseases mediated by neutrophils infiltration such as: chronic obstructive pulmonary disease, bronchiectasis, certain forms of asthma etc. [19].

As it has been shown in our experiment, the combination of variable magnetic field and red light is not necessary to obtain higher decrease of hydrogen peroxide production by neutrophils. Both variable magnetic field and red light cause reduction of DCF fluorescence.

References

[1] Dahlgren C and Karlsson A 1999 Respiratory burst in human neutrophils *J. Immunol. Methods* **232** 3-14

[2] Mayer-Scholl A, Averhoff P and Zychlinsky A 2004 How do neutrophils and pathogens interact? *Curr. Opin. Microbiol.* **7** 62-66

[3] Freitas M, Porto G, Lima JLFC and Fernandes E 2009 Optimization of experimental settings for the analysis of human neutrophils oxidative burst *in vitro Talanta* **78** 1476–83

[4] Bertram A and Ley K 2011 Protein Kinase C Isoforms in Neutrophil Adhesion and Activation *Arch. Immunol. Ther. Exp.* **59** 79–87

[5] Capasso M, DeCoursey T E, Dyer MJS 2011 pH regulation and beyond: unanticipated functions for the voltage-gated proton channel, HVCN1 *Trends in Cell Biology* **21** 1 20-28

[6] Jones RD, Hancock JT, and Morice AH 2000 NADPH OXIDASE: A UNIVERSAL OXYGEN SENSOR? *Free Radic. Biol. Med.* **29** 5 416–24

[7] Babior BM 2004 NADPH oxidase *Curr. Opin. Immunol.* **16** 42–47

[8] Murphy R and DeCoursey TE 2006 Charge compensation during the phagocyte respiratory burst *Biochim. Biophys. Acta* **1757** 996–1011

[9] Musset B, Cherny V V, Morgan D, DeCoursey TE, 2009 The intimate and mysterious relationship between proton channels and NADPH oxidase *FEBS Letters* **583** 7-12

[10] Salmon M D and Ahluwalia J 2009 Relationship between calcium release and NADPH oxidase inhibitionin human neutrophils *Biochem. Biophys. Res. Commun.* **384** 87–92

[11] Bei L, Hu T, Ming Qian Z and Shen X 1998 Extracellular Ca^{2+} regulates the respiratory burst of human neutrophils *Biochim. Biophys. Acta* **1404** 475- 83

[12] Nawrocka H, Poniedziałek B, Jaroszyk F and Wiktorowicz K 2006 Effects of low intensity magnetic fields and red light on respiratory burst of neutrophiles *Polish J. of Environ. Stud.* **15** 4A 28-30

[13] Gomes A, Fernandes E and Lima J 2005 Fluorescence probes used for detection of reactive oxygen species *J. Biochem. Biophys. Methods* **65** 45-8

[14] Jakubowski W and Bartosz G 2000 2,7-dichlorofluorescin oxidation and reactive oxygen species: what does it measure? *Cell Biol. Int.* **24** 10 757-60

[15] Walrand S, Valeix S, Rodriguez C, Ligot P, Chassagne J and Vasson M 2003 Flow cytometry study of polymorphonuclear neutrophiles oxidative burst: a comparison of tree fluorescent probes *Clin. Chim. Acta* **331** 103-10

[16] van Eeden SF, Klut ME, Walker BAM and Hogg JC 1999 The use of flow cytometry to measure neutrophil function *J. Immunol. Methods* **232** 23–43

[17] Chapman GV 2000 Instrumentation for flow cytometry *J. Immunol. Methods* **243** 3–12

[18] Amat A, Rigau J, Waynant R W, Ilev I K and Anders J J 2006 The electric field induced by light can explain cellular responses to electromagnetic energy: A hypothesis of mechanism *J. Photochem. Photobiol. B* **82** 152–60

[19] Morgan MC and Rashid RM 2009 The effect of phototherapy on neutrophils *Int. Immunopharmacol.* **9** 383–88

[20] Fujimaki Y, Shimoyama T, Liu Q, Umeda T, Nakaji S and Sugawara K 2003 Low-level laser irradiation attenuates production of reactive oxygen species by human neutrophils *J. Clin. Laser Med. Surg.* **21** 3 165–70

[21] Lopes-Martins RAB, Albertini R, Martins PSL, Bjordal JM and Castro-Faria-Neto HC 2005 Spontaneous effects of low-level laser therapy (650 nm) in acute inflammatory mouse pleurisy induced by carrageenan *Photomed. Laser Surg.* **23** 4 377–81

[22] Aimbire F, Ligeiro de Oliveira AP, Albertini R, Correa JC, Ladeira de Campos CB, Lyon JP, 2008 Low level laser therapy (LLLT) decreases pulmonary microvascular leakage, neutrophil influx and IL-1β levels in airway and lung from rat subjected to LPSinduced inflammation *Inflammation* **31** 3 189–97

[23] Panagopoulos DJ, Karabarbounis A and Margaritis LH 2002 Mechanism for action of electromagnetic fields on cells *Biochem. Biophy.s Res. Commun.* **298** 95 – 10

[24] Balcavage WX, Alvager T, Swez J, Goff CW, Fox MT, Abdullyava S and King MW 1996 A mechanizm for action of extremely low frequency electromagnetic fields on biological systems *Biochem. Biophys. Res. Commun.* **222** 374 – 78

[25] Berg H 1999 Problems of weak electromagnetic field efects in cell biology *Bioelectrochemistry and Bioenergetics* **48** 355 – 60

[26] Binhi VN, 1998 Interference mechanizm for some biological effects of pulsem magnetic Field *Bioelectrochemistry and Bioenergetics* **45** 73–81

[27] Noda Y, Mori A, Liburdy RP and Packer L 2000 Magnetic fields and lipoic acid influence the respiratory burst in activated rat peritoneal neutrophils *Pathophysiology* **7** 137–41

[28] Sheiko EA, Shikhlyarova AI, Zlatnik E Yu, Zkora GI and Nikipelova EA 2004 Electromagnetic oscillations as a factor modulating blood neutrophil function *Bull. Exp. Biol* . Med **5,** 499 – 502

[29] Varani K, Gessi S, Merighi S, Iannotta V, Cattabriga E, Pancaldi C, Cadossi R and Borea PA 2003 Alteration of A3 adenosine receptors in human neutrophils and low frequency electromagnetic fields *Biochemical Pharmacology* **66**1897–1906

[30] Varani K, Gessi S, Merighi S, Iannotta V, Cattabriga E, Spisani S, Cadossi R and Borea PA 2002 Effect of low frequency electromagnetic fields on A_{2A} adenosine receptors in human neutrophils *Br. J. Pharmacol.* **136** 57-66

[31] Karu T 1999 Primary and secondary mechanisms of action of visible to near-IR radiation on cell *J. Photochem. Photobiol. B* **49** 1 – 17

[32] Cieślar G, Zalejska-Fiolka J, Birkner E, Kasperczyk S and Sieroń A 2004 Influence of long – lasting exposure to weak variable magnetic field on activity of antioxidant enzymes in rats *Physica Medica Supp. 1* **20** 25 – 27

Journal of Physics: Conference Series **329** (2011) 012024 doi:10.1088/1742-6596/329/1/012024

The changes of the frequency specific impedance of the human body due to the resonance in the kHz range in cancer diagnostics

K P Michalak[1], H Nawrocka-Bogusz[1]

[1] Biophysics Department, Poznań University of Medical Sciences

Fredry Street 10, 61-701 Poznań, Poland

E-mail: kmichalak@ump.edu.pl

Abstract. The frequency-specific absorption of kHz signals has been postulated for different tissues, trace elements, vitamins, toxins, pathogens, allergens etc. for low-power (μV) signals. An increase in the impedance of the human body is observed only up to the given power of the applied signal. The highest amplification of the given signal being damped by the body makes it possible to determine the intensity of the given process in the body (e.g. amount of the toxin, trace element, intensity of the allergy) being connected with a given frequency spectrum of the signal.

The mechanism of frequency-specific absorption can be explained by means of the Quantum Field Theory being applied to the structure of the water. Substantially high coincidence between the frequencies of the rotation of free quasi-excited electrons in coherent domains of water and the frequencies being used in the MORA diagnostics (Med-Tronic GmbH, EN ISO 13485, EN ISO 9001) can be observed. These frequencies are located in the proximity of f=7kHz · i (i=1,3,5,7,...). This fact suggests that the coherent domains with the admixtures of the given substances create structure-specific coherent domains that possess frequency-specific absorption spectra.

The diagnostic tool called "MORA System diagnosis" was used to investigate 102 patients with different types and stages of cancer. Many signals were observed to be absorbed by many cancer patients, e.g.: 'Cellular defense system', 'Degeneration tendencies', Manganese, Magnesium, Zinc, Selenium, Vitamin E, Glutamine, Glutathione, Cysteine, Candida albicans, Mycosis. The results confirm the role of oxidative stress, immunological system deficiency and mitochondria malfunction in the development of cancer.

1. Introduction

Today cancer is commonly considered a serious illness being the cause of high mortality in the contemporary society and measures taken in order to treat cancer are found to be rather unsatisfactory. It may come from the fact that the knowledge of biochemical and biophysical processes taking place in a cancer cell is insufficiently considered.

One of the problems connected with biochemical diagnostics of cancer is lack of the methodology of the analysis of different metabolites inside the cells in public medicine. The concentration of e.g. the Krebs Cycle metabolites, amino acids, trace elements and vitamins inside the cells cannot be commonly measured by using the available methods. Concentration of the substances in blood does not reflect its concentrations inside the cells.

One of the possible solutions of this problem is the use of a spectrum-specific increase in impedance of the human body in the area of the kHz. This method is applied e.g. in the MORA System Diagnosis device.

2. The fundamentals of the MORA diagnostics

Franz Morrel and Emil Rasche observed 40 years ago that a human body possesses the ability to increase the impedance in the range of kHz for the signals possessing very small amplitudes and being the representation of emission spectra of different allergens, trace elements, vitamins, amino acids, tissues etc. These signals can be registered by using very sensitive analogue-digital cards if the external electromagnetic fields are diminished by using the Faraday box and by introduction into the AC-card the differential signal between the empty electrode and the electrode on which the given substance is placed. The signal conduction through the body gives the effect of an increase in the impedance by 10%-30% if the pathology being connected with the given substance exists in the body (e.g. allergy for the given substance, the lack of the given trace element, the charge of the organism with the given toxin, the illness of the given tissue). This increase of the impedance is observed, however, only up to the given amplification of the signal power. If the amplification increases over the given level, the effect vanishes. The maximal amplification level for which the increase of the impedance is observed refers approximately to the intensity of the given process. Therefore, it can be concluded that the effect of energy absorption possesses the resonance basis. The amount of energy the tissue is able to absorb by using the resonance effect is limited and the energy exceeding this level goes through the body without absorption.

However, what seems important to stress, trials to register the spectra of the given signals have not been fully successful until now. The Fourier spectra do not possess any peaks for the given frequencies. Every signals look like the noise. Assuming that the spectra are nevertheless specific, three possibilities should be considered as the reason for this effect.

1. Spectra possess many tiny peaks that are not significantly higher than the level of the noise in the Fourier spectrum. The mean absorption of energy for only slightly higher amplitudes than noise gives the total observed effect of an increase in the impedance by 10-30%.
2. The emitted registered signals are periodic but they do not possess the shape of a sine. The peaks of such signals are not directly visible in the Fourier spectrum but are spread over many frequencies. The information is stored especially in the phase relations between the neighbouring and harmonic frequencies. This thesis is promoted by Galle [1]. The exemplary phase spectra presenting the significant phase relations between the successive frequencies in the Fourier spectrum of the specific water solutions are presented by Ludwig [2].
3. The signals that are emitted and registered are non-linear from the viewpoint of the deterministic chaos theory. The fluctuation of energy between different excited states in the complex systems possessing many regulatory elements, as the biological ones, can be chaotic. It means that the signals are at most quasi-periodic and regular only to some degree. The complexity of such signals can be estimated by using e.g. modified Takens-Ellner algorithm [3]. The information is also stored in the mutual phase relations between the frequencies in the Fourier spectrum. Application of these signals into the body would repeat the way of the energy fluctuations between different ground and excited states of the biological systems absorbing the energy in the non-linear way. This hypothesis may be partially supported by the results from the estimation of dimensional complexity in the magnetic field fluctuation of myomas at a frequency range of 1-100Hz [4].

If the reason of the observed increase in impedance depends on the phase relations (like presented in points 2 or 3) then the information can be destroyed by the phase randomization of consecutive frequencies in the Fourier spectrum and this procedure should annihilate the effect of an increased impedance of the signal even if the amplitudes of consecutive frequencies are the same.

Journal of Physics: Conference Series **329** (2011) 012024 doi:10.1088/1742-6596/329/1/012024

2.1. The postulated range of frequency applied in the MORA diagnostics

The Mora device is able to perform the phase-constant amplification of the signal in the frequency range of 1Hz-150kHz. The range of frequency that is responsible for the absorption of energy in the signals representing some allergens was investigated indirectly by using signals filtering. The resonance for the given signals was observed after applying different low- and high-pass filters. The lower frequency range of the resonance effect can be determined by using the gradual increase in the filter frequency of the high-pass filters and looking for the vanishing of the resonance. Similarly, the upper resonance range can be determined by using the gradual decrease in the filter frequency of the low-pass filter applied.

These ranges are described as [5,6]:

- syntetic substances (e.g. formaldehyde, preservatives, heavy metals): 7.6 - 13.5 kHz
- simple living organisms (bacterias, fungs, vires): 22 - 34 kHz
- food products: 49.5 - 53.5 kHz

Spectrum of all signals is postulated to be below 70kHz.

2.2. The possible mechanisms of energy absorption

Wolfgang Ludwig postulated that water is responsible for the energy absorption [2]. Water clusters can take different structures depending on the substances being in contact with water molecules. Not only did Wolfgang Ludwig present the exemplary spectra of the solutions of some substances and low-potency homeopathic drugs that possess the characteristic peaks in the range of 10^2-10^4 Hz. He also suggested that high potency homeopathic drugs possess the resonant frequency in higher frequency range of about 10^4-10^5Hz and the phase relations between neighbouring frequencies are probably the characteristic features of these spectra.

The thesis according to which water is responsible for the absorption of energy in the MORA diagnostics can be supported by the Quantum Field Theory being applied to the water structure by E. del Giudice [7-9]. According to presented calculations, the atoms oscillate between two levels in tune with the non-vanishing electromagnetic field so that the field is trapped within the region whose size is λ. It is the size of the water coherence domain (CD) being equal to about 100nm. The number of water molecules in CD is postulated to be about 10^7. About 13% of electrons in the CD's are quasi-free and they can be put in rotation within the CD. These electrons are in the excited states and the frequency of rotation is postulated to be ~7, 21, 35, 49,.. kHz. Since the motion is frictionless, the lifetime of these excited levels is quite long. One can see that these frequencies correlate strongly with frequencies presented by the authors of the MORA device in the previous subsection. The admixture of the given substances to water CD's or the contact of water CD's with the specific proteins may force the appearance of both specific structure of the CD's and specific level of the excited states of these quasi-free electrons.

The tubular structures in the water corresponding to the ones postulated by del Giudice CD's that should appear in contact with the isolated molecules were photographed by Shui Yin Lo by using the electron microscopy (3-D conformation and packing of IN Crystals mica surface with NanoScope@ Multimode AFM in TappingModeTM. AFM is a high resolution technique for studying surface) [10,11]. This observation also supports the possible role of water as the frequency-specific energy absorber of MORA signals.

On the other hand, Smith proposes that arithmetic and logic operations could be performed on the frequencies imprinted into water [12]. He postulates the existence of the memory in living systems based on the assumption that DNA and water is quantum holographic and syntactic with the information encoded in phase and operations modelled on computer 're-write' systems with a 'nil-potency' rule [13]. The bandwidth of the frequency range being possible to imprint into water is supposed to be from 1 to 10MHz.

What is next important to realize, in the case of trace elements or vitamins both its overdose and deficiency lead to the resonance effect. This fact can be also treated as the resonant effect of the specific water structure. As it is postulated by Froehlich [14,15], large dipoles (e.g. enzymes and co-factors) may attract in the cytoplasm due to common electromagnetic vibration frequency. Water surrounding both the enzyme and co-factor may possess the same frequency of the excited states of quasi-free electrons. If the concentrations of enzymes and co-factors are equal, there are no free enzymes and free co-factors in the cytoplasm and no water structures would occur on the unfitted enzymes or co-factors that could be responsible for the absorption of the energy (see Fig.1). If the concentrations are not equal then the resonance is caused by the water surrounding the overdosed element of this equilibrium.

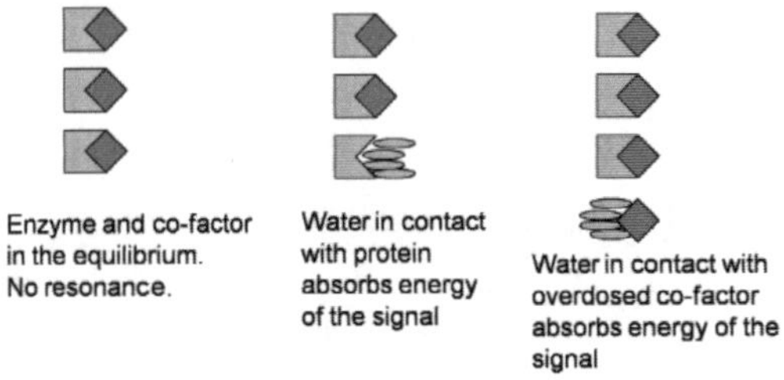

Figure 1. The mechanism shows that both deficiency and overdose of cofactor may lead to the resonance effect. Water being in contact with the overdosed element of the pair enzyme-cofactor may be responsible for the absorption of signal energy.

2.3. The impedance of human body in the kHz range

The problem of the absorption of kHz alternate electric currents (AC) by human body must be analyzed mainly in the context of the permeability of lipid membranes in this frequency range. The capacitance of lipid membranes is equal to about $C=10mF/m^2$. The accurate value depends on the amount of the immersed proteins. This value is relatively high due to very low thickness of the membrane and high electrical permeability of the lipids used for building it. As it was measured, the lipid membranes are to small degree permeable for AC in the kHz range. The signal running through the body is conducted mainly via extracellular fluid and only to some extent through the inside of the cells. The percentage of energy in ~10kHz signal that is absorbed inside and outside the cell is difficult to estimate precisely because of high complexity of the system of lipid membranes that modify the flow of this current. However, on the basis of the total resistance and reactance of the human body, it can be stated that the amount of energy that is absorbed in the cell inside is equal to about 10-30% and in the outside – 70-90%. This observation coincides with an observed increase of the impedance of the signals observed in the MORA diagnostics. On the other hand, assuming that it is possible to transfer water through the lipid membrane without changing its structure, to some extent the extracellular fluid can be also responsible for the absorption of energy. The quantitative analysis of this problem should be performed both in the theory and practice.

3. Biochemical and immunological aspects of the cancer

The Mora diagnosis device is able to find many biochemical aspects of the organism. Some aspects described in the biochemical and cancer journals are mainly focused on the topics that deal with cancer. The current paper compares the knowledge concerning these aspects with the results obtained by using MORA diagnostics.

3.1. Mitochondria and oxidative stress

Otto Warburg was the first researcher who investigated the biochemistry of cancer cells. He discovered that the cancer cells do not use the oxygen even in the existence of the oxygen in the cells

[16]. This phenomenon is called the Warburg effect and it is observed to be a common feature of majority of cancer types. Many contemporary experiments show that the malfunction of the mitochondria [17-27] may be the reason for this effect. The decreased activity of tricarboxylic acid cycle (TCA) enzymes and the oxidative phosphorylation cytochrome chain (OXPHOS) can be found in the specific types of cancers. One of the most common features of cancer is the decreased activity of pyruvate dehydrogenase (PDH) [16,28-30]. This enzyme makes it possible to utilize glucose in the TCA cycle and produce energy. When this enzyme is lacking, it leads to the increased aerobic glycolysis, the formation of large amount of lactate, and finally to the acidosis of the extracellular fluid of tumor that stops the activity of the NK lymphocytes [31]. The restoration of the proper activity of this enznyme by means of dichloroacetate (DCA) is described to result the apoptosis of the cancer cells [32,33].

Disturbed activity of the OXPHOS chain leads in some types of tumors to the leakage of the electrons from this chain. It causes the creation of reactive oxygen species (ROS) and leads to an increase in the oxidative stress in the cell [34]. The increased amount of energy is necessary to restore this pathological state but the production of the energy is disturbed. The cell degenerates gradually. If the mechanisms of the apoptosis do not kill the cell, it evolves into a cancer cell.

3.2. The role of trace elements and vitamins

Five trace elements are directly involved in the energy production in mitochondria, namely: manganese (Mn), magnesium (Mg), zinc (Zn), selenium (Se) and iron (Fe).

Manganese is an activator of three very important enzymes that occur only in mitochondria: pyruvate dehydrogenase (PDH), alfa-ketoglutarate dehydrogenase (KGDH) and mitochondrial dysmutase (SOD_{Mn}) [35-37]. PDH and KGDH are twilling enzymatic complexes immersed in the inner mitochondrial membrane that perform similar decarboxylation reactions; they are necessary for correct activity of the TCA and production of $NADH_2$ for the OXPHOS chain. As it was mentioned, decreased activity of PDH is the common feature of many types of cancer. Decreased activity of KGDH can lead to the decreased concentration gradient alfa-ketoglutarate/succinate that initiates the production of Hypoxia Inducible Factor (HIF) in the cell. An increase of the HIF level is the initial regulatory stimulus for the change of the activity of many enzymes in the cell e.g. decreasing PDH functioning [31,38]. The high HIF level was found in some types of cancer and it seems to be an important element of the metabolic conversion in a cancer cell [24].

Mitochondrial dysmutase (SOD_{Mn}) is an important enzyme removing O_2^- ions that is the most important free radical being the effect of the leakage of electrons from the disturbed cytochrome chain. Its decreased activity appears to be an important factor increasing the oxidative stress and reducing the production of energy in mitochondria.

Magnesium and zinc are the most important co-factors for many enzymes. Each of them activates about 300 enzymes. Magnesium, like Mn, is an activator of PDH and KGDH [36,37,39]. It is also necessary for the glycolysis enzymes and proper activity of the immunological system. Zinc is a co-factor of e.g. cytoplasmatic dysmutase and it plays an important role in the process of cell division [39]. Zinc is postulated to be an important anti-tumor agent in the prostate cancer and other tumors [40].

Selenium is the trace element that is a co-factor of the glutathione peroxidase (GPO) [39]. It restores the oxidized glutathione G-S-S-G to the reduced form G-SH. Thus, it seems vital in the maintenance of the correct level of ROS and in the detoxification processes.

Ferrum is the co-factor of the catalase. This enzyme is the first barieer against ROS. It decomposes H_2O_2 that arises in the amount of about 0.5kg/day/70kg'human in the last step of the OXPHOS chain: translation of 2 electrons from the cytochrome C into O_2. The decreased level of Fe and reduced activity of catalase can lead to the decreased energy production in the cells.

There are three vitamins that are involved in the ROS reduction: A, C and E. Vitamin E is mainly considered as care against the oxidation of the unsaturated fatty acids building the lipid membranes.

Vitamin C reduces the cytoplasmatic free radicals and supports the reduction of the oxidized vitamin E.

The increased ROS level leads to the uncoupling of the cytochrome chain in order to reduce further production of ROS. Thus, deficiency of each of these trace elements can lead to both decreased ROS reduction, increased ROS level, uncoupling of the OXPOX chain, and reduced energy production in the mitochondria, gradual degeneration of the cell and alternative conversion into the cancer cell.

The magnetic resonance spectroscopy has shown that the concentration of many ions inside the cancer cell is changed due to the initial lactate and H^+ production [41]. The concentration of Na^+ and Ca^{2+} is higher and of Mg^{2+} is lower in the cancer cells. The concentration gradient across the membrane of all these ions is lower than in healthy cells due to lower energy production being necessary to shift the ions across the membrane. The negative electrical potential of the cell inside being equal in the normal state to about -90mV serves as an important factor pulling the positive ions into the cell. According to Nernst formula, this potential is able to create the concentration gradient of Me^{2+} ions equal to about 1000x. If this potential drops to e.g. -60mV due to the insufficient energy production, the concentration gradient of Me^{2+} ions decreases to 100x. One can conclude that the concentration of other ions such as: Mn^{2+}, Se^{2+}, Fe^{2+}, Zn^{2+} may be lower inside cancer cells due to its lower membrane potential.

3.3. Glutamine and cachexy phase of the cancer

The other aspect of cancer is overuse of the glutamine in the cachectic phase of cancer [42,43]. The strongly growing tumor utilizes big amount of glutamine to synthesize necessary biochemical compounds like DNA, amino acids and fatty acids. The low level of glutamine in patient's blood stimulates the decay of the proteins in the muscles to restore the glutamine level that is necessary for the proper activity of intestine, kidneys, immunological system and the majority of other organs in the body. The lack of glutamine causes the collapse of metabolic equilibrium, finally it causes death. As it is presented in some experiments, the supplementation of glutamine during the chemotherapy and radiotherapy minimizes the side effects of the treatment [44-54], increases the concentration of some drugs inside the cancer cells [55,56], and decreases its concentration in healthy cells.

The intake of glutamine in the cachectic phase of cancer may stop the cachexy process and in this way elongate patient's life to some extent. This effect can be more efficient if the intake of glutamine is separated in time with the intake of carbohydrates in the diet. The glutamine is mainly the source of $NADPH_2$ that cannot be directly converted to ATP [39] and glucose is mainly the source of ATP and not of $NADPH_2$ [39]. The optimal condition for the growth of cancer is the simultaneously high level of both the ATP and $NADPH_2$. The separation of maximal concentrations of these metabolites in time should slow down the growth of the tumor.

3.4. Glutathione

The glutamine effect is partially caused by its modulation of glutathione equilibrium. Glutathione is the three amino acid peptide composed from the glutamate, glycine and cysteine.

The glutathione synthesis cycle shows that the substrate for glutamate is glutamine. An active place of glutathione molecule is the –SH residue being the part of the cysteine. Glutathione is an important molecule involved in redox equilibrium of the cell. It is able to remove free radicals that arise during the energy production. It is also the detoxification particle that takes part in removing toxins from the cell [57].

Glutathione is involved in a variety of metabolic functions like modulation of calcium homeostasis, activation of transcription factors, DNA repairing, cell cycle regulation, and regulation of enzyme activity. Most of these functions are related to its ability to maintain the reduced cellular environment. [58,59] Malignant diseases are accompanied by reduction of glutathione in the organism, which can be to some degree reversed by the dietary glutamine intake. [60] Other reports suggest that glutathione constituent amino acids, including glutamine, inhibit cancer promotion, at least in part, by their influence on the glutathione metabolism. [61,62]

The other explanation for glutamine supplementation in cancers through the modification of glutathione metabolism was presented by Todorova et al. [58] They show that cancer cells possess higher concentration of reduced glutathione than healthy cells do, which attributes to higher cell proliferation rate and resistance to chemo- and radiotherapy. Therefore, selective tumor GSH reduction is a promising strategy in the treatment of cancer.

3.5. Immunological system

Cellular Defence System serves as a natural care against cancer cells; it consists of NK lymphocytes, macrophages, and neutrophiles that are able to kill the cancer cells [43,63]. The immunological system is a very important barrier against the growth of cancer and its role is underestimated in the current approach to the cancer treatment. The activity of NK cells is strongly lowered in the acidic milieu being a commonly described feature of many types of cancer. The NK cells are able to kill the cancer cells if pH is higher than about 7.0 and the pH of the extracellular fluid in the cancer tumours is equal as a rule to about 6.2-6.8 [31]. It can be concluded that the activity of NK lymphocytes is nearly completely reduced in cancer tumours and especially in tumours producing big amount of lactate in the absence of the activity of PDH.

The other important elements that are analyzed in the current paper and are necessary for the functioning of immunological system are: magnesium, glutamine and glutathione. Mg^{2+} is an activator of many reactions participating in the immunological response and glutamine is the primary fuel for the increased metabolic activity of different immunological cells.

Glutathione regulates the binding, internalization, degradation and T-cell proliferative activity of interleukine-2, thus may affect the growth of cytotoxic T-cells [64].

3.6. Candida albicans and other fungi

Not only is the Cellular Defence System a barrier against the cancer but it also serves as a barrier against the other pathogens, such as: fungi and parasites. It has been observed that the shape and structure of the fungus cells growing in the tight space becomes similar to the cancer cells. It may come from the similar cytoskeleton structure, energy production system and similar kind of growth of both cell types. The fungi are commonly existing pathogens which must be consistently removed from the organism. Generally, fungi produce different enzymes and toxins (antibiotics) that reduce the activity of immunological system and the absorption of the important trace elements in the intestine: Se, Mn, Zn and Fe [65]. Mycosis is also said to be in correlation with allergy [66]. Candida albicans are the most prevalent fungus that exists on the human's mucous membranes. It is the saprophyte only until the mucous membrane is healthy, but after it breaks the mucous barrier it starts to penetrate the submucous area and it becomes the pathogen [66]. Produced toxins inhibit the defence system and acute response of the immunological system (heat, pain, oedema, red mark) is not observed during the gradual fungal infiltration of the organism.

4. Methods and results

The group of 102 cancer patients was investigated by using the MORA device RM10C + Ritu int. The patients were diagnosed to suffer from cancer by using recognized diagnostic methods. The patients possessed different cancer types and different stages of the illness. The age of the patients varied between 3 and 82 and the mean age was 55±16 years.

The MORA diagnostic test was performed to consist of about 100 different signals conducted through the patient's body. The maximal resonating amplification of every signal was remembered for statistical analysis. The Figures present results for most often resonating signals correlating with known biochemical dysfunctions in cancer cells and in the total organism of the patient. The successive figures show the appearance frequencies of the maximal signal amplifications for which the resonance was observed. What seems most important is the proportion between all the colourful areas representing the resonance of the signal and the white area representing the lack of resonance. One should remember that the maximal amplification of the observed resonance is the relative number that cannot be directly converted to any value being represented by SI unit. These amplifications seem to correlate with the intensities of the given processes but, until now, the precisions of these relations have not been analyzed. Approximately, the value R=10 denotes the weak resonance, R=30 - the medium and R>50 - high resonance. The results for the remaining tested signals that present the ratio of the patients with resonating signal to all patients are presented in Table 1.

4.1. Oxidative stress

The signal which represents the existence of oxidative stress is called 'Degenerative tendencies'. The statistics is presented in Figure 2. It can be observed that more than 70% of patients suffered from the oxidative stress in the organism. The intensity of this process was high (A>50) by more than 20%. These result show that the oxidative stress occurs very frequently in the group of cancer patients.

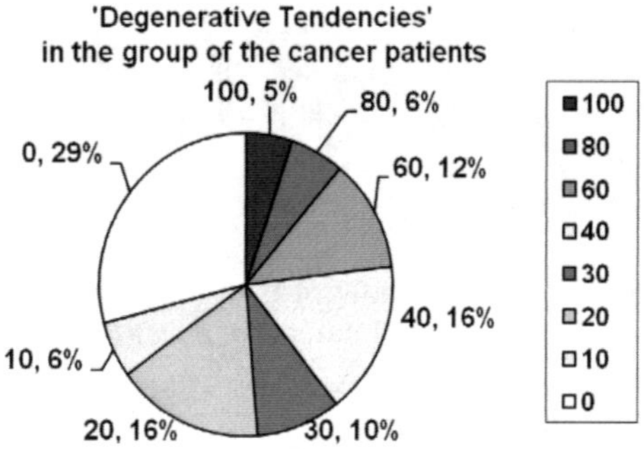

Figure 2. The signal 'Degenerative tendencies' that refers to the oxidative stress has resonated by about 70% cancer patients.

4.2. Trace elements and vitamins

The results for the trace elements: Mn, Mg, Se, Zn, and vitamin E are presented in Figure 3. It is visible that manganese, selenium and vitamin E are most often resonating elements. The other ones were also found in the analyzed group.

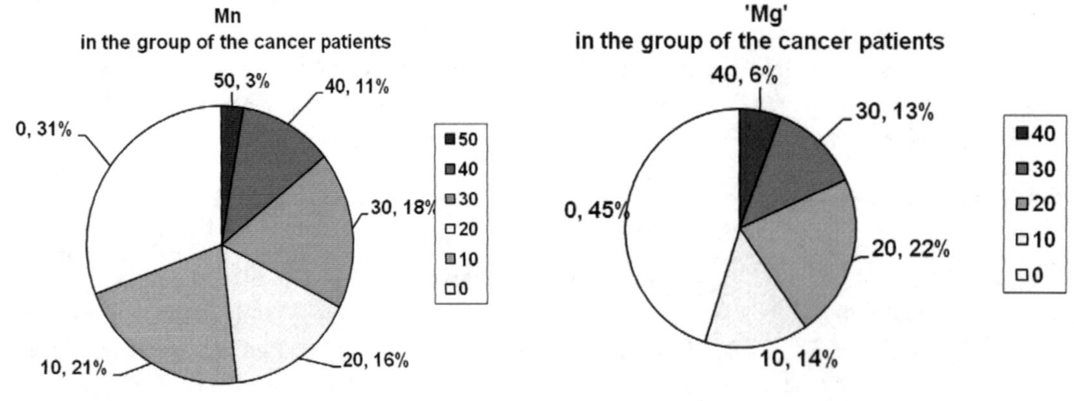

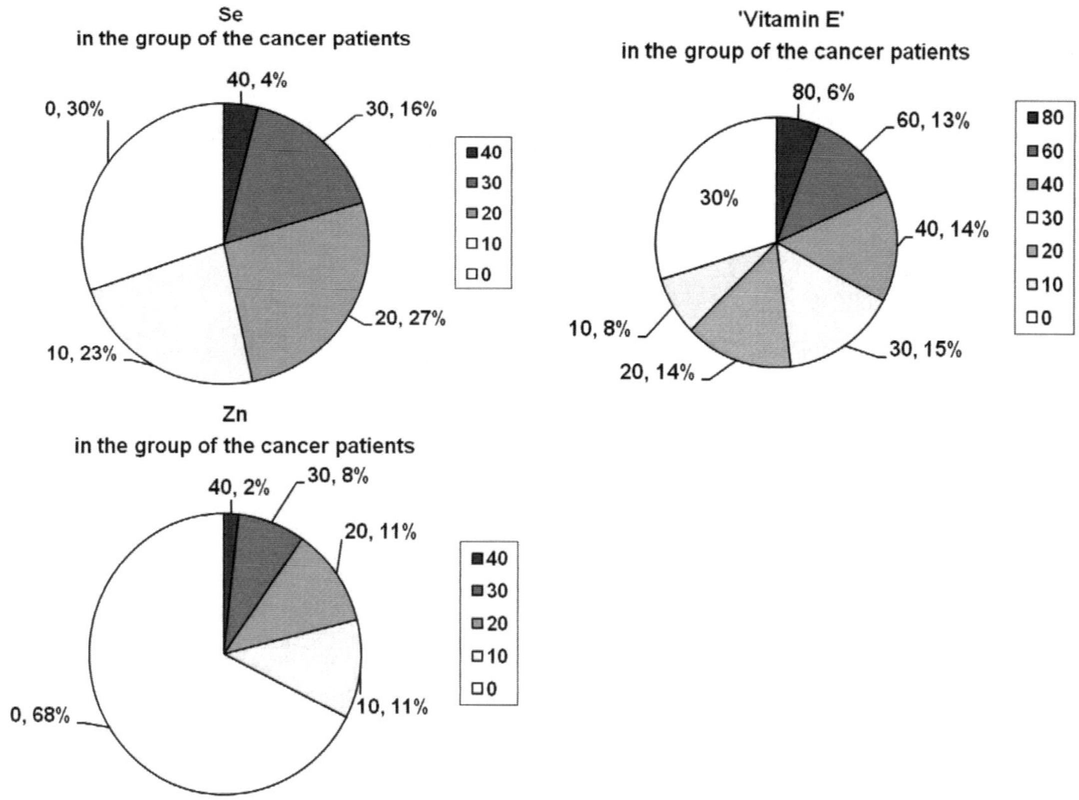

Figure 3. The resonance for the signals: Mn, Mg, Se, Vit E, Zn in the group of cancer patients. Most of them resonated in more than 50% of patients.

4.3. Amino acids

The results for the amino acids are presented in Figure 4. It can be observed that both cysteine and glutathione involved in the cell redox balance and oxidative stress reduction resonated by more than 50% patients. Glutamine that is responsible mainly for the cachexy and indirectly for the oxidative stress reduction resonated in about 35% of patients.

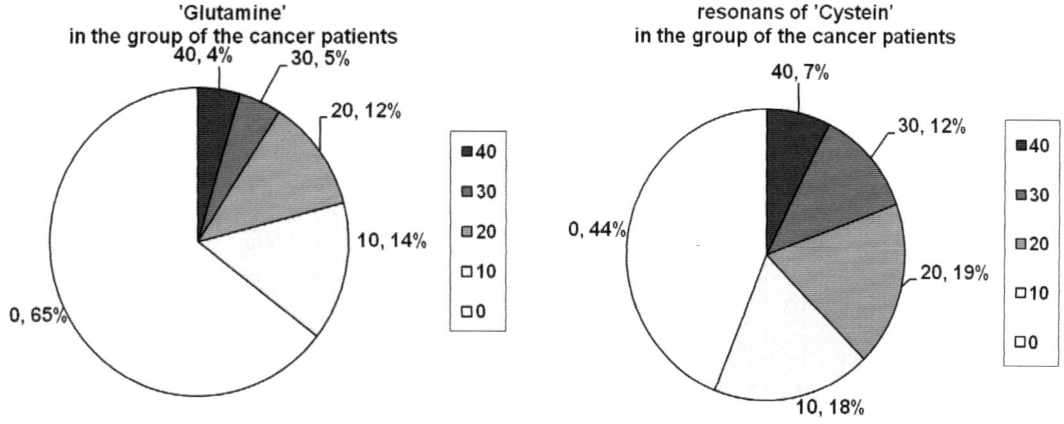

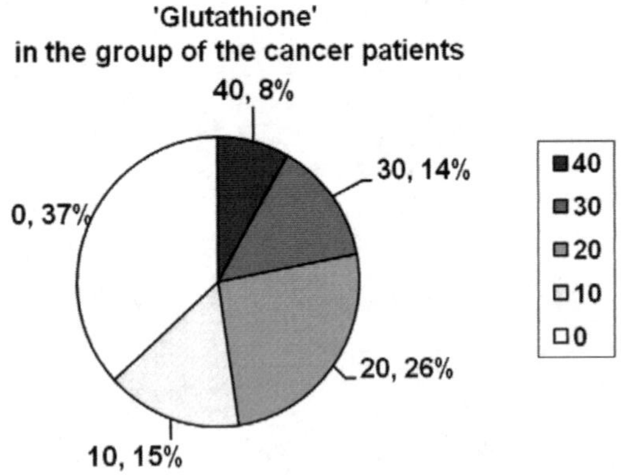

Figure 4. The resonance of the amino acids: cysteine, glutamine and glutathione in the cancer patients group.

4.4. Immunological system and fungi

Figure 5 presents the ratios of the observed resonance of signals "Cellular defence system' Candida albicans' and 'Mycosis'. It can be observed that the deficiency of immunological system and fungal infiltration are common processes occurring in many cancer patients. The coexistence of mycosis and cancer may be connected with insufficient activity of immunological system being a common barrier against both kinds of cells. Mycosis may be considered to be both the reason and effect of the deterioration of immunological system. On the basis of the results obtained, it can be hypothesized that the gradual fungal infiltration may be treated as a potential risk factor for cancer development due to the gradual deterioration of immunological system. The lack of significant symptoms of this infiltration is the reason for insufficient attention paid by the contemporary medicine to the coexistence of fungal infiltration and cancer.

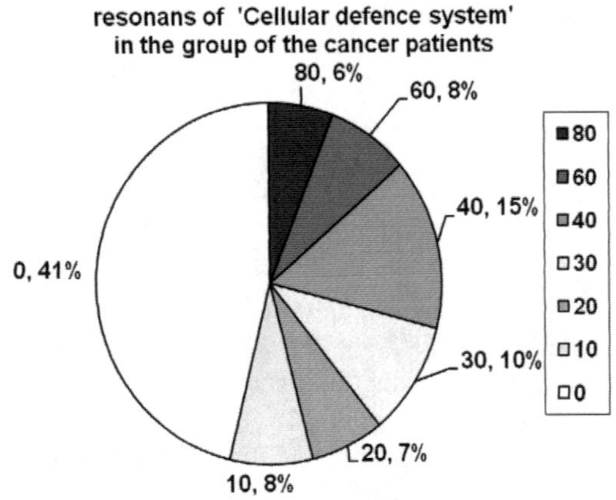

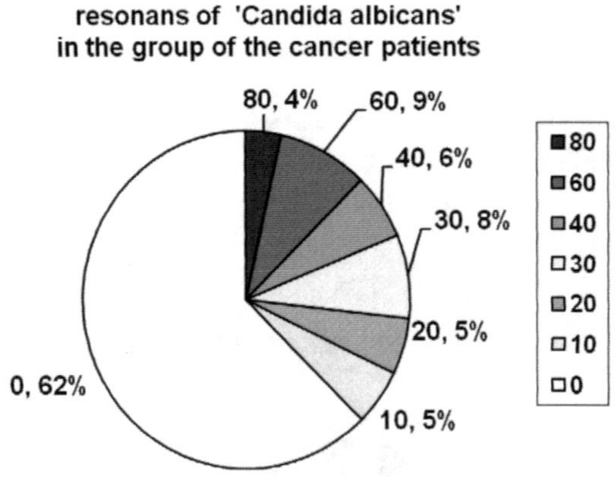

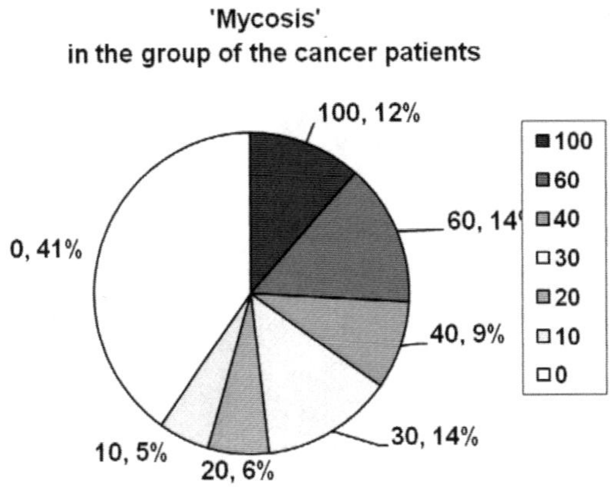

Figure 5. Resonance of the Cellular defence system and mycosis. As the results suggest, the deterioration of cellular defence system and coexisting mycosis are important factors in cancer patients. Mycosis may be considered to be both the reason and effect of the deterioration of immunological system.

4.5. Other signals

Table 1 presents the summary of the frequency of the observed resonance R≥10 for other analyzed signals. These results are presented without additional comment due to the lack of the analysis of the references concerning the individual problems.

It seems important to stress most often observed resonance in the following groups of signals:

1. Many toxic substances (formaldehyde, putrafective and fermentative processes, heavy metals, lipophilic toxines).
2. The main allergens: wheat, milk and egg white, barley.
3. Many other micro-, macro-elements and vitamins, amino acids.
4. The important regulatory systems: intestinal defence system, lymphatic system, reticulo-endothelial system.

Table 1. The other important signals that have resonated in the group of cancer patients.

Signal	percent of patients with resonating signal [%]
Intestinal defence system	60
Wheat	53
Lymphatic system	53
Egg white	45
Vitamin B6	40
Milk	39
Folic acid	39
Putrafective processes	37
Lead	37
Formaldehyde	37
Barley	36
Reticulo-endothelial system	36
Mesenchyme	35
Rye	34
Acidosis	33
Chromium 3+ (trace element)	32

Ferrum	31
Potassium	31
Cystes	30
Alkalosis	29
Arsenic	29
Xyladecor-Mix	28
Fermentative processes	27
Pentachlorophenol	27
Mercury	26
Dioxine	26
Isocyanates	26
Lindane	26
Benign tumours	26
Histamine	25
Cadmium	25
Calcium	24
Vitamin B12	24
Vitamin A	24
Carnitine	24
Beryllium	23
Vitamin B1	23
Solvents-Mix	23
PAK	23
Autoaggressive illnesses	22
Nickel/ Chromium 6+ (toxin)	21
Cuprum	21
PCB	21
Palladium	20
Oats	19
Vitamin B3	18
Pyrethrum	18
Yolk	17
Sodium	17
L-Phenylanaline	16
L-Lysine	14
Molybdenium	13
L-Tryptophan	12

5. Summary

Assuming that analyzing the concentration of different substances inside the cell has been almost impossible in common medicine until now, the MORA diagnostics tool seems a very interesting alternative for the non-invasive estimation of the biochemical processes that take place in the human organism. Many biological aspects of the patient may be analyzed during relatively short time of the measurement giving the broad information about the patient's condition and the initial and secondary causes of the illness. Also, the cost of the examination is low with respect to the amount of information about the patient's condition.

Generally, the results obtained by means of this method correlate with the most recent knowledge that is available for the areas of cancer science, being analyzed in details in the current paper.

The method needs to be validated with respect to its sensitivity and accuracy. Moreover, biophysical fundamentals of the spectrum-specific energy absorption of ultra-weak signals must be explained in details.

References

[1] Galle M, "Klinischer Bericht zu den MORA-Bioresonanzgeraeten," (2008).

[2] Ludwig W, Wasser und Homeopathie (CKH Verlag, 2002).

[3] Michalak K 2011 Modifications of the Takens-Ellner algorithm for medium- and high-dimensional *Phys Rev E* **83** 026206

[4] Kotini A, Anastasiadis AN, Anninos P, Koutlaki N, and Anastasiadis P 2006 Nonlinear analysis of biomagnetic signals recorded from uterine myomas *BioMagnetic Research and Technology* **4**

[5] Cornelissen G, (Promora Polska, 2004).

[6] Cornelissen G, (Promora Polska, 2006).

[7] del Giudice E, in Coherent quantum electrodynamical organization of biochemical reactions, Neuss 08-2007, 2007.

[8] del Giudice E and Vitiello G 2006 Role of the electromagnetic field for the formation of domains in the process of symetry-breaking phase transitions *Physical Review A* **74** 022105

[9] Arani R, del Giudice E, and Preparata G 1995 QED Coherence and the thermodynamics of water *Intl. J. of Mod. Phys. B* **9** 1833-1841

[10] Lo SY and Bonavida B, in Survey of IE clusters, Los Angeles, California, 1997 (Word Scientific).

[11] Lo SY, Li WC, and Huang SH 2000 Water clusters in life *Med Hypotheses* **54** 948-953

[12] Smith CW 2005 Watergates - logic operations in water *Intl J Comp Anticip Syst* **19** 323331

[13] Smith CW 2008 Water - nanoscale to microscale to amcroscale *Intl. J. of Computing Anticipatory Systems* **22** 100-110

[14] Froehlich H 1970 Long range Coherence and the Action of Enzymes *Nature* **228** 1093

[15] Froehlich H 1972 Selective long range dispersion forces between large systems *Physics Letters* **39A** 153-154

[16] Warburg O 1956 On the Origin of Cancer Cells *Science* **123** 309-314

[17] Copeland WC, Wachsman JT, Johnson FM, and Penta JS 2002 Mitochondrial DNA alterations in cancer *Cancer Invest* **20** 557-569

[18] Senthilnathan P, Padmavathi R, Magesh V, and Sakthisekaran D 2006 Modulation of TCA cycle enzymes and electron transport chain systems in experimental lung cancer *Live Sci* **78(9)** 1010-1014

[19] Briere J, Favier J, Gimenez-Roqueplo AP, and Rustin P 2006 Tricarboxylic acid cycle dysfunction as a cause of human diseases and tumor formation *Am J Cell Physiol* **291(6)** C1114-1120

[20] Meierhofer D, Mayr JA, Foetschl U, Berger A, Fink K, Schmeller N, Hacker GW, Hauser-Kronberger C, Kofler B, and Sperl W 2004 Decrease of mitochondrial DNA content and energy metabolism in renal cell carcinoma *Carcinogenesis* **25** 47-61

[21] Isidoro, A., Martinez M, Fernandez PL, Ortega AD, Santamaria G, Chamorro M, Reed JC, and Cuezva JM 2004 Alteration of the bioenergetic phenotype of mitochondria is hallmark of breast, gastric, lung and oesophagal cancer *Biochem J* **378** 17-20

[22] Rossignol R, Gilkerson R, Aggeler R, Yamagata K, Remington JS, and Capaldi RA 2004 Energy substrate modulates mitochondrial structure and oxidative capacity in cancer cells *Cancer Res* **64** 985-993

[23] Bucay AH 2007 The biological significance of cancer: Mitrochondria as cause of cancer and the inhibition of glycolysis with citrate as cancer treatment *Med Hypotheses* **69** 826-828

[24] Hervouet E, Simonnet H, and Godinot C 2007 Mitochondria and reactive oxygen species in renal cancer *Biochimie* **89** 1080-1088

[25] Shaw RJ 2006 Glucose metabolism and cancer *Curr Opin Cell Biol* **18** 598-608

[26] Pathania D, Millard M, and Neamati N 2009 Opportunities in discovery and delivery of anticancer drugs targeting mitochondria and cancer cell metabolism *Advanced Drug Delivery Reviews* **61** 1250-75

[27] Singer S, Souza K, and Thilly WG 1995 Pyruvate utilization, phosphocholine and adenosine triphosphate (ATP) are markers of human breast tumor progression: a 31P- and 13C-nuclear magnetic resonance (NMR) spectroscopy study *Cancer Res* **55** 5140-5145

[28] Unwin RD, Craven RA, Harnden P, Hanrahan S, Totty N, Knowles M, Eardley I, Selby PJ, and Banks RE 2003 Proteomic changes in renal cancer and co-ordinate demonstration of both the glycolytic and mitochondrial aspects of the Warburg effect *Proteomics* **3** 1620-1632

[29] Chen Z, Lu W, Garcia-Prieto C, and Huang P 2007 The Warburg effect and its cancer therapeuthic implications *J Bioenerg Biomembr* **39** 267-274

[30] Faure H, Vigny A, Heddi S, Giraud S, Chautard D, and Stepien G 1996 Expression of oxydative phosphorylation genes in renal tumors and tumoral cell lines *Mol Carcinog* **16** 165-172

[31] Brahimi-Horn MC and Poussegur J 2007 Oxygen, a source of live and stress *FEBS Lett* **581** 3582-3591

[32] Bonnet S, Michelakis E, Archer SL, Allalunis-Turner J, Haromy A, Beaulieu C, Thompson R, Lee CT, Lopaschuk GD, Puttagunta L, Bonnet S, Harry G, Hashimoto K, Porter CJ, Andrade MA, Thebaud B, and Michelakis ED 2007 A Mitochondria-K+ Channel Axis Is Suppressed in Cancer and Its Normalization Promotes Apoptosis and Inhibits Cancer Growth *Cancer Cell* **11** 37-51

[33] Anderson KM, Jajeh J, Guinan P, and Rubenstein M 2009 In vitro effects of dichloroacetate and CO2 on hypoxic HeLa cells *Anticancer Research* **29** 4579-88

[34] Pelicano H, Lu W, Zhou Y, Zhang W, Chen Z, Hu Y, and Huang P 2009 Mitochondrial ysfunction and reactive oxygen species imbalance promote breast cancer cell motility through a CXCL14-mediated mechanism *Cancer Res* **69** 2375-2383

[35] Markiewicz J and Strumiło S 1997 The effect of Mn2+ on the catalytic function of heart muscle 2-oxoglutarate dehydrogenase complex *Biochemical Archives* **13** 127-129

[36] Strumilo S 2005 Short-term regulation of the alpha-ketoglutarate dehydrogenase complex by energy-linked and some other effectors *Biochemistry (Mosc)* **70(7)** 726-729

[37] Czarnecki J and Czygier M 2001 Cooperation of divalent ions and thiamine diphosphate in regulation of the function of pig heart pyruvate dehydrogenase complex *J Nutr Sci Vitaminol* **47** 385-386

[38] Roudier E and Perrin A 2009 Considering the role of pyruvate in tumor cells during hypoxia *Biochim Biophys Acta* **1796** 55-62

[39] Murray RK, Granner DK, Mayes PA, and Rodwell VW, Biochemia Harpera, IV ed. (PZWL, Warszawa, 2001).

[40] Franklin RB and Costello LC 2007 Zinc as anti-tumor agent in prostate cancer and in other cancers *Arch Biochem Piohys* **463** 211-217

[41] Stubbs M, Veech RL, and Griffiths JR 1995 Tumor metabolism: the lessons of magnetic resonance spectroscopy *Adv Enzyme Regul* **35** 101-115

[42] Miller AL 1999 Therapeutic Considerations of L-Glutamine: A Review of the literature *Alternative Medicine Review* **4** 239-248

[43] Klimberg V and McClellan JL 1996 Glutamine, cancer, and its therapy *Am J Surg* **172** 418-424

[44] Souba WW, Klimberg VS, Hautamaki RD, Mendenhall WH, Bova FC, Howard RJ, Bland KI, and Copeland EM 1990 Oral glutamine reduces bacterial translocation following abdominal radiation *J Surg Res* **48** 1-5

[45] Jensen JC, Schaefer R, Nwokedi E, Bevans DW, Baker ML, Pappas AA, westBrook KC, and Klimberg VC 1994 Prevention of Chronic Radiation Entheropaty by Dietary Glutamine *Ann Surg Oncol* **1(2)** 157-163

[46] Klimberg VS, Souba WW, Dolson DJ, Salloum RM, Hautamaki RD, Plumley DA, Mendenhall WM, Bova FJ, Khan SR, Hackett RL, Bland KI, and Copeland EM 1990 Prophylactic glutamine protects the intestinal mucosa from radiation injury *Cancer* **1** 62-68

[47] Klimberg V and all. e 1991 Does glutamine facilitate chemotherapy while reducing toxicity? *Surg Forum* **42** 16-18

[48] Klimberg V, in Oral glutamine protects against cytoxan induced cardiotoxicity and death, Hershey, PA, 1993.

[49] Anderson PM, Schroeder G, and Skubitz KM 1998 Oral glutamine reduces the duration and severity of stomatitis after cytotoxic cancer chemotherapy *Cancer* **83** 1433-1439

[50] Skubitz K and Anderson P 1996 Oral glutamine to prevent chemotherapy induced stomatitis :a pilot study *J Lab Clin Med* **127** 223-228

[51] Cao YH, Kennedy R, and Klimberg VS 1999 Glutamine Protects against Doxorubicin-Induced Cardiotoxicity *J Surg Res* **85** 178-182

[52] Fox A, Kripke SA, De Paula J, Berman JM, Settle RG, and Rombeau JL 1988 Effect of glutamine supplemented enteral diet on methotrexate induced enterocolitis *J Parenter Enteral Nutr* **12** 325-331

[53] Muscaritoli M, Micozzi A, Conversano L, Martino P, Petti MC, Cartoni C, Cascino A, and Rossi-Fanelli F 1997 Oral glutamine in the prevention of chemotherapy-induced gastrointestinal toxicity *European Journal of Cancer* **33** 319-320

[54] Sax HC 1992 Clinical and metabolic efficacy of glutamine-supplemented parenteral nutrition after bone marrow transplantation. A randomized, double-blind, controlled study *J Parenter Enteral Nutr* **16** 589-590

[55] Klimberg VS, Pappas AA, Nwokedi E, Jensen JC, Broadwater JR, Lang NP, and Westbrook KC 1992 Effect of supplemental dietary glutamine on methotrexate concentration in tumors *Arch Surg* **127** 1317-1320

[56] Rubio IT, Cao Y, Hutchins LF, Westbrook KC, and Klimberg VS 1998 Effect of glutamine on methotrexate efficacy and toxicity *Ann Surg* **227** 772-780

[57] Pastore A, Federici G, Bertini E, and Piemonte F 2003 Analysis of glutathione: implications in redox and detoxification *Clinica Chimica Acta* **333** 19-39

[58] Todorova VK, Harms SA, Kaufmann Y, Luo S, Luo KQ, Babb K, and Klimberg VS 2004 Effect of dietary glutamine on tumor gluthatione levels and apoptosis-related proteins in DMBA-induced breast cancer of rats *Breast Canc Res Treat* **88** 247-256

[59] Arrigo AP 1999 Gene expression and the thiol redox state *Free Radic Biol Med* **27** 936-944

[60] Rouse K, Nwokedi E, Woodliff JE, Epstein J, and Klimberg VS 1995 Glutamine enhances selectivity of chemotherapy through changes in glutathione metabolism *Ann Surg* **221** 420-426

[61] Wang T, Chen X, Schecter RL, Baruchel S, Alaoui-Jamali M, Melnychuk D, and Batist G 1996 Modulation of glutathione by a cysteine pro-drug enhances in vivo tumor response *J Pharmacol Exp Ther* **276** 1169-1173

[62] Baruchel S, Wang T, Farah R, Jamali M, and Batist G 1995 In vivo selective modulation of tissue glutathione in a rat mammary carcinoma model *Biochem Pharmcol* **50** 1505-1508

[63] Klimberg VS, Kornbluth J, Cao Y, Dang A, Blossom S, and Schaeffer R 1996 Glutamine suppresses PGE2 synthesis and breast cancer growth *J Surg Res* **63** 293-297

[64] Liang CM, Lee N, Cattell D, and Liang SM 1989 Glutathione regulates interleukin-2 activity on cytotoxic T-cells *Journal of Biological Chemistry* **264** 13519-13523

[65] Michalak K and Jaroszyk F 2010 Biofizyczne podstawy Diagnostyki Systemowej Mora i jej wykorzystanie do oceny zawartości makro- i mikroelementów w organizmie *Journal of Elementology* **15** 66-67

[66] Hofman T 2008 Kandydoza jest głównym czynnikiem torującym alergizację i powstawanie chorób alergicznych *Medical Tribune* **4** 22

Journal of Physics: Conference Series **329** (2011) 012025 doi:10.1088/1742-6596/329/1/012025

On the Effect of Weak Magnetic Field on Solutions of Glutamic Acid: the Function of Water

A De Ninno[1] and A Congiu Castellano[2]

[1]C.R. ENEA Frascati, Dept. UTAPRAD-DIM via E. Fermi 27, 00044 Frascati, Italy ,
[2]University Sapienza, Dept. of Physics,p.le Aldo Moro 2, 00185 Roma, Italy

E-mail: antonella.deninno@enea.it

Abstract. Even though a wide literature is available on the effects of weak Extremely Low Frequency ElectroMagnetic Fields (ELF-EMF) on the biochemical reactions, nevertheless the physical nature of these effects is largely unknown. The main challenge is the mismatch between the infinitesimal amount of the energy carried by the perturbation and the entity of the response of the system. More than 10 years ago, it was claimed that a weak current can be induced in a aqueous solution of Glutamic acid exposed to a weak magnetic field having a certain frequency. We have checked the influence of ELF-EMF on Glutamic acid using the FTIR-ATR spectroscopy in order to observe structural changes induced by the exposure. Samples have been exposed for 10, 20 or 30 minutes to a magnetic field generated by Helmoltz coils and then placed into the spectrometer. After the exposure solutions having different pH values always shown a shift toward the de-protonated species. The effect lasts for minutes after the exposure before the native configuration is restored.

1. Introduction

In the early 1990s, at the Institute of Cell Biophysics in Russia, a cycle of experiments was carried out on the action on aqueous solutions of amino acids of weak low-frequency AC magnetic fields combined with a parallel DC magnetic field. Glutamic acid was selected because it afforded the most evident effect, even if several aminoacids were tested, in that it was observed that a weak current, in order of tens of nA, could be detected in a solution exposed to a combination of a static magnetic field of 50 μT and alternating magnetic field 1000 times weaker, whenever the frequency matched the cyclotron frequency of the Glutamic acid in its ionic form [1]. Experiments such as this followed the discovery in 1985 of an unusual resonance effect caused by a simultaneous weak DC and low frequency AC magnetic field in concentrations of calcium in the nervous tissue [2]. Although controversial, the unexpected Zhadin effect has been confirmed by a number of independent researchers [3, 4, 5]

Two main problems arise in trying to provide physical insights of this phenomenon: 1) the very small amount of energy carried by these EMFs is much lower (by as much as an order of magnitude) than k_BT (k_B being the Boltzmann constant and T room temperature) and thus any information carried is hidden in thermal noise, this is known as the KT paradox; and 2) even if an excited state is induced by the perturbation, according to the general accepted models of molecular dynamics, it should rapidly disappear as a result of the thermal collisions that resume when the perturbation is removed.

Furthermore, the attempt of explaining the role of the cyclotron frequency $\Omega = \frac{1}{2}\frac{e}{m}B$, i.e. the characteristic frequency due to the Lorentz force acting on an ion having charge e and mass m in presence of a magnetic field B, have to face the problem arising from the large cyclotron orbits in comparison with the mean free path of the molecules at room temperature.

In 2002, E. Del Giudice et al. [6] proposed that the peculiar structure of water, depicted with a two-fluids model according to the predictions of Quantum Electrodynamics (QED), may account for the Zhadin effect. In such a picture ions in a static magnetic field move in special orbits around the coherence domains (CD) of water, the radius of these orbits being that of the water CD. The motion of these ions is protected by an energy gap, thus removing the disruptive actions of the thermal motion on the trajectory. The addition of a weak alternating magnetic field of frequency ω adds sideband to the fundamental rotational frequency having frequencies $\Omega - n\omega$. The "Zhadin condition" corresponds to $n=1$ and provides the essential mechanism for extracting ions from the cyclotronic orbits.

A complementary experimental technique, such as the vibrational spectroscopy, may provide useful information on the amino acid structure with a high sensitivity and allows to gain a powerful insight on the modifications of the chemical chains involved in the reactions that occur in the changing environment. We therefore selected FT-IR spectroscopy in order to investigate the mechanisms of interaction between weak fields and Glutamic acid using a very simple experimental layout, in order to minimise the effects of other spurious experimental factors.

2. Materials and Methods

Glutamic acid belongs to the group of amino acids that have a COOH group as a functional group. At low pH values, both the carboxylic and the amine groups are protonated, and deprotonated species appear with increasing pH. The acid dissociation constant of the functional COOH group, pK_a is 2.2 while the acid dissociation constant of side chain carboxylic group, pK_R is 4.3, and the amino group dissociation constant is pK_a=9.7. Above this value, the amino acid is fully deprotonated as can be seen in Fig.1. The values of the assignments of the spectral parameters, such as band position, band width, and absorption coefficient, in different protonation states are widely available [7], and it is therefore possible to detect the modifications induced by the applied perturbation quite clearly.

Several samples of 0.06M solutions of Glutamic acid in water were prepared. The concentration in double distilled water was 9 mg/ml, and the pH was adjusted by adding HCl or NaOH. Each sample was subdivided into two and each subsample decanted into safe-lock Eppendorf polypropylene tubes, each of 1.5 ml, to prevent contamination by air during the measurements. One tube was used as a control and measured as prepared, and the other was exposed to the ELF-EMF for 10, 20 and 30 minutes, removed from the field and then inserted into an IR spectrometer. During the exposure, the control tube was kept far away from the EMF.

The spectra were recorded using an FTIR/410 Jasco Fourier Transform IR spectrometer.

The samples were exposed to the EMF generated by Helmoltz coils that were oriented vertically in order to produce a horizontal field that was almost perpendicular to the local geomagnetic field B_0. The amplitude of the field was adjusted to 45 μT in order to match the amplitude of the geomagnetic field as measured with a Bell-Sypris gaussmeter having a precision of 1 μTesla. The coils were kept in the open air and the space within them was maintained at room temperature T=21±1 °C, as was the desk where the control samples were stored.

In order to test the hypothesis that the effect was related to the cyclotron frequency, part of the test were done with DC current but also sinusoidal signals were applied to the coils whose frequency was determined by the cyclotron frequency $\omega_C = \frac{1}{2\pi}\frac{q}{m}B_0$. Some test have been also performed using a 50Hz frequency magnetic field.

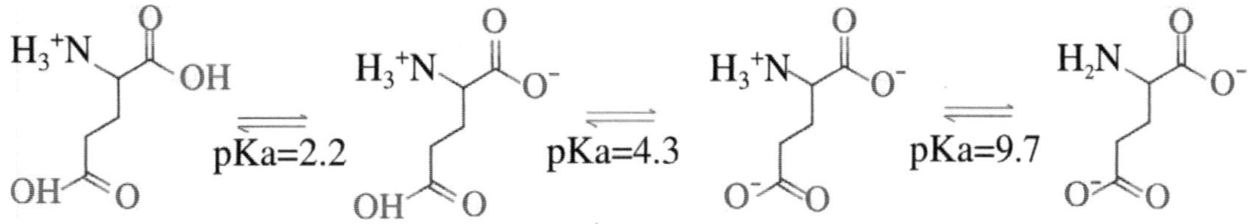

Fig.1 Speciation of Glutamic acid

The electric charge ranges from +1 in the fully protonated form to -2 according to the speciation scheme.

3. Results

The effect of the exposure was checked at different values of pH. The low-pH solution consisted mainly of fully protonated species, whereas the high pH solution mostly contained fully deprotonated species.

The IR spectra at low pH (< 3.2) contain the clear signature of the stretching and bending bands of the protonated COOH especially in the broad band absorbtion centered around 1235 cm^{-1} which is attributable to the coupled C-O stretch and O-H bend of the COOH group. Figure 2 shows a comparison between the spectra of the pH=1.53 Glutamic acid as prepared and after 10 and 20 min of exposure to the EMF. A subsequent measure repeated after 30 min of exposure shows no further modifications with respect to the spectrum acquired after 20 min, and it can therefore been assumed that the modifications of interest take place within 20 min. The overall effect of the field is the modification of the spectra in the regions of absorption of COOH and of the NH_3^+ head group. However, because of the complexity of the spectra, a simple comparison of the spectra of exposed with those of unexposed samples may be misleading, a kinetic analysis using the ratio of the protonated/deprotonated forms is not straightforward and a more sophisticated statistical treatment such as the Principal Component Analysis is needed in order to verify the statistical difference between exposed and not exposed samples. Another interesting effect of the exposure is the splitting of the band centred around 1630 cm^{-1}. Such a shift witness a change in the C-O vibrations which are sensitive to H-bonds.

At intermediate values of pH, the solutions were characterized by very complex IR spectra in which many of the bands were broader and overlapped and the attribution of absorptions to single modes is also more difficult at these pH values.

A straightforward evaluation of the effect is possible at high pH values. Fig.3 shows a comparison between the unexposed and exposed spectra at pH=11.86, at which pH the Glutamic acid should be fully deprotonated; however, the presence of a small band centred around 1650 cm^{-1} reveals the presence of the contribution of the deformation of NH_3^+ in addition to the deformation of NH_2 and the stretching of C=O. Such a band is severely restricted by exposure signalling the shift towards the deprotonated specie.

Similar effects were detected for both the cyclotron frequency of Glutamic acid (i.e., 4.7 and 9.4 Hz respectively for the acidic and basic forms), 50 Hz and DC field. Hence, the hypothesis of the existence of a resonant effect is not supported by these results.

The reversibility of the effect was also investigated by measuring the samples 1 hour after exposure. In all cases the spectra returned to their original, unexposed conditions, which shows that the effect of the EMF is reversible.

A detailed description of the Glutamic acid preparation and measurement together with comprehensive discussion of its IR spectra has been published by the Authors elsewhere [8].

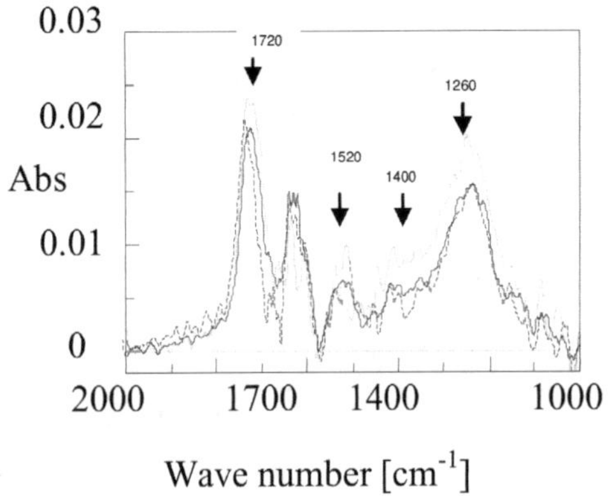

Fig.2 IR spectra of pH=1.53 Glutamic acid in water. Comparison among samples unperturbed (solid line), exposed to the magnetic field for 10 min (dotted) and 20 min (dashed). The major changes in spectra show the absorption for the stretching mode of C=O (1720 cm^{-1}), C-O (1260 cm^{-1}) and of carboxylate (1400 cm^{-1}) and for the bending of NH^+_3 (1520 cm^{-1}).

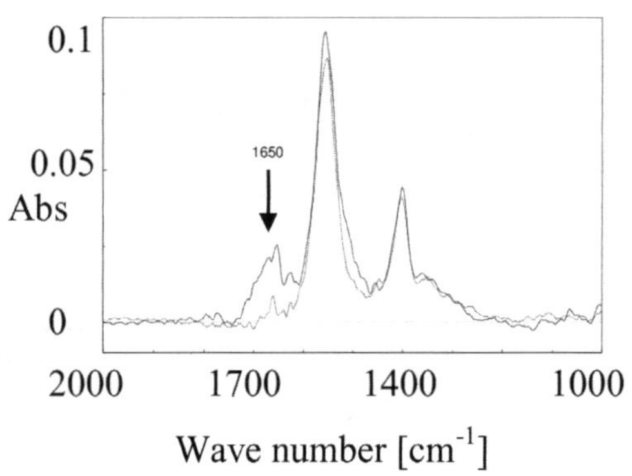

Fig.3. IR spectra of pH=11.86 Glutamic acid in water. Comparison between unperturbed samples (solid line) and samples exposed to magnetic field for 20 min (dotted line).

4. Discussion

The most remarkable characteristic of this effect is the evidence of the existence of a long living state induced by the perturbation, i.e. the effect induced survive to the removal of the perturbation for very long times compared to the characteristic relaxation times of the molecular excited states. The modification of the protonated vs deprotonated population of Glutamic acid in aqueous solution is lasting enough to be revealed by the IR spectroscopy in spite of the very feeble amount of the energy carried by the perturbation.

The energy transfer mechanism described by the semiclassical model of "hopping" along discrete energy levels cannot account for such a figure. It has been shown [9] that long-lived electronic quantum coherence plays an important part in the energy transfer process in biological systems justifying the extreme energetic efficiency of livings. In example, it is known [10,11] that light-absorbing molecules can funnel energy with a near-unit quantum efficiency across a total distance of 20-100 nm. This implies an high degree of correlation among a great number of molecules; the existence of defined phase relationship among molecules can account for a coupling among molecules stronger than to the environment and explain how the energy can be transferred. Nevertheless, the very fast (nanoseconds) relaxing time observed in those systems cannot explain the present work observations, in fact decoherence time as long as tens of minutes should imply a stable mesoscopic coherent structure in our solutions which needs to be identified.

Therefore a new hypothesis may be suggested: the magnetic field effect is mediated by the water structure.

The existence of water in two different grades of co-ordination population was foreseen in the framework of Quantum Electrodynamics (QED), in which the interaction among atoms mediated by electromagnetic fields is explained from first principles [12, 13]. The electrodynamic attraction due to coupling to the electromagnetic fields that exist in condensed matter is responsible for the long-range forces that explain the existence of the condensed state. These forces are additive to the short-range static attractions that come into play only after the molecules have been brought sufficiently close together by the attraction of the long-range force.

Since the beginning of the last century, it has been widely accepted that the attractive forces that act between two neutral atoms or molecules separated by a distance that is large in comparison with their dimensions (Van der Waals Forces) are electromagnetic in origin. A theory of long-range attractive forces in condensed matter was first proposed in 1955 by Lifshitz, using Quantum Field Theory. He proposed that "the interaction between bodies is considered to take place through a fluctuating electromagnetic field" [14]. In consideration of these fluctuations, Walther Nernst suggested the possibility of tuning the fluctuations of all the components of a system together in order to produce the appearance of a common phase. Several decades later, QED allowed it to be shown that fluctuating electromagnetic fields induce the appearance of a coherent state among atoms or molecules that co-resonate, thereby giving rise to spatial regions known as Coherence Domains (CD) that are characterised by components triggered by a fluctuating electromagnetic field having the same frequency and phase.

At temperatures other than T=0, electrodynamic interaction is counterbalanced by thermal collisions, in which components are extracted from a coherent state, and a dynamical distribution between the two phases Fc, Fnc of coherent and non-coherent molecules is thereby established, depending on temperature: $Fc(T) + Fnc(T) = 1$. According to this view, each atom or molecule belongs to one of the two fractions in a dynamical sense.

However, it is quite clear that the boundary between the two phases represents an interesting discontinuity surface that merits a more detailed investigation.

The particular case of water deserves special consideration. As shown in [15, 16] , the excited state of the coherent oscillation of water molecules (12.06 eV) lies just below the ionisation threshold (12.60 eV). This leads to the appearance of an almost delocalised, quasi-free electron for each molecule, which contributes to an easily excitable reservoir of electrons associated with each CD and accessible from outside the surface, see Fig.4.

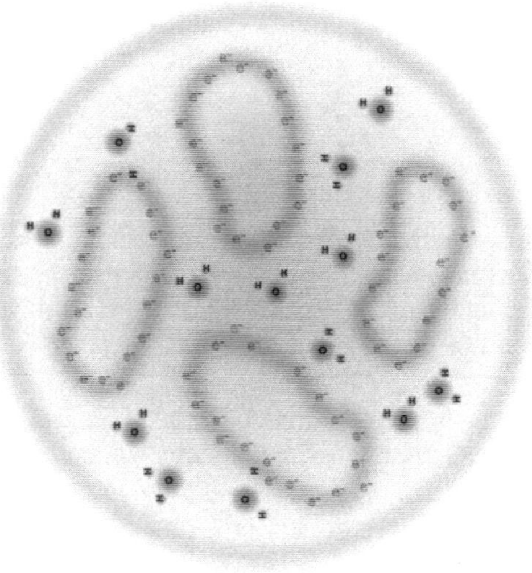

Fig.4 Pictorial representation of the two components model of water. The dimensions of the physical objects shown have just an imaginative meaning.

A number of observations, especially those made by biologists, seem to point to the existence of two kinds of water-water bonds that coexist as microdomains throughout the liquid range of temperatures [17, 18]. Furthermore, the dependence of the Raman [19], IR [20] and x-ray absorbtion spectra [21] on temperature seem also to indicate the existence of HB configurations that are spectrally distinguishable. In particular, IR spectra of water, in Fig. 5, provide a clear evidence for the existence of substructures, each of which may be assigned to a particular kind of water molecule. In particular, the broad band of absorbtion related to the OH stretching mode (ranging from 2800 to 3800 cm^{-1}) may be satisfactorily decomposed into three Gaussians, as can be seen in Fig.6, each corresponding to a stretching mode of a particular energy. Because the OH bond is sensitive to the environment, these energies may easily be ascribed to different degrees of co-ordination of the molecules concerned. We propose that low energy mode may easily be identified with the coherent phase, while the high energy population coincides with the non-coherent fraction. The intermediate energy mode may be classified as the fraction of water molecules located on the surface of the CD, whose extension is a function of the temperature T [22].

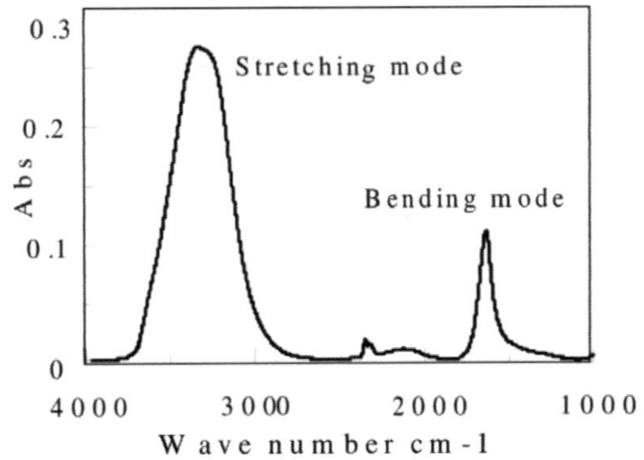

Fig.5 IR spectrum of liquid water at T=25°C.

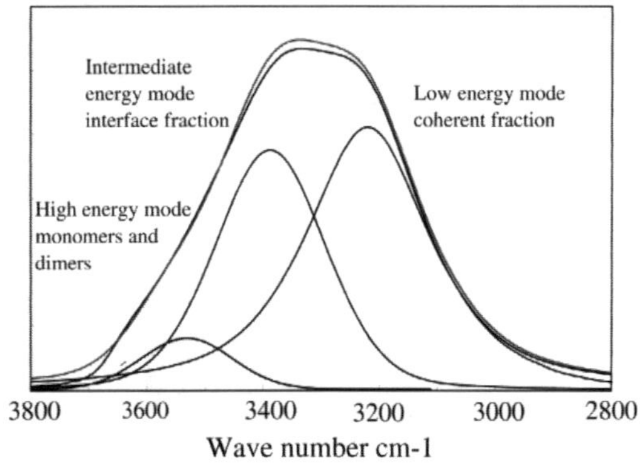

Fig.6 Deconvolution of the stretching mode area. The deconvolution has been done using the Jasco Curve Fitting Analysis software.

It may be noteworthy to ask whether an effect of the magnetic field can be observed also on the structure of liquid water.

We exposed a sample of bi-distilled water to the same EMF used on the Glutamic acid solution using the same experimental procedure. The difference between the spectra of non exposed and exposed samples, even though quite small, is clear and meaningful. In Fig. 7 the spectra are shown together with their difference amplified by a factor of 10 in order to be clearly visible on the same graph. It is interesting to observe that the field affects in a different way the three populations described above. In particular, the larger effect is on the lower energy part of the stretching absorbtion band, which is related to the coherent population which decrease increasing the fraction of the intermediate population while the "free molecules" seem not to be affected by the field in the range of the errors, see Fig. 8. The effect is very small and just a little over the errors, however we believe that it is useful to be cited as example to show the way to understand the origin of the effect of weak magnetic fields on the water solutions of amino acids and ions. Larger effect, as high as the reduction of 20% in the maximum light scattering intensity, has been observed with a pulsed (1-500Hz) low-frequency field having a magnetic field density 20 times greater than which used in our experiment [23].

The role of the magnetic field on water seems actually to be that of a "chaotrope" (disorder-maker) agent.

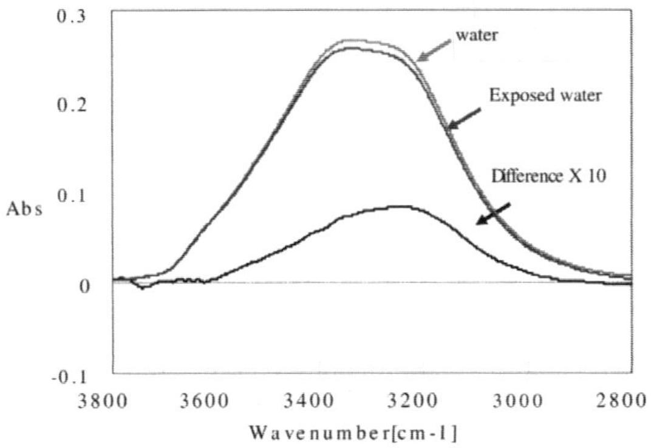

Fig. 7. IR spectra of liquid water, OH stretching band. Comparison between exposed and non-exposed sample (see text for details). The difference is plotted with an amplification factor of 10 in order to be visible on the graph.

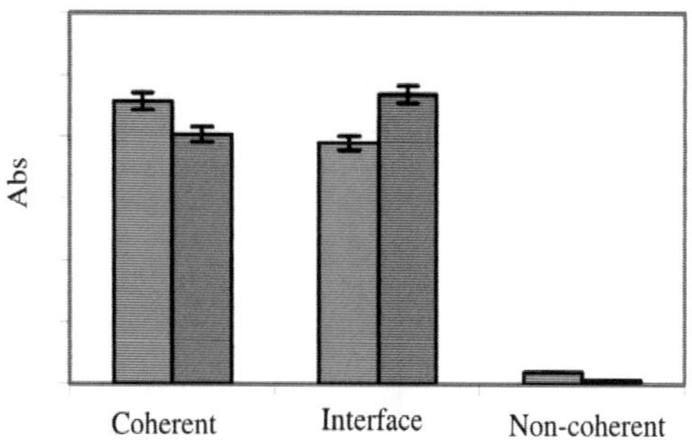

Fig.8. Comparison between the areas below the Gaussians corresponding to the different populations: coherent, intermediate and non-coherent. Blue bars refer to non exposed water; magenta bars to the exposed water

5. Conclusions

The detailed mechanism which leads to the de-protonation of the amino acid is still unclear, however we have observed how the effect of the weak EMF on the solutions of Glutamic acids is mediated by the water: the exposure to the field seems to modify the structure of water in terms of the relative abundance of the coherent and non-coherent fraction.

The amount of water molecules forced out the coherent phase increases the quantity of molecules at the boundary of the CDs, that we have called intermediate population. Such a region is the most reactive place in liquid water because of the abundance of quasi-free electrons and of the existence of the evanescent tail of the electrodynamic field which acts as the source of the ponderomotive force acting on the ions in proportion to the inverse of their masses, thereby inducing the formation of a zwitterion form of water $H \equiv O - H$ [24]. Because of its electrical charges, the zwitterion will be highly soluble in water, in that the hydrogen atoms in the bulk non coherent water immediately surround the negatively charged group of the zwitterion and the oxygen atoms immediately surround the positively charged group of the zwitterion, thereby thoroughly dissolving it. This process could, finally, alters the ionic product of water $[H^+][OH^-]$ and affect the speciation scheme of the Glutamic acid. A detailed exam of the dynamics of this process is needed and will be faced in a next paper.

We have shown that the consequence of a weak EMF on the Glutamic acid has a plausible origin in the disorder-maker result of the field on the coherent structures of water. A fraction of molecules is expelled by the CD under the effect of a EMF and a shift in the quantitative distribution of the Glutamic acid species occur. This provides the essential mechanism for inducing in the solution the temporary injection of ions which may, in due course, give origin to the so called "Zhadin effect".

Acknowledgments

We wish to tank Emilio Del Giudice for many bright and thought-provoking conversations.

References

[1] Zhadin M N, Novikov V V, Barnes F S, Pergola N F 1998 Combined Action Of Static And Alternating Magnetic Fields On Ionic Current In Aqueous Glutamic Acid Solution *Bioelectr.* **19** 41-45

[2] Liboff A R 1985 Geomagnetic Cyclotron Resonance in Membrane Transport *J.Biol. Phys.* **13** 99-102

[3] Pazur A 2004 Characterisation Of Weak Magnetic Field Effects in an Aqueous Glutamic Acid Solution by Nonlinear Dielectric Spectroscopy and Voltammetry *BioMagn. Res. and Tech.* **2** 8

[4] Comisso N, Del Giudice E, De Ninno A, Fleischmann M, Giuliani L, Mengoli G, Merlo F, Talpo G 2006 Dynamics of the Ion Cyclotron Resonance Effect on Amino Acids Adsorbed at the Interfaces *Bioelectr.* **27**(1) 16-25

[5] Giuliani L, Grimaldi S, Lisi A, D'Emilia E, Bobkova N, Zhadin M 2008 Action of Combined Magnetic Fields on Aqueous Solutions of Glutamic Acid: The Further Development of Investigations *BioMagn. Res. and Tech.* **6** 1

[6] Del Giudice E, Fleischmann M, Preparata G, Talpo G 2002 On The Unreasonable Effects of Elf Magnetic Fields Upon a System of Ions *Bioelectr.* **23** 522-30

[7] Barth A 2000 The Infrared Absorption of Amino Acid Side Chains *Prog. in Bioph. & Mol. Biol.* **74** 141–173

[8] De Ninno A, Congiu Castellano A 2011 Deprotonation of glutamic acid induced by weak magnetic field: An FTIR-ATR study *Bioelectr.* **32** 218-255

[9] Engel G S, Calhoun T R, Read E L, Tae-Kyu Ahn T, Mancˇal, Yuan-Chung Cheng R, Blankenship E & G. R. Fleming 2007 Evidence For Wavelike Energy Transfer Through Quantum Coherence In Photosynthetic Systems Nature **446** 782- 785

[10] Scholes G D 2010 Quantum-Coherent Electronic Energy Transfer: Did Nature Think of it First? *J.Phys. Chem. Lett.* **1** 2-8

[11] Sholes G D 2003 Long Range Resonance Energy Transfer In Molecular Systems *Annu. Rev. Phys. Chem.* **54** 57-87

[12] Preparata G QED Coherence in Condensed Matter: World Scientific New Jersey, Singapore, London, 1995

[13] Del Giudice E, Preparata G, Vitiello G 1988 Water as a free electric dipole laser *Phys. Rev. Lett.* **61** 1085

[14] Dzyaloshinskii I E, Lifshitz E M and Pitaevskii M 1961 *Sov. Phys. Uspekhi*, **73** 3-4, 153-176

[15] Arani R, Bono I Del Giudice E, Preparata G 1995 QED coherence and the thermodynamics of water *Int. J. of Modern Phys.B* **9** 1813-1833

[16] Marchettini N, Del Giudice E, Voiekov VL, Tiezzi E 2010 Water: A medium where dissipative structures are produced by a coherent dynamics *J. Theor. Bio.* **265** 511-516

[17] Chaplin M 2001 Water: its importance to Life *Biochem. and Molec. Biol. Education* **29**, 54-59

[18] Wiggins P 2008 Life depends upon two kinds of water PLos ONE, **3**(1), e1406.

[19] Walrafen G E 1972 Water: A comprehensive Treatise (Plenum, New York)

[20] Brubach J B, Mermet A, Filabozzi A, Gershel A, Roy P 2005 Signatures of the hydrogen bonding in the infrared bands of water *Journal of Chem. Phys.* **122**, 184509

[21] Wernet Ph et al. 2004 The Structure of the First Coordination Shell in liquid water *Science* **304** 995-999

[22] De Ninno A, Congiu Castellano A 2011 submitted to *J. of Mol. Struc.*

[23] Vallée P et al. 2005 Effects of pulsed low-frequency electromagnetic fields on water characterized by light scattering techniques: role of bubbles *Langmuir* **21** 2293-2299

[24] Voiekov V, Del Giudice E 2009 Water respiration – The Basis of the Living State *Water* **1** 52-75

Journal of Physics: Conference Series **329** (2011) 012026 doi:10.1088/1742-6596/329/1/012026

Quantum Coherence in (Brain) Microtubules and Efficient Energy and Information Transport

Nick E. Mavromatos
King's College London, Physics Department, Strand, London WC2R 2LS, UK;
CERN. Theory Division, CH-1211 Geneva 23, Switzerland
E-mail: nikolaos.mavromatos@kcl.ac.uk

Abstract. Prompted by recent experimental results in marine algae, indicating quantum entanglement at ambient temperature, with correlations between essential biological units separated by distances as long as 20 Angströms and decoherence times, due to environmental influences, of order 400 fs, I review here a related topic proposed several years ago in connection with the possible rôle of quantum mechanics and/or field theory on dissipation-free energy transfer in (brain) microtubules (MT). The basic assumption was to view the cell MT as quantum electrodynamical cavities, providing sufficient isolation *in vivo* to enable the formation of electric-dipole quantum coherent solitonic states across the tubulin dimer walls. Crucial to this, were argued to be the electromagnetic interactions of the dipole moments of the tubulin dimers with the dipole quanta in the ordered water interiors of the MT, that play the rôle of quantum coherent cavity modes. Quantum entanglement between tubulin dimers was argued to be possible, provided there exists sufficient isolation from other environmental cell effects. Thus, decoherence times as long as $10^{-7} - 10^{-6}$ s could characterise the MT systems. The model was based on certain ferroelectric aspects of MT. In the talk I revisit these decoherence time scales in light of the algae measurements and argue that, even if the environmental decoherence implies short time scales of order of a few hundreds of fs, this is a sufficient time for some kind of quantum computation to take place in (brain) MT, so that within these time scales the cell "quantum calculates" the optimal "path" along which energy and signal (information) are transported most efficiently along the MT.

1. On the rôle of Quantum Mechanics in Biology: fiction or fact?

It is a common perception that Quantum Mechanics (QM) pertains to the *small* (*micro*scopic) and *cold*, whilst classical physics affects the large (*macro*scopic) and complex systems, usually embedded in relatively *hot environments*.

Elementary particles are from the above point of view the best arena for studying quantum effects, and this has lead to important discoveries regarding the structure of our Universe at microscopic scales, of length size less than 10^{-18} m [1]. However, there are important examples from condensed matter physics where quantum effects manifest themselves at relatively *large distances* and/or *high temperatures*. A famous example is the superconductivity phenomenon [2], where formation of electron pairs exhibiting quantum coherence at macroscopic scales of order of a few thousand Angströms (*i.e.* at distances three orders of magnitude larger than the atomic scale) is responsible for electric-current transport virtually dissipation free. Subsequently, high-temperature superconductors, with critical temperatures for the onset of superconductivity up to

140 K (albeit with coherence length of order of a few Angströms), have also been discovered [3]. In ref. [5] the authors describe the *macroscopic* entanglement of two samples of Cs atoms (containing more than 10^{12} atoms) at *room temperature.* Quite recently, it has also been demonstrated experimentally that in certain polymer chains, and under certain circumstances, one may observe quantum phenomena associated with intrachain (but not interchain) coherent electronic energy transport at *room temperature* [4].

A natural question, therefore, which comes to one's mind is whether similar quantum phenomena may occur in biological systems, which are certainly complex, relatively large (compare to atomic physics scales) entities, living at room temperatures. In fact this is an old question, dating back to Schroedinger [6], who in his famous 1944 Book entitled "*What is life*", attempted to argue that certain aspects of living organisms, such as mutations (changes in the DNA sequence of a cell's genom or a virus), might not be explainable by classical physics but required quantum concepts, for instance quantum leaps.

Several years later, H. Fröhlich [7] have suggested that macroscopic quantum coherent phenomena may be responsible for dissipation-free energy and signal transfer in biological systems through coherent excitations in the microwave region of the spectrum due to nonlinear couplings of biomolecular dipoles. The frequency with which such coherent modes are 'pumped' in biological systems was conjectured to be of order

$$t_{\text{coherence Froehlich}} \sim 10^{-11} - 10^{-12} \text{ s} . \tag{1}$$

which is known as Fröhlich's frequency.

Soon after, A.S. Davydov [8], proposed that *solitonic excitation states* may be responsible for dissipation-free energy transfer along the α-helix self-trapped amide in a fashion similar to superconductivity: there are two kinds of excitations in the α-helix: deformational oscillations in the α-helix lattice, giving rise to quantized excitations ("phonons"), and internal amide excitations. The resulting non-linear coupling between these two types of excitations is a Davydov soliton, which traps the vibrational energy of the α-helix and thus prevents its distortion (solitons are classical field theory configurations with finite energy).

In a rather different approach, F. Popp [9] suggested that studies of the statistics of counts of photons in ultra-weak bioluminescence in the visible region of the spectrum point towards the existence of a coherent component linked to the living state. There have also been attempts [10] to link the mechanisms for quantum coherence in Biology suggested by Fröhlich and Popp.

In the 1990's a suggestion on the rôle of quantum effects on brain functioning, and in particular on conscious perception, has been put forward by S. Hameroff and R. Penrose [11], who concentrated on the microtubules (MT) [12] of the brain cells. In particular, they noted that one may view the tubulin protein dimer units of the MT as a quantum two-state system, in coherent superposition. The model of [11] assumes, without proof, that sufficient environmental isolation occurs, so that the *in vivo* system of MT in the brain undergoes *self-collapse*, as a result of sufficient growth that allowed it to reach a particular threshold, namely a critical mass/energy, related to *quantum gravity* (orchestrated reduction method). This type of collapse should be distinguished from the standard *environmental decoherence* that physical quantum systems are subjected to [13]. In this way, the authors of [11] argue that decoherence times of order O(1 s), which is a typical time for conscious perception, may be achieved, thereby deducing that consciousness is associated with *quantum computations* in the mind.

Unfortunately, in my opinion, environmental decoherence, even for *in vivo* MT, cannot be ignored. In a series of works, which I will review below [14, 15], we have developed a *quantum electrodynamics cavity model* for MT, in which electromagnetic interactions between the electric dipole moments of the tubulin protein dimer units and the corresponding dipole quanta in the (thermally isolated) water interiors of the *in vivo* MT, are argued to be the dominant forces, inducing environmental entanglement and eventual decoherence [13] in at most $O(10^{-6} - 10^{-7})$ s.

Such times are much shorter than the required time scale for conscious perception, but have been argued to be sufficient for *dissipation-less* energy transfer and signal transduction along moderately long MT of length sizes of order $\mu m = 10^{-6}$ m. As I will discuss below, the basic underlying mechanism is the formation of appropriate *solitonic* dipole states along the protein dimer walls of the MT, which are reminiscent of the quantum coherent states in the Fröhlich-Davydov approach. We have also speculated [15] that under sufficient environmental isolation, which however is not clear if it can be achieved in *in vivo* MT systems, these coherent states may provide the basis for an operation of the MT as quantum logic and information teleporting gates. At any rate, our main concern in the above works was the search for, and modeling of, possible quantum effects in cell MT which may not be necessarily associated with conscious perception. In fact in this talk I will disentangle the latter from dissipation-free energy and signal transfer in biological matter, which I will restrict my attention to.

All the above are so far mere speculations. Until recently, there was no strong experimental evidence (if at all) to suggest that macroscopic quantum coherent phenomena might have something to do with the living matter. There were of course consistency checks with such assumptions, as is, for instance, the work of [16], which, by performing experiments on the brain of *Drosophila*, provided consistency checks for the cavity model of MT [14, 15], especially on its possible rôle for brain memory function. Nevertheless, despite the interesting and quite delicate nature of such experiments, one could not extract conclusive experimental evidence for quantum aspects of the brain.

To such skepticism, I would also like to add the fact that some theoretical estimates on the environmental decoherence time in brain microtubules, performed by Tegmark [17], in a model-independent way (which, however, as I will discuss below, is probably misleading), place the relevant decoherence MT time scales in the range

$$t_{\rm decoh\ MT\ estimate} \in 10^{-20} - 10^{-13}\ {\rm s}\ , \qquad (2)$$

depending on the specific environmental source. Such estimates made the author of [17] to suggest that there is no rôle of quantum physics in the functioning of the brain. For the purposes of our talk below, I notice that it is the upper limit in (2) that has been proposed in [17] as a conservative estimate on the characteristic decoherence time for MT, the main source of decoherence being assumed to be the Ca^{2+} ions in each of the 13 microtubular protofilaments. Although I would disagree with such estimates, for reasons to be explained below, nevertheless, I will point out later in the talk that, even if a MT decoherence time scale of order 10^{-13} s is realized in Nature, this still allows for quantum effects to play a significant rôle associated with optimization of efficient energy and signal transfer.

At any rate, because of such a distinct lack of experimental confirmation so far, many scientists believed that any claim on a significant rôle of quantum physics in biology constituted science *fiction*.

2. Recent Evidence for Biological Quantum Entanglement at Ambient Temperature? - Photosynthetic Algae

The situation concerning the experimental demonstration of a concrete rôle of quantum physics on basic functions of living matter started changing in 2007, when research work on photosynthesis in plants [18] has presented rather convincing experimental evidence that light-absorbing molecules in some photosynthetic proteins capture and transfer energy according to *quantum-mechanical probability laws* instead of classical laws at temperatures up to 180 K.

Even more excitingly, in the beginning of this year, compelling experimental evidence on quantum effects on living matter at ambient temperatures was provided in ref. [19]. Using photo echo spectroscopy methods on two kinds of light-harvesting proteins, isolated appropriately from

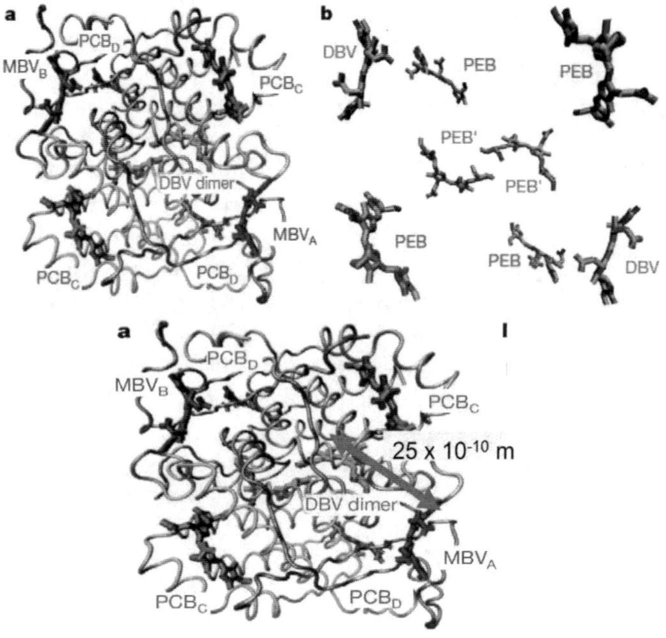

Figure 1. ***Top figures:*** **a.** Structural model of one type of *Cryptophytae Marine Algae* (CMA) protein antenna, PC645 [19]. The eight bilin molecules (Chromophores) responsible for light harvesting are indicated in various colours. **b.** The Chromophores from the structural model of the second type of CMA protein antenna studied in [19], PE545. ***Bottom figure:*** the same as in **a.** above, but with the alleged quantum-entanglement (coherent-wiring) distance of about 25 Angström between bilin molecules indicated by a red double arrow.

cryptophyte marine algae, the authors of [19] have demonstrated that there exist long-lasting electronic oscillation excitations with (quantum) correlations across the 5 nm long proteins, even at *room temperatures* of order 294 K.

More specifically, there are eight light-harvesting molecules (pigments-*Chromophores*, *i.e.* substances capable of changing colour when hit by light as a result of selective wavelength absorption) inside the protein antennae of marine algae (see fig. 1). The authors of [19] studied the electronic absorption spectrum of this complex system, and the results are indicated in fig. 2.

In the experiments, a laser pulse (indicated by a dashed line in fig. 2) of about 25 fs duration is applied to the biological entities, exciting a coherent superposition (in the form of a wave packet) of the protein antenna's vibrational-electronic eigenstates (the relevant absorption bands are indicated by coloured bars in fig. 2). The relevant theory, pertaining to the quantum evolution of a system of coupled bilin molecules with such initial conditions, predicts that the relevant excitation subsequently oscillates in time between the positions at which the excitation is localized, with distinct correlations and anti-correlations in phase and amplitude (*c.f.* fig. 3). Such coherent oscillations last until the natural eigenstates are restored due to *decoherence*, as a consequence of environmental entanglement [13]. The experimental results of [19] confirmed such a behaviour (*c.f.* fig. 4), thereby indicating a quantum superposition of the electronic structure of the bilin molecule dimer DBV (dihydrobiliverdin) for some time, which in the experiments was found to be relatively long, for a room temperature system, of order

$$t_{\rm decoh} = 400 \text{ fs} = 4 \cdot 10^{-13} \text{ s} . \tag{3}$$

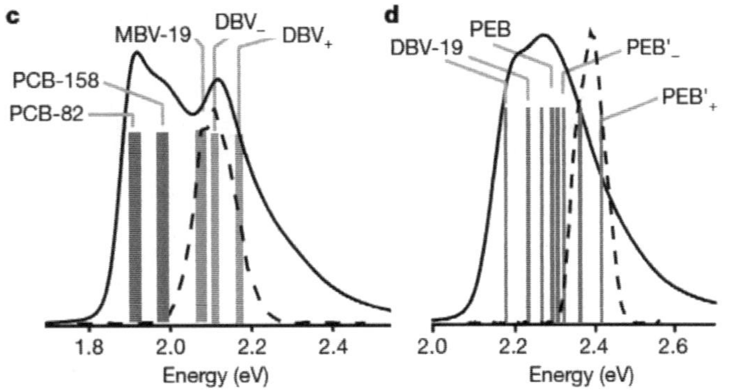

Figure 2. **c.** The approximate electronic absorption energies of the bilin molecules indicated in fig. 1 for the PC645 protein in aqueous buffer at ambient temperatures (294 K). **d.** The same but for the protein PE545, with the same external conditions (pictures taken from [19]). The externally applied laser pulse that excites the system is indicated by a dashed line. Coloured bars denote the absorption band positions.

The quantum oscillations of the DVB molecules were transmitted to the other bilin molecules in nthe complex, at distances 20 Angströms apart, as if these molecules were connected by springs. The authors of [19], therefore, suggested that distant molecules within the photosynthetic proteins are '*entangled*' together by *quantum coherence* ("coherently wired" is the used terminology) for more efficient light harvesting in marine algae. In other words, by exploiting such correlations, the biological cell '*quantum calculates*' – within the decoherence time scale (3) – which is the most efficient way and path to transport energy across macroscopically large distances of order of a few nm (*path optimization*). Some authors would interpret this behaviour as a prototype of 'quantum computation', although personally I believe we are rather far from rigorously demonstrating this.

There is an interesting question as to what guaranteed sufficient environmental isolation of the bilin molecules in the *cryptophytae* antenna studied in [19] so as to have such relatively long decoherence times (3) at room temperatures. The authors speculate that this might be due to the fact that, unlike most of photosynthetic pigments in Nature, the eight bilin molecules in this case are covalently bound to the proteins of the *cryptophytae* antenna complex.

These are, in my opinion, quite exciting experimental results that, for the first time, provide concrete evidence for quantum entanglement over relatively large distances in living matter at ambient temperature, and suggest a rather non-trivial rôle of quantum physics in path optimization for energy and information transport. Given that I have worked in the past on such issues regarding Cell Microtubules (MT), it is rather natural to revisit the pertinent theoretical models, in light of the findings of ref. [19]. It is the point of this talk, therefore, to first review the theoretical models of [14, 15] suggesting quantum coherent properties of MT, discuss the associated predictions as far as the scale of quantum entanglement and decoherence times are concerned, and attempt a comparison with the data of [19].

Caution should be exercised here. Algae light-harvesting antennae and MT are entirely different biological entities. Nevertheless, they are both highly complex protein bio-structures, and the fact that quantum effects may play an essential rôle in energy transfer in algae at room temperature, as seems to be indicated by the recent experiments, might be a strong indication that similar coherent effects also characterize energy and signal transfer in MT *in vivo*, as

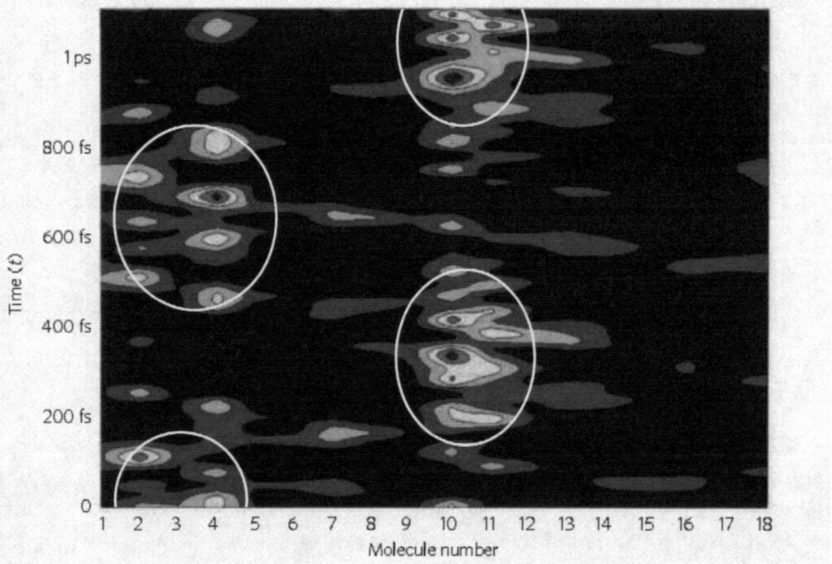

Figure 3. Theoretical Calculations of the electronic excitation dynamics in photosynthetic light-harvesting biological complexes [20]. The example is for the light-harvesting complex of the bacterium *Rhodopseudomonas acidophila*, which contains eighteen pigment molecules, arranged in a circle. Upon an external photo-stimulus, such as a laser pulse, the pigments enter excited electronic states by the absorption of photons, with coherent quantum correlations among the various pigments. The picture shows the probability of such an excitation to reside in a certain position in the complex (the dark-blue regions correspond to zero such probability, while the red-coloured regions indicate maximal probability). The vertical axis is time (in fs), while the horizontal line refers to the pigment molecules numbered 1 to 18. The white circles indicate the (small) number of molecules among which coherent oscillations of the excitation occur. In the model of [20] such oscillations have a half period of 350 fs. Similar oscillatory dynamics is observed in the algae complexes in [19] (*c.f.* fig. 4).

conjectured in [14]. At this point, to avoid possible misunderstandings, I would like to stress once more that even if this turns out to be true, it may have no implications for conscious perception or in general brain functioning [21], although, of course, such exciting possibilities cannot be excluded.

The structure of the remainder of the talk is as follows: in the next section 3, I review the basic features of the MT cavity model and discuss its predictions, especially in the light of the recent findings of [19]. In section 4, I discuss some experimental tests of the model, especially as far as ferroelectric properties are concerned, which unfortunately are not conclusive. I also discuss avenues for future experiments that could confirm some other aspects of the model, including a possible extension of the photo-experiments of [19] to MT complexes. Finally, section 5 contains our conclusions.

3. Cavity Model for Microtubules (MT) revisited: Quantum Coherence and Dissipation-Free Energy transfer in Biological Cells

Microtubules (MT) [12] are paracrystalline *cytoskeletal* structures that constitute the fundamental scaffolding of the cell. They play a fundamental rôle in the cell *mitosis* and are also believed to play an important rôle in the transfer of electric signals and, more general,

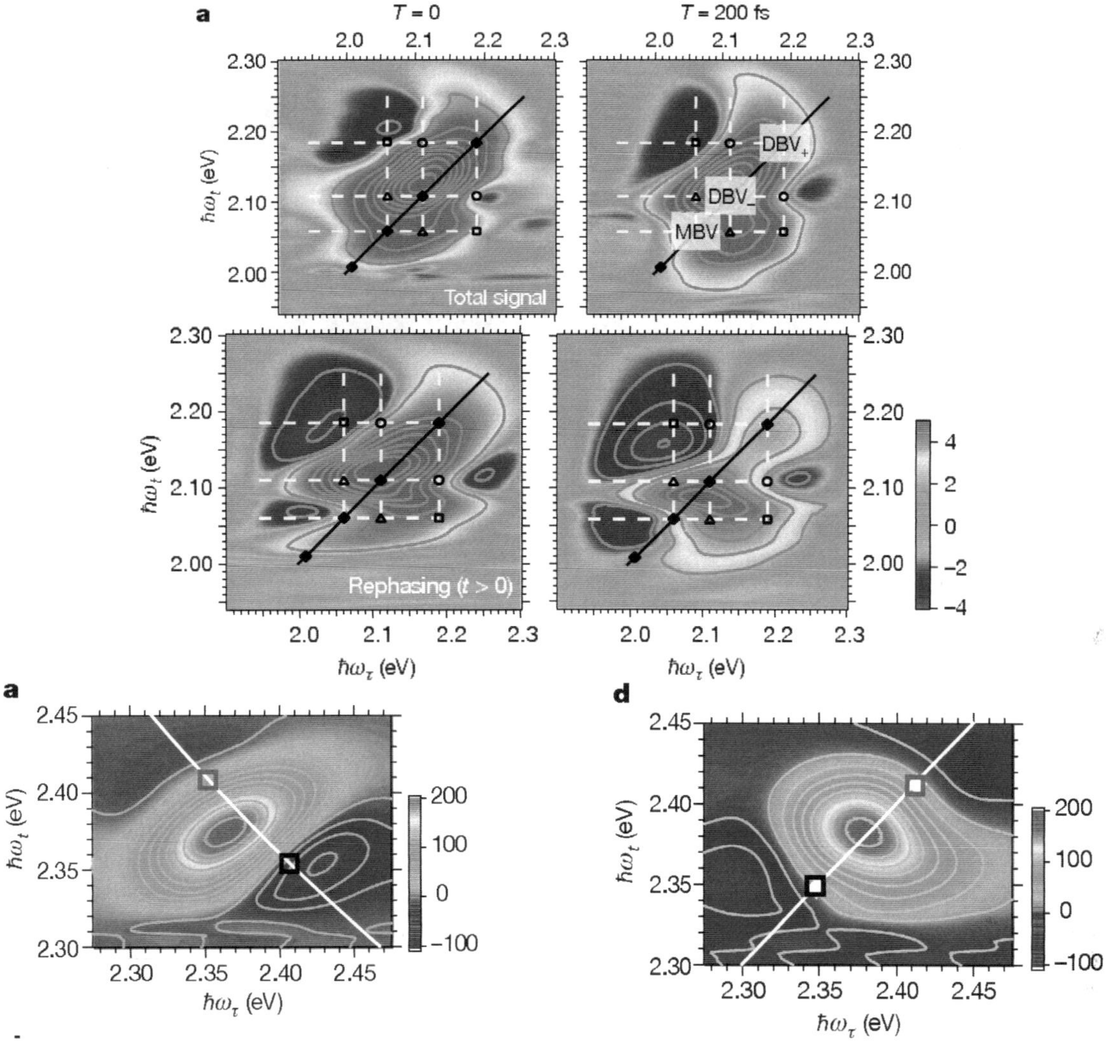

Figure 4. The two-dimensional photon echo (2DPE) data for PC645 protein of photosynthetic marine algae (upper picture) and for PE545 protein (lower picture). In the upper picture **a.**, the right column shows the data for a time of 200 fs since the excitation by the external pulse. The 2DPE data show the signal intensity in an arcsinh scale plotted as a function of the coherence frequency ω_τ and emission frequency ω_t. The lower picture shows data for times 100 fs, left picture **a.** involves re-phasing of the real signal, while in picture **d.** such a re-phasing has not been performed (pictures taken from [19]).

of energy in the cell. They are cylindrical structures (*c.f.* fig. 5) with external cross section diameter of about 25 nm and internal diameter 15 nm. A moderately long MT may have a length of the order of a few $\mu\text{m} = 10^{-6}$ m. Their exterior walls consist of tubulin protein units (*c.f.* fig. 6). The tubulin protein dipers are characterized by two hydrophobic pockets, of length $4 \text{ nm} = 4 \cdot 10^{-9}$ m each (the total length of a dimer being about 8 nm), and they come in two conformations, $\alpha-$ and $\beta-$ tubulin, depending on the position of the unpaired charge of 18 e relative to the pockets.

The tubulin also has an electric dipole moment. A complete electron microscope chartography

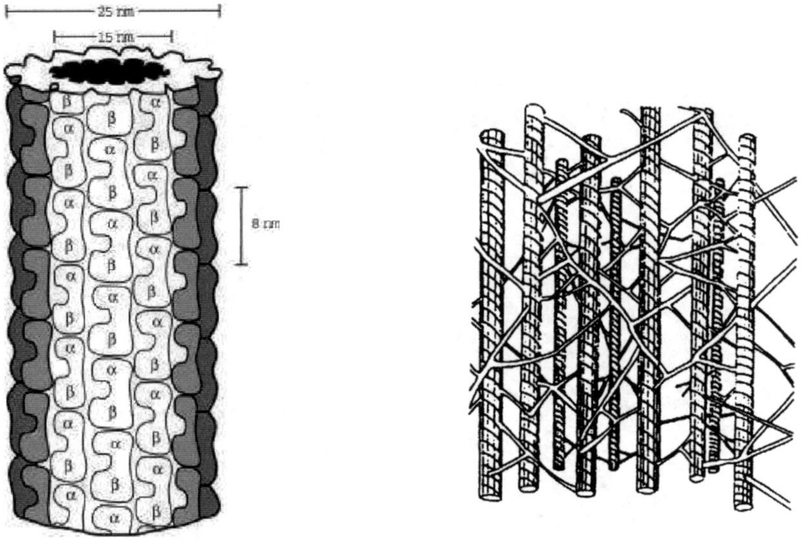

Figure 5. ***Left Picture***: A Microtubule arrangement in the cell. The exterior walls consist of tubulin protein dimers which are arranged in 13 protofilaments (vertical chain-like structures, parallel to the main axis of MT). The interiors are full of ordered water molecules. ***Right Picture:*** A network of Microtubules in a cell. The subtance connecting the various MTs together are the Microtubule Associated Proteins (MAP) (figures from ref. [22]).

of the tubulin protein dimer is available today at 3.5 Angström resolution [23]. This allows for theoretical modelling and computer calculations of the electric dipole moment of the dimers as well as of the entire MT [24, 25]. Current simulations have shown that the bulk of the tubulin's electric dipole moment lies on an axis *perpendicular* to the protofilament axis of the MT and only a fifth of the total tubulin dipole moment lies parallel to it (see also discussion in section 4). However, for the purposes of constructing a (rather simplified) model of MT dynamics [22, 14] that captures the essential features of dissipation-free energy and information transfer, it suffices to observe that the two conformations of the tubulin dimer differ by a relative angle (of about 29^0) relative to the protofilament axis in the monomers orientation. This will have implications on the electric dipole moment of the monomer, as indicated in fig. 6. In this simplified picture, one ignores the components of the electric dipole perpendicular to the protofilament's axis, and concentrates rather on a description of the array of the dipole oscillators along the MT protofilaments by a single effective degree of freedom, namely the projection, u_n, on the MT cylinder's axis of the displacement of the n-th tubulin monomer from its equilibrium position. The strong uniaxial dielectric anisotropy of the MT supports this picture, which enables one to view the MT as one-space dimensional crystals.

As we shall discuss below, this rather simplified geometry captures essential features of the MT, insofar as soliton formation and dissipation-free energy transfer are concerned. It is understood, though, that microscopic detailed simulations of the complete MT, which recently started becoming available [26], should eventually be used in order to improve the theoretical modelling of MT dynamics [22, 14] and allow for more accurate studies of their possible quantum entanglement aspects.

Journal of Physics: Conference Series **329** (2011) 012026 doi:10.1088/1742-6596/329/1/012026

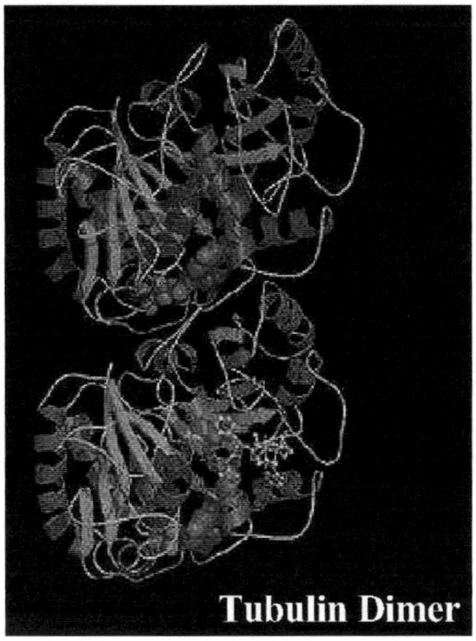

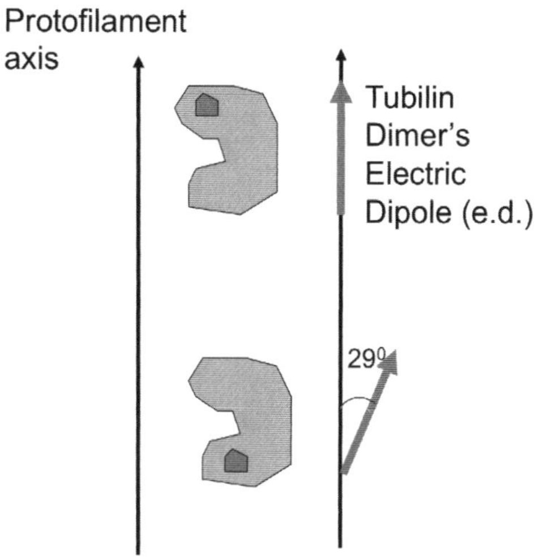

Figure 6. ***Left Picture***: Electron Microscopy chartography of tubulin dimer at 3.5 Angström resolution [23]. ***Right Picture:*** Schematic view of the two conformations α, β and the position (relative to the MT axis) of the electric dipole moment. The two conformations arise from the position of the unpaired electric charge (red polygon) relative to the two hydrophobic pockets of the tubulin dimer [12]. In simplified models, the two monomer conformations differ by a relative displacement of the monomer's electric-dipole orientation by an angle of about 29^0 relative to the protofilament axis. More complicated geometries for the permanent electric dipole moment of the tubulin dimer may occur in realistic systems, according to detailed simulations [24, 25], in which the bulk of it lies in a direction perpendicular to the main symmetry axis of a MT.

3.1. Classical Solitons in MT and dissipation-free energy transfer

Based on such ingredients, the authors of [22] have attempted to discuss a *classical physics* model for dissipationless energy transfer across a MT, by conjecturing ferroelectric properties for MTs at room temperatures, and thus describing the essential dynamics by means of a lattice ferroelectric-ferrodistortive one-spatial dimensional chain model [27].

In the approach of [22], as mentioned previously, the relevant degree of freedom was the displacement vector $\vec{u}$ arising from the projections of the electric dipole moments of the $\alpha-$ and $\beta-$ tubulin conformations onto the protofilament axis. As a result of the inter and intra-chain interactions, this vector may well be approximated by a *continuously* interpolating variable, at a position x along a MT protofilament, which at time t has a value

$$u(x,t) \tag{4}$$

The time dependence is associated with the dipole oscillations in the dimers.

The relevant continuum Hamiltonian, obtained from the appropriate Lattice model [27] reads then [22]:

$$H = kR_0^2(\partial_x u)^2 - M(\partial_t u)^2 - \frac{A}{2}u^2 + \frac{B}{4}u^4 + qEu \ , \quad A = -|\text{const}|\,(T - T_c) \ , \tag{5}$$

where the critical temperature T_c for the on-set of ferroelectric order is assumed at room temperatures. k is a stiffness parameter, R_0 is the equilibrium lattice spacing between adjacent

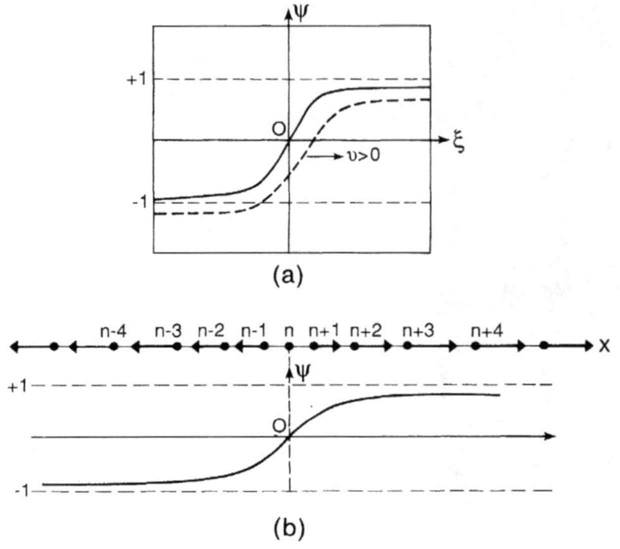

Figure 7. The kink soliton solution (8) of friction equation (9), along a protofilament (x-) axis of a MT, which is a travelling wave with velocity v along the x direction. The points n in the lower picture denote dimer positions in an one-dimensional lattice chain, representing the protofilament tubulin dimers (from ref. [22]).

dimers and the term linear in u is due to the influence of an external electric field of intensity $\vec{E}$, assumed parallel to the protofilament axis (x axis for concreteness in the model we are discussing); $q = 36e$ is the mobile charge (the reader should recall that in MT there is an unpaired charge 18 e in each dimer conformation) and M is the characteristic mass of the tubulin dimers. The simple double-well u^4 non-linear potential terms in (5) have been assumed in [22] to describe inter-protofilament interactions. In [14, 15] we have generalized such potential terms to arbitrary polynomial terms $V(u)$ of a certain degree.

The presence of ordered water in the MT interior (fig. 5) is approximated in this approach by the *addition* to the *equations of motion* derived from the Hamiltonian (5) of a friction term, linear in the time derivative of u, with a phenomenological coefficient γ:

$$M\frac{\partial^2 u(x,t)}{\partial t^2} - kR_0^2\frac{\partial^2 u(x,t)}{\partial x^2} - Au + Bu^3 + \gamma\frac{\partial u(x.t)}{\partial t} - qE = 0. \tag{6}$$

As is well known from mechanics, when friction is present, the standard lagrangian formalism breaks down, unless one enhances the degrees of freedom of the system to include the dynamics of the environment, in order to provide a microscopic dynamical description of the friction. We shall come to this point later on. For the moment, we take into account the presence of the ordered water environment and its effects on the dimers merely by the above-mentioned friction term in eq. (6). This friction should be viewed as an environmental effect, which however does not lead to energy dissipation, as a result of the formation of non-trivial solitonic structures in the ground-state of the system and the non-zero constant force due to the electric field. This is a well known result, directly relevant to energy transfer in biological systems [28]. Indeed, equation (6) admits as a unique bounded solution a *kink soliton*, acquiring the form of a travelling wave [22] (see fig. 7) [1]:

$$\psi(\xi) = a + \frac{b-a}{1+e^{\frac{b-a}{\sqrt{2}}\xi}} , \tag{8}$$

$$\psi \equiv \frac{u(\xi)}{\sqrt{A/B}} , \quad \xi = \alpha(x-ut) , \quad \alpha \equiv \sqrt{\frac{|A|}{M(v_0^2-v^2)}} ,$$

with $\psi(\xi)$ satisfying the equation (primes denote derivatives with respect to ξ):

$$\psi'' + \rho\psi' - \psi^3 + \psi + \sigma = 0 , \qquad \rho \equiv \gamma v[M|A|(v_0^2 - v^2)]^{-\frac{1}{2}}, \qquad \sigma = q\sqrt{B}|A|^{-3/2}E , \tag{9}$$

and the parameters b, a and d are defined as:

$$(\psi - a)(\psi - b)(\psi - d) = \psi^3 - \psi - \left(\frac{q\sqrt{B}}{|A|^{3/2}}E\right) \tag{10}$$

The quantity

$$v_0 \equiv \sqrt{k/M}R_0 . \tag{11}$$

is the "sound" velocity, which is of order 1 km/s for the system at hand [22].

The kink (8) propagates along the protofilament axis with fixed velocity

$$v = v_0[1 + \frac{2\gamma^2}{9d^2M|A|}]^{-\frac{1}{2}} \tag{12}$$

This velocity depends on the strength of the electric field E through the dependence of d on E via (10). Notice that, due to *friction*, $v \neq v_0$, and this is essential for a non-trivial second derivative term in (9), necessary for wave propagation. For realistic biological systems $v \simeq 2$ m/s (although under certain circumstances it may even be of order 20 m/s) [22]).

With a velocity of this order, the travelling waves of kink-like excitations of the displacement field $u(\xi)$ transfer energy along a moderately long microtubule of length $L = 10^{-6}m$ in about

$$t_T = 5 \times 10^{-7}\text{s} . \tag{13}$$

The total energy of the kink solution (8) is easily calculated to be:

$$E = \frac{1}{R_0}\int_{-\infty}^{+\infty} dxH = \frac{2\sqrt{2}}{3}\frac{A^2}{B} + \frac{\sqrt{2}}{3}k\frac{A}{B} + \frac{1}{2}M^*v^2 \equiv \Delta + \frac{1}{2}M^*v^2 \tag{14}$$

and is *conserved* in time. The 'effective' mass M^* of the kink is given by

$$M^* = \frac{4}{3\sqrt{2}}\frac{MA\alpha}{R_0B} \tag{15}$$

[1] In [14, 15] we have generalized the double-well u^4 potential terms to arbitrary polynomial terms $V(u)$ of a certain degree. In the mathematical literature [29] there has been a classification of solutions of friction equations (9) with generalized potentials $V(u)$. For certain forms of the potential [14, 15] the solutions include *kink solitons* with the more general structure:

$$u(x,t) \sim c_1 \left(\tanh[c_2(x - vt)] + c_3\right) \tag{7}$$

where c_1, c_2, c_3 are constants depending on the parameters of the dimer lattice model. For the double-well potential of [22], of course, the soliton solutions reduce to (8).

The first term of equation (14) expresses the binding energy of the kink and the second the resonant transfer energy. In realistic biological models the sum of these two terms, of order of 1 eV, dominate over the third term [22]. On the other hand, the effective mass in (15) is of order $5 \times 10^{-27} kg$, which is about the proton mass ($1GeV$) (!).

This amount of energy (14) is then transferred across a moderately long MT in a time scale (13), virtually *free* from *dissipation*. It should be noted at this stage that the above one-dimensional soliton structures may be embedded in two dimensional *helical* structures, along the MT cylinders, without significant changes as far as the propagation characteristics are concerned.

The above *classical* kink-like excitations (8) have been discussed in [22] in connection with physical mechanisms associated with the hydrolysis of GTP (Guanosine-ThreePhosphate) tubulin dimers to GDP (Guanosine-DiPhosphate) ones.

However, because the two forms of tubulins correspond to different conformations α and β, it is conceivable to speculate that the *quantum mechanical oscillations* between these two forms of tubulin dimers might be associated with a *quantum version* of kink-like excitations in the MT network, in which the solution (8) is viewed as a *macroscopic coherent state*. In such a picture, this state would be itself the result of *incomplete environmental decoherence* [13]. Indeed, it is known that certain quantum systems, with specific Hamiltonian interactions with the environment, undergo *incomplete* localization of their quantum states, in the sense that localization stops before it is complete. In such cases, a quantum coherent ('*pointer*') state is formed as a result of decoherence [30]. This is argued to be the case in the so-called cavity model of MT proposed in [14], which we now come to review.

3.2. Quantization of the Soliton Solutions

Mathematically speaking, a semiclassical quantization of solitonic states of the form (8) (or more generally (7)), has been considered in [14] in a way independent of any microscopic model (for the water-induced friction).

To this end, one assumes the existence of a canonical second quantized formalism for the $(1+1)$-dimensional scalar field $u(x,t)$, based on creation and annihilation operators $a_k^\dagger$, a_k. One then constructs a squeezed vacuum state [31]

$$|\Psi(t)>= N(t)e^{T(t)}|0> \quad ; \quad T(t) = \frac{1}{2}\int\int dxdyu(x)\Omega(x,y,t)u(y) \tag{16}$$

where $|0>$ is the ordinary vacuum state annihilated by a_k, and $N(t)$ is a normalization factor to be determined. $\Omega(x,y,t)$ is a complex function, which can be split in real and imaginary parts as

$$\begin{aligned}\Omega(x,y,t) &= \frac{1}{2}[G_0^{-1}(x,y) - G^{-1}(x,y,t)] + 2i\Pi(x,y,t)\\ G_0(x,y) &= <0|u(x)u(y)|0>\end{aligned} \tag{17}$$

The squeezed coherent state for this system can be then defined as [31]

$$|\Phi(t)>\equiv e^{iS(t)}|\Psi(t)> \quad ; \quad S(t) = \int_{-\infty}^{+\infty} dx[D(x,t)u(x) - C(x,t)\pi(x)] \tag{18}$$

with $\pi(x)$ the momentum conjugate to $u(x)$, and $D(x,t)$, $C(x,t)$ real functions. With respect to this state, $\Pi(x,t)$ can be considered as a momentum canonically conjugate to $G(x,y,t)$ in the following sense

$$< \Phi(t)| - i\frac{\delta}{\delta\Pi(x,y,t)}|\Phi(t)>= -G(x,y,t) \tag{19}$$

The quantity $G(x, y, t)$ represents the *modified boson field* around the soliton. From the expression (18), one may also identify $C(x, t)$ with the dynamical field representing solitonic excitations, which in our case are the quantum-corrected solitons of the dipole configurations of the tubulin dimers in the MT wall,

$$u_q(x, t) = C(x, t) \ . \tag{20}$$

To determine the functions C, D and Ω one applies the *Time-Dependent Variational Approach* (TDVA) [31] according to which

$$\delta \int_{t_1}^{t_2} dt < \Phi(t)|(i\partial_t - H)|\Phi(t) >= 0 \tag{21}$$

where H is the canonical Hamiltonian of the system. This leads to a canonical set of (quantum) Hamilton equations

$$\begin{aligned} \dot{D}(x,t) &= -\frac{\delta \mathcal{H}}{\delta C(x,t)} \qquad \dot{C}(x,t) = \frac{\delta \mathcal{H}}{\delta D(x,t)} \\ \dot{G}(x,y,t) &= \frac{\delta \mathcal{H}}{\delta \Pi(x,y,t)} \qquad \dot{\Pi}(x,y,t) = \frac{\delta \mathcal{H}}{\delta G(x,y,t)} \end{aligned} \tag{22}$$

where the quantum energy functional $\mathcal{H}$ is given by[31]

$$\mathcal{H} \equiv < \Phi(t)|H|\Phi(t) >= \int_{-\infty}^{\infty} dx \mathcal{E}(x) \tag{23}$$

with

$$\begin{aligned} \mathcal{E}(x) = \frac{1}{2}D^2(x,t) &+ \frac{1}{2}(\partial_x C(x,t))^2 + \mathcal{M}^{(0)}[C(x,t)] + \\ +\frac{1}{8} < x|G^{-1}(t)|y > &+ 2 < x|\Pi(t)G(t)\Pi(t)|y > + \frac{1}{2} lim_{x\to y}\nabla_x\nabla_y < x|G(t)|y > - \\ -\frac{1}{8} < x|G_0^{-1}|y > &- \frac{1}{2} lim_{x\to y}\nabla_x\nabla_y < x|G_0(t)|y > \ . \end{aligned} \tag{24}$$

Above we used the following operator notation in coordinate representation $A(x, y, t) \equiv < x|A(t)|y >$, and

$$M^{(n)} = e^{\frac{1}{2}(G(x,x,t) - G_0(x,x))\frac{\partial^2}{\partial z^2}} U^{(n)}(z)|_{z=C(x,t)} \quad ; \quad U^{(n)} \equiv d^n U/dz^n \tag{25}$$

The function U denotes the potential of the original soliton Hamiltonian, H. Notice that the quantum energy functional is conserved in time, despite the various time dependencies of the quantum fluctuations. This is a consequence of the canonical form (22) of the Hamilton equations.

Performing the functional derivations in (22) one obtains

$$\begin{aligned} \dot{D}(x,t) &= \frac{\partial^2}{\partial x^2} C(x,t) - \mathcal{M}^{(1)}[C(x,t)] \\ \dot{C}(x,t) &= D(x,t) \end{aligned} \tag{26}$$

which after elimination of $D(x, t)$, yields a modified soliton equation for the (*quantum corrected*) field $u_q(x, t) = C(x, t)$ (*c.f.* (20)) [31]

$$\partial_t^2 u_q(x.t) - \partial_x^2 u_q(x,t) + \mathcal{M}^{(1)}[u_q(x,t)] = 0 \tag{27}$$

with the notation

$$M^{(n)} = e^{\frac{1}{2}(G(x,x,t)-G_0(x,x))\frac{\partial^2}{\partial z^2}}U^{(n)}(z)|_{z=u_q(x,t)} \ , \quad \text{and} \quad U^{(n)} \equiv d^nU/dz^n \ .$$

The quantities $M^{(n)}$ carry information about the quantum corrections. For the kink soliton (7) the quantum corrections (27) have been calculated explicitly in ref. [31], thereby providing us with a *concrete example* of a large-scale quantum coherent *squeezed* state. The whole scheme may be thought of as a *mean-field-approach* to quantum corrections to the soliton solutions. Again we stress that the above effective one-dimensional models may be embedded into two dimensional helical structures along the MT, without significant changes qualitatively.

Having established these facts, it is interesting to attempt to formulate a microscopic model for MT, leading to the above-described solitonic coherent states. Such a model has been proposed in [14] and elaborated further in [15], where its quantum information processing aspects have been discussed. The basic ingredient of the model consists of viewing the MT as *quantum electrodynamical cavities.* We next proceed to review briefly the model's basic features.

3.3. The Quantum Cavity Model of MT

According to the Quantum Cavity Model of MT, the two conformations α and β, which, as mentioned above, differ by the positions of the unpaired electrons relative to the two hydrophobic pockets of the dimer (fig. 6), are viewed as two different states of a two-state quantum system, excitable by an external stimulus. For instance, applying an external pulse, or an electric field, excites a quantum superposition of these two states, and there are quantum oscillations between the two dimer configurations, which damp out after a finite time, as a consequence of environmental decoherence. This aspect is shared of course with the starting point of the Hemeroff-Penrose model for quantum MT [11]. However the mechanisms underlying the formation of the quantum coherent states and their eventual environmental decoherence are entirely different, as we now proceed to discuss.

In this scenario, Tubulin is viewed as a typical *two-state* quantum mechanical system, where the dimers couple to conformational changes with $10^{-9} - 10^{-11}$s transitions, corresponding to an angular frequency $\omega \sim \mathcal{O}(10^{10}) - \mathcal{O}(10^{12})$ Hz. In [14] we assumed the upper bound of this frequency range to represent (in order of magnitude) the characteristic frequency of the dimers, viewed as a two-state quantum-mechanical system:

$$\omega_0 \sim \mathcal{O}(10^{12}) \text{ Hz} \tag{28}$$

In the quantum Cavity Model, the MT are viewed as *quantum electrodynamical cavities.* In fact the *thermally isolated cavity regions* are thin interior regions of thickness of order of a few Angströms *near* the *dimer* walls, in which the electric dipole-dipole interactions between the dimers and the ordered water molecules overcome thermal losses, and provide the necessary conditions for quantum coherent states to be formed along the tubulin dimer walls of the MT [2].

Indeed, if we consider two electric dipole vectors $\vec{d_i}$, $\vec{d_j}$, at locations i and j at a relative distance r_{ij}, one pertaining to a water molecule, and the other to a protein dimer in the MT

[2] At this point we would like to point out the following. It is known experimentally [32], that in a thin exterior neighborhood of MT there are areas of atomic thickness, consisting of charged ions, which isolate the MT from thermal losses. This means that the electrostatic interactions overcome thermal agitations. It seems theoretically plausible, albeit yet unverified, that such thermally isolated exterior areas can also operate as *cavity regions*, in a manner similar to the areas interior to MT. At this point it is unclear whether there exist the necessary coherent dipole quanta in the ionic areas. Further experimental and theoretical (simulational) work needs to be done regarding this issue.

chain, then the dipole-dipole interaction has the form:

$$E_{dd} \sim -\frac{1}{4\pi\varepsilon}\frac{3(\hat{\eta}.\vec{d}_i)(\hat{\eta}.\vec{d}_j) - \vec{d}_i.\vec{d}_j}{|r_{ij}|^3} \tag{29}$$

where $\hat{\eta}$ is a unit vector in the direction of $\vec{r}_{ij}$, and ε is the dielectric constant of the medium.

First of all we note that, since each dimer has a mobile charge [12]: $q = 18 \times 2e$, e the electron charge, one may *estimate* the electric dipole moment of the dimer roughly as

$$d_{\rm dimer} \sim 36 \times \frac{\varepsilon_0}{\varepsilon} \times 1.6 \times 10^{-19} \times 4 \times 10^{-9}\ {\rm C\,m} \sim 3 \times 10^{-18}\ {\rm C\,m} = 90\ {\rm Debye}\ . \tag{30}$$

where we used the fact that a typical distance for the estimate of the electric dipole moment for the 'atomic' transition between the α, β conformations is of $\mathcal{O}(4\ {\rm nm})$, *i.e.* of order of the distance between the two hydrophobic dimer pockets. We also took account of the fact that, as a result of the water environment, the electric charge of the dimers appears to be screened by the relative dielectric constant of the water, $\varepsilon/\varepsilon_0 \sim 80$. We note, however, that the biological environment of the unpaired electric charges in the dimer may lead to further suppression of d_{dimer} in (30) [3].

Assuming that the medium between the ordered-water molecules in the layer and the MT dimers corresponds exclusively to the tubulin protein, with a typical value of dielectric constant $\varepsilon \sim 10$ [33], and taking into account the generic (conjectural) values of the electric dipole moments for tubulin dimers (30) and water molecules [14] (see discussion below eq. (31)), we easily conclude that the dipole-dipole interactions (29) may overcome thermal losses at room temperatures, $\sim k_BT$, for distances $|r_{ij}|$ of up to a few tenths of an Angström. Notice that for such distances the respective order of the energies is $\mathcal{O}(10^{-2}\ {\rm eV})$. Such thin cavities may not be sufficient to sustain quantum coherent modes for sufficiently long times necessary for dissipation-less energy transfer along the MT.

However, isolation from thermal losses could be assisted enormously by the existence of a *ferroelectric* transition below some critical temperature T_c, for the system of protein dimers in MT. In the models of [22] and ours [14] this is precisely what happens, with the critical temperature for the onset of ferroelectricity near ambient temperature

Ferroelectricity implies an effective dielectric "constant" $\varepsilon(\omega) < 1$ in (29); in such a case, these interactions can overcome thermal losses at room temperatures for up to *a few* Angströms. An additional possibility would be that for this range of frequencies a *negative* (dynamical) dielectric constant arises. This would mean that the dimer walls become *opaque* for the modes in a certain range of frequencies [14] below some critical value, thereby providing concrete support to the idea of MT behaving as isolated 'cavities', by trapping such modes. As explained in [14], such frequencies occur naturally within the range of frequencies of our model.

Inside such thin interior cavity regions, there are *quantum coherent modes*, which in [14] have been argued to be the *dipole quanta*, conjectured to exist in water in ref. [34], and discussed in this conference. These coherent modes arise from the interaction of the electric dipole moments of the water molecules with the quantized radiation of the electromagnetic field [34], which may be self-generated in the case of MT arrangements [22, 14]. The corresponding Hamiltonian interaction terms are of the form [34]

$$H_{ow} = \sum_{j=1}^{M}[\frac{1}{2I}L_j^2 + \vec{A} \cdot \vec{d}_{ej}] \tag{31}$$

[3] The reader is referred to section 4 below for further discussion on detailed simulations of the permanent electric dipole moment of the tubulin dimer. The order of magnitude of our estimates below, however, is not affected by such detailed considerations.

where $\vec{A}$ is the quantized electromagnetic field in the radiation gauge, M is the number of water molecules, L_j is the total angular momentum vector of a single molecule, I is the associated (average) moment of inertia, and d_{ej} is the electric dipole vector of a single molecule, $|d_{ej}| \sim 2e \otimes d_e$, with $d_e \sim 0.2$ Angström. As a result of the dipole-radiation interaction in (31) coherent modes emerge, which in [34] have been interpreted as arising from the quantization of the Goldstone modes responsible for the *spontaneous breaking* of the electric dipole (rotational) symmetry. Such modes are termed '*electric dipole quanta*' (EDQ). It is our view that such modes do not characterize ordinary water, but may well arise in the ordered water in the MT interior, which is a different phase of water. Such quanta play a rôle similar to the coherent modes of quantized electromagnetic radiation in ordinary cavities.

In our cavity model for MT [14] such coherent modes play the role of '*cavity modes*' in the quantum optics terminology [35]. These in turn interact with the dimer structures, mainly through the unpaired electrons of the dimers, leading to the formation of a quantum coherent solitonic state that may extend over the entire MT, and under sufficient isolation even over the entire MT network. As mentioned above, such states may be identified [14] with semi-classical solutions of the friction equations (27). In the model of [14], such coherent, *almost classical*, states are viewed as the result of *incomplete decoherence* of the dimer system due to its interaction/coupling with the water environment [13]. Incomplete decoherence may characterize some systems, in the sense that environmentally induced decoherence stops before complete localization of the quantum state. If this happens, then such partial decoherence time scales could be identified with the time taken for the dimer solitons to form.

In [14] estimates for the *formation time* of such solitons have been given, using conformal field theory methods for the description of the dynamics of the dipoles along the (one-dimensional) protofilaments, represented as Ising spin chains. Admittedly, these are crude approximations, and thus the so-obtained formation time estimates may not be accurate. For completebess we mention that such formation times of the solitons along the dimer walls may be of order 10^{-10} s, although smaller times cannot be excluded. This is the time scale over which solitonic coherent pointer states in the MT dimer system are formed ('*pumped*'), according to our scenario, which is not far from the originally assumed Fröhlich's coherence time scale (1). Eventually, these soliton coherent states will decohere to purely classical configurations. The decoherence scales involved in this second stage can be estimated by applying standard cavity quantum optics considerations to the MT system (see fig. 8).

Indeed, the above-mentioned dimer/water coupling leads to a situation analogous to that of atoms in interaction with coherent modes of the electromagnetic radiation in *quantum optical cavities*. An important phenomenon characterizes such interactions, namely the so-called *Vacuum-Field Rabi Splitting* (VFRS) effect [36]. VFRS appears in both the emission and absorption spectra of atoms [37]. For our purposes below, we shall review the phenomenon by restricting ourselves for definiteness to the absorption spectra case.

Consider a collection of N atoms of characteristic frequency ω_0 inside an electromagnetic cavity. Injecting a pulse of frequency Ω into the cavity causes a doublet structure (splitting) in the absorption spectrum of the atom-cavity system with peaks at [36]:

$$\Omega = \omega_0 - \Delta/2 \pm \frac{1}{2}(\Delta^2 + 4N\lambda^2)^{1/2} \tag{32}$$

where $\Delta = \omega_c - \omega_0$ is the detuning of the cavity mode, of frequency ω_c, compared to the atomic frequency. For resonant cavities the splitting occurs with equal weights

$$\Omega = \omega_0 \pm \lambda\sqrt{N} \tag{33}$$

Notice here the *enhancement* of the effect for multi-atom systems $N >> 1$. The quantity $2\lambda\sqrt{N}$ is called the 'Rabi frequency' [36]. From the emission-spectrum analysis an estimate of λ can

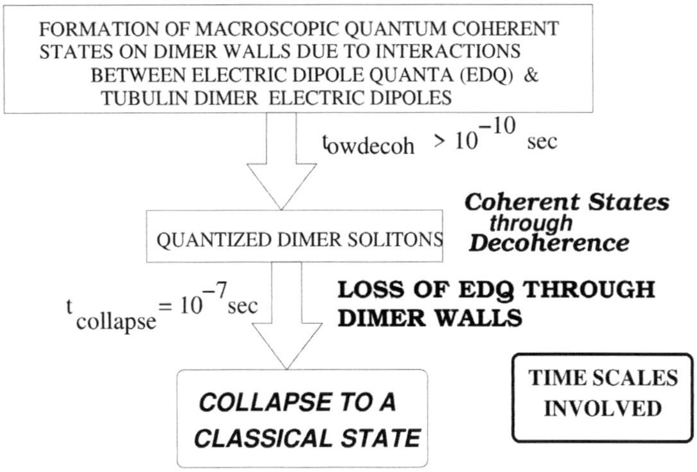

Figure 8. In the Cavity model for MT of [14], there are two stages of quantum decoherence: one involves incomplete decoherence, arising from the rôle of ordered water as an environment to the dimer system, which results in the formation of the soliton coherent state. The second, which lasts much longer, results in the complete collapse of the soliton coherent state, as a result of the loss of dipole quanta and energy through the imperfect dimer walls. Several uncertainties as to the precise origin of the incomplete decoherence exists in the model, which may shorten significantly the first time scale, below the $O(10^{-10})$ s. For dissipationless energy transfer in MT, it is the second (complete) decoherence time scale that is important.

be inferred which involves the matrix element, $\vec{d}$, of atomic electric dipole between the energy states of the two-level atom [36]:

$$\lambda = \frac{E_c \vec{d}.\vec{\epsilon}}{\hbar} \tag{34}$$

where $\vec{\epsilon}$ is the cavity (radiation) mode polarisation, and

$$E_c \sim \left(\frac{2\pi\hbar\omega_c}{\varepsilon V}\right)^{1/2} \tag{35}$$

is the r.m.s. vacuum (electric) field amplitude at the center of a cavity of volume V, and of frequency ω_c, with ε the dielectric constant of the medium inside the volume V. In atomic physics the VFRS effect has been confirmed by experiments involving beams of Rydberg atoms resonantly coupled to superconducting cavities [38].

In the analogy between the thin cavity regions near the dimer walls of MT with electromagnetic cavities, the role of atoms in this case is played by the unpaired two-state electrons of the tubulin dimers [14] oscillating with a frequency (28). To estimate the Rabi coupling between cavity modes and dimer oscillations, one should use (34) for the MT case.

In [14] we have used some simplified models for the ordered-water molecules, which yield a frequency of the coherent dipole quanta ('cavity' modes) of order [14]:

$$\omega_c \sim 6 \times 10^{12} \text{ s}^{-1} . \tag{36}$$

Notably this is of the same order of magnitude as the characteristic frequency of the dimers (28), implying that the dominant cavity mode and the dimer system are almost in resonance.

Note that this is a feature shared by atomic physics systems in cavities, and thus we can apply the pertinent formalism to our system. Assuming a relative dielectric constant of water w.r.t to that of vacuum ε_0, $\varepsilon/\varepsilon_0 \sim 80$, one obtains from (35) for the case of MT cavities:

$$E_c \sim 10^4 \text{ V/m} \tag{37}$$

Electric fields of such a magnitude can be provided by the electromagnetic interactions of the MT dimer chains, the latter viewed as giant electric dipoles [22]. This suggests that the coherent modes ω_c, which in our scenario interact with the unpaired electric charges of the dimers and produce the kink solitons (8), (27) along the chains, owe their existence to the (quantized) electromagnetic interactions of the dimers themselves.

The Rabi coupling for an MT with $\mathcal{N}$ dimers, then, is estimated from (34) to be of order:

$$\text{Rabi coupling for MT} \equiv \lambda_{MT} = \sqrt{\mathcal{N}}\lambda \sim 3 \times 10^{11} \text{ s}^{-1} , \tag{38}$$

which is, on average, an order of magnitude smaller than the characteristic frequency of the dimers (28).

In the above analysis, we have assumed that the system of tubulin dimers interacts with a *single* dipole-quantum coherent mode of the ordered water and hence we ignored dimer-dimer interactions. More complicated cases, involving interactions either among the dimers or of the dimers with more than one radiation quanta, which undoubtedly occur *in vivo*, may affect these estimates. Moreover, the use of more sophisticated models for the description of water, are in need. We hope to be able to come back to such improved analyses in the near future, especially in view of the rapid recent developments on the experimental detection of quantum coherent effects in biological systems [18, 19].

For the time being, we note that the presence of a Rabi coupling between water molecules and dimers provides a *microscopic* description for the *friction* that leads to quantum coherent solitonic states (8), (27) of the electric dipole quanta on the tubulin dimer walls. To estimate the decoherence time we remark that the main source of dissipation (environmental entanglement) comes from the imperfect walls of the cavities, which allow *leakage of coherent modes and energy.* The time scale, T_r, over which a cavity-MT dissipates its energy, can be identified in our model with the average life-time t_L of a coherent-dipole quantum state, which has been found to be [14]: $T_r \sim t_L \sim 10^{-4}$ s. This leads to a first-order-approximation estimate of the quality factor for the MT cavities, $Q_{MT} \sim \omega_c T_r \sim \mathcal{O}(10^8)$. We note, for comparison, that high-quality cavities encountered in Rydberg atom experiments dissipate energy in time scales of $\mathcal{O}(10^{-3})-\mathcal{O}(10^{-4})$ s, and have Q's which are comparable to Q_{MT} above.

Applying standard quantum mechanics of quantum electrodynamical cavities [35, 36, 38], we may express the pertinent decoherence time in terms of T_r, the Rabbi coupling λ and the number of oscillation quanta n included in a coherent mode of dipole quanta in the ordered water. The result is [14]:

$$t_{collapse} = \frac{T_r}{2n\mathcal{N}\sin^2\left(\frac{\mathcal{N}n\lambda^2 t}{\Delta}\right)} \tag{39}$$

The time t represents the 'time' of interaction of the dimer system with the dipole quanta. A reasonable estimate [14] of this time scale can be obtained by equating it with the average life-time of a coherent dipole-quantum state, which, in the super-radiance model of [39], used in our work, can be estimated as [4]

$$t \sim \frac{c\hbar^2 V}{4\pi d_{ej}^2 \epsilon N_w L} \tag{40}$$

[4] In this model the entire interior volume of the MT is taken into account. In our cavity model, however, it is only a thin cylindrical region near the dimer walls that plays the rôle of the cavity. Nevertheless, this does not affect our estimates, since the density of the relevant water molecules N_w/V is assumed uniform.

with d_{ej} the electric dipole moment of a water molecule, L the length of the MT, and N_w the number of water molecules in the volume V of the MT. For moderately long MT, with length $L \sim 10^{-6}$ m and $N_w \sim 10^8$, a typical value of t is:

$$t \sim 10^{-4} \text{ s} . \tag{41}$$

On the assumption that a typical coherent mode of dipole quanta contains an average of $n = \mathcal{O}(1) - \mathcal{O}(10)$ oscillator quanta, and for detuning between the cavity mode and the characteristic frequency of the dimer quanta of order $\Delta/\lambda \sim \mathcal{O}(10) - \mathcal{O}(100)$, the analysis of [14] yields the following estimate for the collapse time of the kink coherent state of the MT dimers due to cavity dissipation:

$$t_{collapse} \sim \mathcal{O}(10^{-7}) - \mathcal{O}(10^{-6}) \text{ s} . \tag{42}$$

This is of the same order as the time scale (13) required for energy transport across the MT by an average kink soliton in the models of [22]. The result (42), then, implies that quantum physics is relevant as far as *dissipationless* energy transfer across the MT is concerned.

In view of this specific model, we are therefore in stark *disagreement* with the conclusions of Tegmark in [17], *i.e.* that *only* classical physics is relevant for studying the energy and signal transfer in biological matter. Tegmark's conclusions did not take proper account of the possible isolation against environmental interactions, which seems to occur inside certain regions of MT with appropriate geometry and properties. The latter can screen decoherening effects in the immediate environment of an MT, for instance those due to neighboring ions used in [17] to arrive at the estimate (2) for the decoherence time of MT. We also note that the arguments of Tegmark have also been criticised in [40] within the framework of the Orchestrated-Reduction model of Hameroff and Penrose [11], which also predicts much longer decoherence times, of order $10^{-5} - 10^{-4}$ s.

However, I must stress once again that the above estimates are very crude. We do not have a good microscopic model for the ordered water in the MT interiors, nor we understand the detailed structure of the water-dimer couplings. Above, we have ignored interactions among dimers, when we considered the dipole quanta-dimer coupling, assuming that there is a single coupling between a dipole quantum and a dimer oscillator. Moreover, there is no fundamental reason why the detuning Δ should have the order assumed. All these complicate accurate estimates of the decoherence time, and the actual situation may be very different. One needs, first of all, to develop theoretical models for the ordered water, and use sophisticated numerical simulations in order to estimate the various quantum optical characteristics of the model, such as vacuum Rabi splittings, dipole electric moments *etc.* This is not done as yet. We hope that such detailed modelling will be available in the near future. Thus, although we disagree with Tegmark's conclusion that quantum effects play no essential role in energy and information process in biological systems, such as the brain, one cannot exclude the possibility that decoherence time estimates as short as (2) may, after all, characterize MT. As I will argue below, though, this is still a long enough time interval to allow for an essential rôle of quantum physics to be played in the functioning of MT.

3.4. MT Cavity Model Parameters Revisited in Light of Algae Experiments

In this subsection I would like to re-analyze the characteristic features of the cavity model of MT in light of the recent experimental evidence on quantum coherent effects in protein antennae of marine cryptophyte algae [19]. This is essential, in the sense that the recent experiments gave us a feeling on the scale and the order of magnitude of the quantum coherent effects that appear to characterize biological complex systems at room temperatures.

First of all, let us remind the reader of the analogies between the cases of the photosynthetic algae and the MT, as far as their quantum aspects are concerned. In algae, one applies an external stimulus in the form of an external laser pulse, which excites quantum superpositions of electronic states in the protein-antennae pigments. There are coherent quantum oscillations and entanglement between pigments spatially separated at distances as long as 25 Angströms, at *room temperatures*. The eventual environmental decoherence, induced by the rest of the algae complex, damps out such oscillations. The collapse of the pertinent wavefunctions occurs within 400 fs time intervals, which sets the time scale (3) as the characteristic time scale for decoherence of this system at ambient temperatures.

In photosynthetic systems like algae, the (quantum) oscillation periods are of order of a few hundreds of fs [20]. One may naively think that such short decoherence time scales have no relevance for any interesting biophysical process. However, the authors of [19], and I tend to agree with them, speculated that such a time period is sufficient for the bio-complex of algae to *quantum compute* the most efficient, *optimal*, way for energy and information transport across the pigments.

The experiments of ref. [19], as well as the relevant simulations of such bio systems [20], constitute strong (experimental) evidence that quantum effects at ambient temperatures, playing a rôle in the system's functioning, are *facts*, which, however, characterize *some* but *not all* photosynthetic biosystems. The cryptophyte algae have the pigments co-valently bound with the rest of the complex, and this was argued in [19] to be the main reason why quantum coherence at ambient temperatures can occur, entangling parts of the complex at distances over 20 Angströms.

Tubulin dimers, on the other hand, have been theorized to be characterized by quantum superporsitions of their two conformations α and β. In this picture, the dimers constitute two-state quantum systems, oscillating between their two energy eigenstates with a frequency (28), corresponding to a period of $\sim$ 6000 fs. For comparison, we mention that the characteristic period of quantum oscillations among the photosynthetic pigments in the calculated dynamics of *Rhodopseudomonas acidophila* [20] is about 500 fs (*c.f.* fig. 3), while in the photosynthetic algae studied in [19], the period of the quantum oscillations is of order 60 fs. Presumably such quantum superpositions in tubulin can also be excited by external stimuli, such as applied (or self-generated [22, 14]) electric and magnetic fields, photon laser pulses *etc.* When the quantum oscillating dimers find themselves bound in the walls of a MT, coherence has been argued to extend over macroscopic scales at distances of order μm, across the entire MT. The cavity model of MT has been invoked in [14, 15] as a plausible microscopic model underlying such properties. The unpaired charge of the tubulin dimer constitutes the two-state quantum system, and the rest of the tubulin protein complex, as well as the ordered water environment of the MT and its C-termini appendices [12], constitute the environment in this case.

If sufficient isolation from other electrostatic effects, such as Ions in MT [17], occurs in MT, then decoherence times as long as $10^{-6} - 10^{-7}$ s can be achieved, (42). As we have seen above, for this to happen [14] one needs strong electric dipole-dipole interactions, between tubulin protein dimers and ordered water molecules inside the thin cavity regions near the dimer walls. This would provide an analogue to the strong co-valence binding of the pigments into the protein antennae of the cryptophyte algae [19].

In view of the fact that *not all* photosynthetic systems exhibit such a strong binding, and that the latter has been linked [19] to the persistence of quantum coherence for relatively long times at ambient temperature, one is tempted to conjecture that such strong dipole-dipole interactions and cavity region realization may *not* characterize *all* MT in *all* biosystems. For instance, it would be interesting from the point of view of conscious perception, if only brain MT exhibited such strong cavity effects. Unfortunately the current experimental evidence towards this is nonexistent, and it is generally believed that there is a universal structure of MT, at least as far as the ordered water environment and the basic chemistry of tubulin dimers are concerned. But one cannot exclude surprises. Moreover, one should not dismiss the possibility that potentially important differences between *in vitro* and *in vivo* properties of biological matter may occur. The latter point of view is supported by the fact that living matter constitutes a non-equilibrium system, as we have also heard in the talk by E. del Giudice in this conference, and hence many concepts and properties we are accustomed to in physics, that pertain to equilibrium situations, may not characterize systems out of equilibrium. Nevertheless, at present such non universal behavior of MT constitutes a mere speculation, lacking any experimental confirmation or evidence.

In view of the *experimental findings* of [19] on quantum decoherence times of order 400 fs, it would be interesting to examine what range of the parameters of the MT cavity model of [14] could lead to such short decoherence time scales. Of course, MT and photosynthetic algae are completely different biological entities, and there is no *a priori* reason why quantum effects, if any observed in MT, should be of the same order as in cryptophyte algae. Nevertheless, the question has a physical meaning, given in particular the fact that such short decoherence scales are of the same order as those induced by the neighboring ions in MT, (2) [17].

In our model of [14], the main source of dissipation is assumed to come from the loss of dipole quanta through the imperfect dimer walls. From (39), (40) we observe that an immediate six-order of magnitude reduction in the decoherence time could be achieved by a corresponding reduction in the time T_r over which the cavity MT dissipates its energy, which in turn would reflect on a much shorter life time of the alleged dipole quanta inside the cavity regions (40), which T_r is identified with. Such a reduction appears unnatural within the super-radiance model of [39], used in our approach. Indeed, the Rabbi coupling λ that could conceivably change, even by several orders of magnitude, given the rough way its magnitude was estimated in our work so far, appears in (39) inside a sinusoidal function, which is always less than unity. Hence, a change in the order of magnitude of λ cannot reduce the decoherence time significantly.

The only parameters of the model that could conceivably change and reduce (39) by six orders of magnitude would be: (i) the density of ordered water molecules near the dimer walls N_w/V, which in our model had so far been assumed uniform inside the MT interior regions, and (ii) the number of coherent dipole quanta. However, we find that changes of an unnatural magnitude in these two quantities would be required in order for (39) to be reduced to the order of O(400) fs. It is hard to imagine, for instance, how we can arrange for $n \sim 10^6$ dipole quanta to be contained in a coherent cavity mode in the model. The cavity model of [14], makes use of mesoscopic cavities, with n=O(10) at most, which is a natural number for dipole quanta inside the conjectured thin interior cavity regions of MT.

On the other hand, such short decoherence times in the quantum oscillations of the dimers, if observed, would indeed point towards the fact that the cavity model of MT is in trouble, although quantum mechanics could still play an important rôle. For instance, as we have already discussed, short decoherence times of order (2) could be provided by the electrostatic effects of the Ca^{2+} and other ions [17] in the C-termini and other neighborhoods of the dimers. In such a case, although the cavity model of MT would be probably invalidated, nonetheless we would still disagree with the conclusions of [17] that quantum effects play no rôle in efficient energy transfer in the (brain) cells. Following the speculations in [19], we may conjecture that such short times may be sufficient for the MT to *quantum calculate* (through the quantum entanglement of (part of) its tubulin subunits) the *optimal direction* along which energy and information would be most efficiently transported. This would be a pretty important function. In this scenario, although energy would still be transported according to classical physics, nevertheless its optimum path would have been decided by a *quantum computation*, lasting at most 10^{-13} s.

Another important aspect of the algae photosynthetic systems was the fact that they involved only a relatively small number of fundamental 'units' (8) that undergo quantum superpositions and are entangled at room temperatures. On the other hand, the MT involve a rather large number of tubulin dimers, more than 100 in each protofilament. It may actually be that in realistic situations *not all*, but only a *small subset* of relatively nearby tubulin dimers in a MT, separated by distances of up to 40 Angström, are actually "coherently wired". This could still be sufficient for the MT to *quantum calculate* the *optimal* direction for energy and signal transfer, as in algae protein antennae [19].

However, it may also be the case that the much longer –as compared to algae – decoherence time scales in the cavity model of MT, can actually be explained, in a rather counterintuitive way, by the fact that a much larger number of fundamental "units" are coherently entangled in MT. Indeed, there are situations in physics where such a phenomenon occurs [15]. As mentioned in the introduction of the talk, in ref. [5] the authors describe the macroscopic entanglement of two samples of Cs atoms at room temperature. The entangling mechanism is a pulsed laser beam and although the atoms are far from cold or isolated from the environment, partial entanglement of bulk spin is unambiguously demonstrated for 10^{12} atoms for $\sim 0.5ms$. The system's resilience to decoherence is in fact *facilitated* by the existence of a *large number of atoms*: even though atoms lose the proper spin orientation continuously, the *bulk* entanglement is not immediately lost. Quantum informatics, the science that deals with ways to encode, store and retrieve information written in qubits, has to offer an alternative way of interpreting the surprising resilience of the Cs atoms by using the idea of "*redundancy*". Simply stated, information can be stored in such a way that the logical (qu)bits correspond to many physical (qu)bits and thus are resistant to corruption of content.

Yet another way of looking at this is given in the work of ref. [41], where the authors have demonstrated experimentally a *decoherence-free quantum memory of one qubit* by encoding the qubit into the "decoherence-free subspace" (DFS) of a pair of trapped Berrilium $^9Be^+$ ions. They achieved this by exploiting a "safe-from-noise-area" of the Hilbert space for a *superposition* of two basis states for the ions, thus encoding the qubit in the superposition rather than one of the basis states. By doing this they achieved decoherence times on average an order of magnitude longer.

Both of the above works show that it is possible to use DFS, error correction and high redundancy to both store information and to keep superpositions and entanglements alive for biologically relevant times in macroscopic systems at high temperature. Thus it nay not be entirely inappropriate to imagine that in biological *in vivo* regimes, one has, under certain circumstances, such as the ones specified above, similar entanglement of tubulin/MT arrangements [15].

I believe that very interesting future experiments can be done with MT, which could shed light on the above aspects of MT as quantum devices, which presently belong to the realm of *science fiction*. Let me now discuss briefly some of such experiments.

4. Experimental Tests of the MT Cavity Model: Past and Future

I proceed now to discuss some experimental tests of the cavity model of MT presented in the previous section. I will concentrate on direct physical tests of the model, rather than expanding on its physiological or quantum information aspects. As I have already mentioned in the beginning of the talk, there are very interesting experiments, for instance, showing consistency of the model as far as its predictions on the memory function of the brain are concerned [16, 21], but I will not touch upon such aspects here. I am also not qualified to discuss experiments to study physiological/biological properties of MT and/or the tubulin dimers in general.

Some of the experiments I will describe below, require for their interpretation detailed theoretical knowledge on the structure of the tubulin dimers and MT, especially as far as their electrical properties are concerned. Computer molecular simulations for the permanent electric dipole moments of the tubulin, to be discussed briefly below, can be found in [24, 25] and are based on the structure of the tubulin dimer provided by electron-microscope studies at 3.5 Angström resolution in ref. [23]. For detailed recent molecular dynamics simulations on the structure of the MT, which may also be useful in future studies and tests of properties of MT, including our cavity model, I refer the reader to the interesting works of ref. [26]. However, for the past experiments I will discuss below, we used the simulations and tubulin structure data available up to the year of the experiments (2005).

4.1. Measurement of Electric Dipole Moments and Ferroelectric Properties Tests

An essential aspect of the model is its ferroelectric features at room temperatures, namely the induction of a permanent electric dipole moment that remains after switching off an externally applied electric field. We have performed such experiments with porcine brain MT in 2005, in collaboration with the molecular biology group of E. Unger [42]. We have examined the effects

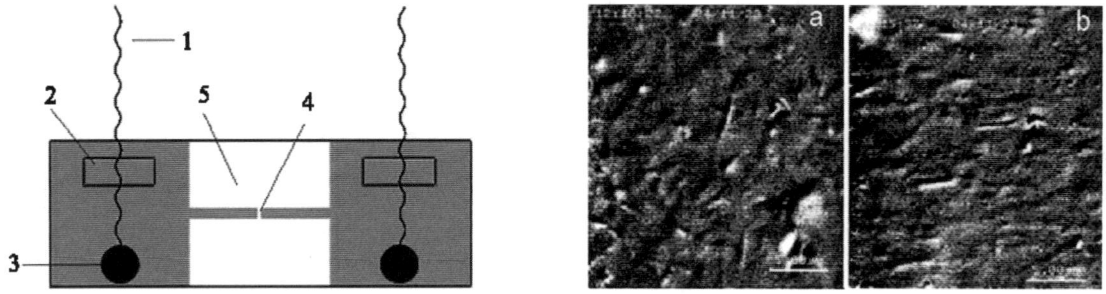

Figure 9. ***Left Picture***: Experimental arrangement for measuring the electron dipole moment of MT in suspension. The picture shows the electrode system used for the measurements in alternating electric fields. The gray areas denote gold layers. 1: thin, flexible wires for the application of the voltage; 2: strips of adhesive tape as pull relief; 3: drops of conductive silver for contacting the wires with the gold; 4: electrode gap filled with the sample of porcine brain MT; 5: gold-free glass areas. ***Middle and Right Pictures:*** **a.(Middle:)** Random positions of MT in the absence of electric field. This state is again realized after switching off the field. **b. (Right):** Alignment of MT in the direction of the externally applied alternating electric field 210,000 V/m and 2 MHz frequency. The visualization of MT has been achieved by video-enhanced DIC Microscopy (Pictures taken from ref. [42]).

of both constant (up to 2 × 103 V/m) and high-frequency alternating fields (up to 2.1 × 105 V/m, with frequencies from 200 kHz to 2 MHz) on suspended porcine microtubules. At pH 6.8 and 120 mM ionic strength, constant fields cause a motion of microtubules toward the anode (*c.f.* fig. 9).

The electrophoretic mobility amounts to 2.6×10^{-4} cm^2/V s, reflecting a negative net charge of approximately 0.2 elementary charges per tubulin dimer. The moving microtubules are randomly space oriented. Alternating high-frequency fields induce electric dipoles and align the microtubules parallel to the field direction. By determining the angular velocity of the turning microtubules, we estimate a dipole moment for the MT roughly

$$p_{\rm MT} = 34,000 \text{ Debye} \tag{43}$$

at 2.1 × 105 V/m and 2MHz frequency. By comparing the potential energy of the dipole in the applied field with the thermal energy of microtubules, we obtained a minimum value of 6,000 Debye, necessary for efficient alignment.

Unfortunately *no evidence* for permanent electric dipole has been found at ambient temperatures, where the experiment has been performed. Thus ferroelectricity has *not* been confirmed as yet. In what follows, I will attempt a comparison of these results with theoretical estimates and seek possible explanations for our inability to observe a permanent electric dipole moment, other than the straightforward dismissal of any ferroelectric properties of MT.

Computer molecular simulations of the permanent electric dipole moment of the tubulin dimers [24], taking into account their detailed structure [23], have shown that the bulk of it is directed in a direction perpendicular to the protofilament axis (x-axis) of the MT, and only about a fifth of the total electric dipole moment is along the x-axis:

$$p_x = 337 \text{ Debye} , \qquad p_y = -1669 \text{ Debye} , \qquad p_z = 198 \text{ Debye} . \tag{44}$$

Taking into account the screening effect of water (whose dielectric constant may be as high as 80) on the dimer charges, yields the following suppressed estimate for the (permanent) dimer dipole moment [25]

$$p_{\text{dimer}} = 90 \text{ Debye} . \tag{45}$$

The unpaired charges,that is, the net charges inside the hydrophobic pockets of the dimers, that appear isolated from their environment, may lead to even further suppression , that is, one may take as an order of magnitude [42]

$$p_{\text{dimer}} = 15 \text{ Debye} . \tag{46}$$

One may speculate that the other components perpendicular to the microtubule axis will be neutralized in the cylindrical microtubular geometry and screened by the environment, thereby leaving the p_x component as the dominant contribution to the total dipole moment of the microtubule. The reader is reminded at this point that such an assumption also characterizes the simplified ferroelectric one-dimensional lattice model for MT, discussed in [22] and adopted in our studies in [14].

If this is the case, then taking into account that in moderately long microtubules of an average length L = 3.5 μm with 12 protofilaments each (as used in the experiments of [42]), there are about $\mathcal{N}$ = 5280 tubulin dimers of average length 8 nm each, one would arrive at the most optimistic estimate for the total dipole moment (all dimer dipoles contributing equally to the x-direction) in the range

$$p_{\text{total}} = \mathcal{N} \, p_{\text{dimer}} = 79,200 \text{ Debye} . \tag{47}$$

This is likely to be further suppressed if details on the water and other environments and geometry are properly taken into account.

Such a suppression of the total dipole moment may provide an explanation for our inability to observe an alignment of microtubules in constant electric fields up to 2 × 103 V/m. As discussed in [42], estimates of the interaction energy of the supposed permanent dipoles in the electric fields show that this energy is too small to overcome the influence of the thermal energy. Safety requirements prevented the use of higher intensity constant fields in our experiments. On the other hand, as we already mentioned, the application of alternating fields with intensities as high as 2 × 105 V/m, at 2 MHz frequency, induce alignment of the MT along the direction of the field (*c.f.* fig. 9), from which we estimated the dipole moment (43).

The results of this experiment, therefore, probably imply that a possible permanent part of the dipole moment does not play a role for the orientation at high frequencies because the dipole cannot follow the changes of the field. Only the induced part of the dipole is responsible for the orientation, because the torque does not depend on the field direction. Comparing the potential energy of the dipole in the applied field with the thermal energy of microtubules, shows, as mentioned above, that a minimum value of the dipole moment is necessary for a successful orientation.

Thus the non-observation of ferroelectric properties in our experiment does not falsify the ferroelectric-ferrodistortive models of MT [22] on which the cavity approach is based [14, 15]. Further, refined, experiments in this direction are certainly due.

4.2. Quantum-Optics-Inspired Experiments for the Cavity Model of MT: are they feasible?

One of the most important ingredients in the approach of [14] was the rôle of the coherent cavity modes of the electric dipole quanta. Their (Rabi) dipole coupling with the dipoles of the tubulin dimers provides the necessary "friction" environment, as we discussed above, which is responsible for the soliton (coherent) states that transfer energy across the MT in a dissipation-free manner. One of the decisive tests of the MT cavity model, therefore, would be to observe such coherent modes/Rabi couplings, adapting properly the pertinent quantum optics/cavity electrodynamics experiments [38, 35] in the MT situation. For completeness, below I will discuss these experiments, with a view to see whether there is a possibility of applying them to the biological systems at hand.

In cavity quantum electrodynamics, the easiest way to observe the Rabi splitting of atoms [38], discussed briefly in subsection 3.3 above, (32), (33), is through the experimental set up indicated in fig. 10. Some two-state Rydberg atoms (or, more generally (*c.f.* fig. 10), atoms properly prepared so as to resemble two-state systems) are injected through a wave guide into a cylindrical superconducting cavity, which in the experiments of [38] is cooled down to 1.7 K. Before entering the cavity the atoms are prepared in some configuration of the hyperfine splitted $39s_{1/2}$ F=2, and F=3 states. The cavity mode couples the $39s_{1/2}$ F= 3 state to the $39p_{3/2}$ state. As a result of this coupling, upon the application of the external electric field of frequency ν, the cavity mode does not resonate at the bare atom frequency but exhibits instead two peaks in the corresponding absorption spectrum, as indicated in the lower part of fig. 10. From these peaks one clearly can compute the Rabi coupling, using (32) or (33).

However, experimentally (*c.f.* fig. 10) there are more features and structures than the two symmetric absorption peaks theory predicts: the two observed Rabi main peaks appear asymmetric and there is a much weaker third peak at frequencies around $(\omega_0 + \Delta)/2\pi$. The latter feature is clearly associated with the non-resonant coupling of the $39s_{1/2}$ F=2 atoms to the cavity modes, since $(\omega_0 + \Delta)/2\pi$ is the frequency of the $39s_{1/2}$ F=3 $\rightarrow$ $39p_{3/2}$ transition (*c.f.* fig. 10). The former are associated with the motion of the atoms, as well as the fluctuations in the atom number inside the cavity. Numerical simulations of such effects confirm the experimental results and thus provide convincing explanations of the Rabi-splitting phenomenon in realistic systems [38].

Such measurements are then used for the classification of atoms that are employed in quantum-optics experimental demonstrations of environmentally-induced decoherence effects in atomic physics [43]. In these experiments, two-state atoms in a quantum superporition are sent through isolated cavities filled by microwaves. The corresponding Rabi couplings of the atoms to the cavity coherent modes cause a shift in the phase of the microwave field, by different amounts. The experiment involves Rydberg atoms that interact one at a time with the few photon coherent modes $(O(1 - 10))$, trapped in the cavity. In this way, the field in the cavity is also put in a superposition of two states. The exchange of energy of the field with its environment, and the loss of photon coherent cavity modes through the (imperfect) cavity walls, imply decoherence and therefore the eventual collapse of the field superposition into a single definite state. The authors of [43] have observed experimentally this decoherence, while it unfolded, via the study of correlations between the energy levels of pairs of atoms sent through the cavity with various time delays between the atoms.

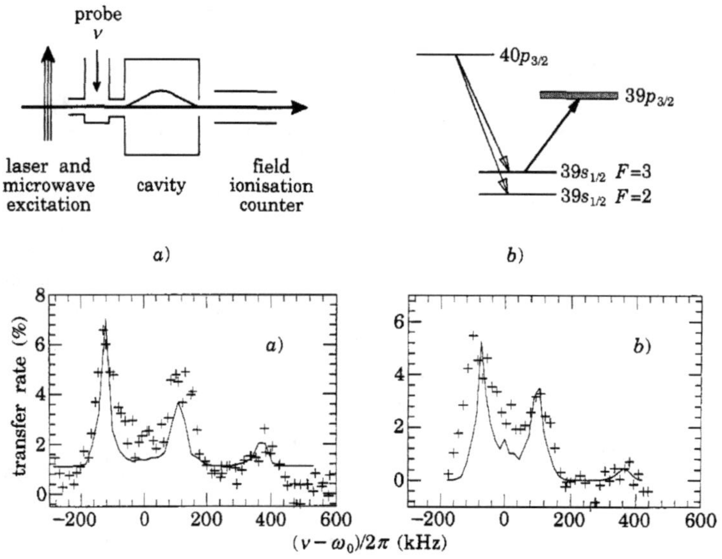

Figure 10. ***Upper Figures***: **a)** Experimental set up for the measurement of the Vacuum Field Rabi splitting in atomic physics. The atomic beam is represented by a horizontal arrow. The atoms are prepared in some initial configurations ($39s_{1/2}$ with hyperfine two-state splitting $F = 2, 3$, indicated by arrows pointing downwards) by laser and microwave excitation. The probing electric field (with a frequency ν and sine arch time variation of its amplitude, indicated in the figure by a continuous curve inside the cavity region) is injected into the cavity through the same wave guides as the atoms. **b.)** The cavity mode couples the $39s_{1/2}$ F=3 hyperfine level to the $39p_{3/2}$ levels (their hyperfine structure remains unresolved in the conditions of the experiment). This is indicated by an arrow pointing upwards. ***Lower Figures:*** The observed Rabi spectra, corresponding to the following number of atoms in the cavity: **a.)** N = 10, **b.)** N = 5. Crosses indicated experimental points, continuous lines are theoretical simulations. $\omega_0/2\pi = 68.38$ GHz is the frequency of the $39s_{1/2}$ F=3 $\rightarrow 39p_{3/2}$ transition. The transition $39s_{1/2}$ F=2 $\rightarrow 39p_{3/2}$ has a shifted frequency $(\omega_0 + \Delta)/2\pi$, $\Delta/2\pi = 320$ KHz (Pictures and results taken from ref. [38]).

In the case of the cavity model of MT, there are formal similarities between the quantum superpositions of the two energy levels of the two-state atoms used in the above-described quantum optics experiments and those of the two conformations of the tubulin dimers. However, from the technical point of view, the situation is much more complicated. Unlike atoms, the MT complexes contain many entities in their environment, exhibiting motion, vibrations (due to (room) temperature effects) *etc*, which complicate simple tests in search of Rabi-couplings between the dimer excitations, playing the rôle of the atoms, and the electric-dipole quanta inside the cavity regions of the MT, playing the rôle of the quantized cavity field modes in the atomic physics experiments. Such couplings would lead [14] to frequency splittings, with the characteristic peaks (*c.f.* fig. 10) (32), (33), in the appropriate absorption spectra of MT (see discussion in subsection 3.3).

Unfortunately, I am not qualified to discuss in detail the feasibility of such experiments, I can only make speculations which, from an experimental point of view, may not be realizable in practice. Nevertheless, as this is a talk, I am free to speculate, so here are my thoughts/suggestions on such experiments: one needs, first of all, to have isolated microtubules, probably in suspension. One should apply laser fields through them and study the absorption spectrum. Since the conformational changes of the tubulin dimers in MT are expected theoretically [14] to be in the thousand GHz region (28), one should arrange for the externally applied fields to have frequencies near such values. According to the model of cavity MT, reviewed above, the characteristic coherent modes of the cavity regions of the MT, excited by the field, are almost in resonance with the dimer oscillations, and hence standard Rabi-splitting phenomenology (33) should be expected, if the model correctly describes nature.

The induced Rabi splitting in the frequencies of the absorption spectra of MT, if observed, would then constitute compelling evidence for the existence of *both* isolated cavity regions in MT interiors and coherent modes inside such regions, which in the model of [14] would be the dipole quanta of the water molecules, suggested in [34]. It goes without saying, that in view of the complicated MT structure, and the associated motions and vibrations of the various biological entities in the tubulin protein or the C-termini appendices, there would be non resonant couplings of atoms and ions in the dimers to the cavity electron dipole quanta, in addition to the simple dipole Rabi couplings examined so far. These would complicate the resulting spectra, however one should still expect to see pronounced Rabi peaks, in analogy with the atomic physics experiments of fig. 10.

Unfortunately, observations of such effects alone (even if they are realized in nature) would not constitute a proof of the quantized nature of the dimer excitations of the MT. Although the latter is a plausibility, and according to such interpretations, the (conjectural) thin interior cavity regions of MT would entail vacuum fluctuations of the ordered-water dipole quanta that would split the resonance line of the dimers by an amount proportional to the collective dimer-water-cavity coupling (a sort of induced dynamical Stark Effect on the quantum superpositions of the dimer states), alternative explanations of the Rabi splitting phenomenon [38] exist. The latter point towards an interpretation of the phenomenon as a consequence of the fact that the MT dimer *medium* behaves as a *refractive* one with a *classical complex* (*i.e* containing imaginary parts) index of refraction that splits the cavity mode into two components.

To demonstrate, therefore, unambiguously the existence of quantum coherent effects in MT one needs to explicitly observe experimentally the quantum oscillations of the dimers and measure their environmentally induced decoherence. A straightforward extension of the atomic physics experiments, observing quantum decoherence in electrodynamical cavities [43] may, unfortunately, not be feasible in the case of MT. Indeed, in the case of the atomic physics experiments of ref. [43], a Rydberg atom beam in a quantum superposition is sent through a cavity, *exits* from it and is eventually counted in one or the other Rydberg state by appropriate ionizing detectors, so that only decoherence of the atom-cavity system ("atom plus measuring apparatus") is measured in the experiment. In contrast, in the case of MT, the "cavity" regions are attached to the "atoms", being part of the MT structure. One cannot separate the quantum oscillations of the dimers from the rest of the MT and its ordered water interiors. Thus, the methods of [43], that could in principle measure quantum decoherence of the field-dimer (and hence of the entire MT) system, as the latter unfolds, appear inapplicable.

On the other hand, it seems to me that Photon echo absorption data, like the ones in *Cryptophyte Algae* [19], appropriately adapted to the much more complex case of MT, might be a way forward, in order to observe the decoherence of the coupled "field-dimer" system inside the MT, but presently I am agnostic as to the feasibility of such measurements, lacking the appropriate qualifications. Nevertheless, I find the prospect of performing such experiments very exciting, and I am sure there are ways one can proceed along these lines in the near future.

5. Conclusions and Outlook

The exciting experimental developments in the light-harvesting *Cryptophyte Algae* provided compelling evidence on an important rôle of quantum effects in biological systems at room temperature. Specifically, on exciting by photon pulses certain dimer pigments (DVB) of the photosynthetic protein antennae of the algae into a quantum superposition of appropriate electronic states, one observes experimentally, by means of two-dimensional photon echo (absorption) data [19], the quantum oscillations between the two electronic states of the DVB dimers, as well the quantum entanglement of the DVB molecules with the other pigments, at distances of order 20 Angstrom away. In this way, the entanglement is responsible for "coherently wiring" pigment molecules across the entire protein antennae. The experiment has been performed at ambient temperatures (294 K).

The situation is reminiscent of coupled oscillators through extended springs. Such action at a distance is the result of quantum correlations between the quantum states of the pigments. The eventual decoherence of the relevant oscillations, induced by the complex environment of the protein antennae, has been observed to last for about 400 fs. Although this time scale is relatively short, nevertheless, the authors of ref. [19] have argued that it may be sufficiently long for the protein antenna to *quantum calculate* in which direction energy and information would be transported more efficiently. Thus the observed coherent 'wiring' across the entire antenna complex is thereby linked with energy transfer optimization in photosynthetic algae.

Not all photosynthetic proteins exhibit such a behavior, which in [19] has been attributed to the fact that the pigments are covalently bound in the protein complex, unlike the case of other light-harvesting biosystems. This strong bounding is probably responsible for the rather long (from the point of view of information transfer optimization) decoherence time scales at room temperatures.

Such exciting developments brought back to life, at least in the mind of the speaker (!), a rather old model of biological microtubules (MT) viewed as quantum electrodynamical cavities, developed in [14] and elaborated further in [15]. In this model, quantum coherent oscillators in the dimer walls was the result of environmental entanglement of the tubulin protein dimers (viewed as a two state quantum system, oscillating between the two conformational states, α and β - tubulin) with quantum dipole quanta inside thin interior regions of the MT, extending from the dimer walls inwards, for a few Angströms, which are full of ordered water. These regions have been argued to play the rôle of thermally isolated cavity regions, as a result of strong dipole-dipole interactions between the electric dipoles of the dimers and those of the dipole quanta in the ordered water. The latter arise as a result of the interaction of water molecules with quantized electromagnetic fields inside the cavity regions.

Electromagnetic fields in MT can be either externally applied or self generated. In this talk, it has been argued that strong experimental evidence in favor of the existence of dipole quanta in the interior of MT would be the observation of characteristic Rabi splitting in the photon absorption frequency spectra of tubulin dimers, under the influence of externally applied electromagnetic fields of definite frequency.

The dipole-dipole water/dimer coupling is held responsible [14] for inducing macroscopic quantum coherence across the entire MT, of some μm$=10^{-6}$ m long, which would manifest itself as a dimer-dipole kink soliton, responsible for dissipation-free energy and signal transfer. It has been estimated that, within some natural range of the parameters of the model of [14], the decoherence time scale, within which such solitonic states collapse to a classical state, could be as long as

$$t_{\rm MT\ decoh}\mathcal{O}\left(10^{-6} - 10^{-7}\right)\ \mathrm{s}\ . \tag{48}$$

under the assumption that the dominant source of decoherence is the loss of dipole quanta through the imperfect cavity walls. Such long decoherence times may be achieved if there is ferroelectric behaviour of MT complexes at room temperature. They are argued to be sufficiently long for dissipation-free energy and signal transfer by the kink soliton along the MT protofilaments. Our mechanism for the emergence of coherent states in MT arrangements

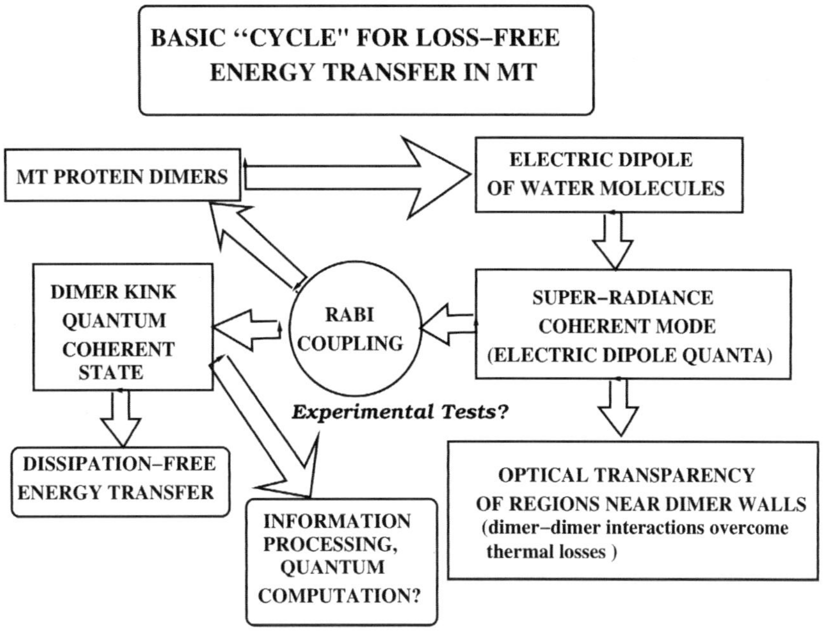

Figure 11. Suggested basic cycle, summarizing functioning and properties of the quantum cavity model for Microtubules [14], and its rôle in dissipation-free energy and information transfer.

therefore inside the cell can be summarized by the 'basic cycle' shown in figure 11.

Much shorter decoherence times of the quantum excitations of the dimers are expected if the ferroelectric behaviour and the strong dipole-dipole water/dimer coupling do not exist in Nature. In particuler, as explained in [17], decoherence times of order of a few hundreds of fs may be the result of Ca^{2+}-ion induced decoherence in MT. Although, such generic estimates have been argued in [17] to eliminate any essential rôle of quantum physics in biological MT, in particular brain MT, nevertheless, in view of the findings of [19], such conclusions should be revised.

Decoherence times, of order of a few hundreds of fs, *may be* sufficiently long for optimization of the direction along which energy and information transfer in MT would be more efficient, even if the latter occurs by means of classical physics (mechanics) laws of transport rather than quantum. Thus, even if the cavity model of MT turns out *not* to provide the correct description of natural MT, some sort of quantum computation in energy and signal optimization could still take place inside the biological cells, in the way conjectured above.

All such features would leave the realm of *science fiction* and become *realistic possibilities* or even realities, once some experimental evidence on the basic properties of models entailing quantum dynamics of MT emerges. The latter can come in many ways, from a direct observation of quantum oscillations of dimers in an MT, using photon echo data as in [19], assuming that such a technique is feasible in MT, and indirect verification of the existence of cavity regions in their interior, through the observation of Rabi-splitting peaks in appropriate absorption spectra, to an explicit demonstration of ferroelectric properties of MT at room temperatures, through the measurement of permanent dipole moments and other phenomena characteristic of ferroelectricity [44]. Unfortunately, at present there is no such experimental evidence. However, the situation can easily change in the near future.

In fact, a quite interesting development has been presented in this conference by A. Bandyopadhyay [45], who claims the observation of helical solitons along MT cylinders, using sophisticated nanotechnology experiments. If confirmed, this would be a significant step towards an understanding of the mechanisms for energy transfer along these fundamental cell structures. Moreover, one is hopeful that with the advancement of technology at the nanoscale, the experimental testing of the ideas concerning the cavity nature of the MT, described above, via the observation of Rabi splitting-like phenomena in the absorption or emission spectra, may become a reality in the not-so-distant future, thereby opening up the way for the falsification (or verification!) of macroscopic quantum entanglement phenomena in the Microtubular arrangements of (brain) cells.

I would like to close this talk by stating something that physicists tend to forget, though, when experimenting with living matter. *In vitro* (Laboratory) situations may be entirely different from actual *in vivo* situations. Living organisms are non equilibrium systems, as emphasized in this conference by E. del Giudice [46], hence, their behaviour may be entirely different from dead *in vitro* samples in suspension. It is then possible that some quantum aspects of *in vivo* Microtubules are not easily (if at all) revealed in the laboratory. More sophisticated than usual techniques might be needed to bring forward such aspects.

I sincerely believe that the emergence of quantum phenomena in light-harvesting algae will open new and exciting avenues for research in this direction, and, if such coherent phenomena prove to characterize also the fundamental building blocks of the cell, the Microtubules, this may eventually lead to establishing a possibly important rôle of quantum physics on basic functions of the cell. Time will probably tell whether all these constitute *science fiction* or *facts*...

Acknowledgements

I wish to thank the organisers of the 9th International Fröhlich symposium EDALC 2011 (Academy of Sciences of the Czech Republic, Prague, July 1-3 2011) for the invitation in this interesting and thought stimulating conference.

References

[1] See, *e.g.*: C. Amsler *et al.* [Particle Data Group], *Review of particle physics*, Phys. Lett. B **667**, 1 (2008).
[2] H. K. Onnes, Commun. Phys. Lab. Univ. Leiden **12**, 120 (1911); W. Meissner and R. Ochsenfeld, Naturwiss. **21** (44): 787 (1933) [doi:10.1007/BF01504252]; F. London and H. London (1935) Proc. R. Soc. London **A 149** (866): 71 (1935) [doi:10.1098/rspa.1935.0048]; V. L. Ginzburg and L.D. Landau (1950) Zh. Eksp. Teor. Fiz. **20** (1064), (1950); J. Bardeen, L. N. Cooper and J. R. Schrieffer, Phys. Rev. **106** (1): 162 (1957); *ibid.* **108** (5): 1175 (1957).
[3] J.G. Bednorz and K.A. Mueller, Z. Phys. **B64** (2) 189 (1986); for a review see: Anthony Leggett, Nature Physics **2**: 134 (2006) [doi:10.1038/nphys254].
[4] E. Collini and G. D. Scholes, Science **323**, 369 (2009).
[5] B. Julsgaard, A. Kozhekin and E. Polzik, Nature **413**, 400 (2001).
[6] E. Schrödinger, *What is Life* (Cambridge University Press, 1944, reprinted 1992 [ISBN 0521427088]).
[7] H. Frohlich, Int. J. Quantum Chem., **2**, 641 (1968); Nature **228**, 1093 (1970); Phys. Lett. **39A**, No 2, 153 (1972); also in *Bioelectrochemistry* (F. Guttman and H. Keyzer eds., Plenum, New York 1986).
[8] A.S. Davydov, Journal of Theoretical Biology **66** (2): 379 (1977).
[9] F. A. Popp, in: *Electromagnetic Bio- Information* (F.A.Popp, G.Becker, H.L.Konig and W.Peschka, eds., Urban & Schwarzenberg, Munich- Vienna-Baltimore (1979)).
[10] see: J. Swain, arXiv:physics/0603137 (2006) and references therein.
[11] S. Hameroff and R. Penrose, J. Conscioussness Studies **2** (1995), 98; ibid. **3**, 36 (1996); Mathematics and Computers in Simulation **40**, 453 (1996); also in *Towards a science of Consciousness*, The First Tucson Discussions and Debates, eds. S. Hameroff *et al.* (MIT Press, Cambridge MA 1996), 507. See also: R. Penrose, *The Emperor's New Mind* (Oxford Univ. Press 1989); *Shadows of the Mind* (Oxford Univ. Press 1994).
[12] See, *e.g*: P. Dustin, *MicroTubules* (Springer, Berlin 1984); Y. Engleborghs, Nanobiology **1** (1992), 97.
[13] W. H. Zurek, Rev. Mod. Phys. **75**, 715 (2003), and references therein.

[14] N. E. Mavromatos and D. V. Nanopoulos, Int. J. Mod. Phys. B **11**, 851 (1997) [arXiv:quant-ph/9512021]; Int. J. Mod. Phys. B **12**, 517 (1998) [arXiv:quant-ph/9708003]; Adv. Struct. Biol. Vol. **5**, 283 (1998) [arXiv:quant-ph/9802063].

[15] N. E. Mavromatos, Bioelectrochemistry and Bioenergetics **48**, 273 (1999); N.E. Mavromatos, A. Mershin, and D.V. Nanopoulos, Int. J. Mod. Phys. B **16** (2002) 3623-3642 [arXiv:quant-ph/0204021].

[16] A. Mershin, D. V. Nanopoulos and E. M. C. Skoulakis, arXiv:quant-ph/0007088; A. Mershin, E. Pavlopoulos, O. Fitch, B. C. Braden BC, D. V. Nanopoulos and E. M. Skoulakis, Learn Mem. **11** (3), 277 (2004); see also A. Mershin *et al.* in [21] below, and references therein.

[17] M. Tegmark, Phys. Rev. **E 61**, 4194 (2000) [arXiv:quant-ph/9907009].

[18] G.S. Engel, *et al*, Nature **446**, 782 (2007); H. Lee, Y.C. Cheng and G.R. Fleming, Science **316**, 1462 (2007); I.P. Mercer, *et al.*, Phys. Rev. Lett. **102**, 057402 (2009).

[19] E. Collini *et al.*, Nature **463**, 644 (2010).

[20] V. I. Novoderezhkin, D. Rutkauskas and R. van Grondele, Biophys. J. **90**, 2890 (2006); R. van Grondele and V. I. Novoderezhkin, Phys. Chem. Chem. Phys. **8**, 793 (2006); also in Nature **463**, 614 (2010), *news and views* on the results of ref. [19].

[21] For potential implications of quantum effects in brain functioning and the associated experimental tests, the reader may be referred to the collection of relevant articles in: *The Emerging Science of Consciousness* (J.A. Tuszynski ed., The Frontiers Collection, Springer, ©Springer-Verlag Berlin, Heidelberg 2006); ISBN 3-540-23890-5), and references therein.

[22] M.V. Satarić, J.A. Tuszyński and R.B. Zakula, Phys. Rev. **E 48**, 589 (1993). For an updated analysis see: M.V. Satarić and J. A. Tuszynski, Journal of Biological Physics **31**, 487 (2005).

[23] J. Löwe, H. Li, K.H. Downing, E. Nogales, J Mol Biol. **313**, 1045 (2001). See also: E. Nogales, S. G. Wolf, K. H. Downing, Nature **391**, 199 (1998).

[24] see: J.A. Tuszynski, J.A. Brown, E. Crawford, E.J. Carpenter, M.L.A. Nip, J.M. Dixon and M.V. Sataric: Mathematical and Computer Modelling **41**, 1055 (2005) and references therein; also: J.A. Tuszynski, T. Luchko, S. Portet and J.M. Dixon: European Journal of Physics E: Soft Condensed Matter **17** (No.1), 29 (2005).

[25] A. Mershin, A.A. Kolomenski, H.A. Schuessler, D.V. Nanopoulos, Biosystems **77**, 73 (2004).

[26] D.B. Wells and A. Aksimentiev, Biophys. J. **99** , 629 (2010).

[27] M.A. Collins, A. Blumen, J.F. Currie, J. Ross Phys. Rev. **B19** 3630 (1979); *ibid.* 3645 (1979).

[28] P. Lal, Physics Letters **111A**, 389 (1985).

[29] M. Otwinowski, R. Paul and W.G. Laidlaw, Phys. Lett. **A128**, 483 (1988).

[30] W. H. Zurek, S. Habib and J. P. Paz, Phys. Rev. Lett. **70**, 1187 (1993).

[31] Y. Tsue and Y. Fujiwara, Progr. Theor. Phys. **86**, 469 (1991).

[32] D.L. Sackett, *Structure and Function in the "Tubulin Dimer and the Role of the Acidic Carboxyl Terminus"* in Subcellular Biochemistry, Vol. **24**. Proteins: Structure, Function and Engineering (Biswas B.B., Roy, S. Eds., Plenum Press, New York (1995)).

[33] J. A. Brown and J. A. Tuszynski, Phys. Rev. **E 56**, 5834 (1997).

[34] E. Del Giudice, G. Preparata and G. Vitiello, Phys. Rev. Lett. **61**, 1085 (1988); For a discussion on the rôle of non-equilibrium field theory in living matter, in a more general setting see: E. Del Giudice, S. Doglia, M. Milani and G. Vitiello, Nucl. Phys. **B251** (FS 13), 375 (1985); *ibid* **B275** (FS 17), 185 (1986).

[35] S. Haroche and J.M. Raimond, *Cavity Quantum Electrodynamics*, ed. P. Berman (Academic Press, New York 1994), p.123, and references therein.

[36] J.J. Sanchez-Mondragon, N.B. Narozhny and J.H. Eberly, Phys. Rev. Lett. **51**, 550 (1983).

[37] G.S. Agarwal, Phys. Rev. Lett. **53**, 1732 (1984).

[38] F. Bernardot *et al.*, Europhysics Lett. **17**, 33 (1992).

[39] M. Jibu, S. Hagan, S. R. Hameroff, K. Pribram, K. Yasue, Biosystems **32**, 195 (1994).

[40] S. Hagan, S. Hameroff and J.A. Tuszynski, Phys. Rev. **E 65**, 061901 (2002).

[41] D. Kielpinski, V. Meyer, M.A. Rowe, C.A. Sackett, W.M. Itano, C. Monroe adn D.J. Wineland, Science **291**, 1013 (2001).

[42] K. Böhm, N. E. Mavromatos, A. Michette, R. Stracke and E. Unger, Electromagnet. Biol. and Medicine **24**, 297 (2005).

[43] M. Bruen *et al.*, Phys. Rev. Lett. **77**, 4887 (1996).

[44] N. E. Mavromatos, D. V. Nanopoulos, I. Samaras and K. Zioutas, Adv. Struct. Biol. Vol. **5**, 127 (1998) [arXiv:quant-ph/9803005]; J.A. Tuszynski, T.J.A Craddock and E.J. Carpenter, Journal of Computational and Theoretical Nanoscience, Volume **5**, Number 10, 2022 (2008).

[45] See talk by Anirban Bandyopadhyay, these proceedings.

[46] See talk by Emilio del Giudice, these proceedings.

Systems Information Therapy and the central role of the brain in allostasis

Alberto Foletti*° and Settimio Grimaldi°

*Lugano, CH 6900, Switzerland and Institute of Translational Pharmacology, CNR, Rome,Italy ,
°Institute of Translational Pharmacology, CNR, Rome, Italy.

E-mail: contact@albertofoletti.ch

Abstract. This work arose from the necessity to up date and clarify some basic concepts in contemporary medical practice such as those of health, disease, therapy and prevention. According to this perspective the work starts with a general epistemological review and goes on with an epistemological revision of biology and medicine. The concept of adaptation and the central role of the brain is then analysed and stated as the base to next consideration and deepening from a biophysical perspective. Physio-pathology of adaptation is assumed as a key concept in the definition and in the understanding of health and disease. A huge amount of endogenous and external stimuli has to be processed and response to them may lead to increase, stability or decrease of coherence in agreement with Frohlich's pioneering ideas. In this framework, the concept of stress, allostasis and allostatic load are outlined. Allostasis is defined as the capability of keeping stability through dynamic changes. A particular attention is paid to the emerging paradigms in biology and medicine especially those of system biology and system medicine trying to integrate the concept of complexity and hierarchical organization of the information flow in living organisms and in humans. In this framework biophysical signalling could play a significant role in modulating endogenous dynamics and in mediating external interactions. Additionally biophysical mechanisms could be involved in biological systems inner communication and could be responsible for the maintenance of systems inner coherence. The integration of the biophysical paradigm into contemporary medical practice is leading from one side to a better understanding of many pathways in physiopathology and from the other side to some new effective clinical applications. System Information Therapy is, for instance, is rising as a suitable and coherent tool in the application of thise concept being able to restore the self regulation and self regeneration capabilities both at the local and at the system level operating with endogenous and external electromagnetic signals in the range of the extremely low frequency electromagnetic signals. Some practical applications are described such as the clinical detection and treatment of fluctuating asymmetry by Vega Select 719. Fluctuating asymmetry, as well known, is related to the presence of an allostatic load and its disappearance after a biophysical treatment is a good clinical evidence of restoring of allostasis mediated by the brain at systemic level presumably through a biophysical repatterning in which we assume a key role should be played by membranes, cytoskeleton and especially by microtubules.

1. Introduction.

The living human body achieves his own systems stability, in spite of the continuous fluctuations in its external and internal environment, by an ongoing brain-coordinated process of both systemic and local physiological adjustment called allostasis.In other words the term allostasis describe the process of maintaining stability through dynamic changes.It is different from the usual term of omeostasis employed so far because it includes the concept of a dynamic process finalized to the aim of maintaining the alive's dynamic as "a continuous flow of matter, energy and information" [1] as embedded in a biofield.

Being alive means to keep far from the thermodynamic equilibrium and maintain a dissipative dynamic in agreement with Prigogine's description of the living organisms: in one word being alive means to keep in allostasis.

Adaptation to a current stressor is referred to as allostatic accommodation. The number and the nature of stressors to whom we are exposed at any time during all our life span could be very huge. Any input coming from outside or from inside the organism must be translated from different language, depending on the different channel of the input (sight, hearing, feel, taste, smell, balance, pallaesthesia,...) into a unique code, the so called organic code [2,3], in order to process these signals as flow of information and translating them into specific meanings. It is a continuous and dynamic process of signals transduction in order to allow the computational process of the different stimuli.

The adaptive process is mainly a computational process aimed to maintain allostasis. It is not allowed not to adapt so it is not allowed not to provide output to any single input coming from the external or from the internal environment (both real or virtual, because the brain only processes its own representation of the world). Nevertheless the outcome is very different in respect to the different kind of output delivery. At this point the difference that makes difference is, uniquely, the kind of output namely the kind of adaptive response delivered. The main criterion to take in account is related to the coherence of the adaptive response to the nature of the adaptive stimulus in order to achieve the allostasis. This criterion was first described as related to biological systems by Herbert Fröhlich [4]. In this framework we can distinguish: Output coherent to the nature of the input: this is the suitable response, delivered at the suitable moment and maintained for the suitable time. This output is defined as the functional adaptive response. And output not coherent to the nature of the input: this is the not suitable response, delivered at the not suitable moment and maintained for the not suitable time. This output is defined as the dysfunctional adaptive response.

The dysfunctional response may be temporary without leading to damages or lead to a progressive biological damage leading in that case to the so called pathologic adaptive response. In This case we can register and measure the damages at the biological level evaluating the possibility to stop and reverse the process or otherwise the progressive lose of the function with no possibility to restore allostasis.

The cumulative amount of wear and tear, the continuous effort, coming from the continuous allostatic mechanism could represent the concept of allostatic load [5, 6, 7] .This process is spread over all our lifespan beginning from the conception till to the death. As allostatic load accumulates , the body's capability for effective allostatic adaptation decreases, leading to a downward spiral of increased vulnerability to new stressors, and ultimately to impaired global health both in biological mechanisms (immune system, hormonal system, autonomic nervous system, postural response) and in psycho neuro emotional behavioural mechanism. From the above description the following consideration seem to be of interest.

The possibility to estimate the presence of the allostatic load during a clinical examination could not be neglected in the light of the recent finding related to ill health dynamics [8]. The possibility of managing allostatic load in the clinical activity is consequently one of the main aims of any therapeutic intervention.

Moreover reducing allostatic load is the greatest deal of any preventive hints in medicine because of the evidence still available regarding the reduction of all risk of mortality by reducing allostatic load [9, 10].

At this point the following question has to be answered:

- Is it available some clinical method to detect the presence of the allostatic load at first as a general feature and secondly as detailed feature?
- Is it available a clinical method to manage allostatic load in order to restore allostasis?

Some papers showed a relationship between the presence of the allostatic load and the presence of fluctuating asymmetry [11, 12, 13, 14, 15]. Fluctuating asymmetry is defined as the asymmetric activation of symmetric group of muscles due to a loose of coherence into the inter hemispheric synchrony at the brain level. As recently established optimal information transfer in the brain is related to synchronization [16]. Evolutionistic and behavioural biological sciences have extensively proved that human beings develop forms of "Fluctuating Asymmetry" in response to environmental stress. Fluctuating Asymmetries has been defined as the loss of perfect symmetry, which also involve symmetric body segments, induced by stressful environmental factors and is the expression of the presence of allostatic load. Fluctuating asymmetry is most probably related to the lose of synchronization between the two brain hemipheres due to allostatic load.
It is well established that the brain is the key organ in managing stress coping and in keeping allostasis through activation of self regulation and self recovery properties of the system [17, 18] both in event of general adaptation(GAS) syndrome and in local adaptation syndrome(LAS) according to Selye's former description [19]. Therefore the clinical detection of a fluctuating asymmetry through semeiotic evaluation is a good model to asses the presence of allostatic load as referred to a general reaction of the whole organism like in case of General Adaptation Syndrome mediated through the brain. Clinical evaluation of fluctuating asymmetry is reliable and objective, quick to learn and easy to perform, it is also not invasive and comfortable to the patient. Clinical evaluation of fluctuating asymmetry can be integrated in the current clinical evaluation so resulting even time saving and cost effective. Clinical evaluation of fluctuating asymmetry is a very good tool to detect allostatic load and to follow its evolution at different time.

In the last decays a true change of paradigm as began involving at first biology and later on medicine. The progressive fall of reductionism has lead to the rise of the concept of systems biology and only very recently to the concept of systems medicine. In the framework of systems medicine it is clear that some features and behaviour of the system as a whole are not the sum of the single parts but arises from emergent characteristics due to the complexity of the human organism. All the adaptive dynamics we described earlier considering the brain as the central switchboard has therefore to be understood within the System Medicine approach [20, 21, 22]. As a consequence biology and medicine has to deal nowadays from one side with the need to understanding biological complexity and from the other side with the need to manage biological complexity with new and appropriate tools both in experimental biology and in clinical practice. The comprehension of complexity of biological systems is a challenge that requires any effort to build bridges between different sciences such as chemistry, biochemistry, molecular biology, genetics, epigenetic, physics, biophysics, information theory, signals theory, networks theory, bioinformatics, and communication theory, physiopathology of adaptation, evolution theory, and epistemology. Many evidences has accumulated on how cells, tissues and organisms usually generate electromagnetic signals in different frequency ranges and on how living organisms could use electromagnetic signals as an additional mechanism to synchronize specific physiological activities especially at the level of brain, nerves, hearth, muscles and most probably at any level of cell and tissue function both in adaptive allostatic response leading to resilience and in maladaptive adaptive response leading to defiance through allostatic load.

The managing of complexity in clinical medical practice at the light of these new biophysical knowledge is still at is beginning nevertheless thanks to the bridging between Systems Medicine, biophysics, electronics we can useful employ some electro-medical devices that allow to manage different symptoms and diseases especially those of dysfunctional origin coherently with the aim of improving quality of life and reducing allostatic load. This new clinical approach is called Systems Information Therapy (SIT). Systems Information Therapy is based on the concepts of a bio-informational effects of physical factors (pattern of endogenous or exogenous signals within specific range) and assume the physiopathology of adaptation as a basis for the comprehension of the dynamics that underlay to health and disease. The restore of the coherence of response to internal or external stimuli and the restoring of the self regulation and self regeneration abilities locked by appearance of a dysfunction in the biophysical pathways is the main aim of Systems Information

Therapy by biophysical means. In one word the activation of the built in self-recovery ability. We previously demonstrated that electromagnetic signals in the extremely low frequency range generated by a commercially available electro-medical device (Vegaselect 719) were able to transfer specific information pattern to living cells triggering stem cells differentiation [23], epithelial cell [24] and neuronal cell differentiation [25, 26] as well as human neuroblasoma cell differentiation toward a normal cell type by electrically transferring the molecular signals of retinoic acid [27].

And now the answer to the question if a clinical method to manage allostatic load in order to restore allostasis is available. According to our previous experimental finding at cellular level especially those regarding the effect of extremely low frequency electromagnetic signals on neuronal cells [26] we decide to employ the same commercially available electro-medical device (Vega Select 719) in order to asses if it was able to induce clinical effect in managing fluctuating asymmetry. We found out that a Systems Information Therapy approach is effective in erasing the clinical patterns of fluctuating asymmetry leading therefore to the clinical possibility of a useful management of allostatic load.

2. Materials and methods

A total of 300 patients clinically assessed in search for evidence of fluctuating asymmetry were found positive. Patients with evidence of fluctuating asymmetry whit previous history of congenital anomalies, major traumas, fractures or surgical procedures performed at the hips or legs levels was not included into the trial in order to select homogeneous parameters of evaluation. These selective criteria lead to exclusion of 12 patients. As inclusion criteria only patients showing a fluctuating asymmetry of at least one centimetre, or more, was enrolled in order to increase the reliability of the differential evaluation. At the end 288 of them was enrolled and dived into two groups. The first group of 144 patients was treated by Vega Select 719 and the second of 144 was enrolled as control group. The treatment was delivered according to the following experimental design distinguishing treated and non treated group:

- **Treated group**: patients were lying on a usual clinical bed and the electro medical device placed at the left side. Input electrode was placed bilaterally at the level of the forehead by an adhesive while output electrode was placed at the level of the ears and hold in place by the patient. The setting of the device's parameters was as follow. Currying wave swinging from 4.0 to 10.0 Hz, modulating waves fixed 7.0 Hz, intensity of the delivered electromagnetic field was 35% of the maximum of 5 μT available. The entire procedure lasted 10 minutes.

- **Control group**: patients were lying on a usual clinical bed and the electro medical device placed at the left side as for the treated group. Input and output electrodes were placed as for the treated group. The device was simply switched on but started therefore not running. The entire procedure lasted 10 minutes as well.

Each patient of both treated and control groups were asked to walk in the examination room for 10 minutes after the end of the experimental procedure. At this time a new clinical examination was carried out focused on fluctuating asymmetry with the same criteria of first one.

3. Results

In the **treated group**, 144 patients, behaved as follow:

- complete disappearance of fluctuating asymmetry was reported in 116 patients, 81,25%,

- partial reduction of fluctuating asymmetry was reported in 24 patients, 15,97%,

- same amount of fluctuating asymmetry was reported in 4 patients, 2,78%,

In the **control group**, 144 patients, behaved as follow:

- complete disappearance of fluctuating asymmetry was reported in 0 patients, 0%,
- partial reduction of fluctuating asymmetry was reported in 17 patients, 11.8%,
- same amount of fluctuating asymmetry was reported in 127 patients, 88,2%,

None of the patients reported side effects during or after the therapeutic procedure.
Patients of both group reported a significant increase of feeling of relaxation.

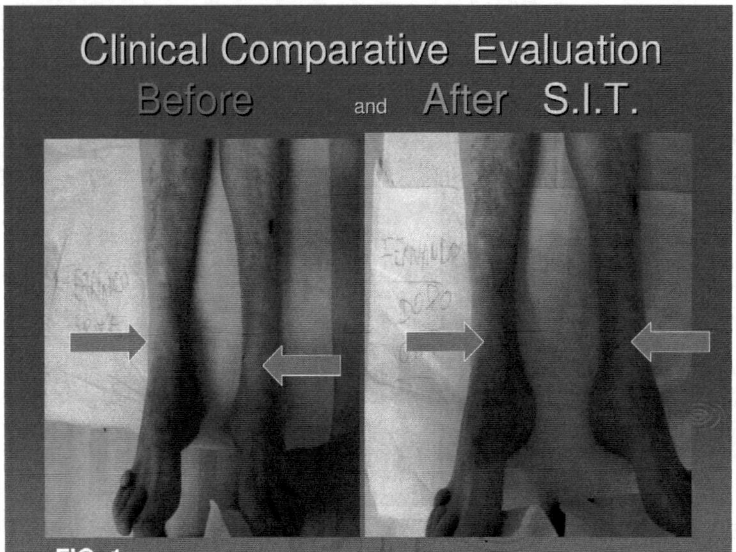

FIG. 1 Evidence of asymmetry before and its disappearance after B.IT.

Fig. 1 shows a clear demonstration of the presence of fluctuating asymmetry according to the picture taken at the first clinical evaluation and comparatively, in the same patient, the positive effect of System Information Therapy(S.I.T.) in re-establish completely the symmetry at the level of the ankles.

4. Discussion

A first astonishing consideration comes from this preliminary, but systematic, clinical assessment of the presence of fluctuating asymmetry: fluctuating asymmetry is almost ubiquitous even if it shows a different amount in its measure. These data are consistent with the framework of allostasis and allostatic load that propose that both allostasis and allostatic load are a direct, and someway, unavoidable consequence of being alive. In this perspective live mean be under a number of stressors, coming both from outside and from inside the organism. Each input stressor is traduced into signals suitable to be processed and translated to the brain. The brain has a central role in processing each signals referring to inner or external inputs [28, 29, 30]. This computational process done by the brain allows to translate this flow information into different meaning according to specific codes [2, 3] and give rise to signals able to deliver the adaptive response as final output. Most of these process are out of the level of consciousness, only a very few amount of these continuous adaptive process occur at the conscious level. Adaptive response involves at once, both the neurological psycho-emotional behavioural side and the biological side including hormonal system, immune system, autonomic nervous system and muscle-skeletal system. Actually any adaptive response is a true complex pattern involving in different time and in different blended amount each of the above mentioned system. In this framework when an input stressor appears two possible outcomes are outlined as possible: functional, namely adaptive responses, or dysfunctional, namely maladaptive responses.

While, on one side, functional adaptive response sustains allostasis and improves resilience increasing health and, by an upward spiral, salutogenesis. On the other side, dysfunctional maladaptive response sustain allostatic load and improve defiance increasing disease potential and, by a downward spiral, pathogenesis. As a consequence heath and disease could now be considered as the two side of the same life-lasting adaptive effort and could not be longer considered separately but rather as a dynamic process aimed to survival and personal evolution.

According to the literature evidence of fluctuating asymmetry has to be considered a as a clinical marker of allostatic load due to impairment of the physiologic inter hemispheric synchronization leading to the lose of symmetry between the two side of the body that , as well known, are controlled each one from the opposite side of the brain hemisphere. This impairment in keeping lateral symmetry is the effect of the dysfunctional adaptive response involving muscle-skeletal system at the body level as peripheral expression of the central control output delivered at the brain level. Therefore the clinical evaluation of fluctuating asymmetry is as easy and reliable tool to evaluate the presence of allostatic load involving the brain. Consistently with these considerations it does not so astonish the ubiquitous presence of fluctuating asymmetry because allostatic load, even if in different amount, is someway underpinning any life process. Moreover it is of interest focus on the great clinical utility of a first, even if general, criterion of assessment and monitoring of allostatic load also in the growing field of preventive medicine. Disappearance of fluctuating asymmetry could be a marker of a disappearance or at least of a not negligible decrease of the amount of the allostatic load of the examined subject. That is the reason for which in this preliminary study we assume significant reduction or disappearance of fluctuating asymmetry as a good clinical response as well as expression of an effective management of allostatic load. The Systems Information Therapy approach was performed in this study by a commercial available electro medical device (Vega Select 719) of which the authors have reported some interesting evidence of effect at cellular level on different cell culture especially differentiation of human pituitary neurons in vitro [26]. These evidences lead to the clinical application of the same device in order to asses any possible useful effect in humans in agreement with some previous report of similar devices operating in the extremely low frequency range of the electromagnetic signals by means of a resonance effect [31, 32, 33, 34, 35, 36] by external signal. We integrated the use of endogenous signals recorded at the level of the forehead bilaterally. We settled the currying wave at 7 Hz, in agreement with our previous finding on neurons [26], and set the modulating frequency within the range from 4 to 10 Hz swinging very slowly from 4 up to 10 Hz and backward in agreement with the usual range of functioning of the brain during rest and deep relaxation. Finally the magnetic field delivered was settled at 35% of the 5 μT (micro Tesla) available. We speculate that these setting could trigger the self-regulation and self-regeneration capability at the level of neurons and promote the inter hemispheric synchronization leading to the fast recovery of central symmetric activation of symmetric muscles chains thus restoring symmetry at any peripheral level as at the knees and at the ankles.

5. Conclusions

Coherence of the adaptive response to the nature of the external or internal stimuli is the suggested criterion to take in account defining health and disease dynamics in the emerging framework of Systems Information Therapy that arises from bridging Systems Medicine and physio-pathology of adaptation. The central role of the brain could be taken in account also from a biophysical and computational perspective in which microtubules are assumed to play an important role [37, 38]. Adaptive dynamics are assumed as a key concept in definition and understanding of health and disease including the evolutionary perspective [39]. Lots of endogenous and external stimuli, namely stressors, have to be processed and the adaptive response given to them may lead to increase, stability or decrease of coherence in agreement with Frohlich's description [4]. In this framework, the concept of stress, allostasis and allostatic load has been outlined until to the definition of a true "price of adaptation" [40] referred to the consequences of allostatic load.

The emerging paradigms in biology and medicine especially those of system biology and system medicine try to integrate the concept of complexity and hierarchical organization of the information flow in living organisms and in humans stating the brain as the interface between inner and external stressors. In this framework biophysical signalling could play a significant role in modulating endogenous dynamics and in mediating external interactions through brain's computational activity. Biophysical mechanisms are involved in biological systems inner communication including brain function and could be responsible for the maintenance of systems inner coherence. I particular some evidence tribute to microtubules a role in the maintenance of synchronicity [41]. This preliminary report leads us to some interesting closing remarks.

Systems Information Therapy showed a very significant effect in reducing or erasing fluctuating asymmetry. In the control group none of the patient showed a complete recovery, only 11,8% showed some partial reduction of the previous amount of fluctuating asymmetry and 88,2% showed no variation from the initial record. Surprising in the treated group a complete recovery was reported in 81,25% with a complete restore of symmetry, partial reduction was reported in 15,97% and no variation only in 2,78%. It is a well stated and current opinion that fluctuating asymmetry is a clinical correlate of allostatic load. Therefore we could argue that the disappearance of fluctuating asymmetry has to be considered as a clinical evidence of a significant reduction of allostatic load achieved through a restore of coherence at the brain level involving peripheral changes.

These preliminary results are very promising and indicate a Systems Information Therapy approach as an effective and safe tool suitable to manage allostatic load in clinical practice. Future studies should be designed in order to expand the number of patients, to study which other clinical parameters should undergo some significant useful modification and in order to achieve a better definition of the possible use in managing allostatic load and in contributing to management of dysfunctional disease and to improve health quality moreover contributing to successful aging.

References

[1] Brizhik L S, Del Giudice E, Popp F-A, Maric-Oeler W, Schlebusch K P 2009 *Electromagn. Biol. Med.* **28** 28
[2] Barbieri M 2003 *Hist. Philos Life Sci.* **25** 243
[3] Barbieri M 2004 *Riv. Biol.* **97** 91
[4] Fröhlich H, editor 1988 *Biological coherence in response to external stimuli* (Berlin: Springer Verlag)
[5] McEwen B S, Stellar E 1993 *Arch. Intern. Med.* **153** 2093
[6] McEwen B S 1998 *Ann. N.Y. Acad. Sci.* **840** 33
[7] McEwen B S, Wingfield J C 2003 *Horm. Behav.* **43** 2
[8] Fava G A, Guidi J, Semproni F, Tomba E, Sonino N 2010 *Psychother. Psychosom.* **79** 280
[9] Karlamangla A S , Singer B H , McEwen B S , Rowe J W, Seeman T E 2002 *J. Clin. Epidemiol.* **55** 696
[10] Karlamangla A S , Singer B H , Seeman T E 2006 *Psychosomatic Med.* **68** 500
[11] Parson P A 1990 *Biol. Rev. Camb. Philos. Soc.* **65** 131
[12] Livshits G ,Kobyliansky E 1991 *Hum Biol.* **63** 441
[13] Kowner R 2001 *Br. J. Psychol.* **92** 447
[14] Al-Elisa E, Egan D, Wassersung R 2004 *Evolution and Human Behavior.* **25** 31
[15] Brown W M, Price M E, Kang J, Pound N, Zhao Y, Yu H 2008 *Proc. Natl. Acad. Sci. USA* **105** 12938
[16] Buehlmann A, Deco G 2010 *PloS Comput. Biol.* **6** e1000934.
[17] McEwen B S 2007 *Physiol. Rev.* **87** 873
[18] McEwen B S, Gianaros P J 2010 *Ann .N.Y. Acad. Sci.* **1186** 190
[19] Selye H 1956 *The stress of life* (New York: McGraw-Hill Book Co.)

[20] Dhar P K , Zhu H , Mishra S K 2004 *IEEE Trans. Nanobioscience* **3** 144
[21] Bateson P 2005 *J. Biosci.* **30** 31
[22] Federoff H G, Gostin L O 2009 *J.A.M.A.* **302** 994
[23] Lisi A, Ledda M, De Carlo F, Pozzi D, Messina E, Gaetani R, Chimenti I, Barile L, Giacomello A, D'Emilia E, Giuliani L, Foletti A, Patti A, Vulcano A, Grimaldi S 2008 *Electromagn. Biol. Med.* **27** 127
[24] Lisi A, Ledda M, De Carlo F, Foletti A, Giuliani L, D'Emilia E, Grimaldi S 2008 *Electromagn. Biol. Med.* **27** 230
[25] Foletti A, Lisi A, Ledda M, De Carlo F, Grimaldi S 2009 *Electromagn. Biol. Med.* **28** 71
[26] Foletti A, Ledda M, De Carlo F, Grimaldi S, Lisi A 2010 *Electromagn. Biol. Med.* **29** 63
[27] Foletti A, Ledda M, D'Emilia E, Grimaldi S, Lisi A 2011 *J. Altern. Complement. Med.* **17** 701
[28] Goldstein D S, McEwen B S 2002 *Stress* **5** 55
[29] Goldstein D S 2007 *Stress* **10** 109
[30] Goldstein D S 2008 *Ann. N. Y. Acad. Sci.* **1148** 223
[31] Rossi E, Corsetti M T, Sukkar S, Poggi C 2007 *Electromagn. Biol. Med.* **26** 277
[32] Vallesi G, Raggi F, Rufini S, Gizzi S, Ercolani E, Rossi R 2007 *Electromagn. Biol. Med.* **26** 283
[33] Ciafaloni A 2007 *Electromagn. Biol. Med.* **26** 299
[34] Crescentini F 2007 *Electromagn. Biol. Med.* **26** 305
[35] Mancuso M, Ghezzi V, Di Fede G 2007 *Electromagn. Biol. Med.* **26** 311
[36] Raggi F, Vallesi G, Rufini S, Gizzi S, Ercolani E, Rossi R 2008 *Electromagn. Biol. Med.* **27** 325
[37] Hameroff S R, Watt R C 1982 *J. Theor. Biol.* **98** 549
[38] Dayhoff J, Hameroff S, Beltra R L, Swenberg C 1994 *Eur. Biophys.* J. **23** 79
[39] Nesse R M, Williams G C 1998 *Scientific American* **279** 86
[40] Seeman T E, Singer B H, Rowe J W, Horwitz R I, McEwen B S 1997 *Arch. Intern. Med.* **157** 2259
[41] Insinna E M 1992 *Nanobiology* **1** 191

Journal of Physics: Conference Series **329** (2011) 012028 doi:10.1088/1742-6596/329/1/012028

Multiscale Modelling for investigating single molecule effects on the mechanics of actin filaments

Deriu Marco A. (1), Bidone Tamara C. (1), Carbone Laura (1), Bignardi Cristina (1), Montevecchi Franco M. (1), and Morbiducci Umberto (1)

(1) Department of Mechanics, Politecnico di Torino, Turin, Italy

marco.deriu@polito.it

Abstract. This work presents a preliminary multiscale computational investigation of the effects of nucleotides and cations on the mechanics of actin filaments (F-actin). At the molecular level, Molecular Dynamics (MD) simulations are employed to characterize the rearrangements of the actin monomers (G-actin) in terms of secondary structures evolution in physiological conditions. At the mesoscale level, a coarse grain (CG) procedure is adopted where each monomer is represented by means of Elastic Network Modeling (ENM) technique. At the macroscale level, actin filaments up to hundreds of nanometers are assumed as isotropic and elastic beams and characterized via Rotation Translation Block (RTB) analysis. F-actin bound to adenosine triphosphate (ATP) shows a persistence length around 5 μm, while actin filaments bound to adenosine diphosphate (ADP) have a persistence length of about 3 μm. With magnesium bound to the high affinity binding site of G-actin, the persistence length of F-actin decreases to about 2 μm only in the ADP-bound form of the filament, while the same ion has no effects, in terms of stiffness variation, on the ATP-bound form of F-actin. The molecular mechanisms behind these changes in flexibility are herein elucidated. Thus, this study allows to analyze how the local binding of cations and nucleotides on G-actin induce molecular rearrangements that transmit to the overall F-actin, characterizing shifts of mechanical properties, that can be related with physiological and pathological cellular phenomena, as cell migration and spreading. Further, this study provides the basis for upcoming investigating of network and cellular remodelling at higher length scales.

1. Introduction

The actin filament is a structural component of the cell cytoskeleton, undergoing various dynamics and mechanical transitions related with cell remodeling. Ion influx and filament treadmilling induce respectively magnesium/calcium exchanges from the high affinity binding site of the actin monomer (G-actin) and dephosphorylation of adenosine triphosphate associated with actin polymerization. These phenomena lead to a cascade of events that, starting from the molecular rearrangements of G-actin, affect the F-actin flexibility. The underlying reasons for these concurrent phenomena rely on the multiscale organization of F-actin. F-actin is a right-handed double helix built up from actin monomers (Figure 1a). Each G-actin is tightly bound to one adenosine nucleotide, ATP or ADP, and one tightly-bound metal cation, Ca^{2+} or Mg^{2+}. Nucleotides and cations are freely exchangeable, depending upon the cycle of polymerization/depolymerization and the local ion concentrations. The nucleotide is situated in the inter-domain cleft of G-actin, where it interacts, through its phosphate oxygens, with the divalent cation positioned at the bottom of the cleft (Figure 1b).

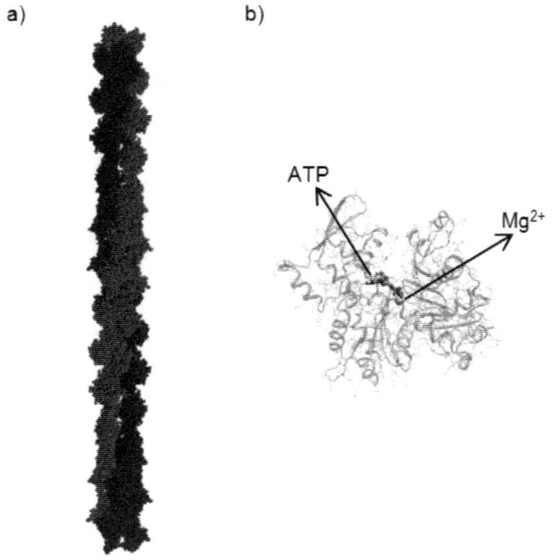

Figure 1: Double helical assessment of F-actin (a); molecular model of G-actin (b): the nucleotide binding cleft is situated in central position and contains the nucleotide (in licore representation) and the cation (in yellow, Van der Walls representation).

The exchanges of nucleotide and cation can be concomitant or take place once at a time, but both result in variations of the kinetics of polymerization for F-actin and of the affinities towards different binding proteins: in addition to these changes in the dynamic properties of F-actin variations of the mechanical properties occur.

The flexibilities of F-actin polymerized with Mg^{2+}-G-actin and Ca^{2+}-G-actin were measured using electron microscopic techniques [1]: an higher flexibility was found for Mg^{2+}-Factin (persistence length, l_p, of about 1.6 μm) than for Ca^{2+}-F-actin (l_p of about 6 μm). To obtain information about the structural changes at the basis of these changes in flexibility, limited proteolysis [2] was employed and the conformation of specific sites in the actin monomers in Ca- and Mg-F-actin was probed, revealing that, depending on whether the high-affinity site for cation is occupied by Ca^{2+} or Mg^{2+}, there are differences in the molecular structure of F-actin. In particular, upon substitution of Ca^{2+} for Mg^{2+} localized regions of G-actin undergo rearrangements, that result in variations of the intra- and inter-monomer contacts. In particular, the bound cation modulates the molecular topologies of the surface loop 38-52, including the DNase-I binding loop, those of the internal segment 61-69, in proximity of the so-called sensor loop (residue 73), and of residues 364-372, corresponding to the COOH-terminal segment. Direct E.L.I.S.A. [3] revealed that the region around residue 201 undergoes a structural rearrangement induced by the type of bound cation. The positions of these loops are illustrated in Figure 2.

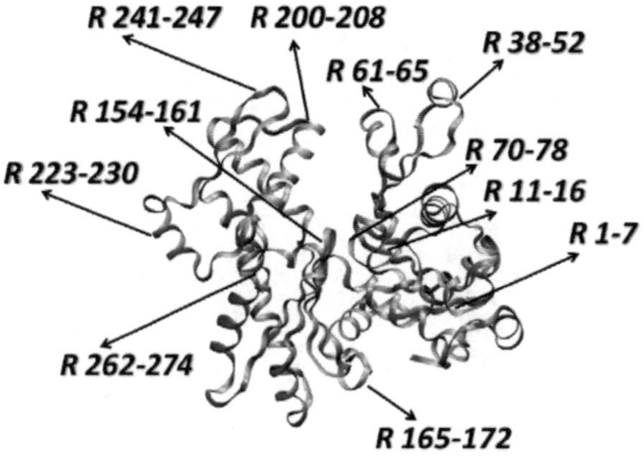

Figure 2: Residue loops subjected to conformational rearrangements upon nucleotide and/or cation exchanges.

Similarly, ATP hydrolysis leads to a softening of F-actin [4-6]. Experimental analysis of F-actin thermal fluctuations confirmed that ATP-F-actin is more rigid than ADP-F-actin with values of persistence length around 12-15 μm and 7-10 μm, respectively [7]. When ADP is bound, the conformation of the DB loop within the subdomain 2 of G-actin is structured as an α-helix and the nucleotide binding cleft is open. On the contrary, when ATP is bound, the DB loop is unstructured and the nucleotide binding cleft is closed [8]. Thus, a transition from a coil to an helix for the DB loop and a closure of the nucleotide binding cleft occur upon ATP hydrolysis [9, 10]. Molecular mechanisms explicable as loop rerrangments and fluctuations drive the modifications in the mechanical properties from ATP-F-actin to ADP-F-actin.

However, the sequence of molecular transitions that are induced by the combinatory effect of cation and nucleotide and that further modulate the elastic properties of F-actin remain to be elucidated. For this reason, we employ a series of simulation techniques each referring to a different length and time scale and evaluate what molecular rearrangements of G-actin induced by a nucleotide and by couples of cation/nucleotide reflect on the biomechanics of F-actin. In particular, we detect what loops of residues in G-actin undergo the highest fluctuations in presence of difference combinations of nucleotide/cation and relate them with the variations in the F-actin flexibility. We start by applying Molecular Dynamics (MD) simulations on ATP-G-actin and ADP-G-actin, then we study the same nucleotide bound monomers with bound Mg^{2+} or Ca^{2+}, exploiting all possible combinations of nucleotide/cation: ADP-Ca^{2+}-, ADP-Mg^{2+}-, ATP-Ca^{2+}-, and ATP-Mg^{2+}- Then, using the MD-refined G-actins as building units, several systems of F-actin are assembled, with lengths varying from about 50 to 500 nm. Each F-actin is further modelled as an Elastic Network (EN), where the positions of the alpha carbons of the system is taken into account and the collective motions induced from thermal fluctuations are analyzed. Finally, Normal Mode Analysis (NMA) is performed and the mechanical properties of F-actin are estimated, in terms of bending stiffness k_f (i.e. flexural rigidity), torsional rigidity, k_t, Young's modulus, E_n, and persistence length, l_p, on the basis of the identified bending, stretching and torsional modes. Specific aims of this work are to investigate: *i)* what inter- and intra-monomer rearrangements of ADP-G-actin and ATP-G-actin are induced by a tightly-bound Ca^{2+} and Mg^{2+}; ii) how these rearrangements reflect on the overall mechanical properties of F-actin.

2. Methods

The atomic structures of ATP-G-actin and ADP-G-actin were taken from the RCSB Protein Data Bank: PDB entry 1ATN.pdb [11] and 1J6Z.pdb [8], respectively. A metal ion, Mg^{2+} or Ca^{2+}, was added at the high affinity binding site, for a total of 4 G-actin systems: Mg^{2+}-ATP-G-actin, Ca^{2+}-ATP-G-actin, Mg^{2+}-ADP-G-actin, and Ca^{2+}-ADP-G-actin. G-actin monomers were arranged in a rectangular box (of about 5nm × 8nm × 7nm) and fully solvated (Figure 3c and Figure 3d), using the Simple Point Charge (SPC) explicit water model [12, 13], neutralized by sodium ions. The total number of atoms of each system was around 26000 (about 4000 atoms for the protein and 22000 for the solvent). Periodic boundary conditions were applied and the total energy was minimized by means of 500 steps of Steepest Descent algorithm [13, 14]. Then, 150 ns equilibrium MD simulation was performed at 300 K in a NVT canonical ensemble (i.e., with a constant number N of atoms, constant volume V and constant temperature T). Particle Mesh Ewalds (PME) method [15] was used to model long range electrostatic interactions, with a short range cut-off set at 1.2nm and the force-field GROMOS 53a6, implemented in the code GROMACS 4 [13], was adopted [16, 17].

The analysis of the results was performed with the software packages PROCHECK [18] and VMD [19]. PROCHECK was used to check the "stereochemical quality" of G-actin at the end of the MD simulations [20, 21]. Instead, VMD was used for drawing the MD trajectories of G-actin and for the rendering of the structures. Technically, the analysis of the trajectories of G-actin was conducted in terms of Root Mean Square Deviation (RMSD) and Root Mean Square Fluctuation (RMSF).

The outcome of the atomistic simulations, i.e. the G-actin monomers with atomic coordinates renewed from the fluctuations at equilibrium, were used to assembly actin filaments: ATP-F-actin, ADP-F-actin, Mg^{2+}-ATP-F-actin, Mg^{2+}-ADP-F-actin, Ca^{2+}-ATP-F-actin, and Ca^{2+}-ADP-F-actin. The correct orientation of the monomers was achieved by docking G-actin on the first monomer of a validated structure of F-actin (PDB entry 3B5U.pdb [22]). Then, other G-actins were added longitudinally, each one rotated of 166° and translated of 2.75 nm, according to Holmes et al. [6, 23]. Several models of F-actin with lengths varying from about 50 nm (corresponding to 6500 aminoacids, or alpha carbons C_α) to about 500 nm (corresponding to 68000 C_α) were assembled.

The atomic structure of each filament was modelled as an elastic network (ENM) composed of nodes (in the positions of the C_α) and springs (harmonic potentials of 1 kcal/molA^2) [24-26]. The total potential energy, E_p, was expressed in the anisotropic network model formulation [27] as follows:

$$E_p = \Delta\mathbf{R}^T \mathrm{H} \Delta\mathbf{R} \quad (1)$$

where $\Delta\boldsymbol{R}$ is a $3N$ dimensional vector describing the fluctuations $\Delta\boldsymbol{R}_i$ of the position vector $\boldsymbol{R}_i$ with respect to the equilibrium position ($1 \leq i \leq N$, where N is the total number of nodes), and $\boldsymbol{H}$ is the Hessian matrix ($3N$x$3N$).

The Hessian matrix for F-actin, even when the filament is assumed to be made only of C_α atoms, is dramatically large for direct diagonalization. For this reason, the C_α atoms corresponding to the 4 functional subdomains of G-actin were clustered into rigid blocks.

Using the Rotation Translation Block (RTB) approach [28-31], F-actin was modelled as made of rigid blocks with interactions derived from atomistic interactions. The vibrational properties of F-actin were computed performing the Normal Mode Analysis (NMA) [31]. Trajectories of the filaments were analyzed (for modes of motion identification) by employing the package VMD.

Assuming F-actin as an isotropic and elastic beam [32-34], the first mode of bending (Figure 2c), of stretching (Figure 2d) and of torsion (Figure 2e) were directly related to the bending stiffness (k_f), stretching stiffness (k_s) and torsional stiffness (k_t) [30, 31, 35], respectively, following the same approach used in recent papers by Deriu and co-workers [31, 36]. The effect that MD-based refinements of ATP-G-actin and ADP-G-actin bound to different ions have on actin filaments was evaluated in terms of the computed mechanical properties of the corresponding F-actin.

3. Results

All G-actins, throughout the whole equilibrium MD simulations, resulted stable in aqueous environment in terms of tertiary and secondary structure, as no major unfolding took place. The root mean square deviation (RMSD) at 150 ns resulted around 0.4 nm (Figure 3a). The trend of the RMSD curves was similar for ADP- and ATP- G-actin. It is worth noticing that between 50 and 100 ns these structures seemed to reach a stable RMSD value around 0.3 nm. However in the range of 100 ns a structural change can be observed. As shown by Figure 3b, this change corresponds to the onset of high fluctuations of residues 223-230, with a RMSF peak of 0.6 nm in the interval 90-120 ns: these fluctuations do not correspond to a secondary structure transition of the region, within this time interval.

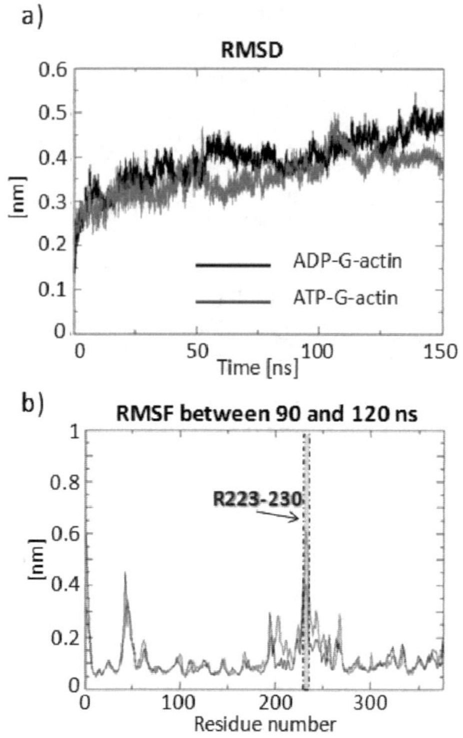

Figure 3: RMSD of ATP-G-actin and ADP-G-actin (a); RMSF of ATP-G-actin and ADP-G-actin between 90 and 120 ns (b): enhanced are the highest peak.

In general, the RMSF values of C_α atoms of ATP-G-actin are higher than those of ADP-G-actin (Figure 4), although the overall average displacements, in terms of RMSD, is lower for ATP-G-actin than for ADP-G-actin (Figure 3a).

The highest values of RMSF for both nucleotide bound forms of the monomer range from 0.4 to 0.8 nm and are distributed as follows: one peak is relative to few residues of subdomain 1 (part of the N- terminal region, residues 1-7); one peak pertain to subdomain 2 (the DB loops, residues 38-52); one peak is relative to a group of residues of subdomain 4 (residues 223-230), as shown in Figure 4a. At equilibrium, RMSF values show that ATP-G-actin is more stable than ADP-G-actin (Figure 4b). Two chain regions of ADP-G-actin are characterized by RMSF values higher than 0.2 nm, i.e., residues of subdomain 1 (part of the N- terminal region, residues 1-7) and the DB loop in subdomain 2 (residues 38-52).

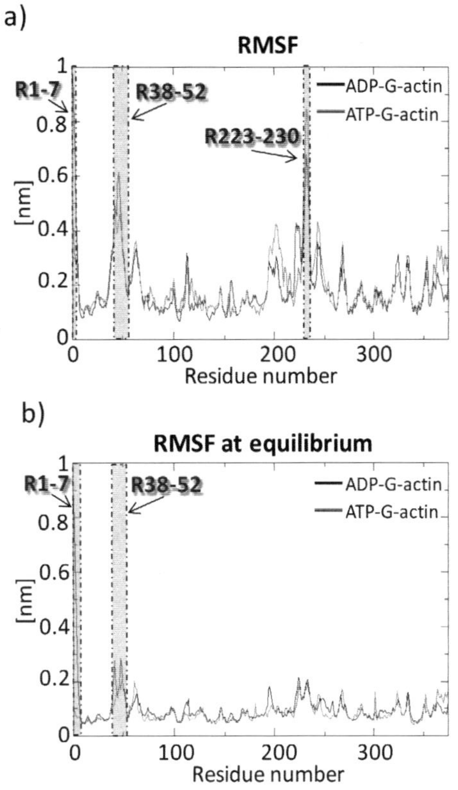

Figure 4: RMSF of ATP-G-actina and ADP-G-actin computed from the beginning of the simulations (a); RMSF of the same monomers at the equilibrium (b).

By comparing the monomers bound to the cation, ATP-G-actin bound to Ca^{2+} was slightly more stable than ATP-G-actin bound to Mg^{2+}, with lower average values of RMSD (Figure **5**a). On the contrary, ADP-G-actin bound to Mg^{2+} showed lower average values of RMSD than ATP-G-actin bound to Ca^{2+} (Figure **5**b) throughout the overall simulation time.

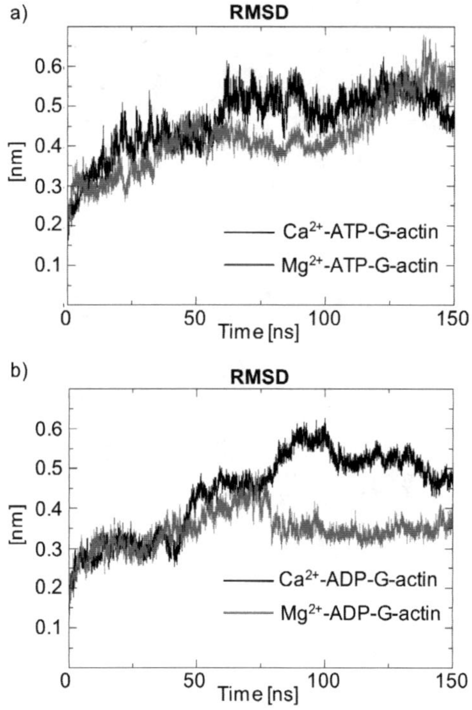

Figure 5: RMSD of Ca^{2+}-ATP-G-actin and Mg^{2+}-ATP-G-actin (a); RMSD of Ca^{2+}-ADP-G-actin and Mg^{2+}-ADP-G-actin (b).

By computing the Root Mean Square Fluctuations (RMSF) of the 4 cation containing monomers, similar trends can be observed: higher fluctuations were shown for Mg^{2+}-ATP-G-actin than for Ca^{2+}-ATP-G-actin (Figure 6a), while the fluctuations were generally higher for Ca^{2+}-ADP-G-actin than for Mg^{2+}-ADP-G-actin (Figure 6b). In particular, 6 groups of residues presented values of RMSF higher than 0.4 nm: residues 38-52, 61-69, 200-208, 223-230, 230-247 and 262-274.

Journal of Physics: Conference Series **329** (2011) 012028 doi:10.1088/1742-6596/329/1/012028

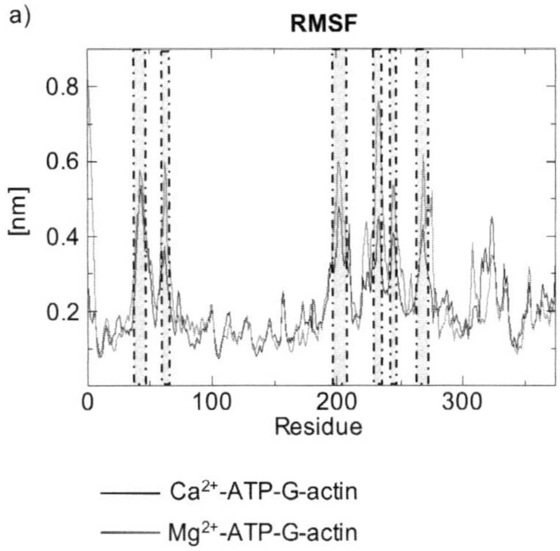

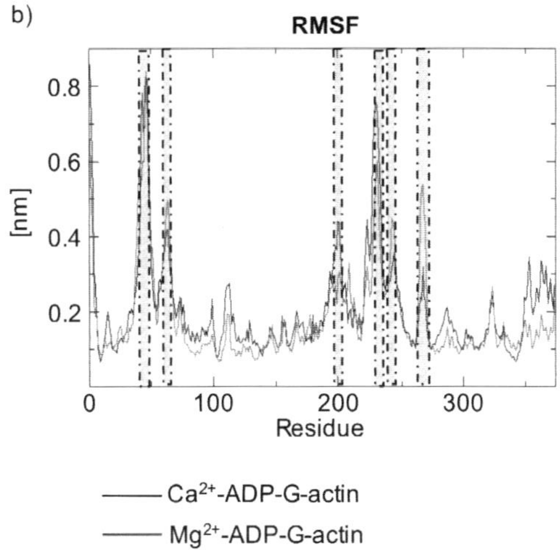

Figure 6: RMSF of Ca^{2+}-ATP-G-actin and Mg^{2+}-ATP-G-actin (a); RMSF of Ca^{2+}-ADP-G-actin and Mg^{2+}-ADP-G-actin (b).

In order to investigate the effect of Ca^{2+} on G-actin, the RMSF of the Ca^{2+} bound forms of ATP-G-actin and ADP-G-actin was computed: 2 peaks resulted higher for ADP- than for ATP- G-actin (Figure 6a), namely residues 38-52 and 223-230, with values of RMSF around 0.8 nm. A similar analysis was carried out for the case of Mg^{2+} bound forms of G-actin: 5 peaks resulted slightly higher for ATP-G-actin in comparison with ADP-G-actin (Figure 6b), namely residues 61-69, 200-208, 223-230, 230-247, and 262-274, with RMSF between 0.6 and 0.7 nm.

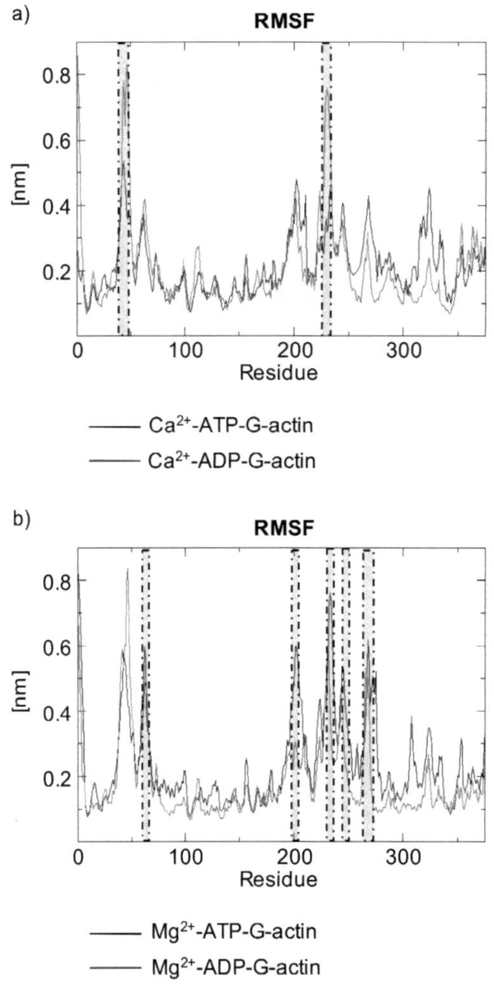

Figure 7: RMSF of Ca^{2+}-ATP-G-actin and Ca^{2+}-ADP-G-actin (a); RMSF of Mg^{2+}-ATP-G-actin and Mg^{2+}-ADP-G-actin (b).

These results obtained at the lowest scale allowed us to relate the molecular rearrangements to the mechanical properties of the actin filament. The average frequencies of vibration of ATP-F-actin (ADP-F-actin) range from about 0.25 cm^{-1} (0.22 cm^{-1}) in the case of the first bending mode, up to about 1.99 cm^{-1} (1.68 cm^{-1}) and 4.45 cm^{-1} (4.10 cm^{-1}) for the first torsion and stretching modes, respectively. Independent of the nucleotide-bond, the computed values of k_f, k_s, and k_t of F-actin filaments result to be length free. The average values of elastic stiffness obtained for ATP-F-actin are: k_f = 1.89±0.08 10^{-26} Nm^2; k_s = 7.33±3.57 10^{-2} Nm^{-1}; k_t = 4.11±0.05 10^{-27} Nm^2. The average elastic stiffness values obtained for ADP-F-actin are: k_f = 1.45±0.07 10^{-26} Nm^2; k_s = 6. 23±0.03 10^{-2} Nm^{-1}; k_t = 2.92±0.04 10^{-27} Nm^2.

Under the assumption of homogeneous, isotropic beam , the Young's modulus E_n (coincident with the bending modulus E_b) is found to be not dependent on the chain length, with values of about 0.4-0.5 GPa for ATP-F-actin and of about 0.6-0.7 for ADP-F-actin (Figure 8).

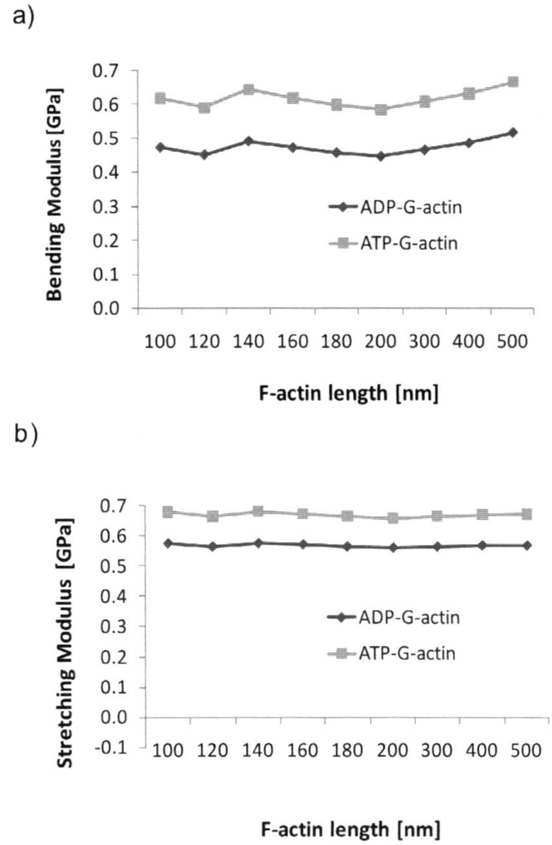

Figure 8: Young's modulus computed from the 1st modes of bending (a) and from the 1st modes of stretching (b) as functions of different lengths for the systems of F-actin assembled after equilibrium MD of ATP-G-actin and ADP-G-actin.

The computed persistence length of the filament is found to be of about 4 μm (Figure 9a). From our study, F-actin bound to ATP results to be slightly stiffer than F-actin bound to ADP, independent of the length of the filament (Figure 9b).

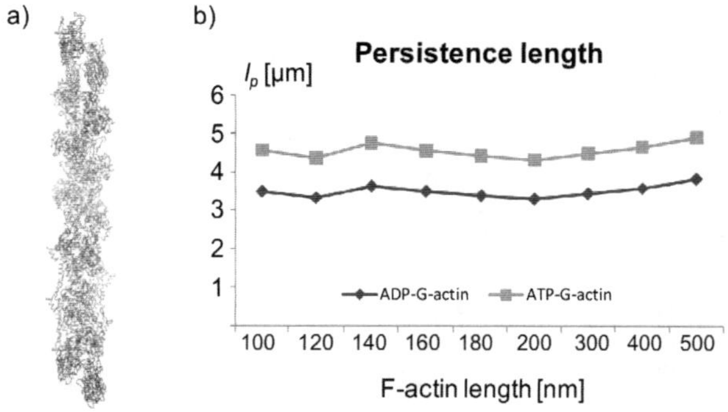

Figure 9: All-atom representation of F-actin (a); Persistence length as a function of different lengths for ATP-F-actin and ADP-F-actin (b).

Also in the cases of the cation bound filaments, the computed values of k_f, k_s, and k_t showed to be length free. The average values of elastic stiffness's obtained for Mg^{2+}-ATP-F-actin were: k_f = 1.55±0.09 10^{-26} Nm^2; k_s = 6.45±0.04 10^{-2} Nm^{-1}; and k_t = 3.26±0.04 10^{-27} Nm^2. The average values of elastic stiffness's obtained from Mg^{2+}-ADP-F-actin were: k_f = 1.15±0.02 10^{-26} Nm^2; k_s = 5.14±0.03 10^{-2} Nm^{-1}; and k_t = 2.11±0.41 10^{-27} Nm^2. In the case of Ca^{2+}-ATP-F-actin, the following average values were found: k_f = 1.55±0.96 10^{-26} Nm^2; k_s = 6.45±0.04 10^{-2} Nm^{-1}; k_t = 3.26±0.38 10^{-27} Nm^2. For Ca^{2+}-ADP-F-actin, the average values of elastic stiffness's resulted: k_f = 1.38±0.11 10^{-26} Nm^2; k_s = 7.09±0.05 10^{-2} Nm^{-1}; k_t = 2.93±0.69 10^{-27} Nm^2.

Under the assumption of a homogeneous and isotropic beam, the Young's modulus E_n (coincident with the bending modulus E_b) resulted independent from the chain length, with a value from 0.3 GPa to 0.6 GPa (Figure 10a) and a persistence length of roughly 3 μm (Figure 10b).

a)

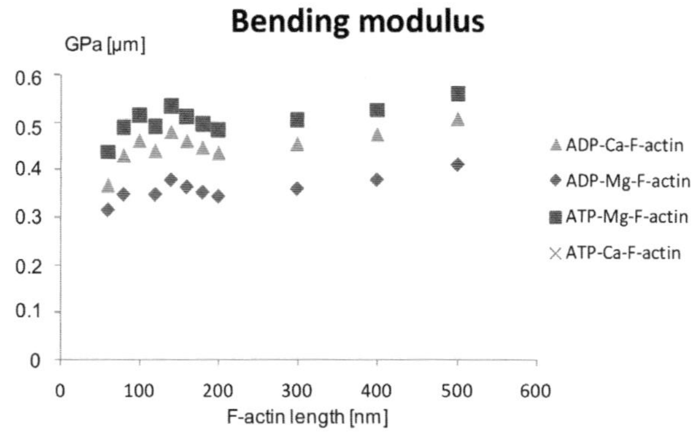

b)

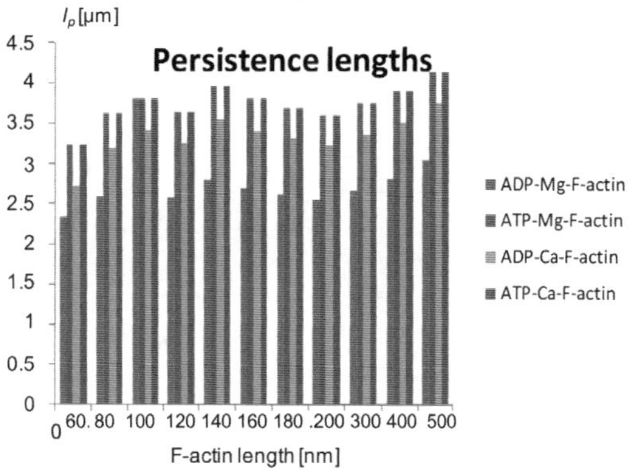

Figure 10: Bending modulus as a function of the filament length for the 4 nucleotide/cation systems of G-actin (a); persistence length as a function of the filament length for the 4 nucleotide/cation systems of G-actin (b).

4. Discussion

This work represents a systematic investigation of the effects of different combinations of a nucleotide and a cation on the biomechanics of F-actin. The model employed for the analysis of the effects of ADP/Ca^{2+}, ADP/Mg^{2+}, ATP/Ca^{2+} and ATP/Mg^{2+} on the mechanics of F-actin is able to maintain the entire number of residues of the building G-actins and to reproduce the intramonomer flexibilities, as explained in [30]. This modeling strategy makes full use of a computational technique, without introducing experimental parameters. Moreover, the direct comparison with cation free systems (ATP-G-actin and ADP-G-actin) allows isolate single contributions from nucleotide and cations and elucidate what molecular reasons are at the basis of changes on flexibility for both cation bound and only nucleotide bound F-actin systems.

With reference to the local rearrangements of G-actin, Mg^{2+}-G-actin resulted slightly more stable than Ca^{2+}-G-actin, as demonstrated by the lower RMSD (Figure 5) values. This result can be related with the fact that the ionic conditions within living cells favor the binding of G-actin to Mg^{2+} rather than Ca^{2+} [37]. However, at the end of our simulation, Mg^{2+}-ATP-G-actin presents higher values of RMSD than Ca^{2+}-ATP-G-actin, thus resulting more instable. The driving force that makes Mg^{2+}-ATP-G-actin faster in polymerizing with respect to Ca^{2+}-ATP-G-actin [2] might be due to such instability: the monomer might need to find a more stable configuration given by the assembly into a filament. Other than the higher RMSD for Mg^{2+}-ATP-G-actin than Ca^{2+}-ATP-G-actin, specific groups of residues result more fluctuating in for Mg^{2+}-G-actin than Ca^{2+}-ATP-G-actin (Figure 6a): residues 38-52, 61-69, 200-208, 223-230, 230-247 and 262-274. Some of these residues are involved in longitudinal contacts among nearby monomers of the same strand in the double helical assessment of F-actin, while others are involved in lateral interactions among nearby monomers of opposite strands.

In addition, most of these residues pertain specifically to subdomains 2 (residues 33-69) and 4 (residues 181-269) and are involved in the closing of the nucleotide binding cleft, upon polymerization into F-actin. At a higher scale of investigation, the different molecular structures of Ca^{2+}-ATP-G-actin, Ca^{2+}-ADP-G-actin, Mg^{2+}-ATP-G-actin and Mg^{2+}-ADP-G-actin reflect on the biomechanics of the filament. The ATP- bound forms of F-actin result stiffer than the ADP- bound forms at all lengths (Figure 8 and Figure 10). This stiffening is caused by the rearrangements of residues 38-52 and 223-230 (Figure 4a), that strengthen longitudinal and lateral interactions, respectively. In particular, high molecular rearrangements are carried out from residues 223-230, involved in lateral interactions with Cys 374 of the opposite monomer [38]. Results from our MD simulations has also confirmed the observations of Dalhaimer et al. [39] that subdomain 2 of G-actin is highly flexible regardless the nucleotide state: no secondary structure of the DNase I binding loop has formed in either cation forms of ATP-G-actin, nor ADP-G-actin during the 150 ns of MD simulations. The reason behind this tendency is the characteristics of hydrophobicity of this loop, that tend to rearrange itself to find a more stable arrangement far from the solvent. We think that, when ADP-G-actin is arranged into a filament, the DB loop is enclosed within the residues of nearby monomers and is therefore able to fold into an alpha helix, but this effect was not seen during our MD simulations of ADP-G-actin in isolation, since the DB loop was directly in contact with the polar solvent. However, Mg^{2+}-ATP-F-actin, if compared with the cation-free form of ATP-F-actin, thought the high rearrangements of residues 1-7 and 223-230, resulted more flexible, in agreement with the spectroscopic measurements of Yanagida et al. [40].

Further results of this study confirm that molecular rearrangements and consequent variations of the overall filament flexibility occur after hydrolysis of ATP in ADP. ADP-F-actin results more flexible than ATP-F-actin at all lengths (Figure 9), confirming that a decrease in flexibility occurs from the ATP- to the ADP- bound forms of the filament. By considering the presence of a bound cation in the ADP form of F-actin, Mg^{2+}-F-actin result more flexible than Ca^{2+}-F-actin while an equal flexibility can be seen for the ATP bound forms of the filament. The results of this study also confirm that the substitution of Ca^{2+} for Mg^{2+} in ADP-F-actin influences the surface loop 39-51, including the DNase-I binding loop, and segment 61-69 (Figure 6). Our simulations did not highlight any difference in the rearrangements of residues 364-372, corresponding to the COOH-terminal segment, between the Mg and Ca bound forms of ADP-G-actin, in contrast with previous experimental results of Strzeleckagolaszewska et al. [41]. We assume that further rearrangements of residues 364-372, not visible in our simulation, may occur upon incorporations of G-actin into F-actin. However, some of the regions that this study recognized as the most fluctuating are the same loops individuated by experimental analyses of Mejean et al. [3] and Tirion et al. [38]. Indeed, upon the substitution of Ca^{2+} for Mg^{2+}, in addition to residues 39-51 and 61-69, high fluctuations were carried out from a region around residue 200, from residues 223-230, from the so-called "hydrophobic plug", among residues 230-247, and from residues 262-274. The orientation and displacements of the hydrophobic plug reduce or increases the portion of exposed surface to the neighbour monomer on the opposite strand. Taken together, these molecular rearrangements led to values of torsional rigidity for Ca^{2+}-ADP-F-actin ($2.93\ 10^{-27}$ Nm^2) of about the same values of those of Mg^{2+}-ADP-F-actin ($2.11\ 10^{-27}$ Nm^2). Also the flexural rigidity of Ca^{2+}-ADP-F-actin ($1.38\ 10^{-26}$ Nm^2) was smaller than that of Mg^{2+}-ADP-F-actin ($1.15\ 10^{-26}$ Nm^2).

The result that the flexural rigidity is sensitive to the bound metal is also in contrast with the results of Isambert *et al.* (1995), but confirms the results of Yasuda et al. [34].

Summarizing, although global analysis of the actin structures did not reveal relevant changes in conformation (RMSD) or dynamics (RMSF), the minor fluctuations of these several localized regions show a specific conformational dependence on the state of the bound cation and nucleotide and induce variations in the overall F-actin mechanics. The work points out the attention on how single molecule effect at the nanoscale level can be transferred to higher scales affecting the overall mechanics of entire actin filaments.

References

[1] Orlova A, and Egelman E H 1993 J. Mol. Biol. 232 334.
[2] StrzeleckaGolaszewska H, Wozniak A, Hult T, and Lindberg U 1996 Biochem. J 316 713.
[3] Mejean C, Hue H K, Pons F, Roustan C, and Benyamin Y 1988 Biochem. Biophys. Res. Commun. 152 368.
[4] Blanchoin L, and Pollard T D 2002 Biochem. 41 597.
[5] Pollard T D, Goldberg I, and Schwarz W H 1992 J. Biol. Chem. 267 20339.
[6] Chu J W, and Voth G A 2005 Proc Natl Acad Sci U S A 102 13111.
[7] Isambert H, Venier P, Maggs A C, Fattoum A, Kassab R, Pantaloni D, and Carlier M F 1995 J. Biol. Chem. 270 11437.
[8] Otterbein L R, Graceffa P, and Dominguez R 2001 Science 293 708.
[9] Chu J W, and Voth G A 2006 Biophys. J. 90 1572.
[10] Pfaendtner J, Branduardi D, Parrinello M, Pollard T D, and Voth G A 2009 P Natl Acad Sci USA 106 12723.
[11] Kabsch W, Mannherz H G, Suck D, Pai E F, and Holmes K C 1990 Nature 347 37.
[12] Berendsen H J C, Postma J P M, Vangunsteren W F, Dinola A, and Haak J R 1984 J. Chem. Phys. 81 3684.
[13] Van Der Spoel D, Lindahl E, Hess B, Groenhof G, Mark A E, and Berendsen H J 2005 J. Comput. Chem. 26 1701.
[14] Hess B, Kutzner C, van der Spoel D, and Lindahl E 2008 J. Chem. Theory Comput. 4 435.
[15] Darden T, York D, and Pedersen L 1993 J. Chem. Phys .98 10089.
[16] Oostenbrink C, Soares T A, van der Vegt N F, and van Gunsteren W F 2005 Eur. Biophys. J. 34 273.
[17] Oostenbrink C, Villa A, Mark A E, and van Gunsteren W F 2004 J. Comput. Chem. 25 1656.
[18] Laskowski R A, Rullmannn J A, MacArthur M W, Kaptein R, and Thornton J M 1996 J. Biomol. NMR 8 477.
[19] Wriggers W, and Schulten K 1997 Proteins 29 1.
[20] Laskowski R A, Rullmann J A C, MacArthur M W, Kaptein R, and Thornton J M 1996 J. Biomol. Nmr 8 477.
[21] Laskowski R A, Macarthur M W, Moss D S, and Thornton J M 1993 J. Appl. Crystallogr. 26 283.
[22] Cong Y, Topf M, Sali A, Matsudaira P, Dougherty M, Chiu W, and Schmid M F 2008 J. Mol. Biol. 375 331.
[23] Holmes K C, Popp D, Gebhard W, and Kabsch W 1990 Nature 347 44.
[24] Atilgan A R, Durell S R, Jernigan R L, Demirel M C, Keskin O, and Bahar I 2001 Biophys. J. 80 505.
[25] Chennubhotla C, Rader A J, Yang L W, and Bahar I 2005 Phys. Biol. 2 S173.
[26] Yang L W, and Chng C P 2008 Bioinform. Biol. Insights 2 25.
[27] Doruker P, Atilgan A R, and Bahar I 2000 Proteins 40 512.
[28] Tama F, Gadea F X, Marques O, and Sanejouand Y H 2000 Proteins 41 1.
[29] Philippe Durand, Georges Trinquier, and Yves-Henri Sanejouand 1994 Biopolymers 34 759.
[30] Deriu M A, Soncini M, Orsi M, Patel M, Essex J W, Montevecchi F M, and Redaelli A 2010 Biophys. J. 99 2190.
[31] Deriu M A, Bidone T C, Mastrangelo F, Di Benedetto G, Soncini M, Montevecchi F M, and Morbiducci U 2011 J. Biomech. 44 630.
[32] Kojima H, Ishijima A, and Yanagida T 1994 Proc. Natl. Acad. Sci. U S A 91 12962.
[33] Tsuda Y, Yasutake H, Ishijima A, and Yanagida T 1996 Proc. Natl. Acad. Sci. U S A 93 12937.
[34] Yasuda R, Miyata H, and Kinosita K, Jr. 1996 J. Mol. Biol. 263 227.
[35] Bahar I, and Rader A J 2005 Curr. Opin. Struct. Biol. 15 586.
[36] Deriu M A, Soncini M, Orsi M, Patel M, Essex J W, Montevecchi F M, and Redaelli A 2010 Biophys. J. 99 2190.

[37] Lange K, and Brandt U 1996 Febs Lett. 395 137.
[38] Tirion M M, Benavraham D, Lorenz M, and Holmes K C 1995 Biophys. J. 68 5.
[39] Dalhaimer P, Pollard T D, and Nolen B J 2008 J. Mol. Biol. 376 166.
[40] Yanagida T, Nakase M, Nishiyama K, and Oosawa F 1984 Nature 307 58.
[41] Strzeleckagolaszewska H, Moraczewska J, Khaitlina S Y, and Mossakowska M 1993 Eur J Biochem 211 731.

9th International Fröhlich's Symposium IOP Publishing
Journal of Physics: Conference Series **329** (2011) 012029 doi:10.1088/1742-6596/329/1/012029

Chemotherapy Drugs Thiocolchicoside and Taxol Permeabilize Lipid Bilayer Membranes by Forming Ion Pores

Md. Ashrafuzzaman[1], **M Duszyk**[2], **J A Tuszynski**[1, 3]
Department of Oncology[1], Cross Cancer Institute, University of Alberta, Edmonton, Canada
Department of Physiology[2], University of Alberta, Edmonton, Alberta, Canada
Department of Physics[3], University of Alberta, Edmonton, Alberta, Canada

E-mail: ashrafuz@ualberta.ca

Abstract. We report ion channel formation by chemotherapy drugs: thiocolchicoside (TCC) and taxol (TXL) which primarily target tubulin but not only. For example, TCC has been shown to interact with $GABA_A$, nuclear envelope and strychnine-sensitive glycine receptors. TXL interferes with the normal breakdown of microtubules inducing mitotic block and apoptosis. It also interacts with mitochondria and found significant chemotherapeutic applications for breast, ovarian and lung cancer. In order to better understand the mechanisms of TCC and TXL actions, we examined their effects on phospholipid bilayer membranes. Our electrophysiological recordings across membranes constructed in NaCl aqueous phases consisting of TCC or TXL under the influence of an applied transmembrane potential (V) indicate that both molecules induce stable ion flowing pores/channels in membranes. Their discrete current versus time plots exhibit triangular shapes which is consistent with a spontaneous time-dependent change of the pore conductance in contrast to rectangular conductance events usually induced by ion channels. These events exhibit conductance (~0.01-0.1 pA/mV) and lifetimes (~5-30 ms) within the ranges observed in e.g., gramicidin A and alamethicin channels. The channel formation probability increases linearly with TCC/TXL concentration and V and is not affected by pH (5.7 - 8.4). A theoretical explanation on the causes of chemotherapy drug induced ion pore formation and the pore stability has also been found using our recently discovered binding energy between lipid bilayer and the bilayer embedded ion channels using gramicidin A channels as tools. This picture of energetics suggests that as the channel forming agents approach to the lipids on bilayer the localized charge properties in the constituents of both channel forming agents (e.g., chemotherapy drugs in this study) and the lipids determine the electrostatic drug-lipid coupling energy through screened Coulomb interactions between the drug molecules and lipids. The strength of this electrostatic energetic coupling determines the stability of the drug-induced ion pores. These findings may elucidate cytotoxic effects of chemotherapy drugs and aid in the development of novel drugs for a broad spectrum of cancers and other diseases.

Introduction

Colchicine has a long history of use in immune-system diseases (1, 2). In 2009, colchicine won FDA approval as a stand-alone drug for acute gout and familial Mediterranean fever. It inhibits leukocyteendothelial cell adhesion (3) and T-cells activation (4) by binding to tubulin dimers, which prevents their polymerization into microtubules (MTs) (5). Due to increased rate of mitosis cancer cells are more vulnerable to colchicine poisoning than normal cells. The potential usefulness of colchicine in cancer chemotherapy is limited by its toxicity against normal cells. Colchicine shifts the dynamic equilibrium in MT's towards disassembly by sequestering most of tubulin (6) and inducing slow disassembly of MT's (7). This may account for the gradual change of the membrane action potential (decrease) or threshold (increase) and the resting potential which, changed at modest concentrations of colchicine by only 2.5 mM, at higher concentrations (10 mM) was reduced by up to ~ 5 mV (6). Colchicine is also found to bind in the nuclear periphery which effects a constriction of the pore-complexes with concomitant inhibition of RNA egress and disordering of the nuclear membrane phospholipid bilayers (8). Since its FDA approval in 1992 taxol has been used for ovarian, lung, breast cancer, etc. (9) due to its stabilization of MTs (10) which makes it suitable for use in combination therapy (11). However, taxol's poor solubility is a serious problem that requires conjugation with cremophor or albumin. A taxol-phospholipid liposome construct increases taxol's antitumor efficacy by delaying tumor progression compared to the effectiveness of free taxol administered in cremophor (12). Taxol is also found to inhibit endosomal-lysosomal membrane trafficking by favorably inducing small densed vesicles over larger ones (13). This suggests that taxol affects the structural integrity of membranes by perhaps changing the lipid curvature profiles. Taxol's effects on liposomes were elucidated using fluorescence, circular dichroism, differential scanning calorimetry, fluorescence polarization and X-ray diffraction (14) showing that taxol incorporated into liposomes penetrates into the acyl chain domain which alters physical properties e.g., the range of the phospholipid phase transition, lipid order parameter, fluidity etc. of both artificial and biological membranes. Both MT-stabilizing taxol and MT-disrupting colchicine were found not to affect the human platelet membrane lipid fluidity (15). The evidence of strong interactions of colchicine with lipids (16) also suggests that colchicine is a poorer candidate as an anti-cancer drug by being weakly accessible to tubulin. Although the anti-cancer efficacy of colchicine and taxol has been found to be affected by lipids, there is still a substantial lack of information about the effects of these molecules on cell membranes, especially in regard to the membranes transport properties which depend on physical, geometrical and structural aspects of membranes. Studying these effects will shed important light on off-target interactions of these chemotherapeutic agents and hence may lead to an improved drug design in the future.

MT's are hollow cylinders constructed from a heterodimer of two 55 kDa proteins known as α and β tubulin. The tubulin dimer locks a single molecule of GTP within the non-exchangeable nucleotide binding site of α tubulin (17) and an additional molecule of GTP binds the β-subunit but it can hydrolyze into GDP following polymerization (Figure 1) providing the conformational flexibility required during the polymerization-depolymerization cycle (18). MTs observed *in vitro* are composed of 11 to 18 individual protofilaments, whereas *in vivo* they are made up of only 13 protofilaments (19). MTs are involved in numerous critical, cellular processes including mitosis, cell motility, maintenance of cellular morphology, and the activity of cell surface receptors. (20, 21). MTs exhibit a process of constant polymerization and depolymerization called dynamic instability, allowing them to efficiently search for attachment sites (e.g. to the kinetochore of a chromosome (22)). Paclitaxel and its analog docetaxel, some of the most successful cancer chemo-therapy agents, bind to a luminal location of an MT and stabilize it (23-25). Hydrogen/deuterium exchange coupled with mass liquid chromatography demonstrated that increased rigidity in paclitaxel stabilized MTs was distinct from stabilization as a result of GTP-induced polymerization (26). A different drug site within β-tubulin was shown to bind colchicine, a water-soluble alkaloid (27) that binds only to free tubulin and not to MTs. Colchicine has extremely strong anti-mitotic activity but its use as a cancer treatment is limited since it binds tubulinindiscriminately.

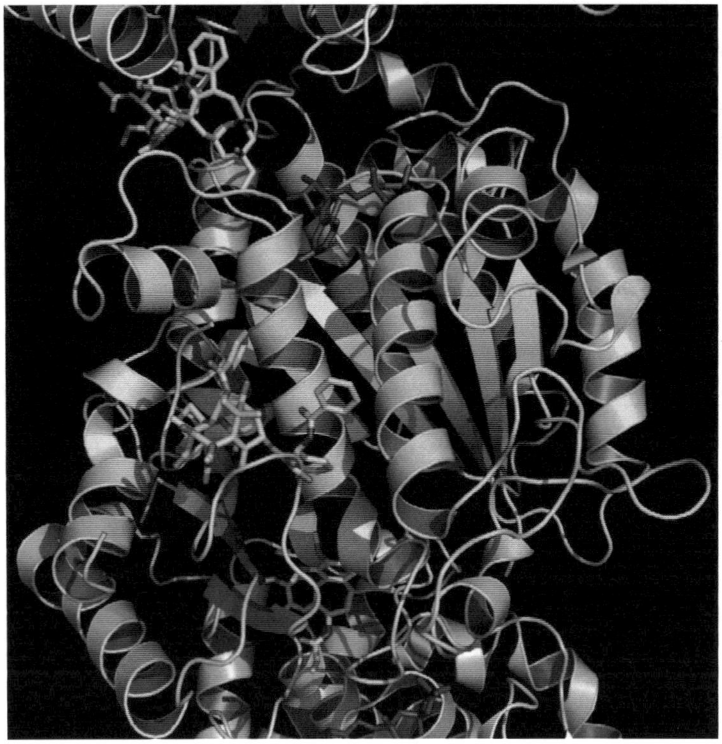

Fig. 1: Paclitaxel, colchicine and Vinca binding sites on α/β tubulin where drug molecules are in green. From bottom to top, colchicine, paclitaxel and vinblastine from PDB files: 1SA0, 1JFF and 1Z2B have been superimposed and fit back onto the 1SA0 structure to obtain the relative positioning of each drug within the β tubulin monomer (cyan) and two α tubulin monomers (yellow) at the top and bottom of the frame. The GTP at the non-exchangeable and GDP at the exchangeable site are colored purple.

Although many studies investigated different aspects of the effects of the chemotherapy drugs in the inner cellular level, none addresses their function in the cellular wall. All the drug effects on the MT assembly depend highly on the fractional cellular intake of the drugs. The intermediate membrane wall and hydrophilic/hydrophobic boundaries in the drug pathways undoubtedly influence the ultimate effectiveness and cytotoxic effects of the anticancer drugs. Effects of the drugs on the membrane's transport properties and the interaction between drug molecules and the membranes constituents are important but remain largely unexplored. The relatively small size of the drug molecules primarily suggests that their mechanisms are distinct from antimicrobial mechanism(s) of membrane permeabilization, e.g. formation of well structured long-lived protein-lined (28-30) ion channels. Otherwise, their amphiphile type effects would modulate the bilayer physical properties like e.g., triton X-100 and capsaicin effects (31) or transient defects disordering in the lipid layers as observed for the antimicrobial peptide gramicidin S (GS) (32). The goal here was to evaluate the two drug molecules thiocolchicoside (TCC) and taxol (TXL) (Fig. 2) for their effects on lipid model membranes using electrophysiology. Our results suggest that these drug molecules do not only permeabilize lipid bilayer membranes but also create certain structures through reorganization between lipid molecules and themselves. We provide evidence for the formation of stable ion pores/channels inside lipid model membranes involving these molecules whose characteristics are unconventional. That is, the discrete conductance events do not exhibit rectangular shapes in their current versus time plots (as seen in most antimicrobial peptides/membrane proteins-induced channels) but instead exhibit novel triangular shape effects. We therefore propose that these chemotherapy drug-induced pores/channels create toroidal-type channel structures (33-37).

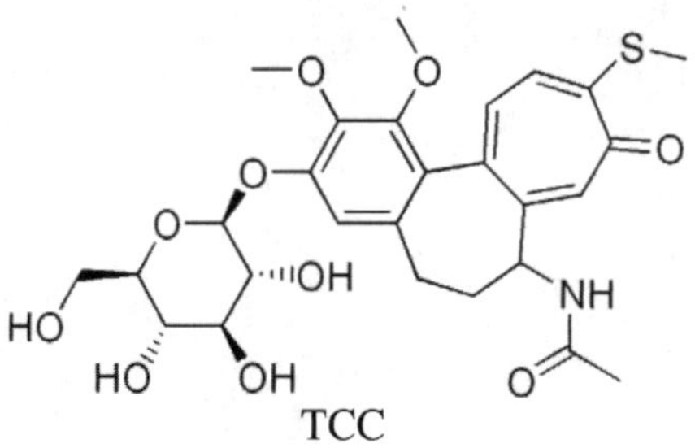

TCC

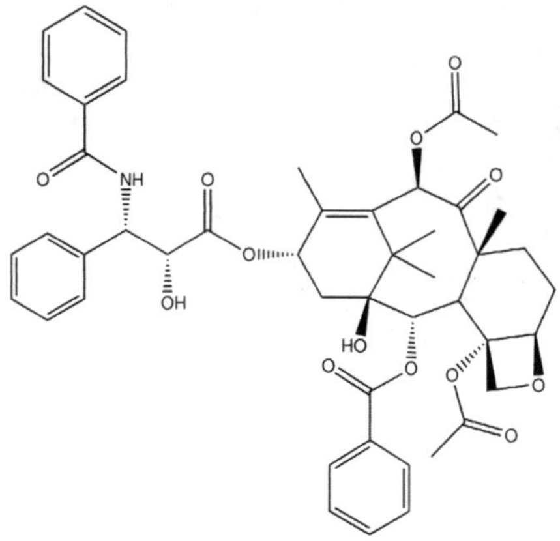

TXL

Fig. 2: Structure of TCC and TXL (Taxol-A).

Materials

Phosphoethanolamine (POPE), phosphatydyleserine (POPS) and phosphatidylcholine (POPC) were from Avanti Polar Lipids (Alabaster, AL, USA), were used without further purification. *n*-Decane was 99.9% pure from ChemSampCo (Trenton, NJ, USA). TCC was from ChemRoutes (Edmonton, Canada) and TXL was from Vector Lab (Berlington, Canada). Dimethylsulfoxide (DMSO) was from Burdick and Jackson (Muskegon, MI).

Methods

Planar lipid bilayers were formed by applying the lipid cocktail of POPE: POPS: POPC (5:3:2, v/v/v)/n-decane. The painting method was used to form membranes over a 150 μm septum of a bilayer cuvette. The working volume of bilayer cuvettes was 1 mL in both the *cis* (recording electrode) and *trans* (reference electrode) chambers. TCC and TXL stock solutions were prepared in DMSO (4 mg/mL), and were then further diluted (1 mg/mL) in standard buffer (0.5 M NaCl, 10 mM HEPES, pH 7.4) to be used in doing experiments. For pH experiments, the pH buffer was changed from HEPES to 2-(N-morpholino) ethanesulfonic acid (MES) for solutions at pH 5.7. TAPS (2-hydroxy-1,1-bis(hydroxymethyl)ethyl)amino-1-1 propanesulfonic acid)) was used to make the buffer (NaCl solution) at pH 8.4. After formation of the bilayer, we waited for about 1 hr and then tested the stability of the bilayer by applying transmembrane potential V=400 mV across it using both polarities. TCC or TXL stock solution was added to the *cis* chamber with continuous stirring for 10 min. The observed ion channel activity at reasonable V was recorded as traces.

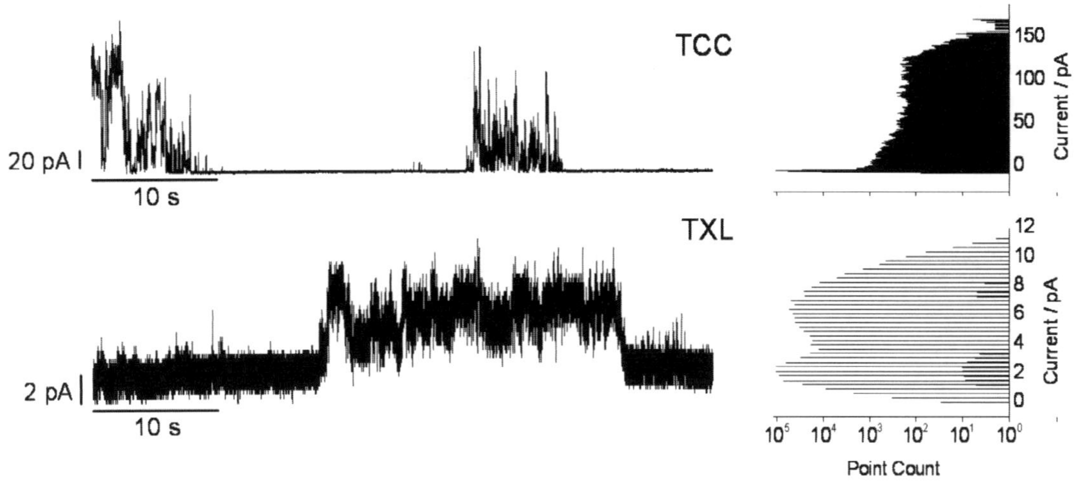

Fig. 3: TCC and TXL (both at 90 μM) permeabilize lipid bilayer membranes by inducing nonzero current events. Here, pH=5.7, V=100 mV. Both traces were filtered at 20kHz but the lower one shows higher noise due to its presentation (current axis) at amplified scale. We also observed conductance events (data not shown) at a much lower range and for more biologically relevant (to cancer therapy) concentrations e.g. at drug concentration c_D=10 μM for both TCC and TXL but at V~200 mV or higher.

Results

Fig. 3 shows considerable conductance events observed when a V is applied across a lipid bilayer membrane doped with TCC or TXL. A lipid membrane itself is generally nonconducting to ions but in the presence of TCC or TXL different current levels were seen. The long-time records indicate that time independent/random appearance of current events was induced by both compounds. The stability of different current levels varied significantly. The point count plot shows that the conductance events do not correspond to discrete current level(s) with any discrete values of their conductance. The current levels appeared with all possible conductance which is similar to the GS induced conductance events (32). Although no major qualitative difference between TCC and TXL-induced membrane permeabilization was observed, TXL-induced current levels cover relatively lower conductance range than those for TCC. Fig.4 shows the change of activity (A) with the change of V or TCC/TXL concentration (c_D) in the aqueous phase bathing the membranes. A linear progression relation between A and V or c_D was seen before A approached 1.0 at very high V and c_D. Although at very high V and c_D the membrane seems to show spontaneous nonzero resistance against ion flow, there are no standard lower or upper cut off values of both V and c_D to initiate or saturate respectively the appearance of conductance events. The lower/upper cut off values for V can be increased/decreased by decreasing/increasing the value of c_D and vice versa.

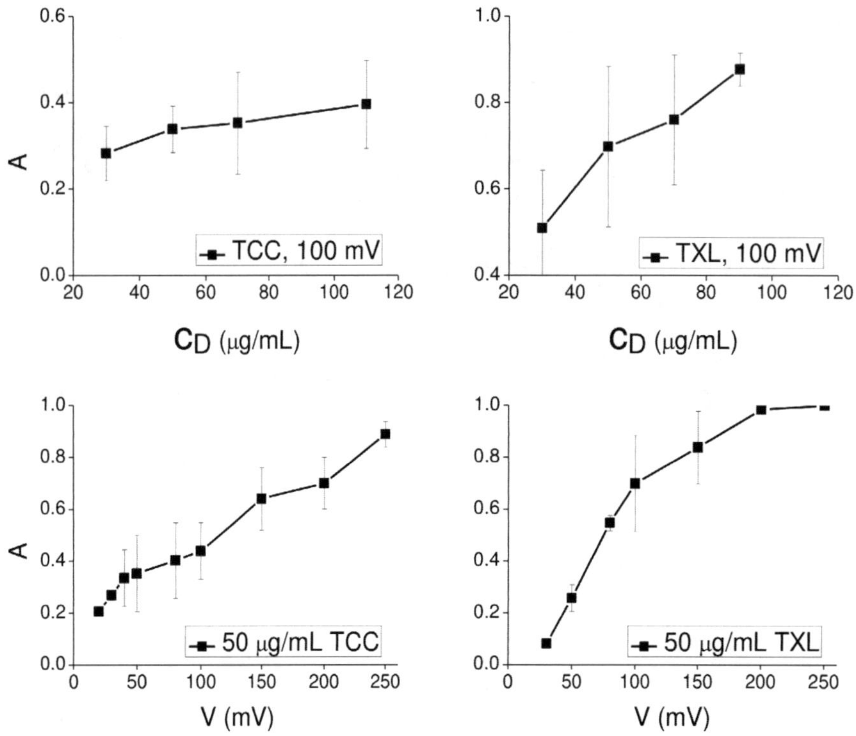

Fig. 4: TCC or TXL-induced channel activity A increases proportionally with both the V and drug concentration c_D. Here Activity ($A=\sum A_i/(\sum A_i+A_{nc})$) is defined as the ratio of the total point count in all conducting phases relative to the sum of the total point counts of both conducting phases and the nonconducting phase. A_i and A_{nc} are total point counts in the conducting phase i and the nonconducting phase. pH 5.7.

In Fig.5 short-time records of drug-induced current events show the appearance of independent conductance events. Any current event can be due to independent conductance event or due to a combination thereof. Current event therefore does not represent any controlled or reproducible parameter to address the nature of the induced conducting pore across membranes. To understand it we need to find independent conductance events, few of which are shown in Fig.5. Note that the independent conductance events appear with triangular shapes meaning conductance of any single event is not constant but it increases/decreases spontaneously over the time interval comparable to the 'lifetime' of any specific conductance event. To the best of our knowledge, the appearance of triangular conductance events is unique. The heights of the peak points of different conductance events are also different. We observed random spontaneous transitions between different current levels within a discrete conductance event during its lifetime. The discrete conductance events detected here were found to be roughly characterized by conductance ~0.01-0.1 pA/mV and lifetimes ~5-30 ms. For comparison, we have also shown the nature of conductance events observed in two other well studied channels namely gramicidin A (gA) and alamethicin (Alm). Like conductance events in channels formed by integral membrane proteins or antimicrobial peptides both gA and Alm, conductance events appear with rectangular shapes meaning transition between discrete current levels is transient or extremely abrupt. In gA channels only a back-and-forth transition between two current levels due to gA dimer/monomer states while in Alm channels many transitions between different conductance states consisting of varied numbers of monomers participating in the formation of channels are found. Importantly in both of these structurally different channels the transition between different current levels corresponding to discrete states is totally transient. The spontaneous transition between different random current levels in TCC/TXL-induced conductance events is therefore an important novel finding.

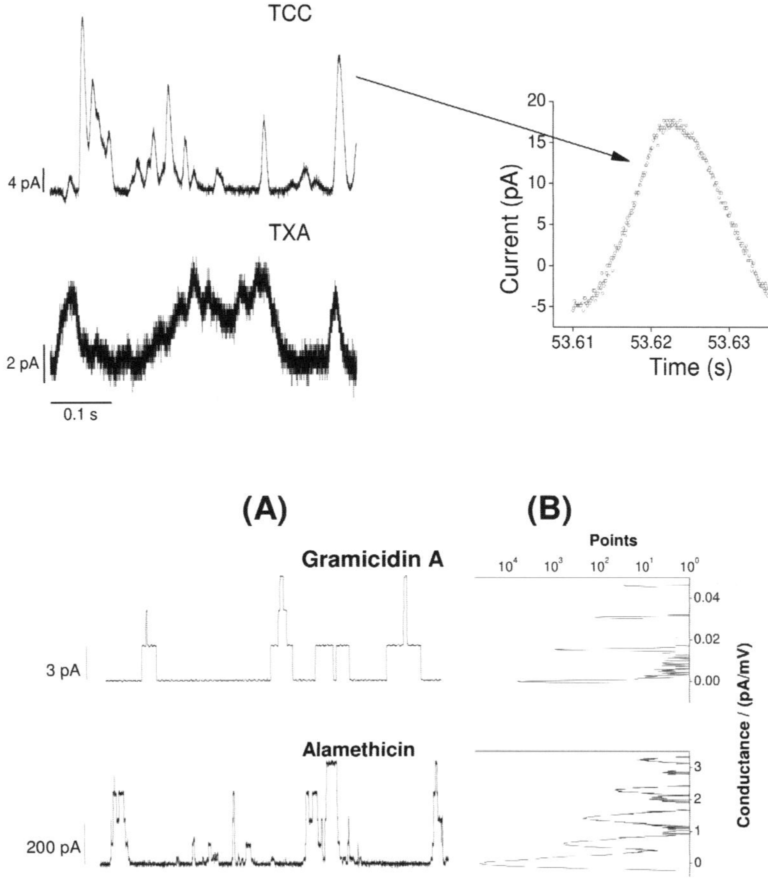

Fig. 5: The upper panel shows triangular-shape conductance events induced by TCC and TXL, both at 90 μM. pH=5.7, V=100 mV. Both traces were filtered at 20 kHz but the lower one shows higher noise due to its presentation (current axis) at an amplified scale. In a high resolution plot (shown in the right side of the arrow) of a single event only with showing individual points (in Origin 8.5 plot) we observe all points (open circle) with increasing and decreasing, respectively, corresponding values of conductance at both left and right lateral sides of the chemotherapy drug induced triangular conductance events. The lower panel **(A)** illustrates rectangular-shape conductance events in gA and Alm channels (32). gA channel activity was recorded at 200 mV and Alm at 150 mV. Traces

representing gA and Alm channel activities in phospholipid bilayers were recorded at filter frequencies 2 kHz and 20 kHz, respectively. A lower filter frequency for traces representing gA channel activity is alright because of the channel's relatively higher stability. In **(B)** the point count plots of the current traces through gA and Alm channels peak at discrete values of conductance.

A

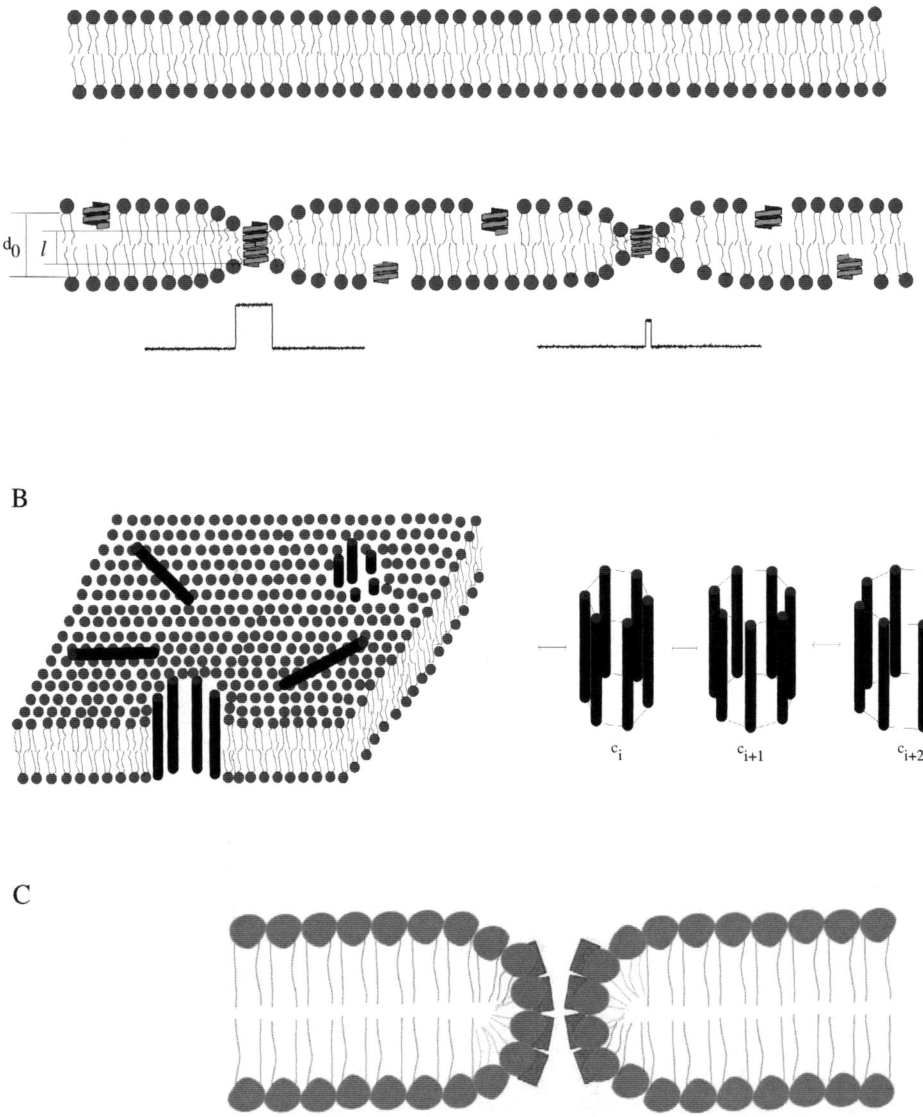

Fig. 6: A, gA channels deform lipid bilayer's resting thickness at the cost of a bilayer deformation energy originating from electrostatic interaction between lipids and gA monomers to compensate the hydrophobic mismatch (d_0-l) between bilayer thickness (d_0) and gA channel length (l). With channel formation bilayer conducts a current pulse with an average pulse width (channel lifetime $\tau \sim$ ms) and height (channel conductance ~ pA) depending on gA monomers forming channels and lipids forming bilayers. Two types of monomers have schematically structured two different channels here. B, In Alm channels cylindrical rods represent monomers on a 'barrel stave' pore. Transition between

different Alm conductance pores by addition/release of monomer(s) from/to the surrounding vicinity is shown in lower panel. C, Chemotherapy drugs TCC and TXL form toroidal-type ion channels which raises the possibility of spontaneous change of the pore radius due to the complex interaction and/or hydrophobic coupling between the lipids and the pore inducing drug molecules (TCC, TXL) as observed experimentally by us.

Discussion and conclusions

Two chemotherapy drugs (TCC and TXL) were investigated for their effects on membrane's electrical conductance due to the formation of special structures inside or across membranes. Erdal et al. has recently reported that a large number of potential anticancer drugs induce p53 protein-independent apoptosis and that lysosomal membrane permeabilization is a mediator of many such responses (38). Our discovery of both membrane permeabilization and the formation of ion pores/channels due to TCC and TXL sheds new light on the cytotoxic effects and other interactions exhibited by these chemotherapy drugs. Stimulation of conductance events due to chemotherapy drugs may also lead to the discovery of a novel type of ion channel formed by compounds whose primary mode of action involves cytoplasmic proteins. All lipid membrane permeabilizing membrane proteins or antimicrobial peptides have been found to form ion channels in which transitions between different current levels happen instantaneously. Only the drugs TCC and TXL induced single conductance events with a current versus time plot exhibits triangular shapes. This indicates that the induced pore radius spontaneously changes back-and-forth between smaller and bigger pores with all possible cross sectional areas in between. The spontaneous change of the TCC/TXL pore conductance immediately rules out the possibility of channel formation which might look like linear gA type, barrel stave Alm type, etc. channels. Although the point count plots of the long time current traces (Fig.3) resemble those observed for GS which induce defects inside membranes (32), a fine tuning of the single conductance events (Fig.5) suggests that the events are long-lived, not transient like those observed for GS (32). No such back-and-forth spontaneous transitions between different non-zero current levels as observed in a chemotherapy drug-induced discrete conductance event (Fig.5) are found in GS-induced defects which shows instantaneous/transient conductance events between zero conducting states only. TCC/TXL events appear with somewhat similar characteristics as Alm conductance events with back-and-forth transitions, though instantaneous, between different non-zero current levels (Fig.5). Our results, however, can neither rule out nor support the possibility of a mechanism which is somewhat similar to the so-called in-plane diffusion model of Bechinger (39, 40). Here, the bilayer disrupting molecules have been found inserted into the phospholipid bilayer disordering the hydrocarbon chains of the adjacent phospholipid molecules, locally thinning the bilayer, leading to local disturbances in bilayer packing and an eventual formation of toroidal type pores (Fig.6). TCC/TXL induced conductance events do not appear to be random defects (e.g., like GS-induced defects) or like other protein lined pores/channels (e.g, gA induced linear β-helix, Alm induced barrel-stave pore etc.), but rather more like events which correspond to a special structure induced inside membranes by forming a TCC/TXL and lipid molecule complex. Here we assume that lipids are forced to line across pore openings (Fig.6) and TCC/TXL molecules sit behind the head group region near the hydrocarbon chains of the lipids (14). Moderate dependence of channel forming activity on c_D and V (linear) (which is much lower than the 2.6 power dependence on the identical parameters in the cases of protein-lined channels (41)) also provides support for the TCC/TXL channel (Fig. 6). This allows the membrane to be doped with higher amounts of the channel forming molecules at a moderate energy cost as the molecules just penetrate into the lipid headgroup-hydrocarbon chain interface. Only this type of broken membrane structure can ensure the spontaneous change of the pore's cross sectional area which leads to the formation of triangular conductance events with spontaneous change of pore conductance. In the case of protein-lined channels the membrane's resting thickness never vanishes, rather shows some kind of bilayer deformation from its resting thickness as was predicted in another mechanosensitive channels study (42). The chemotherapy drug induced channel formation effects were also investigated changing the pH of the aqueous phase and unlike other protein-lined channels e.g. gA channels (43) and Alm channels, which report substantial phenomenological effect of pH changes, we observe almost no considerable effect of pH (Fig. 7). This suggests two main conclusions: (i) formation mechanism of chemotherapy drug induced channels is different from other conventional protein-lined channels and (ii) the investigated chemotherapy drug molecules appear to be equally effective in both normal cellular membranes and cancer cell membranes with different pH environments (44). Both of these points together make proper sense considering that the chemotherapy molecules are, on a broader scale, charge neutral while other channel forming peptides are usually charge bearing with specific (helical, beta sheet etc.) structures.

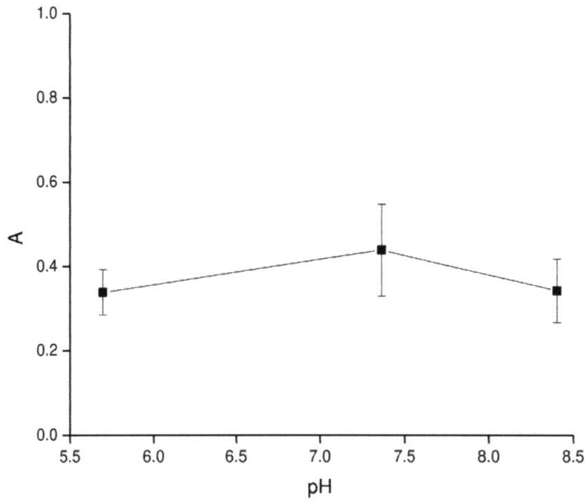

Fig. 7: TCC pore/channel activity A is independent of pH of the membrane bathing aqueous phase. c_D=50 μg/mL, V=100 mV.

Energetics behind the creation of chemotherapy drug induced membrane transport events ion pores

A theoretical explanation on the causes of chemotherapy drug induced ion pore formation and the pore stability can be found by analyzing the recently discovered general pictures of energetics behind the functions of protein-lined ion channels gA and Alm (45). gA and Alm are the two most studied ion channels which form β-helical (46-48) and barrel-stave (28, 29) pores, respectively, across lipid bilayer membranes (see Figure 6). In both cases the channels couple with the membrane's two adjacent monolayers at two longitudinal edges of the channels. We shall here address how the channel bilayer energetic coupling regulates integral channel functions (45).

In bilayer-spanning channel formation the association of two trans-bilayer gA monomers is governed by the dimerization coefficient: $K_D=[D]/[M_g]^2= k_1/k_{-1}=\exp\{-(\Delta G^0_{prot}+\Delta G^0_{def})/k_BT\}$, where $[M_g]$ and $[D]$ are monomer and dimer concentrations; and k_1 and k_{-1} are rate constants (43). Here, k_B and T are the Boltzmann constant and absolute temperature, respectively. Since the bilayer deformation energy ΔG^0_{def} is sensitive to the hydrophobic mismatch (d_0-l) between bilayer thickness (d_0) and gA channel length (l), the bilayer responds to its deformation by imposing a restoring/channel-dissociation force F_{dis} on the edges of a channel. Increasing F_{dis} is reflected in a decreasing τ and channels become molecular force transducers (43). Within limits, the channel structure is invariant when the lipid bilayer thickness is varied (49), meaning that the gA channels are more rigid than the host bilayer. All-atom molecular dynamics simulations of gA in bilayers (50) show how lipid head groups organize themselves in the region of hydrophobic free length d_0-l. Potential-of-mean-force calculations (51) suggest that trans-membrane protein interactions are regulated by a hydrophobic mismatch equivalent to d_0-l. The calculation of F_{dis} has been a long-standing challenge. Based on the so called theory of elastic bilayer deformation (52) $\Delta G^0_{\mathrm{def}}$ has been found to be approximately changing as a quadratic function of d_0-l (52, 53) but subsequent developments (54-58) resulted in the introduction of the lipid intrinsic curvature c_0 (whose positive and negative changes correspond to increases and decreases of the hexagonal lipid phase, respectively, and any such local curvature profile generally controls the lipid packing energy profiles in bilayers) (details in ref. (59)) into the expression for $\Delta G^0_{\mathrm{def}}$ which is now considered to be changing as a quadratic function of d_0-l and intrinsic curvature c_0 ($\Delta G^0_{\mathrm{def}} = H_B.(d_0\text{-}l)^2 + H_x.(d_0\text{-}l).c_0 + H_c.c_0^2$), consequently, F_{dis} is found to be linearly dependent on d_0-l and c_0 [43, 60]: $F_{\mathrm{dis}} = -(-(\partial/\partial(d_0 - l))\Delta G^0_{\mathrm{def}}) = 2H_{\mathrm{B}}\cdot(d_0 - l) + H_{\mathrm{X}}\cdot c_0$, where H_{B}, H_{X} and H_c are phenomenological elastic constants. Using specific 'elastic parameters' in a fluid-like membrane is a good first-order approximation that works well within the limitations of a linear theory. However, in order to extend the applicability of the theory to a nonlinear regime, we propose to use the screened Coulomb interaction approximation (45). I shall briefly describe this published screened Coulomb interaction model which perfectly describes the general bilayer regulation of membrane protein functions. This publication (45) also correctly describes why the well accepted so called theory of elastic bilayer deformation model (52-58) is totally unacceptable due to its inability to explain important general phenomena related to the membrane regulation of membrane protein functions. A brief explanation on this issue will also automatically come here while describing the screened Coulomb interaction model.

In the screened Coulomb interaction model (45), considering $l<d_0$, the channel extends its Coulomb interaction towards lipids sitting on the bilayer's nearest resting thickness. A gA channel directly interacts with a nearest-neighbor lipid by Coulomb forces and this lipid interacts directly with the next-nearest-neighbor lipid but this second lipid's interaction with the channel is screened by the channel's nearest-neighbor lipid. The interaction between the third-nearest neighbor and the channel is screened by the lipids in between. An assumption has been made regarding all lipids participating in this chain that their interactions exist in an identical dielectric environment. The general form of the screened Coulomb interaction is

$$V_{sc}(\vec{r}) = \int d^3k Exp\{i\vec{k}.\vec{r}\} V_{sc}(\vec{k}) \quad (1)$$

whose Fourier transform is (61)

$$V_{sc}(\vec{k}) = \frac{V(\vec{k})}{1 + \frac{V(\vec{k})}{2\pi k_{\mathrm{B}} T} n} \quad (2)$$

where $V(k)=(1/\varepsilon_0\, \varepsilon_r) q_g q_L / k^2$ is the direct Coulomb interaction between gA monomer (c harge q_g) in a channel and the nearest-neighbor lipid. $k \approx 2\pi/r_{\mathrm{LL}}$, r_{LL} is the average lipid-lipid distance (62) which has been assumed to also be the distance between the channel's longitudinal edge and the nearest lipid head group and has been considered here (for simplicity) to correspond to a certain lipid type only. In reality this may also change due to many parameters e.g., variations in the membrane's electrical conditions, the presence of hydrocarbons within, etc., n is lipid density ~1/60 Å^2. Obviously, $k_{\mathrm{B}}T \approx 1.38\ 10^{-23}$ Joule/K (300 K). Here, ε_0 is the dielectric constant in vacuum and ε_r (~2) is the relative dielectric constant inside the membrane (63).

The strong binding energy between two monomers with identical charge profiles in a gA channel inside membranes ($U_{g,g}$) is due to the Lennard-Jones and Coulomb potentials which is supported by earlier work on the derivation of an effective attractive interaction potential between charges of the same type (contrary to the generally expected repulsive interaction) in solution (64) and due to arguments presented in ref. (45). A change of gA channel stability is mainly due to the change of gA channel bilayer coupling energy ($U_{g,bilayer}$) even though the total binding energy is given by $U(r) = U_{g,g} + U_{g,bilayer}$. Here, $U_{g,bilayer}$ is a 1st, 2nd, etc. order term in the expansion of $V_{sc}(r)$ for the hydrophobic mismatch to be filled by single, double etc. lipids representing 1st, 2nd, etc. order screening, respectively. ΔG^0_{prot} and ΔG^0_{def} are proportional to $U_{g,g}$ and $U_{g,bilayer}$, respectively. F_{dis} therefore originates from mechano-electrical properties of membrane and gA monomers although at $d_0 \approx l$ ($\Delta G^0_{def} \sim 0$) any fluctuation in ΔG^0_{prot} may appear as a channel function regulator, too. A complicated computational program using Mathematica 7 has revealed that $\Delta G^0_{def} \approx \mathrm{Exp}\{d_0 - l\}$ which suggests that $F_{dis} \approx \mathrm{Exp}\{d_0 - l\}$, different from that found in the elastic bilayer consideration that is $F_{dis} \approx d_0 - l$ (mentioned earlier). The gA channel lifetime $\tau \sim \mathrm{Exp}\{-\Delta G^0_{def}/k_B T\} \sim \mathrm{Exp}\{-\lambda . F_{dis}/k_B T\}$ (for details see ref. (43)). where λ is the distance two gA monomers move apart to reach the dimer/monomer transition state (64). Slight differences in the bilayer thickness gA channel length mismatch dependence of the theoretical trend of gA channel lifetime appear to depend on whether we use the expression for F_{dis} from the screened Coulomb model ($\sim \exp(d_0 - l)$) or the elastic bilayer model ($\sim (d_0 - l)$) in the case when c_0 is assumed to be unchanged. We have clearly addressed that screened Coulomb interaction model is certainly needed over elastic bilayer model to add the much needed anharmonic terms into the bilayer deformation energy (45). The interaction energy originating from the localized charge properties of the lipids and peptides on the channels mainly contribute to the cause of bilayer channel coupling and the bilayer elastic property originated harmonic energetic term helps the bilayer-channel coupling to be stabilized. Screened Coulomb interaction model serves both of the purposes. The consideration of the bilayer channel harmonic coupling due to the bilayer elasticity can nohow represent the main bilayer channel coupling energetics rather serves some secondary support. This has been wrongly interpreted by many groups during almost three decades doing research in membrane protein functions.

The screened Coulomb interaction model also explains the lipid curvature and charge effects on the regulation of membrane protein functions. Negative c_0 is found from this model to linearly destabilize gA channels, which agrees with experimental observations (45). The model also shows that $\Delta G^0_{def} \sim (q_L/q_g)^s$, s=1, 2, etc. for 1st, 2nd, etc. order screening, respectively, suggesting that channel formation is harder in bilayers containing charged lipids.

Besides gA channel the screened Coulomb interaction model also explains all general aspects of membrane regulation of Alm channel functions. Interested readers should find this analogy from refs. (45).

The above explanations generally suggest that the energetic coupling between channel forming agents and the lipids on the bilayer due to primarily their localized electrical properties regulate the stability and other functions of ion channels/pores. The localized charges emerge perhaps due to the mutual polarizing effects of the localized charges in the atomic level when both lipids and pore forming agents approach to each other even though sometimes the overall net charges of both may be zero. Like some of the zwitterionic lipids chemotherapy drugs despite their having net zero charges therefore are likely to show localized charges as they approach to each other to form ion pore (Figure 6C). In this situation any empirical calculation of the force field finds partial charges on each atom in the drugs (65-68). Each drug then falls under the influence of screened Coulomb interactions with the lipid head group chain. The sum of all these screened interactions between all drugs and all lipids in the drug-bilayer interaction region make the effective binding energy which should follow the general picture of energetics as applied to both gA and Alm channels. Our ongoing molecular dynamics simulations suggest that there exist strong distance dependent electrostatic and van der Waal's interactions (~kcal/mole) between peptides (gA, Alm) or chemotherapy drugs (TCC, TXL) and membrane constructing phospholipids (manuscript in preparation by Ashrafuzzaman, Tseng and Tuszynski). These simulation results also validate our discovery that there exists strong bilayer channel electrostatic energetic coupling arising apparently from the effective localized charge profiles (equation (1)) of both lipids and channel forming peptides (45) or biomolecules like chemotherapy drugs. Other platforms will be used to present those groundbreaking discoveries.

In conclusion, chemotherapy drugs TCC and TXL both permeabilize lipid bilayer membranes by forming ion channels with a new type of conductance events with spontaneously changing channel conductance. This study sheds new light on the molecular mechanism of chemotherapy drugs both across the cell's membrane and in the cell's cytoplasm. This understanding may help find novel drugs to treat cancer and other diseases occurring both in the inner cellular and membrane regions.

Acknowledgment J.A.T. acknowledges funding from NSERC, the Allard Foundation and the Alberta Cancer Foundation. All authors are grateful to Drew Nahirney.

References

[1] Seidemann P, Fjellner B, Johannesson A (1987) Psoriatic arthritis treated with oral colchicine. *J. Rheumatol* 14: 777–779.

[2] Callen JP (1985) Colchicine is effective in controlling chronic cutaneous vasculitis in lupus erythematosus. *J. Am. Acad. Dermatol.* 13: 193–200.

[3] Rosenman SJ, Ganji AA, Gallatin WM (1991) contact dependent redistribution of cell surface adhesion and activation molecules reorganization *FASEB J.* 5: 1603.

[4] Mekory YA, Baram D, Goldberg A, Klajman A (1989) Inhibition of delayed hypersensitivity in mice by colchicines: Mechanism of inhibition of contact sensibility in vivo . *Cell. Immunol.* 120: 330–340.

[5] Borisy GO, Taylor EW (1967) The mechanism of action of colchicine: Colchicine binding to seaurchin eggs and the mitotic apparatus. *J. Cell. Biol.* 34: 533–548.

[6] Matsumoto G, Sakai H (1979) Microtubules inside the Plasma Membrane of Squid Giant Axons and their Possible Physiological Function. *J. Membrane Biol.* 50:1-14.

[7] Haga T, Kurokawa M (1975) Microtubule formation from two components separated by gel filtration of a tubulin preparation. *Biochim. Biophys. Acta* 392: 335.

[8] Agutter PS, Suckling KE (1982) Effect of colchicine on mammalian liver nuclear envelope and on nucleo-cytoplasmic RNA transport. *Biochim. Biophys. Acta* 698: 223-229.

[9] Holmes FA, Kudelka AP, Kavanagh JJ, Huber MH, Ajani JA, Valero V (1995) Current Status of Clinical Trials with Paclitaxel and Docutaxel. In *Taxane Anticancer Agents: Basic Science and Current Status;* Georg, G. I., Chen, T. C., Ojima, I., Vyas, D. M., Eds.; ACS Symposium Series No. 583; American Chemical Society: Washington, DC: 31-57.

[10] Schiff PB, Fant J, Horwitz SB (1979) Promotion of Microtubule Assembly *in vitro* by Taxol. *Nature* 277: 665-666.

[11] Fisherman J, McCabe M, Hillig M (1992) Phase I study of taxol and doxorubucin (Dox) with G-CSF in previously untreated metastatic breast cancer. *Proc. Am.SOC. Clin. Oncol.*: 1175A.

[12] Sharma A., Straubinger, RM (1994) Novel taxol formulations: Preparation and characterization of taxol-containing liposomes. *Pharm. Res.* 11: 889-896.

[13] Sonee M, Barron E, Yarber FA, Hamm-Alvarez SF (1998) Taxol inhibits endosomal-lysosomal membrane trafficking at two distinct steps in CV-1 cells. *Am. J. Physiol. Cell Physiol.* 44: 1630-1639.

[14] Balasubramanian SV, Straubinger RM (1994) Taxol-lipid interactions: taxol-dependent effects on the physical properties of model membranes. *Biochemistry* 33: 8941-8947.

[15] Shiba M, Watanabe E, Sasakawa S, Ikeda Y (1988) Effects of taxol and colchicines on platelet membrane properties. *Thromb Res.* 52: 313-323.

[16] Mons S, Veretout F, Carlier M, Erk I, Lepault J, Trudel E, Salesse C, Ducray P, Mioskowski C, Lebeau L (2000) The interaction between lipid derivatives of colchicines and tubulin:

Consequences of the interaction of the alkaloid with lipid membranes. *Biochim. Biophys. Acta* 1468: 381-395.

[17] Weisenberg RC, Deery WJ, Dickinson PJ (1976) Tubulin-nucleotide interactions during the polymerization and depolymerization of microtubules. *Biochemistry* **15**, 4248-4254.

[18] Wang HW, Nogales E (2005) Nucleotide-dependent bending flexibility of tubulin regulates microtubule assembly. *Nature* **435**, 911-915.

[19] Unger E, Bohm KJ, Vater W (1990) Structural diversity and dynamics of microtubules and polymorphic tubulin assemblies. *Electron Microsc Rev* **3**, 355-395.

[20] Pichichero ME, Avers CJ (1973) The evolution of cellular movement in eukaryotes: the role of microfilaments and microtubules. *Subcell Biochem* **2**, 97-105.

[21] Hyams JS, Stebbings H (1979) The mechanism of microtubule associated cytoplasmic transport. Isolation and preliminary characterisation of a microtubule transport system. *Cell Tissue Res* **196**, 103-116.

[22] Derry WB, Wilson L, Khan IA, Luduena RF, Jordan MA (1997) Taxol differentially modulates the dynamics of microtubules assembled from unfractionated and purified beta-tubulin isotypes. *Biochemistry* **36**, 3554-3562.

[23] Wani MC, Taylor HL, Wall ME, Coggon P, McPhail AT (1971) Plant antitumor agents. VI. The isolation and structure of taxol, a novel antileukemic and antitumor agent from Taxus brevifolia. *J Am Chem Soc* **93**, 2325-2327.

[24] Schiff PB, Horwitz SB (1980) Taxol stabilizes microtubules in mouse fibroblast cells. *Proc Natl Acad Sci U S A* **77**, 1561-1565.

[25] Ringel I, Horwitz SB (1991) Studies with RP 56976 (taxotere): a semisynthetic analogue of taxol. *J Natl Cancer Inst* **83**, 288-291.

[26] Xiao H et al. (2006) Insights into the mechanism of microtubule stabilization by Taxol. *Proc Natl Acad Sci U S A* **103**, 10166-10173.

[27] Ravelli RB et al. (2004) Insight into tubulin regulation from a complex with colchicine and a stathmin-like domain. *Nature* **428**, 198-202.

[28] Boheim G (1974) Statistical analysis of alamethicin channels in black lipid membranes. *J. Mem. Biol.* 19: 277-303.

[29] He K, Ludtke SJ, Huang HW, Worcester DL (1995) Antimicrobial peptide pores in membranes detected by neutron in-plane scattering. *Biochemistry* 34: 15614-15618.

[30] Andersen OS (1983) ION MOVEMENT THROUGH GRAMICIDIN A CHANNELS Studies on the Diffusion-controlled association step, *Biophys. J.* 41: 147-165.

[31] Ashrafuzzaman M, Andersen OS (2007) Lipid bilayer elasticity and intrinsic curvature as regulators of channel function: a single molecule study. *Biophys. J.*: 421a.

[32] Ashrafuzzaman M, Andersen OS, McElhaney RN (2008) The antimicrobial peptide gramicidin S permeabilizes phospholipid bilayer membranes without forming discrete ion channels. *Biochim. Biophys. Acta* 1778**:** 2814-2822.

[33] Matsuzaki K, Murase O, Tokuda H, Fujii N, Miyajima K (1996) An antimicrobial peptide, magainin 2, induced rapid flip-flop of phospholipids coupled with pore formation and peptide translocation. *Biochemistry* 35: 11361–11368.

[34] Ludtke SJ, He K, Heller WT, Harroun TA, Yang L, Huang HW (1996) Membrane pores induced by magainin. *Biochemistry* 35:13723–13728.

[35] Yang L, Harroun T, Weiss TM, Ding L, Huang HW (2001) Barrel-stave model or toroidal model? A case study on melittin pores. *Biophys. J.* 81: 1475-1485.

[36] Sobko AA, Kotova EA, Antonenko YN, Zakharov SD, Cramer WA (2006) Lipid dependence of the channel properties of a colicin E1-lipid toroidal pore. *The J. of Biol. Chem.* 281: 14408-16.

[37] Allende D, Simon SA, McIntosh TJ (2005) Melittin-induced bilayer leakage depends on lipid material properties: Evidence for toroidal pores. *Biophys J.* 88: 1828-1837.
[38] Erdal H, Berndtsson M, Castro J, Brunk U, Shoshan MC, Linder S (2004) Induction of lysosomal membrane permeabilization by compounds that activate p53-independent apoptosis. *Proc Natl Acad Sci USA* 102: 192-197.
[39] Bechinger B (1997) Structure and functions of channel-forming peptides: magainins, sercopins, melittin and alamethicin. *J. Membr. Biol.* 156: 197-211.
[40] Bechinger B (1999) The structure, dynamics and orientation of antimicrobial peptides in membranes by multidimensional solid-state NMR spectroscopy. *Biochim. Biophys. Acta* 1462: 157-183.
[41] Latorre M, Alvarez O (1981) Voltage-dependent channels in plannar lipid bilayer membranes. *Physiol. Rev.* 61: 77-150.
[42] Perozo E, Cortes DM, Sompornpisut P, Kloda A, Martinac B (2002) Open channel structure of MscL and the gating mechanism of mechanosensitive channels. *Nature* 418: 942-948.
[43] Ashrafuzzaman M, Lampson MA, Greathouse DV, Koeppe II RE, Andersen OS (2006) Manipulating lipid bilayer material properties using biologically active amphipathic molecules. *J. Phys.: Condens. Matter* 18: S1235-S1255.
[44] Gagliardi LJ (2005) Electrostatic Considerations in Mitogenesis. *Proceedings ESA Annual Meeting*: 227-241.
[45] Ashrafuzzaman M, Tuszynski J (2011) Regulation of channel function due to coupling with a lipid bilayer. *J. Comput. Theor. Nanosci.*: in press (accepted).
[46] Arseniev AS, Barsukov IL, Bystrov VF, Ovchinnikov YA (1986) Spatial structure of a gramicidin A transmembrane ion channel. NMR analysis in micelles. *Biol. Membr.* 3: 437-62.
[47] Ketchem RR, Roux B, Cross TA (1997) High-resolution polypeptide structure in a lamellar phase lipid environment from solid state NMR derived orientational constraints. *Structure* 5: 1655-69.
[48] Townsley LE, Tucker WA, Sham S, Hinton JF (2001) Structures of gramicidins A, B, and C incorporated into sodium dodecyl sulfate micelles. *Biochemistry* 40: 11676-11686.
[49] Wallace BA, Veatch WR, Blout ER (1981) Conformation of gramicidin A in phospholipid vesicles: circular dichroism studies of effects of ion binding, chemical modification, and lipid structure. *Biochemistry* 20: 5754-5760.
[50] Allen TW, Andersen OS, Roux B (2004) Energetics of ion conduction through the gramicidin channel. *Proc. Natl. Acad. Sci.* 101: 117-122.
[51] F. de Meyer and B. Smit (2009) Comment on "cluster formation of trans-membrane proteins due to hydrophobic mismatching". *Phys. Rev. Lett.* 102: 219801
[52] Huang HW (1986) Deformation free energy of bilayer membrane and its effect on gramicidin channel lifetime. *Biophys. J.* 50: 1061-1071.
[53] Helfrich P, Jakobsson E (1990) Calculation of deformation energies and conformations in lipid membranes containing gramicidin channels. *Biophys. J.* 57: 1075-1084.
[54] Durkin JT, Providence LL, Koeppe II RE, Andersen OS (1993) Energetics of heterodimer formation among gramicidin analogues with an NH2-terminal addition or deletion consequences of missing a residue at the join in the channel. *J. Mol. Biol.* 231: 1102-1121.
[55] Nielsen C, Goulian M, Andersen OS (1998) Energetics of inclusion-induced bilayer deformations. *Biophys. J.* 74: 1966-1983.
[56] Nielsen C, Andersen OS (2000) Inclusion-induced bilayer deformations: effects of monolayer equilibrium curvature. *Biophys. J.* 79: 2583-2604.
[57] Lundbæk JA, Birn PHAJ, Søgaard R, Nielsen C, Girshman J, Bruno MJ, Tape SE, Egebjerg J, Greathouse DV, Mattice GL, Koeppe II RE, Andersen OS (2004) Regulation of sodium

channel function by bilayer elasticity. The importance of hydrophobic coupling. Effects of micelle-forming amphiphiles and cholesterol. *J. Gen. Physiol.* 123: 599-621.

[58] Lundbæk JA, Birn P, Tape SE, Toombes GE, Søgaard R, Koeppe II RE, Gruner SM, Hansen AJ, Andersen OS (2005) Capsaicin regulates voltage-dependent ssodium channels by altering lipid bilayer elasticity. *Mol. Pharmacol.* 68:680–689.

[59] Gruner SM (1985) Intrinsic curvature hypothesis for biomembrane lipid composition: A role for nonbilayer lipids. *Proc. Natl. Acad. Sci. USA* 82: 3665-3669.

[60] Andersen OS, Koeppe II RE (2007) Bilayer thickness and membrane protein function: an energetic perspective. *Annu. Rev. Biophys. Biomol. Struct.* 36: 107-130.

[61] Ashrafuzzaman M, Beck H (2004) in Vortex dynamics in two-dimensional Josephson junction arrays, (University of Neuchatel, http://doc.rero.ch/record/2894?ln=fr), ch 5, p 85.

[62] Harper PE, Mannock DA, Lewis RNAH, McElhaney RN, Gruner SM (2001) X-Ray diffraction structures of some phosphatidylethanolamine lamellar and inverted hexagonal phases. *Biophys. J.* 81: 2693-2706.

[63] Parsegian A (1969) Energy of an Ion crossing a low dielectric membrane: solutions to four relevant electrostatic problems. *Nature* 221: 844-846.

[64] Grønbech-Jensen N, Mashl RJ, Bruinsma RF, Gelbart WM (1997) Counterion-induced attraction between rigid polyelectrolytes. *Phys. Rev. Lett.* 78: 2477-2480.

[65] Goulian M, Mesquita ON, Fygenson DK, Nielsen C, Andersen OS (1998) Gramicidin channel kinetics under tension. *Biophys. J.* 74: 328-337.

[66] Duan Y, Wu C, Chowdhury S, Lee MC, Xiong G, Zhang W, Yang R, Cieplak P, Luo R, Lee T (2003) A point-charge force field for molecular mechanics simulations of proteins based on condensed-phase quantum mechanical calculations. *J. Comput. Chem.* 24: 1999–2012.

[67] Lee MC, Duan Y (2004) Distinguish protein decoys by using a scoring function based on a new Amber force field, short molecular dynamics simulations, and the generalized Born solvent model. *Proteins* 55: 620–634.

[68] Jakalian A, Bush BL, Jack DB, Bayly CI (2000) Fast, efficient generation of high-quality atomic charges. AM1-BCC model: I. Method. *J. Comput. Chem.* 21: 132–146.

[69] Jakalian A, Jack DB, Bayly CI (2002) Fast, efficient generation of high-quality atomic charges. AM1-BCC model: II. Parameterization and Validation. *J. Comput. Chem.* 23: 1623–1641.

Journal of Physics: Conference Series **329** (2011) 012030 doi:10.1088/1742-6596/329/1/012030

Disturbances of electrodynamic activity affect abortion in human

A Jandová[1], M Nedbalová[2], J Kobilková[3], A Čoček[4], A Dohnalová[2], M Cifra[1] and J Pokorný[1]

[1]Institute of Photonics and Electronics, Academy of Sciences of the Czech Republic, Chaberska 57, 182 51 Prague, Czech Republic
[2]Institute of Physiology, First Faculty of Medicine, Charles University, Albertov 5, 128 00 Prague, Czech Republic
[3]Department of Gynecology and Obstetrics, First Faculty of Medicine, Charles University, Apolinářská 18, 128 51 Prague, Czech Republic
[4]Otorhinolaryngological Department, Third Faculty of Medicine, Charles University, Šrobárova 50, 100 34 Prague, Czech Republic

E-mail: jandova@ufe.cz

Abstract. Biochemical research of biological systems is highly developed, and it has disclosed a spectrum of chemical reactions, genetic processes, and the pathological development of various diseases. The fundamental hypothesis of physical processes in biological systems, in particular of coherent electrically polar vibrations and electromagnetic activity, was formulated by H. Fröhlich; he assumed connection of cancer process with degradation of coherent electromagnetic activity. But the questions of cellular structures capable of the coherent electrical polar oscillation, mechanisms of energy supply, and the specific role of the endogenous electromagnetic fields in transport, organisation, interactions, and information transfer remained open. The nature of physical disturbances caused by some diseases (including the recurrent abortion in humans and the cancer) was unknown. We have studied the reasons of recurrent abortions in humans by means of the cell mediated immunity (using immunologic active RNA prepared from blood of inbred laboratory mice strain C3H/H2K, infected with the lactate dehydrogenase elevating virus–LDV) and the cytogenetic examination from karyotype pictures. The recurrent abortion group contained women with dg. spontaneous abortion (n = 24) and the control group was composed of 30 healthy pregnant women. Our hypothesis was related to quality of endometrium in relation to nidation of the blastocyst. The energetic insufficiency (ATP) inhibits normal development of fetus and placenta. We hope that these ideas might have impact on further research, which could provide background for effective interdisciplinary cooperation of malignant and non-malignant diseases.

1. Introduction

The hypothesis of potential risk of multiplication of associations of acrocentric chromosomes for gravidity and, further, of a potential relationship of this phenomenon with the LDH-elevating virus infection was derived on the basis of the following findings: An enhanced associability of chromosomes and, at the same time, an increased susceptibility of leukocytes to the LDH-elevating virus antigen were recorded in women with disordered fertility of an unknown etiology as compared with control groups. Through a statistic evaluation of results obtained via sectional as well as experimental investigation, a relationship was demonstrated by us between the presence of a LDH-elevating virus agent and an altered behavior of chromosomes on mitosis.

A cohort studied consisted of 24 women aged 25-40 years with a load of a spontaneous abortion in 1st trimester in their anamneses. The women became pregnant over 6 to 40 months after the latest abortion. The mean number of spontaneous abortions, parturitions of healthy infants and interruptions made for social reasons are as follows: 1,5 (1 to 3), 0,3 (0 to 1), and 0,3 (0 to 1), respectively. Neither targeted clinical nor laboratory examinations revealed a potential cause of their infertility.

Information as regards the clinical state of women patients and newborns were obtained from case histories and parturition records in the [2]nd Clinic of Gynaecology and Obstetrics, 1st Faculty of Medicine, Charles University, Prague.

2. Materials and Methods

2.1. Leukocyte adherence inhibition test

The leukocyte adherence inhibition test was used. The leukocyte adherence inhibition (LAI) assay is an in vitro quantitative technique to study and monitor the cell-mediated immunity based on observation of adherence of T lymphocytes to substrate in the presence or absence of antigens [1].
LDH virus antigen is an immunologically functional fraction prepared from the serum of inbred laboratory mouse strain C3H/H2K infected with the virus (LDH virus) enhancing the lactate dehydrogenase isoenzyme [2].

The T lymphocytes were prepared from venous blood of women with recurrent abortions and of healthy women.

The number of T cells in suspension was counted before and after 60 min sedimentation period and the difference – the number of nonadherent cells – was determined.

2.2. Cytogenetic examination from karyotype pictures

Complete sterile sets for one examination using quasi micro methods were used. The sets contained all technological components (i.e. solutions and tools for taking of blood samples, cultivation) prepared for manufacturing. The leukocytes were taken from periphery blood at the beginning of pregnancy. The method of chromosome deviations, specially acrocentric associations was used. Number of mitoses (30) with acrocentric associated chromosomes examined and evaluated. Karyotype picture were prepared in the cytogenetic lab of the 2nd Gynecological and Obstetrics Clinic in Prague [3].
The significance of differences between mean values was being determined using the t-test.

3. Results

A cytogenetic examination as carried on the basis of the culture of leukocytes from the peripheral blood taken at the beginning of pregnancy provided a possibility to define a count of mitoses with associated chromosomes, such a count serving as a rate of their associability. The associations were evaluated always in 30 mitoses according to criteria developed by [4].

The susceptibility of leukocytes to the LDH-elevating virus was being ascertained using the leukocyte adherence inhibition test, and the antigen was obtained based on the above mentioned methods.

The control group for cytogenetic and immunologic examination consisted of 30 women bearing a healthy child at term after a physiological gravidity and having neither gynaecologic nor other serious organ diseases in their histories.

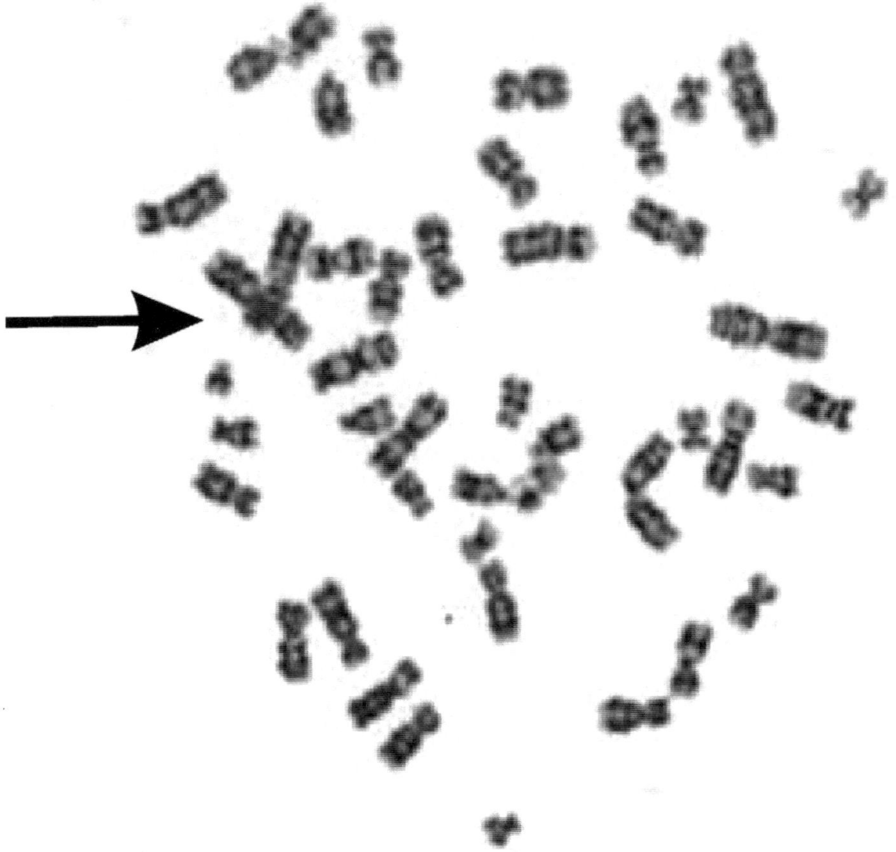

Figure 1. Karyotype containing an acrocentric cluster of chromosomes prepared from T lymphocyte of periphery blood (the arrow points to an acrocentric cluster).

Table 1. Results of evaluation of the associability of acrocentric chromosomes.

Cohort	Number	% of mitoses with associations of acrocentric chromosomes
Control group	30	39.6
Infertile women	24	60.7**

Legend. **................. P<0,01

Table 2. Immune response to LDH-elevating virus by women with disordered fertility of an unknown etiology.

Cohort	Number	% of non-adhered cells
Control group	30	36.8
Infertile women	24	70.7**

Legend. **................. P<0,01

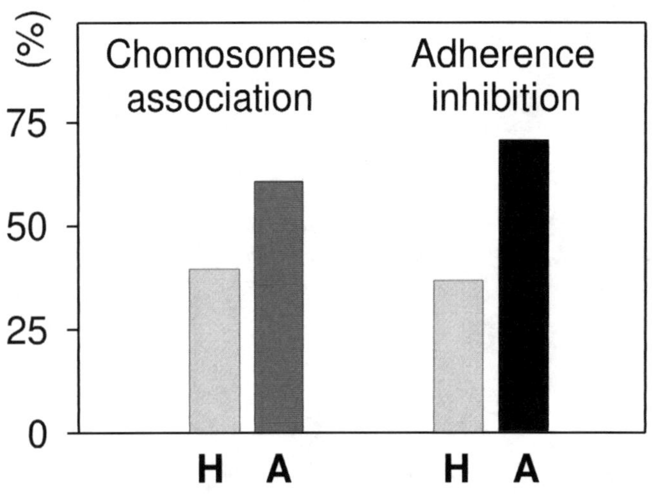

Figure 2. Comparation of associated acrocentric chromosomes and immune response to LDH virus
H – women with normal delivery (healthy group), N = 30
A – women with recurrent abortions (abortion group), N = 24

Twenty women of the abortion group studied (83.3%) bore a live foetus, one woman (4.2%) bore a dead child and three women (12.5%) aborted spontaneously again. 13 women (65%) were hospitalized in the course of their gravidity (8x for bleeding, 2x for EPH gestosis, respectively 3x for signs of a premature parturition).

4. Discussion

The immune response to the LDH-elevating virus decreases with time which has elapsed from the latest abortion, which indicates that the reaction of the nucleus apparatus on the noxa seems to be slower.

It is evident from the above mentioned that in 24 infertile women an increased associability of acrocentric chromosomes and an enhenced susceptibility to the LDH-elevating virus antigen were ascertained in the subsequent gravidity by means of the cytogenetic examination. Pathologic deviations were exhibited both by the course of pregnancies and the health status of newborns [5].

Results of this clinical study, which could not encompass a higher number of women patients due to financial reasons, support the hypothesis that the increased associability is a risk factor for pregnancy and the state of the newborn as well. Its determination along with determination of an immune response of the organism to the LDH-elevating virus in women with a disorder of fertility of an unknown origin is considered suitable in order to optimize determination of the best time for a new conception.

It seems that mutual attraction of chromosomes by a small particle with specific properties as a consequence of defective distribution of electrodynamic activity in chromosomes–generation of a strong field in a preferential direction (may be a result of conformation or structural changes, changed energy supply) are a good hypothesis for future physical research: decreased electrodynamic cellular attraction forces–decreased power and coherence of cellular oscillations [6,7,8,9,10].

In the presented work we like to focus on the cell metabolism and its dependence on the electrodynmic field under the influence of exogenous RNA.

It turns out, that the electrodynamic field is proportional to the rate of ATP production and uptake, the alteration of which affects the cytoplasm, the cell membrane and finally the nearby extracellular space, where it can interfere with the electrodynamic field of the surrounding tissue

[11,12]. We are discussing whether the outlined interaction might play some role in the intercellular signaling and intercellular information exchange.

5. Conclusion

Investigation of abortions from unknown reason revealed that:
Chromosomes from mother's T lymphocytes display acrocentric associations,
Mother's T lymphocytes are sensitive to RNA of LDH virus.
Abortions from unknown reason may belong to a general group of diseases containing pathological consequences of LDH-elevating virus infection.

6. References

[1] Jandová A, Kobilková J, Čoupek J, Vejbora O, Postupa J and Pokorný J 1990 Method of determination of leukocyte adherence inhibition. *Authors Certificate No. 265601*, Prague
[2] Jandová A, Čoupek J, Sanitrák J, Motyčka K, Mach O and Pekárek J 1989 Method of preparation of immunologically functional fraction from the inbred strain mice blood. *Authors Certificate No. 259704*, Prague
[3] Mejsnarová B, Jandová A, Novotná J, Motyčka K, Heyberger K, Sanitrák J and Čoupek J 1979 Karyotyp, toxoplasmatické vyšetření a LAI testu žen se spontánními potraty. Čs. gyn. **44** 741-42
[4] Zeng L, Godeny EK, Methven SL and Brinton MA 1995 Analysis of simian hemorragic fever virus (SHFV) subgenomic RNAs, junction sequences, and 5′ leader. Virology **207** 543-48
[5] Nedbalová M, Čoček A and Jandová A 2010 Study of immunological methods in relation to the energy metabolism of cancer cells I. *Atherosklerosa (Prague, Czech Republic, 8-10 September 2010)* ed S Eichlerová and E Tvrzická (First Faculty of Medicine, Charles University) p 31
[6] Motyčka K, Jandová A and Pezlarová J 1979 Studium přenosu infekce LDH virem z rodičů na potomstvo u myší kmene H. (Study of infection transfer by LDH virus from parents to offsprings in mouse strain H.) *Sb. lék.* **78** 176-80
[7] Nedbalová M, Chválová S, Jandová A and Trojan S 1997 Immunological response of CD4 lymphocytes on LDV antigen in pigs and human parents and children. *Arch. Physiol. Biochem.* **105** 240
[8] Říha J, Jandová A and Polášek M 1993 The effects of lactate dehydrogenase virus (LDV) on reproductive performance and health of bovine females. *Živočišná výroba* **38** 223-38
[9] Riley V 1974 Biological Contaminants and Scientific Misinterpretations. *Cancer Res.* **34** 1752-54
[10] Jandová A, Nedbalová M and Trojan S 1997 LDH virus - parasite or energetic system of animals? *Arch. Physiol. Biochem.* **105** 239
[11] Pokorny J 2009 Fröhlich′s coherent vibrations in healthy and cancer cells. *Neural Netw. World* **19** 369-378
[12] Jandová A, Pokorný J, Kobilková J, Trojan S, Nedbalová M, Dohnalová A, Čoček A, Mašata J, Holaj R, Tvrzická E, Zvolský P, Dvořáková M and Cifra M 2009 Mitochondrial Dysfunction. *Neural Netw. World* **19** 379-91

In vitro T lymphocyte adherence capabilities under the influence of lower induction values (0.1 – 0.01 mT) of 50 Hz external magnetic fields

A Čoček[1], A Jandová[4], A Hahn[1], J Mártonová[1], M Ambruš[2], A Dohnalová[3], M Nedbalová[3], J Pokorný[4]

[1]3rd Faculty of Medicine, Charles University in Prague, Department of otorhinolaryngology, Faculty Hospital Královské Vinohrady, Prague, Czech Republic, Šrobárova 50, 100 34, Prague 10, Czech Republic
[2]3rd Faculty of Medicine, Charles University in Prague, Department of radiotherapy and oncology, Faculty Hospital Královské Vinohrady, Prague, Czech Republic, Šrobárova 50, 100 34, Prague 10, Czech Republic
[3]Physiological Institute, 1st Medical Faculty, Charles University, Albertov 5, 128 00 Prague 2, Czech Republic
[4]Institute of Photonics and Electronics, Academy of Sciences, Chaberská 57, 182 51, Prague 8, Czech Republic

Abstract. Our research thus far has concerned the impact of external magnetic fields (50 Hz) and low (0.01-10 mT) induction on adherence capabilities of T lymphocytes obtained from the blood of patients with head and neck tumors. We know that the in vitro adherence capability of T lymphocytes towards surfaces in cancer patients is less than that of control. Previously, we have found that exposure to magnetic fields (50 Hz / 0.01-10 mT) increases the capability of T lymphocytes, in larynx/pharynx cancer patients, to adhere in vitro to surfaces, achieving almost physiological values, in not only pre-treatment patients but also those receiving treatment in the course of follow-up. The capability of T lymphocytes in controls (voluntary blood donors) to adhere to surfaces was also increased (50 Hz / 0.01-0.5 mT). The present study concentrates on the significance of the level of magnetic field induction in order to determine whether low induction values can restore T lymphocytes adherence capabilities. Testing a set of 20 patients showed a statistically significant difference ($p < 0.05$) in the in vitro adherence capacity of T lymphocytes between both 0.01 and 0.05, and 0.1 mT induction levels. In the control group (patients diagnosed with chronic sensorineural hearing loss) there was even a statistically significant difference between induction values of 0.05 and 0.01 mT. Therefore, we concluded that lower induction values resulted in a more biologically significant response.

1. Introduction

Our research is concerned with the impact of external magnetic fields of power frequency (50 Hz) and low (0.01-10 mT) magnetic induction on the inhibition of T lymphocyte adherence. The research was carried out on blood obtained from patients with head and neck carcinomas. We know that the in vitro adherence capability of T lymphocytes towards surfaces in cancer patients is less than that seen in controls [1]. We have found that exposure to magnetic fields (50 Hz / 0.01 – 10 mT) increases the capability of T lymphocytes from larynx / pharynx cancer patients, to adhere, in vitro, to surfaces, achieving almost physiological values, not only pre-treatment patients but also in those receiving treatment in the course of follow-up [2]. Also, T lymphocyte adherence capabilities in a control group of voluntary blood donors were previously shown to be increased by exposure to magnetic fields (EMF).

The present paper concentrates on the significance of the level of magnetic field induction (0.1 – 0.01 mT) to determine if low induction values can significantly influence T lymphocytes adherence capabilities relative to higher inductions values.

2. Materials and Methods

T lymphocyte adherence inhibition tests were simultaneously carried out with and without an EMF, using comparable inductions, such that T lymphocyte capabilities of individual patients could be compared, i.e. comparing the adherence capability with and without magnetic fields of 0.1 mT, 0.05 mT, and 0.01 mT (n = 20; 13 males, mean age 61 years, range 45-76 years, 7 females, mean age 68 years, range 44-80 years). The control group consisted of patients with chronic sensorineural hearing loss. The group was matched for age and morbidity (n = 30, 13 males, mean age 59 years, range 40-68 years, 17 females, mean age 69 years, range 52-71 years).

2.1. Preparation of T lymphocytes

A venous blood sample (8 ml) was transferred to a test-tube containing 500 units of heparin, and incubated for 30 minutes at a temperature of 37 °C in thermostat. The leukocyte-rich plasma was transferred to another test-tube using a dropper. The contents were diluted with redistilled water back to the original volume. The tube was then centrifuged at 2,500 rpm for 20 minutes. The supernatant was removed and leukocyte sediment diluted with 1 ml of physiological saline (NaCl 0.9 %) and bubbled with a Pasteur pipette. This achieved the immediate lysis of the remaining erythrocytes. Diluted leukocyte sediment was filtered through a nylon filter column to achieve a pool containing 92% T lymphocytes and 8% other cells [3]. The filtrate was transferred to a Bürker counting chamber and diluted to obtain 4 x 10^6 cells per one ml of solution [4]. This process did not affect T lymphocytes.

2.2. Preparation of organ-specific antigen (immunoactive fraction of malignant tumors)

Organ-specific antigen (an immunoactive fraction of a malignant tumor) was obtained from peroperative malignant tumor tissue. The tissue was stored frozen with dry nitrogen. After thawing, the tissue was homogenized and the product was cryolyzed (−60 and +40 °C) three times. Thereafter, the homogenized product was centrifuged at 3.5 x 10^3 G, at + 40 °C, for 60 min. Subsequently, the supernatant containing proteins, nucleoproteins, nucleic acids and low-molecular weight components was extracted for 15 minutes using phenol (1:1 ratio) in the presence of 1% sodium dodecyl sulfate and 0.01% 8-hydroxyquinoline to achieve a pH of 7 – 7.5.

The solution was then centrifuged at 3 x 10^3 G. Afterwards the water phase was removed and shaken with phenol. The water phase – a mixture of cytoplasmic ribonucleic acids – was separated in a cylinder filled with oligo-dT-cellulose (enriched with deoxythymidine), which has an affinity towards polyadenylate i-RNA residues. The residues obtained in the cylinder were rinsed with water. The eluate was subsequently purified using high-pressure gel chromatography. The cylinder was filled with Separon-Hema-300 glc® gel, containing a suspension of 2-hydroxyethyl-methacrylate with ethylene dimethacrylate with an exclusion limit of 3 x 10^5 kDa.
Detection was performed using an R 043 refractometer and a variable wavelength UV spectrophotometer. A saline solution was used as the mobile phase. Immunologically active and organ-specific antigens were obtained in the 25 – 50 cm^3 fraction evaluated at a 340 nm wavelength.

2.3. Preparation of nonspecific antigen (LDV antigen – immunoactive fraction from inbred mice serum)

An immunologically active fraction was obtained from the serum of inbred mice from the C3H/H2K strain infected with the lactate dehydrogenase virus (LDH), which produces lactate dehydrogenase isoenzymes (NAD+ 1.1.1.27 oxidoreductase). The plasma, obtained from heparinized blood, was twice centrifuged for 20-30 minutes at 3 – 3.5 x 10^3 G. The separated supernatant was subjected to further centrifugation for one to two minutes at 9 – 11 x 10^3 G. The sediment was then suspended in buffer (pH of 7 – 7.4) consisting of tris (hydroxymethyl)-aminomethane-NaCl and the dinatrium salt of ethylene-diamintetra acetate acid in a minimal volume (max = 1 ml). The sediment was divided, using a saccharose density gradient, and the area corresponding to 32 – 38% of the saccharose mass, with a maximum of 35%, was retrieved. After dilution in the above-mentioned buffer, in a 1:0.5 – 1.5 volume ratio, the sample was centrifuged for 50 – 70 minutes at 50 – 100 x 10^3 G. Samples were stored after obtaining the last sediment, at 2 – 6 °C. The obtained sediment was diluted in 1.5 – 3 ml of physiological saline and subjected to high-pressure gel chromatography. Fractions corresponding with a molecular mass of 3.0 x 10^5 to 1.5 x 10^4 were immunologically active. Level centrifugation was applied for 180 – 210 minutes at 150 and 180 x 10^3 G. The stationary phase of high-pressure chromatography utilized Sepharon-Hema-glcR, which is a suspension copolymer of 2-hydroxyethyl-methacrylate with ethylene dimethacrylate with a column exclusion limit expressed by a molecular mass of 3.0 x 10^5, with a gel particle size of 32 – 40 μm. Detection was performed using a differential refractometer and an UV spectrometer with variable wavelengths. A physiological saline was employed as a mobile phase and the fraction was entrapped at a wavelength of 340 nm. This resulted in an immunologically pure fraction.

2.4. Sinusoidal magnetic field (EMF)
A EMF was generated in a cylinder with a diameter of 30 cm and height of 33 cm. Since the magnetic field is maximum (100%) at the center of the cylinder, test tubes (green SiAl glass) containing the T lymphocyte suspensions were positioned at the center of the cylinder. The magnetic field thus generated had induction levels of 0.1 mT, 0.05 mT and 0.01 mT. The non-homogeneity of the field within the test tube containing T lymphocytes was in the vicinity of 1%. The oscillating magnetic field had a sinusoidal form with small higher harmonics components, as verified with an oscilloscope. Field exposure lasted 45 minutes, during which samples were collected; the temperature during the experiment was 37 °C. An Immune response was triggered by mixing with a specific or nonspecific (viral) antigen. The antigen was introduced to the T-lymphocyte suspension immediately before exposure to the magnetic field.

2.5. Measurement and evaluation of results
The T lymphocyte suspension, prepared as described above, was mixed with an organ-specific or viral antigen in the following ratio: four parts T lymphocyte suspension to one part antigen. The suspension was left to sediment in a green SiAl glass test tube. Sixty minutes later the number of free non-adhering T lymphocytes (NAL), i.e. those that had not adhered to the glass test tube, were counted. The results were recorded in % (100% represents the absence of T lymphocyte adherence). Physiological NAL values in healthy people can be as high as 33%; values topping 41% are considered pathological [5].

2.6. Statistical evaluation
To assess whether statistically, the NAL values were significantly different (using low-induction field exposure (0.1, 0.05 or 0.01 mT)) compared to those without exposure, as well as between different inductions, we used variance analysis for repeated measurements and the t test for two dependent selections ($p < 0.05$) with a Bonferroni correction to test the pairs. We also evaluated the control set (patients with chronic sensorineural hearing loss) in the same way. In order to determine if statistically, the control NAL values were significantly different from the tested set for the same magnetic induction values, or in the absence of an EMF, we used a t test for two independent sets ($p < 0.001$).

3. Results
We examined a single batch of patients for their adherence capability to in vitro T lymphocytes without exposure to an EMF and with exposure to an EMF (induction levels of 0.1 mT, 0.05 mT and 0.01 mT). A statistical description of the set is available in Table 1.

Table 1. The basic statistical characteristics of values of adherence of T lymphocytes of patients with larynx/pharynx cancer without and after magnetic field exposure (0.1 mT, resp. 0.05 mT, 0.01 mT) and result of ANOVA test (n=20)

	induction (mT)	n	average (NAL)	standard deviation	standard error	min. value	max. value	ANOVA p
LDV	0	20	94.7	14.7	3.3	41.7	100	0.001
	0.1	20	61.0	19.5	4.4	33.1	83.3	
	0.05	20	37.6	16.9	3.8	16.6	68.3	
	0.01	20	34.2	15.0	3.4	6.6	68.3	
ORG	0	20	90.7	14.7	3.3	34.8	100	0.001
	0.1	20	57.6	17.2	3.9	24.4	91.4	
	0.05	20	32.3	12.4	2.8	14.6	55.6	
	0.01	20	29.7	12.1	2.7	6.6	55.6	

LDV nonspecific (LDV) antigen
ORG organ-specific (tumor) antigen
NAL non-adhering T lymphocytes

The results of mutual T tests with Bonferroni corrections, using induction pairs, is presented in Table 2. (graphically represented in Figure 1.).

Table 2. The results of mutual t test with Bonferroni corrections between pairs with different magnetic inductions (n=20)

test with	Without EMF			0.1 mT		0.05 mT
	0.1 mT	0.05 mT	0.01 mT	0.05 mT	0.01 mT	0.01 mT
LDV	x	x	x	x	x	NS
ORG	x	x	x	x	x	NS

EMF magnetic field
LDV nonspecific (LDV) antigen
ORG organ-specific (tumor) antigen

X statistically significant difference at a 0.05 level ($p < 0.05$)
NS statistically nonsignificant difference

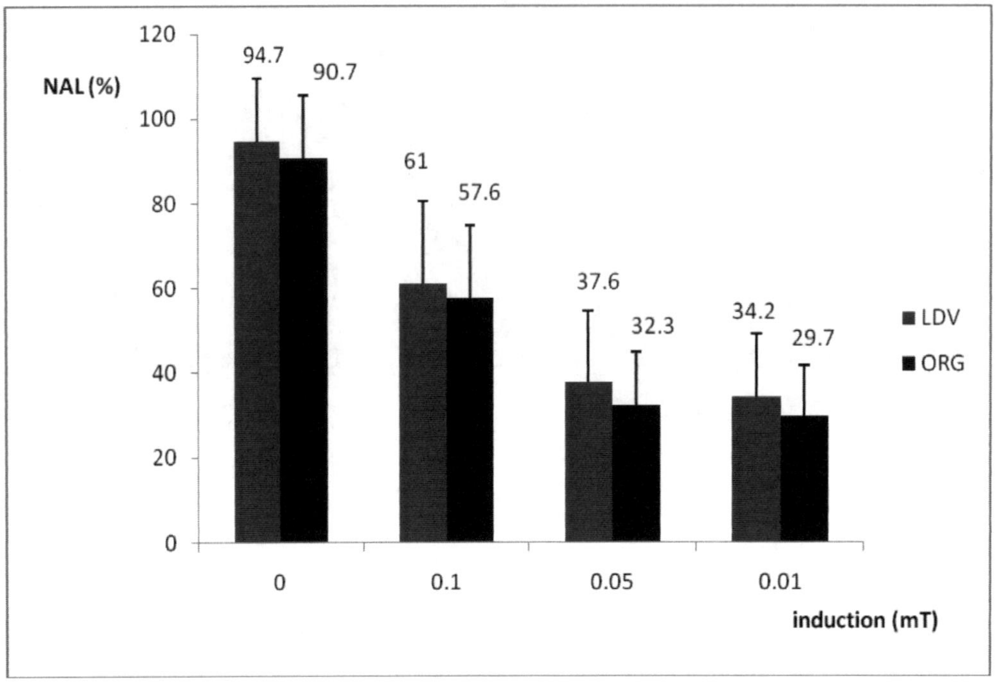

Figure 1. The average values of non-adhering of T lymphocytes of patients with larynx/pharynx cancer before and after magnetic field exposure (0.1 mT, resp. 0.05 mT, 0.01 mT; n=20)

NAL non-adhering T lymfocytes
LDV nonspecific (LDV) antigen
ORG organ-specific (tumor) antigen

Control group consists of patients with chronic sensorineural hearing loss (n = 30). We also ascertained, within this control group, whether the monitored magnetic induction values of 0.1 mT, 0.05 mT or 0.01 mT had a statistically significant impact on increasing T lymphocytes surface adherence capabilities (compared to adherence without an EMF, of course only with respect to a nonspecific antigen). Statistical characteristics are shown in Table 3. and the results of mutual tests with the Bonferroni correction between pairs are shown in Table 4.

Table 3. The statistical characteristics of adherence of T lymphocytes in presence of nonspecific (LDV) antigen without and with magnetic field exposure (0.1 mT, 0.05 mT, 0.01 mT). Result of ANOVA (control group – patients with sensorineural hearing loss), (n=30).

	induction (mT)	n	average (NAL)	standard deviation	standard error	min. value	max. value	ANOVA p
LDV	0	30	27.2	9.8	1.8	15.9	55.9	0.001
	0.1	30	14.5	5.5	1.0	6.6	30.3	
	0.05	30	6.9	2.8	0.5	3.2	13.9	
	0.01	30	6.1	2.6	0.5	2.7	12.9	

LDV nonspecific (LDV) antigen
NAL non-adhering T lymphocytes

Table 4. The results of mutual t tests with Bonferroni corrections between pairs with different magnetic inductions (control group – patients with sensorineural hearing loss), (n=30).

test with	Without EMF			0.1 mT		0.05 mT
	0.1mT	0.05 mT	0.01 mT	0.05 mT	0.01mT	0.01mT
LDV	x	x	x	x	x	x

EMF magnetic field
LDV nonspecific (LDV) antigen
X statistically significant difference at 0.05 level ($p < 0.05$)

In conclusion, we examined whether statistically, T lymphocyte adherence values significantly differed between cancer patients and those in the presence of a nonspecific antigen with EMF exposure (inductions of 0.1 mT, 0.05 and 0.01 mT) and without EMF exposure. A basic statistical characterization of pre- and post-EMF exposure for both the control group and patients is given in Table 5 and graphically presented in Figure 2.

Table 5. The basic statistical characterization of adherence of T lymphocytes in presence of nonspecific (LDV) antigen pre- and post-magnetic field exposure (0.1 mT, 0.05 mT a 0.01 mT) for both the control and the patient group and result of t test between the control and the patient group

induction (mT)		n	average (NAL)	standard deviation	t-test	p
0 mT	control	30	27.2	9.8013	-19.5	0.001
	tumor	20	94.7	14.7138		
0.1 mT	control	30	14.5	5.4657	-10.4	0.001
	tumor	20	61.0	19.4938		
0.05 mT	control	30	6.9	2.7584	-8.05	0.001
	tumor	20	37.6	16.874		
0.01 mT	control	30	6.1	2.6363	-8.28	0.001
	tumor	20	34.2	15.0155		

Control control group (patients with sensorineural hearing loss)
Tumor group of patients with larynx/pharynx cancer
NAL non-adhering T lymphocytes

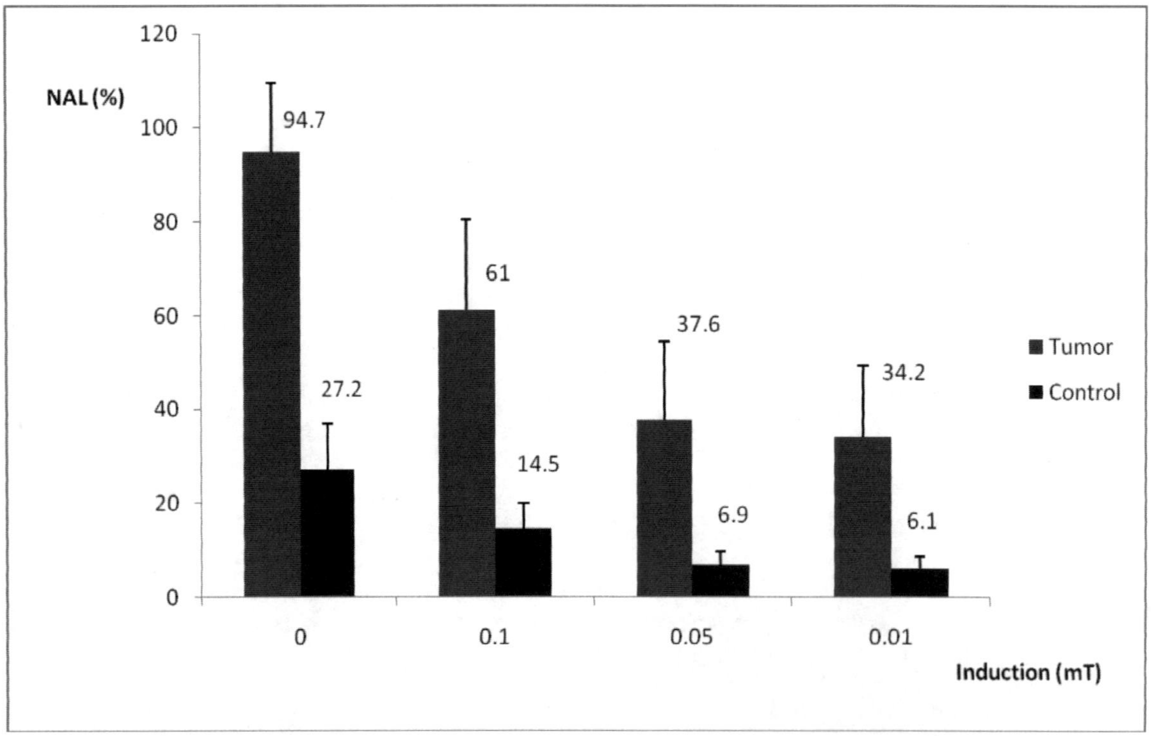

Figure 2. The average number of non-adhering T lymphocytes without and with magnetic field exposure (0.1 mT, 0.05 mT a 0.01 mT) in both tumor (n=30) and control group (n=20) in the presence of nonspecific (LDV) antigen

NAL	non-adhering T lymphocytes
Control	control group (patients with sensorineural hearing loss)
Tumor	group of patients with larynx/pharynx cancer

4. Discussion

T lymphocytes transfer information about mitochondrial dysfunction from target cells. The effects of specific and the nonspecific antigens on T lymphocytes are assumed to reflect mitochondrial dysfunction [6]. Mitochondrial dysfunction is a hallmark of cancer processes [7,8,9].

The relationship between mitochondria and cancer was the reason a nonspecific LDH antigen was used in this research. Riley and Wroblewski (1960) [10], and Rowson and Mahy (1975) [11] analyzed the effect of LDH virus infections and the origin of tumors. They found a correlation between the fast-growing Ehrlich carcinoma, in mice, and LDH activity in the serum – specifically increased levels of isoenzyme 2 and 3 in the initial phase, following tumor implantation, and isoenzyme 4 and 5 during the fast growth phase of the tumor. The lactate hydrogenase virus (LDV, LDH virus, Riley virus) is characterized as a slow virus. It is an ideal cellular parasite. At the start of infection, in mice, it can be detected as dark particles associated with mitochondria. Then the RNA replicates, and when leaving the cell, ceases to be alien to the body, and becomes indigenous. It parasitizes the energy production system of the infected cell [6].

Cellular adhesion capability is an important cell property, crucial to the organization of tissues. A loss of this cell property among cancer cells leads to their proliferation; therefore an adherence inhibition test was used.

We found statistically significant differences in the adherence capabilities in response to all induction levels vis-à-vis adherence capabilities in the absence of an magnetic field ($p < 0.001$). At a statistical significance level of $p < 0.05$, there was a difference between the adherence capabilities of T lymphocytes not exposed to an magnetic field and T lymphocytes exposed to an magnetic field, at all measured induction levels. Significant differences were also found between induction levels, except for between 0.05 mT and 0.01 mT. In other words, even micro-induction levels of 0.05 mT and 0.01 mT influence the adherence capability of T lymphocytes more than levels as low as 0.1 mT. The results were identical for nonspecific and organ-specific antigens.

An identical test was carried out with the control group (T lymphocytes of sensorineural hearing loss patients without current or an anamnestic malignity). The EMF, at all induction levels (0.1 mT or 0.05 mT and 0.01 mT), increased in vitro T lymphocytes adherence capabilities in the presence of a nonspecific antigen at a statistical significance level of $p < 0.001$ compared to cells tested in the absence of an EMF. While testing pairs (without an EMF and individual magnetic induction levels) we managed to demonstrate an even more pronounced effect regarding lower vs. higher magnetic induction values. Without residual values at the defined statistical significance level, the control group proved that the lower induction level, the greater the achieved biological efficiency (i.e. increased capability of T lymphocytes to adhere in vitro).

Finally we tested the differences of T lymphocyte adherence in 20 larynx / pharynx cancer patients against controls ($n = 30$) in the presence of a nonspecific antigen with (0.1 mT, 0.05 mT, and 0.01 mT) and without exposure to an EMF. We found that the T lymphocyte adherence capabilities of controls was, under all conditions, always higher than that of cancer patients at a statistical significance level of $p < 0.001$. It follows from the above that T lymphocyte surface adherence capability in ailing patients is lower than that of T lymphocytes in controls. Under the influence of an magnetic field with defined magnetic induction levels, the adherence capability of the both groups increased; however, control T lymphocytes always had greater in vitro surface adherence capabilities.

This work is the first research on the influence of defined EMFs on T lymphocyte in vitro surface adherence capabilities. T lymphocytes in vitro surface adherence capability is a very complicated problem that is dependent on many factors. EMFs can influence adherence in two ways: (1) they can directly improve T lymphocyte surface adherence capabilities and (2) they can indirectly improve adherence capabilities by disrupting processes that inhibit adherence. Even though various physical mechanisms between interactions of electromagnetic and magnetic fields and biological systems have been proposed, a detailed assessment that offers a complete explanation for the observed biological effect is still missing [12,13]. Fröhlich emphasized that his theory was only a hypothesis, which needed to be clarified and verified experimentally. So far it is impossible to explain why lower EMF induction values have greater biological effects.

5. Conclusion

The results of the present study demonstrated that an magnetic field, with extremely low magnetic induction values (0.01 – 0.1 mT), improves the in vitro adherence capabilities, to solid surfaces, of T lymphocytes taken from patients with larynx / pharynx cancer. Improvement was also demonstrated with controls (patients diagnosed for chronic sensorineural hearing loss).

Testing a group of 20 patients showed a statistically significant improvement in adherence with induction levels of 0.05 and 0.01 mT compared to 0.1 mT. In the control group, a statistically significant difference occurred even between induction values of 0.05 and 0.01 mT. Therefore our paper has corroborated the literary assumption of lower induction values lead to biologically more significant responses.

Although it is not possible to assume that the exposure to an external electromagnetic and magnetic fields represents treatment for cancer transformed cells, this project demonstrates that an external magnetic field can affect a specific phenomenon, in this case, in vitro T- lymphocyte adherence.

References

[1] Čoček A, Hahn A, Jandová A, Nedbalova M and Dohnalova A 1997 Long-term monitoring of an immune response to tumor and LDH viral antigen in patients with carcinoma of otorhinolaryngologic origin. *Med. Sci. Monit.* **3** 657-60

[2] Čoček A, Hahn A, Ambruš M, Dohnalova A, Jandova A and Pokorny J 2008 Changes of Leukocyte Adherence Ability Under the Influence of Magnetic Field in the Course of Treatment of Patients with Laryngeal and Pharyngeal Carcinoma. *Electromagn. Biol. Med.* **27** 1-12

[3] Tlaskalová-Hogenová H, Čoupek J, Pospíšil M, Tučková L, Kamýnková J and Mancal P 1980 Affinity chromatography of human lymphocytes on Spheron immunoadsorbent columns. *J. Polymer. Sci.* **68** 89-95

[4] Jandová A, Hurych J, Pokorný J, Čoček A, Trojan S, Nedbalová M, Dohnalová A 2001 Effects of Sinusoidal Magnetic Field on Adherence Inhibition of Leukocytes. Electro- and Magnetobiology **20** 397 – 413

[5] Čoček A, Hahn A, Nedbalová M and Jandová A 1995 Assessment of the immune reaction to the Riley virus (lactate-dehydrogenase virus, LDH virus, LDV) in head and neck oncology. *Otolaryngology* (Prague) **44** 40-43

[6] Jandová A, Pokorný J, Kobilková J, Trojan S, Nedbalová M, Dohnalová A, Čoček A, Mašata J, Holaj R, Tvrzická E, Zvolský P, Dvořáková M and Cifra M 2009. Mitochondrial Dysfunction. *Neural Netw. World* **19** 379-91.

[7] Warburg O 1956 On the Origin of Cancer Cells. *Science* **123** 309

[8] Cuezva J, Krajewska M, López de Heredia M, Krajewski S, Santamaría G, Kim H, Zapata JM, Marusawa H, Chamorro M and Reed JC 2002 The bioenergetic signature of cancer: A marker of tumor progression. *Cancer Res.* **62** 6674-81

[9] Carew J and Huang P 2002 Mitochondrial defects in cancer. *Mol. Cancer* **1** 9-20

[10] Riley V and Wroblewski F 1960 Serial Lactic dehydrogenase activity in plasma of mice with growing or regressing tumors. *Science* **132** 151-52

[11] Rowson and Mahy 1975 *Lactic dehydrogenase virus* vol 13, ed S Gard and C Hallauer (Wien, New York: Springer) p 121

[12] Markov MS 2007 Magnetic Field Therapy: A Review. *Electromagn. Biol. Med.* **26** 1-23
[13] Simkó M and Mattsson MO 2004 Extremely low frequency electromagnetic fields as effectors of cellular responses in vitro: possible immune cell activation. *J. Cell. Biochem.* **93** 83-92

Journal of Physics: Conference Series **329** (2011) 012032 doi:10.1088/1742-6596/329/1/012032

Doubts about Nonthermal Effects of MM Radiation Have no Scientific Foundations

Yu P Chukova
Krasnopresnenskiy Ecological Fund,
Malaya Gruzinskaya St 6 – 42, 123242, Moscow, Russian Federation

E-mail: y.Chukova@mtu-net.ru

Abstract. All causes of doubtings nonthermal effects of MM radiation are reviewed in details, including problem of reproducibility of outcomes. On the basis of the thermodynamic theory of electromagnetic radiation energy conversion is shown, that they are fine confirming of prognoses of the thermodynamics for conversion of energy in the Rayleigh - Jeans region.

Discovering of nonthermal effects of MM radiation

The first bieffects of radiofrequency radiation were observed more than 100 years ago, but their systematic study was begun much later, in the middle of 20th century with focus on the thermal bioeffects of microwaves. On Jan. 17-18. 1973, there was a scientific session of the Division of General Physics and Astronomy of the USSR Academy of Science in Moscow. At this session, Academician N.D. Devyatkov reported for the first time frequency-dependent nonthermal biological effects in the frequency range from 39 GHz to 60 GHz [1]. These effects were observed at the different levels of living matter organization (molecules, cells, organ, and organism). Results were arrived by different experimentalists in laboratories of different institutes under direction of Acad. Devyatkov.

It was found for the biological objects studied :

- the existence of irradiation effects which depend strongly on frequency, sometimes in a resonant manner,
- the existence of a threshold intensity necessary for induction of such effects, and that over an intensity range of several orders of magnitude above the threshold , the induced effects do not vary with intensity,
- the effects are observed to depend on duration of irradiation.

The results obtained are of great scientific and practical interest. For example, it was established that a vital activity of microorganisms is induced by millimetre wave radiation of low intensity.

The effect may be as both positive and negative one, depending on the particular conditions of irradiation and system characteristics. Different systems and many tests were studied, especially cell division under MM radiation. Here some experiments will be outlined.

Smolanskaya and Vilenskaya [2] reported that under MM wave radiation intracellular systems are responsible for lethal synthesis in bacteria, i.e. the synthesis of substances that result in the death of the cell. The colicinogenic (col) factor of E. coli was chosen as the test. The col-factor is an extrachromosomic genetic element. The functional activity of this element is normally repressed, but a number of physical and chemical influences can induce its activity. Under MM radiation the col-factor results in synthesis of a special protein substance known as colicin. The cell then perishes. Colicins are proteins coded in a small extrachromosomal plasmid called the *col* factor. The colicin that it produces has an antibacterial action with respect to other bacteria of the same and similar species.

The activity of the colicin synthesis was determined by the method of lacunas. It was found that the number of colicin-synthesizing cells increased sharply upon irradiation of the colicinogenic strain with millimetre waves of certain wavelengths. Thus, the number of cells that synthesized colicin increased by an average of 300% upon irradiation at wavelengths of 5.8; 6.5 and 7.1 mm. At the same time, neighbouring wavelengths (6.15 and 6.57 mm) showed no such effect. The results obtained were reproduced with high regularity. Thus the effect of *E. coli* to produce colicin was induced only at particular frequencies.

Smolanskaya and Vilenskaya also studied the influence of the power density of radiation on induction of the colicin synthesis. Variation of the power density by a factor of 100, from 0.01 to 1.00 mW/cm^2, had no influence on the induction coefficient, and only a further reduction of power density down to 0.001 mW/cm^2 resulted in a sharp decrease in the biological effect (Fig.1).

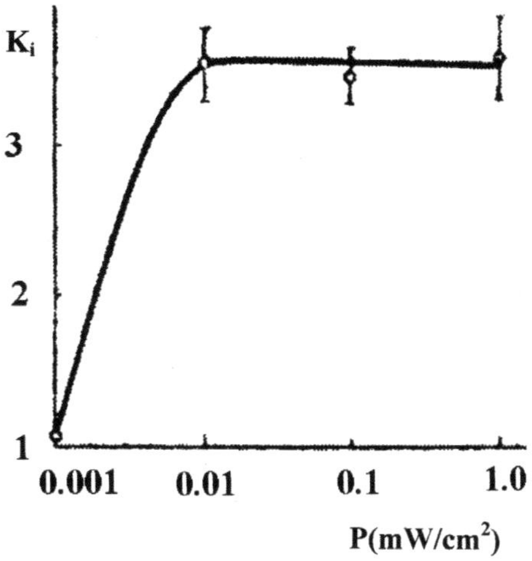

Figure1. Induction coefficient of colicin synthesis as a function of power flux density

That the effect does not depend on power density was a strong argument in favour of the nonthermal effects of millimetre waves, since all thermal effects depend primarily on intensity of irradiation.

The results of the scientific session of the USSR Academy of Science were not pilot results, because beginning around 1965, a number of organizations in the USSR began under the direction of N.D. Devaytkov systematic researches on the effects of low-power millimetre waves on biological objects.

Various yeast cultures were irradiated. A resonant effect of MM-waves irradiation on the division rate of the cells being irradiated was observed. Thus, for example, irradiation of a culture of *Rhodotorula rubra* at some wavelengths showed a sharp frequency dependence.

Irradiation of a *Candida culture* caused a marked change in the nature of cell division as compared with the control. It was revealed an increase in production of proteases with fibrinolytic activity by the fungus *Asp. aryzal*, effects of irradiation on staphylococci etc.

It was established for a wide variety of microorganisms and tests that, beginning at a certain threshold power flux density of about 0.01 mW/cm^2, the effects vary weakly over several (two and five) orders of magnitude when a marked thermal effect is already beginning to make its appearance.

The scientific session of the USSR Academy of Science was a culmination of several years of research and scientific discussions at the highest level. At this scientific session, specialists working in the fields of biophysics, microbiology, biochemistry, and medicine, were invited. Even some the most well-wishing scientists have met the report with doubt. These doubts were called by basic characteristics of nonthermal effects of MM radiation:

- Narrow frequency resonance has had the Q-factor, which one in some experiments was suspiciously high
- Sharp threshold with following plateau on power axis, as in the Smolyanskaya experiment, did not meet to the experimenters earlier
- In first part of the 20th century they thought that nonthermal effects may be only in the region, where $h\nu \gg kT$, and are impossible in region, where $h\nu \ll kT$, as quantum energies $h\nu$ is essentially insufficient to call any chemical transformations in materials.

The brief contents of some papers were published in Soviet journal "Uspekhi physicheskikh nauk ". It has English translation "Soviet Physics – USPEKHI", which was published in 1974 [1, 2]. Therefore scientists of other countries could know the soviet results and try to replicate some of them.

Reproducibility of outcomes

Quickly it has been found, that researchers in this discipline have had great difficulty in replicating results obtained by others. And what is more, the same scientists can observed the effect or can not.

The most strange history was with colicin induction by *E. coli*. As noted above, for the first time this effect has been reported by Smolyanskaya and Vilenskaya in 1973. This effect has been confirmed by M.L Swicord, T.W. Athey, F.L. Buchta and B.A. Krop at the 19th General assembly of URSI in Helsinki in 1978 [3]. However, in 1979, at the Bioelectromagnetics Symposium in Seattle Athey T.W. (Bureau of radiological health, USA) reported: "We devoted about a year to the colicin induction experiment. In the pilot study using a temporary experimental system, we seemed to be getting some positive results; but after refining our experimental system we never again saw any increased colicin induction" [4]. In contrast to Athey, at the Seattle Symposium Motzkin S.M. (Polytechnic Institute of New York) reported results in support of the Smolyanskaya and Vilenskaya original observation. She said:

"Preliminary observation that W3110Col E1 *E. coli* can be induced to produce colicin by CW irradiation at 37°C for 1hr, at 5.8 mm and a power density of 0.5 mW/cm^2"[5]

The Bioelectromagnetics Symposium in Seattle on June 22. 1979 [6] was the most serious discussion of this problem. At this symposium L.M. Partlow (University of Utah, USA) remembered in addition, that Hill et al. [7] have been unable to replicate the frequency-specific effects previously reported by Berteaud et al. [8] in 1975 on the rate of growth of *E.coli*. This bacteria is depressed by radiation of 71 and 73 GHz.

Similarly, Baranski et al. [9], Ismailov [10], Melnick [11], Motzkin [12] have all reported that microwave radiation had an nonthermal effect on the permeability of red blood cells (RBC's). In contrast Liu and Cleary [13], Hamrick and Zinkl [14], Peterson, Partlow and Gandhi [15] have demonstrated that erythrocyte membrane permeability is equally affected by microwave radiation and conventional heating.

In 1978 W. Grundler [16,17] studied the growth behavior of yeast cultures under coherent microwave irradiation. To avoid thermal effects of irradiation he chose to use a yeast cell culture in stirred aqueous suspension, as this assures an efficient thermal exchange between the cells and the surrounding medium. The temperature of the suspension (32±1°C) was continuously monitored. Similar to a yeast experiment described by Devyatkov, frequencies were chosen in a range near 42 GHz. When the cultures were irradiated by CW microwave fields of a few mW/cm^2 the growth rate either stayed constant or was considerably enhanced or reduced depending on the frequencies varying by no more than a few megahertz in the 42 GHz range. A spectral fine structure with a width of the order of 5 -10 MHz was observed. Careful temperature monitoring excludes a trivial thermal origin of this effect.

In 1979 Webb published a paper [18], where the induction of lambda prophages in *E. coli* under irradiation by millimetre microwaves was chosen as the test object. The effect was distinctly frequency-dependent and has dependence on power density at frequency 70.4 GHz. The effect was observed at 35°C after 24 h incubation.

Experiments were conducted at the University of Utah in U.S.A. to obtain nonthermal effects of MM radiation[19 – 21]. Experiments performed at this University focused on the replication of the experiments reported by Grundler et al [16,17]. The University of Utah team used a more perfect instrumentation. No differences greater than +/- 4% were detected in the growth rates of yeast for any of the 15 frequencies chosen to coincide with the " resonance frequencies" reported by Grundler et al.

In 1991 in Moscow there was the International symposium "Millimeter waves of nonthermal intensity in medicine" . Motzkin S. (USA) has represented the abstract with the title "Low power continuous wave millimetre irradiation falls to produce biological effects in lipid vesicles, mammalian muscle cells and *E. coli*" [22]. The last words of this abstract were such: " To date we have not observed any statistically significant alteration in *E. coli* proliferetion, colicin induction, membrane fluidity or variation in resting potentials or miniature end plate potentials. Available data do not conclusively define low level mm waves as a biological hazard".

In early 1990s Ziskin and colleagues tried to look for nonthermal frequency-dependent effects of millimeter wave electromagnetic radiation [23]. Similar to the millimeter wave sources used for Soviet experiments, Ziskin and colleagues also used backward-wave oscillators in the frequency bands 37.5-53.57 GHz and 53.57-78.33 GHz , respectively. It was not possible to find out required nonthermal effects.

It was not surprise, that after such numerous negative results the American experimentalists were sure that nonthermal frequency-dependent bioeffects are not in reality and finished researches in this discipline.

But European researches were continued the study, and new interesting results were obtained. In 1996 the effect of milimeter waves on the genome conformational state of *E. coli* AB1157 cells was studied by the method of anomalous viscosity time dependencies in the frequency range of 51.64 – 51.85 GHz and power density in the range from 10^{-19} to $3x10^{-3}$ W/cm^2 [24]. The power density dependence of the MM waves effect at one of these resonance frequencies (51.674 GHz) differed markedly from the corresponding dependence at the 51.755 GHz resonance, the power window occurring in the range from 10^{-16} to 10^{-8} W/cm^2. The Fig.2 gives the maximum relative viscosity as a function of logarithm of power density (W/cm^2).

In up-to-date practice of scientific publications, it is not done to report about unsuccessful results, but in the case of nonthermal frequently-dependent MM wave bioeffects we have an exception to the rule because one group of experimentalists have shown the lack of nonthermal effects obtained by other group of scientists. The pioneers of nonthermal MM waves bioeffects ignored completely the negative results of American scientists. Furthermore, they developed this discipline in direction of medical uses and introduced the new methods of MM therapy in medical practice of treatment many internal diseases: about 40 in Russia and about 60 in Ukraine [25, 26]. In Moscow new methods of MM therapy are in the list of insurance medicine.

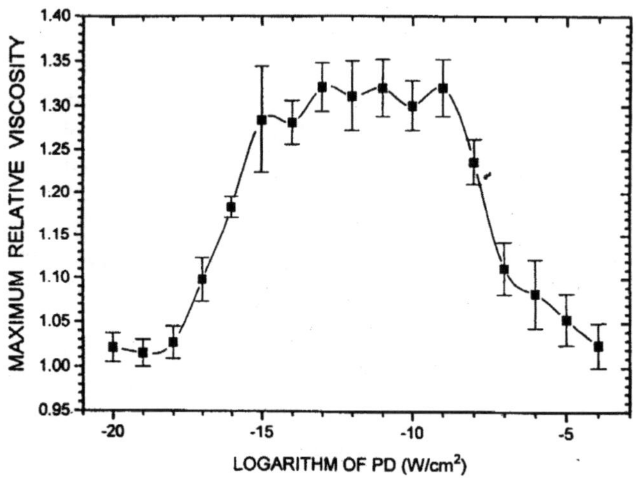

Figure 2. Power density dependence of maximal relative viscosity in cell lysates after *E. coli* cell exposure at the 51.674 GHz. Cell concentration during exposure was $4x10^7$ cells/ml.

The described situation is surprising, because for 40 years the experimenters could not solve it by their forces, despite of a very high level of experimental skill. It means, that the science has met a very great and new problem. In a history of electromagnetic radiation, already there were the striking problems, and all of them were solved by methods of thermodynamics. It is of interest, what is given by thermodynamics for comprehension of non-thermal processes of electromagnetic radiation

Thermodynamic methods of the solution of problems of electromagnetic radiation.

In 19 century efforts of forward physicists were focused on research of the laws of equilibrium thermal radiation. Many theorists attempted to calculate the universal function of G. Kirchhoff $\varepsilon_{\nu,T}$

The German physicist W. Wien was the first to use the thermodynamic method to find a solution of this problem. He propagated concepts of absolute temperature and entropy to heat radiation. In the total W. Wien formulated two laws which described a short-wave wing of equilibrium heat radiation in good conformity with experiment (W. Wien region). For this discovery, he received the Nobel Prize in 1911. However, the formulas of W. Wien were in a drastic contradiction with outcomes of experimental outcomes for a long-wave wing of heat radiation.

The problem of the theoretical description of the long-wave wing of heat radiation was resolved by Lord Rayleigh on the basis of the laws of classic thermodynamics. Later on his outcome was updated by Jeans. The Rayleigh – Jeans formula is conformed to the realm of radio-frequency radiation and extremely low frequencies (the Rayleigh – Jeans region) But the Rayleigh – Jeans formula was not suit at all for a short-wave part of equilibrium radiation. On memory of this fact, the expression «ultra-violet catastrophe» was coined.

Problem of the theoretical description of the universal function of Kirchhoff $\varepsilon_{\nu,T}$ was completely resolved in ingenious paper of M. Planck, in 1900. While using the thermodynamic method, M. Planck introduced, for the first time, the concept of the quantum of energy and formulated nowadays well-known formula, which gives the full description of the Kirchhoff function $\varepsilon_{\nu,T}$ for any wavelength

$$\varepsilon_{\nu,T} = 2\pi h\nu^3 c^{-2} [\exp(h\nu/kT) - 1]^{-1} \qquad (1)$$

For this work, M. Planck received the Nobel Prize in 1918.

For the solution of our problems, I believe it is expedient to copy as

$$c^2\ \varepsilon_{\nu,T} / 2\pi h\nu^3 = [\exp(h\nu/kT) - 1]^{-1} \qquad (2)$$

In 1905. A. Einstein entered a representation about quantum structure of light (Nobel Prize in 1921). In 1924 S. N. Bose elaborated quantum statistics for particles with the integer spin and, on this basis, received the Planck law for heat radiation. Then the physical sense of the exponent in the Planck formula (2) has become clear. The particles with the integer spin are called bosons.

In the beginning of the 20th century M. Planck [27] wrote the book, in which all aspects of equilibrium radiation are reviewed in details. One of the main concepts of this book is the concept of a flow of entropy of equilibrium radiation $\dot{S}$, which is calculated according the following formula

$$\dot{S} = 2\pi k c^{-2} \int \nu^2 [(1+\rho)\ln(1+\rho) - \rho \ln\rho] d\nu \qquad (3)$$

where c is the speed of light, k is the Boltzmann constant, the distribution function ρ is given by the formula (2) for heat radiation

$$\rho^* = c^2\ \varepsilon_{\nu,T} / 2\pi h\nu^3 \qquad (2a)$$

What is the physical sense of this value? It is very simple: the value ρ^* demonstrates number of photons at a definite level of energy.

Many Russian theoretical physicists are sure of, that the formula for flow of entropy (3) was deduced for the first time by the outstanding Soviet theorist, Nobel prize winner L. Landau. This opinion is a mistaken one, because M Planck had already deduced this formula long before the birth of L. Landau . However, L. Landau has the most direct relation to the formula (3).

The fact is that M. Planck had deduced this formula for equilibrium radiation. Einstein and Бose had also considered a system of equilibrium bosons, and L. Landau showed that the formula (3) is justified for non-equilibrium electromagnetic radiation [28]. In this case ρ is determined not by the formula (2a), but formula (4)

$$\rho = c^2 E_\nu / 2\pi h\nu^3 \qquad (4)$$

in which instead of the Kirchhoff function $\varepsilon_{\nu,T}$ (spectral density of equilibrium heat radiation) is used E_ν (spectral density of non-equilibrium electromagnetic radiation). The regularity of Landau's conclusion was confirmed by Western scientists [29, 30].

This theoretical refinement had a large practical significance in the 20th century and allowed to answer some questions. For example, Landau has made the concluding about a capability of light efficiency greater than unity and capability of optical cooling. His conclusion was so radical and unexpected, that only few theoretical physicists-contemporaries could believe it [31 – 33].

However, experimenters obtained the optical cooling in experiment with different physical samples and systems very successfully. In 1997, the Nobel Prizes on physics were awarded to S. Chu, C. Cohen-Tannoudji and W.D. Phillips for optical cooling by laser light.

What is the relation of the foregoing to the explanation of problems generated by the biological effects of MM radiation? It is a direct one: MM radiation and other electromagnetic radiation are subject to the Bose - Einstein statistics. It means that the flow of entropy of this radiation should be

Journal of Physics: Conference Series **329** (2011) 012032 doi:10.1088/1742-6596/329/1/012032

calculated according to formula (3). The graphic dependence of log $\dot{S}$ on log ρ is shown in Fig.3.

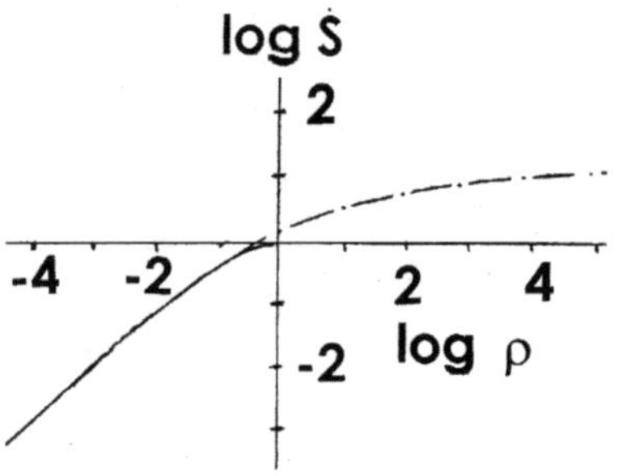

Figure 3. Logarithm of entropy flux of bosons as function of log ρ

For a flow of bosons, the different degree of density of population of energy levels is possible. By a way of a straight submission of numerical characteristics of equilibrium or non-equilibrium radiation of the W. Wien region in the formula (2a) and (4) it was received, that for this realm the condition $\rho << 1$ is realized.

The calculation for the Rayleigh -Jeans region gives other ratio $\rho >> 1$. The bosons, which are submitted to the condition $\rho >> 1$, are called as the Bose condensate because the large number of quanta is condensate at the lowest level. The Bose condensate in Fig. 3 is shown by the dot-and-dash line.

For a better visualization of a situation it is possible to give Tab.1. In Tab.1, the outcomes of concrete direct calculation of number of photons of equilibrium radiation at one level ρ^* are given at room temperature (300 K) and at the temperature of Sun radiation (5800 K).

Table 1. ρ^* as a function of frequency ν for two temperatures (300K and 5800K) of equilibrium radiation

T =300 K

ν	10^{10}	**$4,6\ 10^{10}$**	10^{11}	10^{12}	**$4,3\ 10^{12}$**	10^{13}	$5\ 10^{13}$	10^{14}
ρ*	625	**135**	62	6	**1**	0,25	$3\ 10^{-4}$	10^{-7}

T =5800 K

ν	10^{10}	**$4,6\ 10^{10}$**	10^{11}	10^{12}	$5\ 10^{13}$	**$9\ 10^{13}$**	$5\ 10^{14}$	$9,5\ 10^{14}$
ρ*	12500	**2610**	1204	120	1,9	**1**	$1,6\ 10^{-2}$	$4\ 10^{-4}$

Journal of Physics: Conference Series **329** (2011) 012032 doi:10.1088/1742-6596/329/1/012032

In this table two frequencies are selected. One characterizes frequency, at which ρ* = 1, and the second frequency (4,6 10^{10} Hz) is a frequency of maximum of colicine induction by E. coli in experiment of Smolyanskaya and Vilenskaya [2]. The values ρ* equal to 1, or close to 1, are in the infrared range. At room temperature, in equilibrium radiation on frequency 4,6 10^{10} Hz there are 135 quanta of MM radiation. The intensity of non-equilibrium MM radiation is always higher than equilibrium intensity. The calculation for the Smolyanskaya and Vilenskaya experiment demonstrates, that ρ* in conditions of a beginning of a reversible effect (threshold) is equal to 1100. We shall discuss these in the following section.

At this point, it is useful to recollect the problem κT. As is evident from the foregoing, such a problem simply does not exist for living systems, which exist at room temperature or close to room temperatures. At these temperatures the presence at a level of energy only one quantum of MM radiation is impossible. There are always many, as a matter of fact, lots of them. The ensemble of bosons is very large. One quantum of MM radiation at a level of energy can be received only at a very low temperature. I shall not calculate it, because life does not exist at such low temperature. All foregoing can call bewilderment of the reader, because the concept of Bose condensate in our representation is uniquely connected to the superconductivity of metals and not related to living systems.

This paradox has appeared as follows. When the superconductivity of metals was discovered in experiment at the beginning of the 20th century, the theorists could not find explanation for this phenomenon very long. Metals are subject to the Fermi-Dirac distribution, which forbids presence at one level more than one electron ($\rho < 1$). But for explanation of superconductivity it was necessary to admit $\rho > 1$.

At first, this phenomenon was called as Fermi degeneration and only later investigators understood, that at very low temperatures, the system changes the statistics and passes over from the Fermi –Dirac distribution to the Bose-Einstein distribution, which enables the presence of many electrons on the lower layer, i.e. Bose condensation. The problem of superconductivity was such widely discussed in the 20 century, that the impression was added up for many, that the Bose condensate is by all means metal in conditions of the lowest cooling. Only after obtaining the Nobel Prize in 2003, sensation around of the Bose condensate in metals was abated.

On arena of strife for the Nobel Prizes other exotic phenomenon has appeared. It is Bose condensation of excitons. The exotic phenomena draw away scientific attention from ambient life. Between that, how it is obvious from the foregoing, radio-frequency radiation and extremely low frequencies (the Rayleigh – Jeans region) are the Bose condensate of electromagnetic radiation. The radiation of the Wien region is not such.

While great physicists struggle for Nobel Prize, they do not have time to look on ambient life and to inform ordinary physicists and population of the planet, that they live in Bose condensate of man-made radio-frequency electromagnetic radiation. Its intensity rises permanently as an index of technical advance. Now it is not known, than it threatens to mankind and other forms of life in nature.

At our life there is no understanding that all of us live in a field of the Bose condensate. In these conditions is not surprising almost, that scientific investigating of bioeffects of MM radiation and other radio-frequency radiation, has not had any right direction, and almost 40 years, scientists attempted to troubleshoot, which actually does not exist.

The paradoxical situation with the problem κT in nonthermal effects of MM radiation resembles very much the situation of the hero in a play of the French comedy-writer Moliere. This hero has said: «But I did not know, that I speak prose! » Now each one of us can say: "I did not know, that I live in the sea of the Bose condensate of radio-frequency electromagnetic radiation."

Thermodynamic consideration of biomedical effects of MM electromagnetic radiation

In second half of the 20th century as prolongation of activities of M. Planck [27], L. Landau [28] and M.A. Weinstein [31] the thermodynamic theory of isothermal conversion of energy of electromagnetic radiation in open systems was considered. Fig. 4 shows the system under consideration, which has been described in many journal publications and books of Chukova [34 – 46] and Landsberg & Tonge [33]. Here are represented the following: energy flow and entropy flow of the absorbed electromagnetic radiation ($\dot{W}_a$ and $\dot{S}_a$) and reemission ($\dot{W}_L$ and $\dot{S}_L$), similar characteristics of reactants ($\dot{U}_r$ and $\dot{S}_r$) and products ($\dot{U}_p$ and $\dot{S}_p$) and heat flow ($\dot{Q}$ and $\dot{Q}/T$). $\dot{S}_I$ is energy generation rate due to irreversibility of process.

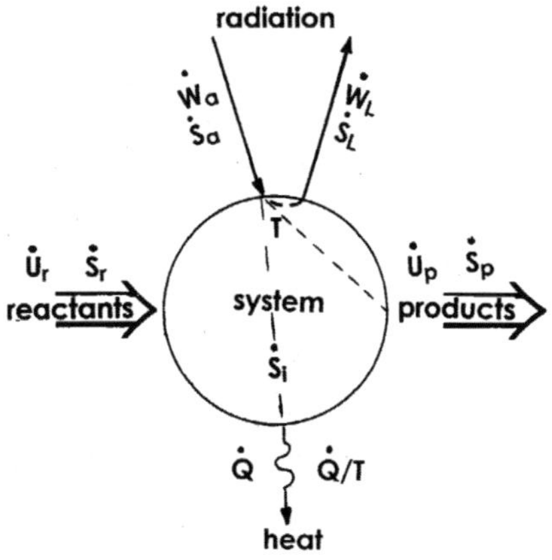

Figure 4. Fluxes of energy and entropy in thermodynamic open system

This system was utilized for explanation of outcomes of rather different experiments. Now conversion of energy of electromagnetic radio-frequency radiation (mainly MM radiation) into the energy of chemical bonds (into the Helmholtz free energy) is the main interest for us. This efficiency η is calculated according to the formula

$$\eta = 1 - T(\dot{S}_a + \dot{S}_i)/\dot{W}_a \qquad (5)$$

which is obtained on the basis of the energy conservation law and the equation of entropy balance in the stationary state of a system without any individual suppositions. Therefore this formula is of a very general nature. It has already been reviewed in detail in many previous publications. For this reason, only the total outcome will be shown here.

Fig. 5 shows the full scale of electromagnetic frequencies, which are now investigated in biomedical experiments. The Wien region and the Rayleigh – Jeans region and - between them - IR (infrared) radiation are represented. The ratio of quantum of thermal oscillations kT and quantum of electromagnetic radiations hν are given, as well as is the ranges of ρ. On the lowest part, are shown the representative dependence of efficiency η* of reversible conversion of electromagnetic radiation energy into the Helmholtz free energy on log of spectral density of absorbed radiation E_ν .

Both figures are given to the semilogarithmic scale: the abscissa axis is the log of spectral density of electromagnetic radiation, and the ordinate is the thermodynamic-limit efficiency η^*. In the Wien region we see the well-known logarithmic relation (the Weber – Fechner law), and in the left-hand part we see step, which was observed by the Soviet scientists. Thus, the relation, obtained by the Soviet scientist, the step-dependence on absorbed energy is not surprising at all: it is dictated by the laws of physics. More precisely, it occurs as a consequent of quite definite dependence of a flow of entropy of electromagnetic radiation on the spectral density of absorbed energy.

In Fig. 3 it is evidence, that the entropy of a flow of bosons increases variously depending on ρ. There are two regions: $\rho << 1$ for the Wien region and $\rho >> 1$ for the Rayleigh –Jeans region (the Bose condensate region). This difference determines essentially different efficiency of conversion of energy of electromagnetic radiation in the Wien region and in the Rayleigh – Jeans region. Thus, the second problem of bioeffects of MM radiation actually is not a problem.

Now concerning the high Q-factor of bio-resonant effects of MM radiation: there is no problem, because the main cause of broadening of absorption (or radiation) line is the energy dissipation on phonons. But this process is possible only for the Wien region and there is no place in the Rayleigh – Jeans region. The outstanding American theorist C. Kittel paid special attention to these in his fine book [47].

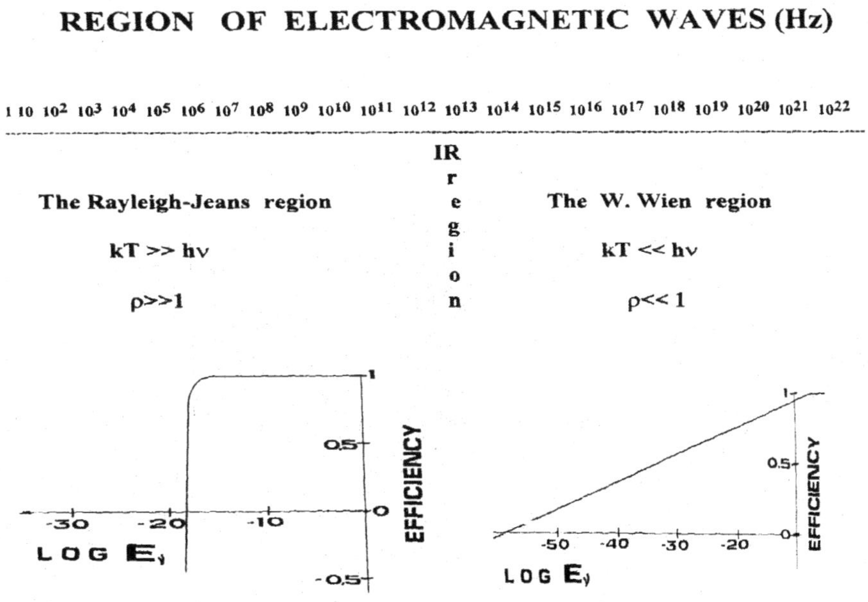

Figure 5. Outline of main characteristics of electromagnetic radiation

Nevertheless, there is a question: «Why do the outcomes of biomedical effects of MM radiation have a surprising poor reproducibility?» The answer was given a long time ago, and it is simple.

In living systems under MM radiation there are processes, which can flow in a direction of increase of the Helmholtz free energy (endergonic processes) or in a direction of decrease (exergonic processes). Nowadays studied effects of MM radiation are endergonic processes [38 - 45]. The position of boundary between them for the reversible process depends on frequency and spectral density of absorbed radiation. This boundary is shown in Fig. 6. Here the light field is an area of endergonic processes, and the area of exergonic processes is hatched (shaded). Only some exergonic processes have been studied to present time [48 – 51].

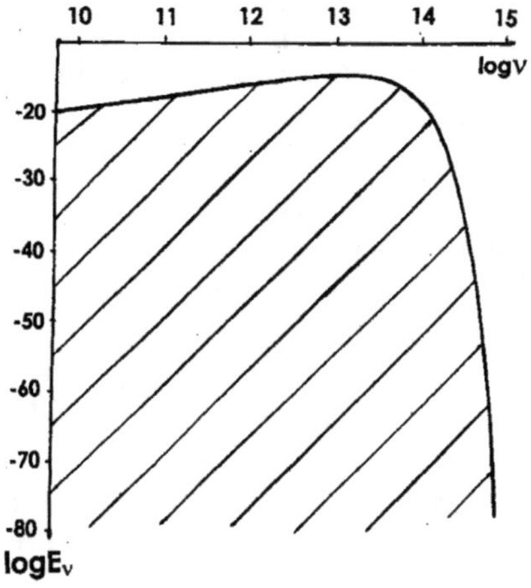

Figure 6. The regions of exergonic processes (hatched area) and endergonic ones (light area) for conditions of thermodynamic reversibility

As is known, reversible processes does not happen in the nature. All processes are irreversible. For describing the irreversibility, the Nobel prize winner I. Prigogine [52] introduced the concept of entropy generation rate $\dot{S}_i$.

$\dot{S}_i$ increases linearly with increase of absorption in a system, if an effect is small, and superlinearly, if effect is strong

$$\dot{S}_i = \alpha \dot{W}_a + \beta \dot{W}_a^n \qquad (6)$$

The linear increase $\dot{S}_i$ gives the shift η^* along the axis of absorbed energy. For already above mentioned synthesis of colicine in E. coli, this shift is submitted in Fig. 7. Here η^* is thermodynamic limit efficiency, the points are experimental data of Smolyanskaya and Vilenskaya, η is real efficiency, obtained by thermodynamic method.

Here it is useful to recollect the already mentioned large value of number of bosons at a level of energy in this classic experiment. If this process is reversible, in point of the threshold of effect the number of bosons would be about 1100. At the point of actual threshold of process of colicine synthesis the number of photons are 2.10^7 . In the last experimental point the number of photons are $2.\ 10^{10}$

Such linear irreversibility of processes under the Rayleigh - Jeans radiation is characteristic for all investigated processes.

The main cause of poor reproducibility of bioresonant effects is connected with the sharp dependence of efficiency on absorption of energy near the point of a threshold of effect. The smallest change of value of absorption is capable to change the sign of effect: endergonic process passes in exergonic or vise versa. In this situation a procedure of averaging, which is used by the experimenters in biological experiments, easily can result in zero outcome. In experiments with microorganisms, where there is an automatic averaging on an ensemble, only an experimenter of the

Journal of Physics: Conference Series **329** (2011) 012032 doi:10.1088/1742-6596/329/1/012032

highest proficiency does not lose the effect, and the experimenters of lower proficiency lose it, even if it was seen in the pioneer experiments.

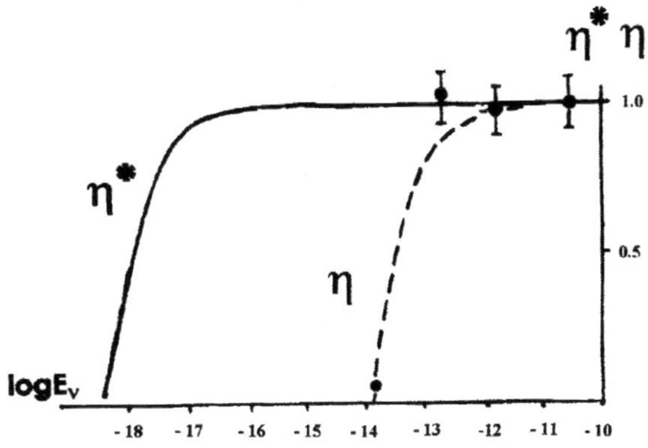

Figure 7. Efficiency of colicine synthesis in E. coli for reversible (η*) and for irreversible real processes (η)

In medical practice the reproducibility of outcomes is higher, than in biological experiments because the doctor works with the separate patient, and the procedure of averaging is eliminated.

The second cause of poor reproducibility of outcomes is that entropy generation rate in a system is non constant. It changes depending on many factors such as age of individual, satisfied or hungry individual, the concentration of different chemical agents in medium, where he/she is and others. Therefore, in the same system the outcome is not always reproduced. The third and last cause is connected to superlinear increase of entropy generation rate at strong absorption of energy. Sometimes, the experimenters increase the absorption, hoping to increase effect. But they reach opposite outcome. At high absorption and strong deviation of a system from equilibrium, the endergonic processes decrease and their efficiency reaches zero point. These have been reviewed in details [41 - 45, 53].

The Soviet pioneers-scientists were successful because they could find some living systems with rather low linear irreversibility and very low superlinear irreversibility.

Conclusion

Thermodynamics does not prohibit isothermal processes in conditions hν << kT (the Rayleigh-Jeans region), but thermodynamics asserts, that such processes have the characteristics much distinct from those, which have been studied by experimenters up to 1973.

For the first time these characteristics were discovered by the Soviet scientists under leadership of Acad. Devyatkov.

For successful studying of biomedical processes under the Rayleigh-Jeans radiation (hν << kT), it is necessary another planning an experiment and other handling of data processing.

The generally accepted norm of averaging of outcomes is extremely unwanted in the Rayleigh-Jeans region, because in conditions of sharp transition from exergonic process to endergonic process at very small change of absorption, microbiological ensembles will give zero total outcome with high probability. Therefore at the 20 century the medical effects of radio-frequency influence has been fixed easier, than microbiological one. It has as result a broad development of MM - therapy in a number of countries, while the number of accurate and stringent physical experiments is rather limited up to the present.

References

[1] N.D. Devyatkov at al., 1974 *Soviet Physics USPEKHI*, p.568
[2] A.Z. Smolyanskaya and R.L. Vilenskaya 1974 *Soviet Physics USPEKHI*, p.571
[3] M.L. Swicord, Athey T.W., Buchta F.L., Krop B.A. 1978 *Abstracts of scientific paper open symposium on the biological effects of electromagnetic waves, 10 General assembly of URSI*, Helsinki, p.35
[4] T.W. Athey 1980 *Proceedings of IEEE*, **68** 116
[5] S.M. Motzkin 1980 *Proceedings of IEEE* .**68** 116.
[6] D.L. Jaggard, J.L. Lords 1980 *Proceedings of IEEE* **68** 114
[7] Hill D.W., Hagman M.J., Riazi A., Gandhi O.P., Partlow L.M., Stensaas L.J. 1978 *Abstracts of scientific papers open symposium on the biological effects of electromagnetic* waves *19 General assembly of URSI,* Helsinki, Finland p.35
[8] J.Berteaud, M. Dardalhon, N. Rebeyrotte and D. Averbeck. 1975 *Compt. Rend. Acad. Sci.* Paris **281** 843
[9] Baranski S., Ludwicka H., Szmigielski S. 1971 *Medycyna Latnicza* Z, **39** 75-79.
[10] Ismailov E.S. 1971. Biol. Nauki USSR, **3** 58
[11] Melnick at al. 1980 URSI *Proceedings of the IEEE, **68** 461*
[12] Motzkin S., Birenbaum L., Rosenthal S., Rybenstein C., Davidow S., Remily R., Melnick R. 1979 . *National Radio Science Meeting Bioelectromagnetics Symposium,* , Seatlle , W.A., 460
[13] Liu L.M., Cleary S.F. 1977 *Abstracts for the International Symp Biol. Effects of Electromagnetic Waves,* 103
[14] Hamrick P.E., Zinkl J.G. 1975 *Radiat Res.* **62** 164
[15] Peterson D.J., Partlow L.M., Gandhi O.P. 1979 *IEEE Trans. Biomed. Eng.*, **26** 428
[16] W. Grundler, F.Keilmann, 1978 *Z. Naturforsh.*, **33** 15
[17] W.Grundler, U.Jentzsch, F.Keilmann, V. Putterlik. 1988 In: *Biological Coherence and Response to external stimuli.* Ed. Herbert Frolich. Springer-Verlag, Berlin, Heidelberg, New York, London, Paris, Tokyo. 65
[18] S. J. Webb 1979 *Physics letters* **73A** 145
[19] L.M. Partlow, L.G. Bush, L.J.Stensaaas, D.W.Hill, A.Riazi and O.P. Gandhi 1981 *Bioelectromagnetics* **2** 123
[20] L.J.Stensaas, L.M.Partlow, L.G.Bush, P.L.Iverson, D.W.Hill, M.J.Hagmann and O.P.Gandhi 1981 *Bioelectromagnetics* **2** 141
[21] L.G.Bush,D.W.Hill, A.Riazi, L.J.Stensaas, L.M.Partlow and O.P. Gandhi 1981 *Bioelectromagnetics* **2** 151
[22] S. M. Motzkin 1991 *Digest of papers International symposium "Millimetre waves of non-thermal intensity in medicine"* Moscow **2** 367
[23] E.P. Khizhnyak and M.C. Ziskin 1994 *IEEE Trans.Biomedical Engineering*, **41(9)** 865.
[24] Belyaev I.Y., Shcheglov V.S., Alipov Y.D., Polunin V.A. 1996 *Bioelectromagnetics* **17** 312
[25] *Biological aspects of low intensity millimetre waves.* 1994 Eds. Deviatkov N.D., Betskii O.V. Moscow, 336 p. (in English).
[26] O.V. Betskii , N.N. Lebedeva *Bioelectromagnetic medicine.* Edited by Paul J. Rosch and Marko S. Markov. Marcel Dekker, Inc. New York, Basel, 741
[27] M. Planck 1914 *The theory of heat radiation* - N.Y.
[28] L.D. Landau. 1946 *J. Phys. USSR*, **10** 503
[29] Rosen P. 1954 *Phys. Rev.* **96** 555
[30] Ore A. 1955 *Phys. Rev.* **98** 887

[31] Weinstein M.A. 1960 *Journal of the Optical Society of America,* **50** 597
[32] Chukova Yu.P. 1969 *JETP Lett* **10** 294
[33] P.T. Landsberg , G.Tonge 1980 *J. Appl. Phys* **51** p.R1 - R20
[34] Chukova Yu.P. 1990 *Russian Journal of Physical Chemistry,* **64**
[35] Chukova Yu.P. 1993 *Transactions 2-nd Congress of the European Bioelectromagnetic Association,* Bled-Slovenia. 50
[36] Chukova Yu.P. 1995 *10 Russian symposium (with participation of foreign scientists) "Millimeter waves in medicine and biology".* Digest of papers. Moscow 149
[37] Chukova Yu.P. 2000 *Proceedings of the Millennium workshop on "Biological effects of electromagnetic fields" October 2000, Crete, Greece* 338
[38] Chukova Yu.P. 2001 *Advances in nonequilibrium thermodynamics of systems under electromagnetic radiation.* Khrizostom, Moscow, 118 p., ISBN 5-7508-9285-X.
[39] Chukova Yu.P. *Low-level influence effects. Thermodynamic, experimental (medical and biological), social, legislative, international and philosophical aspects of the problem.* 2002, Moscow, ISBN 5-89047-013-2
[40] Chukova Yu.P. 2002 *In "Biological effects of EVFs",* ed. P. Kostarakis, Demokritos Publishers, Rhodes (Greece), **1** 318
[41] Chukova Yu.P. 2004 *In Proceedings 3rd International workshop "Biological effects of EVFs",* ed. P. Kostarakis, Demokritos Publishers, Kos (Greece), **1** 399
[42] Chukova Yu.P. 2004 *In Proceedings 3rd International workshop "Biological effects of EVFs"*, ed. P. Kostarakis, Demokritos Publishers, Kos (Greece), **2** 775
[43] Chukova Yu.P. 2006 In: *Biological effects of EMFs. 4th International workshop. Proceedings.* Ed. P.Kostarakis. Crete. Greece..129
[44] Chukova Yu.P. 2006 In: *Biological effects of EMFs. 4th International workshop. Proceedings.* Ed. P.Kostarakis. Crete. Greece. 383
[45] Chukova Yu.P. 2008 *5th International workshop on biological effects of electromagnetic fields. Palermo-Terrasini.* CD-R, ISBN 978-88-95813-03-5.
[46] Chukova. Yu.P. 2009 *Electromagnetic Biology and Medicine,* **28** 41
[47] Kittel Ch.. 1969 *Thermal Physics,* John Wiley and Sons, Inc. New York..
[48] Reiter R. J. 1995 *Integrative Physiological and behavior science,* **30** 314
[49] Reiter R.J. 1997 *Advances in electromagnetic fields in living systems,* ed. By J.C. Lin . Plenum Press, N.Y., **2** 1-27.
[50] Reiter R.J., Tan D.X., Poegeler B., Kavet R.. 1998 Bioelectromagnetics, **19** 318
[51] Reiter R. J. 1999 *Modern Radio Science,* ed. By M.A. Stuchly. Oxford University Press, 287
[52] Prigogine I. 1962 *Introduction to nonequilibrium thermodynamics* N.-Y.:Wiley-Interscience..
[53] Chukova Yu.P. 1999 *Bioelectrochemistry and bioenergetics,* **48** .349

Journal of Physics: Conference Series **329** (2011) 012033 doi:10.1088/1742-6596/329/1/012033

Resonance properties of the biological objects in the RF field

E Cocherova, P Kupec and V Stofanik

Institute of Electronics and Photonics, Slovak University of Technology, Ilkovicova 3, 812 19 Bratislava, Slovak Republic

E-mail: elena.cocherova@stuba.sk

Abstract. Irradiation of people with electromagnetic fields emitted from miscellaneous devices working in the radio-frequency (RF) range may have influence, for example may affect brain processes. The question of health impact of RF electromagnetic fields on population is still not closed. This article is devoted to an investigation of resonance phenomena of RF field absorption in the models of whole human body and body parts (a head) of different size and shape. The values of specific absorption rate (SAR) are evaluated for models of the different shapes: spherical, cylindrical, realistic shape and for different size of the model, that represents the case of new-born, child and adult person. In the RF frequency region, absorption depends nonlinearly on frequency. Under certain conditions (E-polarization), absorption reaches maximum at frequency, that is called "resonance frequency". The whole body absorption and the resonance frequency depends on many further parameters, that are not comprehensively clarified. The simulation results showed the dependence of the whole-body average SAR and resonance frequency on the body dimensions, as well as the influence of the body shape.

1. Introduction

Nowadays, the presence of nearly ubiquitous radio-frequency and particularly microwave (MW) electromagnetic fields is given mainly by continuously increasing utilization of these fields in communication, industrial and domestic appliances. Too large absorption of RF field energy may cause an increase of temperature in tissues, e. g. inside the head and may affect processes in that tissues [1], [2]. Variety of neurological effects of RF fields has been observed [3], e. g. influence on processes, related to attention and manipulation of information in working memory [4], [5].

To protect tissues from possibly detrimental thermal effect, an amount of absorbed RF energy in tissues must be determined.

2. RF dosimetry

The rate of RF energy absorption in tissues is described by an important dosimetric quantity, the specific absorption rate (SAR). Value of SAR [W/kg] depends on material properties of tissue and induced internal electric field in the body according to the following relation [2]:

$$SAR = \frac{\sigma E^2}{\rho} \tag{1}$$

where E is the rms value of electric field intensity [V/m], σ is the tissue electrical conductivity [S/m] and parameter ρ is the mass density [kg/m^3] of biological tissue.

Journal of Physics: Conference Series **329** (2011) 012033 doi:10.1088/1742-6596/329/1/012033

Even though the SAR is defined as the local (point) quantity, the average values of SAR are commonly used for practical purposes:

$$SAR_{avg} = \frac{1}{V}\int_V SAR\, dV \qquad (2)$$

where V is the volume [m^3] of biological tissue with the mass [kg] over which is SAR averaged:

- in the case of the whole-body (total) average SAR, the volume V is volume of whole-body;
- in the case of the head average SAR, the volume V is volume of head;
- in the case of the 10 g (or 1 g) SAR, the volume V is volume of 10 g (or 1 g respectively) of tissue.

Absorption in the human body depends on many parameters of the field (frequency, orientation of field, etc.) and many parameters of irradiated body [2], [6]. For far-field exposure conditions (plane wave), absorption differentiates markedly for three main polarizations: E-, H- and P-polarization, when electric field vector $\vec{E}$, magnetic field vector $\vec{H}$ and Poynting vector of wave propagation $\vec{P}$ is parallel to longitudinal axis of the body (figure 1).

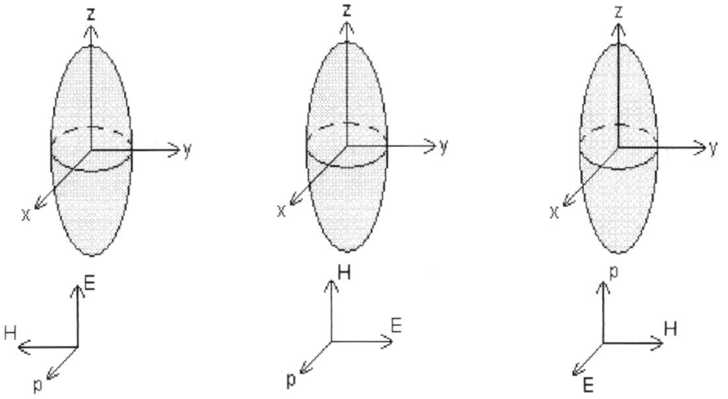

Figure 1. Polarization of plane wave with respect to orientation of longitudinal axis of rotational ellipsoid (z-axis) representing human body.

With regard to character of the field absorption in a human body, the RF range can be subdivided into the four regions [2], shown in figure 2 (for E-polarization):

- The sub-resonance range; frequencies less than 30 MHz. In this range, the whole-body average absorption increases rapidly with frequency.
- The resonance range, that is from 30 MHz to about 300 MHz for the whole-body absorption resonance, and to higher frequencies for partial body resonances.
- The "hot-spot" range, which is extending from about 400 MHz up to about 3 GHz. Localized energy absorption can occur in this range.
- The surface absorption range, for frequencies greater than about 3 GHz. The energy absorption and temperature elevation is localized at the surface of the body.

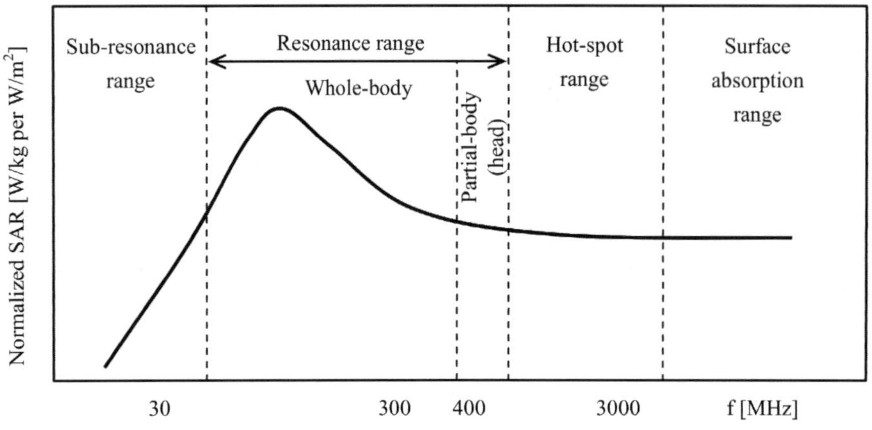

Figure 2. Normalized whole-body average SAR dependence on frequency in the case of E-field vector parallel to longitudinal axis of adult human body.

3. Model and simulation parameters

The electromagnetic field absorption in human body in RF frequency range depends on many parameters, like frequency, model shape and material parameters of tissues [2]. In this article, the spatial distributions of the specific absorption rate (SAR) are evaluated for the models of different shapes: spherical, cylindrical, realistic shape (Sam model of the head, Laura model) and for different heights of models, that represented the head or whole body of children and adult persons.

Simulations were performed in the time-domain by means of the Finite Integration Technique (FIT) in CST Microwave Studio program environment [7]. In the simulations, the models of the head and whole body were exposed to the plane wave (power density of 1 W/m^2) representing far-field exposure conditions. The surrounding space around the models was vacuum and the boundary of computational space (open conditions) was in distance of at least 0.2 m from the model in all directions.

3.1. The head models

The two polarizations were chosen: E-polarization (electric field vector $\vec{E}$ parallel to longitudinal axis of the model) and P-polarization (the Poynting field vector $\vec{P}$ parallel to longitudinal axis of the body). Since RF absorption depends nonlinearly on frequency, reaching usually the maximum for the E-polarization at so called "resonance frequency", the whole-body average SAR was evaluated in wide frequency range from 100 MHz to 1 000 MHz.

The homogeneous models consist only of materials with electrical parameters of brain grey matter (BGM) or brain white matter (BWM): the frequency dependent parameters (permittivity and electrical conductivity) were set according to [8], relative magnetic permeability $\mu_r = 1$ and mass density of 1030 kg/m^3.

3.2. The whole body models

For assessment of energy absorption in the human body we used several homogeneous model types: the ellipsoidal, cylindrical and anthropomorphic model of woman body (Laura model, see figure 3). Simulations were performed for different model dimenstions: e. g. an ellipsoidal model had the constant height to width ratio of 3:1, i. e. the ellipsoids with height $h = 0.5$ m, 0.6 m, 0.9 m, 1.2 m and 1.8 m had width $w = 0.17$ m, 0.2 m, 0.3 m, 0.4 m and 0.6 m, respectively, that simulated the body of new-born, younger and older child and adult person. The height of Laura model was $h = 1.8$ m.

Journal of Physics: Conference Series **329** (2011) 012033 doi:10.1088/1742-6596/329/1/012033

As the body model material was chosen a muscle tissue material. Values of frequency dependent parameters (permittivity and electrical conductivity) were taken from [8], mass density of tissue material of 1030 kg/m^3 was assumed and relative magnetic permeability was $\mu_r = 1$.

Models were exposed to the plane wave with power density of 1 W/m^2 for E-polarization. Laura model was irradiated from side. Simulations have been performed at the frequencies from range of 20 MHz to 500 MHz in CST Microwave Studio program environment [7].

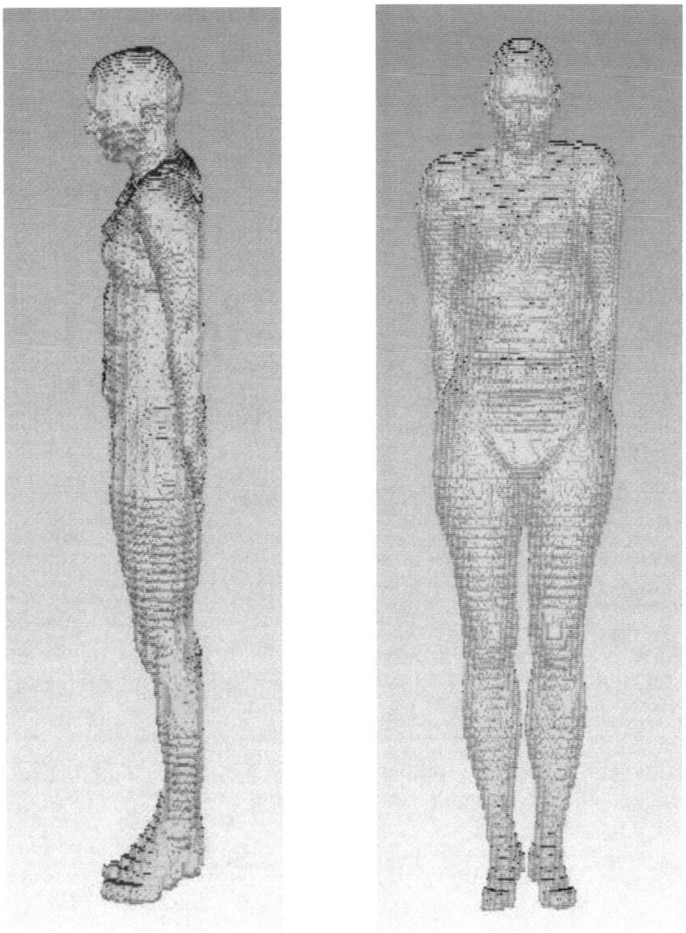

Figure 3. Realistic model of woman body (Laura model).

4. Simulation results

4.1. The head models

The simulated spatial distributions of SAR (10 g average) for spherical and cylindrical model of human head at selected frequencies are shown in figure 4 and figure 5. The simulated spatial distributions of SAR (10 g average) for the SAM model of human head at selected frequencies are presented in figure 6.

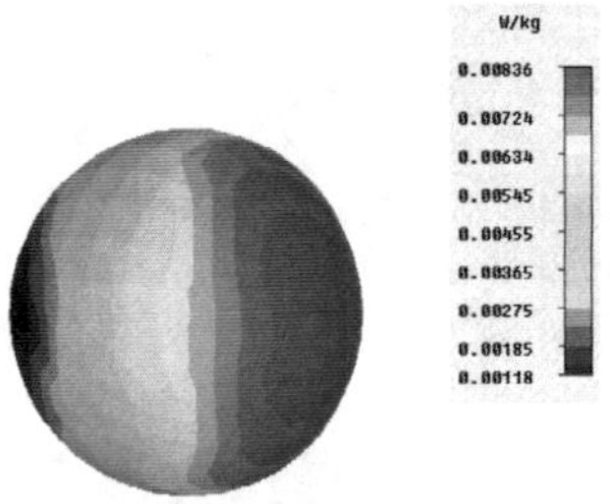

Figure 4. The distribution of SAR (10 g) in the spherical model of height $h = 20$ cm at 450 MHz, resonance frequency.

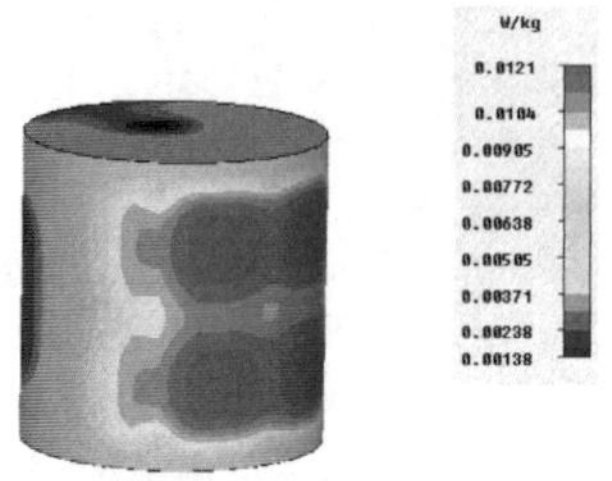

Figure 5. The distribution of SAR (10 g) in the cylindrical model of width and height of 20 cm at 350 MHz, resonance frequency.

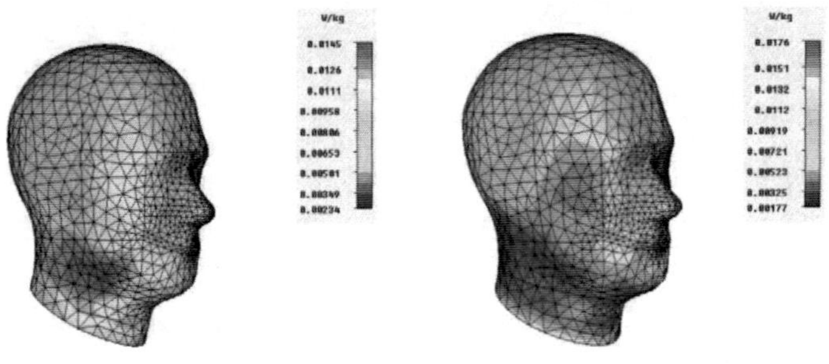

Figure 6. The distribution of SAR (10 g) in the SAM model of height $h = 20$ cm at 450 MHz and 700 MHz.

The frequency dependences of whole-body average (total) SAR for cylindrical model of human head simulated for two different materials and for two polarizations are shown in figure 7 (simulated values are indicated by markers).

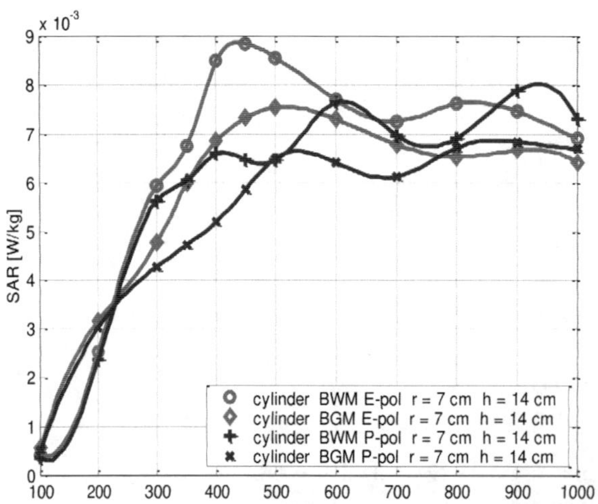

Figure 7. Frequency dependence of whole-body average SAR for the cylindrical model of width and height of 14 cm for two different materials (BWM and BGM) and for two polarizations.

For realized simulation, absorption is higher for brain white matter (BWM) than for brain grey matter (BGM) material for the major part of RF frequency range. The resonance frequency (first clear maximum of absorption) is distinctive only for E-polarization, while for P-polarization the absorption is increasing wavily and the first local maximum is not the highest one.

Summarized frequency dependences of whole-body average SAR (simulated values are indicated by markers) for different head models of various body dimensions are shown in figure 8. The simulation results show that the body resonance frequency depends on the model shape, as well as on the model height. The body resonance frequency and the whole-body average SAR at resonance frequency are approximately inversely proportional to the body height.

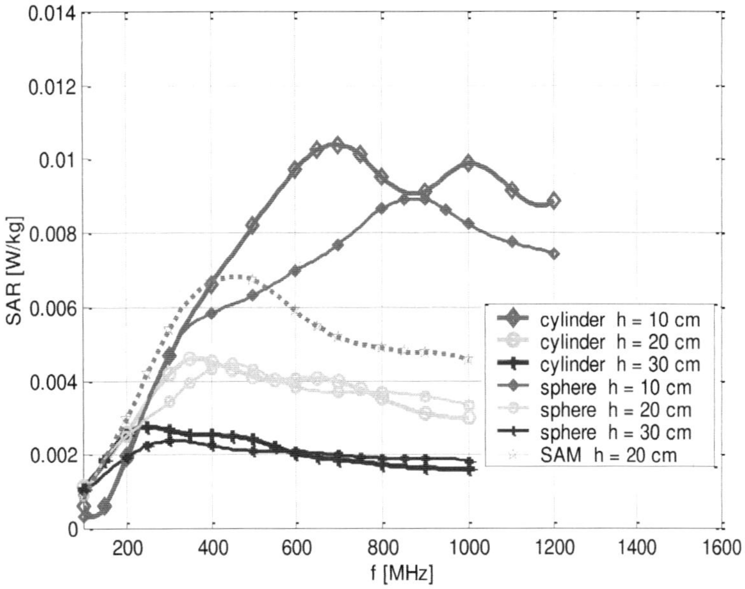

Figure 8. Frequency dependence of whole-body average (total) SAR for the spherical, cylindrical (diameters are equal to heights) and SAM model of BGM material for E-polarization for different model heights..

4.2. The whole body models

The spatial distribution of SAR (10 g average) for ellipsoidal model (of height $h = 0.5$ m, and length : width ratio of 3:1) and cylindrical model (of width 0.4 m and height 1.8 m) are shown in figure 9 and figure 10, respectively.

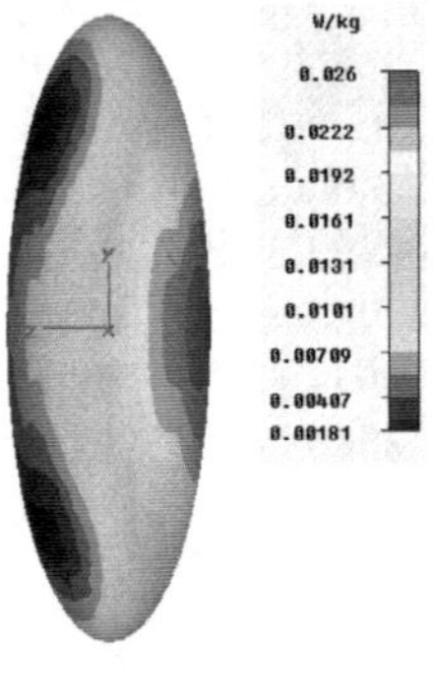

Figure 9. The distribution of SAR (10 g) in the ellipsoidal model for $h = 0.5$ m and ratio 3:1 (length : width) at 247 MHz (resonance frequency).

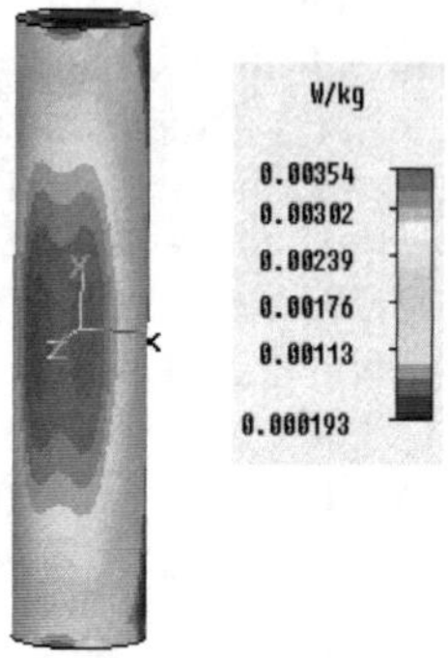

Figure 10. The distribution of SAR (10 g) in the cylindrical model of width $w = 0.4$ m and height $h = 1.8$ m at frequency 32 MHz.

Frequency dependences of the whole-body average SAR for ellipsoidal models of the constant height to width ratio of (3:1) with heights $h = 0.5$ m, 0.6 m, 0.9 m, 1.2 m and 1.8 m are shown in figure 11. It is obvious, that the smaller is the model height, the higher is the resonance frequency, and the higher is the total SAR at resonance frequency. This means, that the resonance frequency of children body is higher than the resonance frequency of adult human bodies and also the total SAR at resonance frequency is higher for children than for adult people.

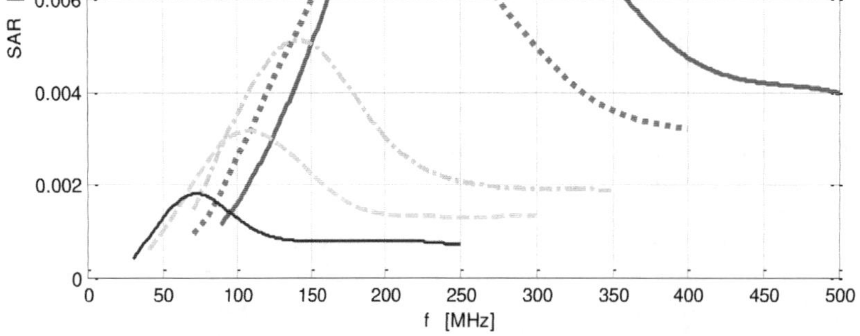

Figure 11. Frequency dependence of whole-body average (total) SAR in the ellipsoidal model of human body for different heights h (constant height to width ratio of 3:1).

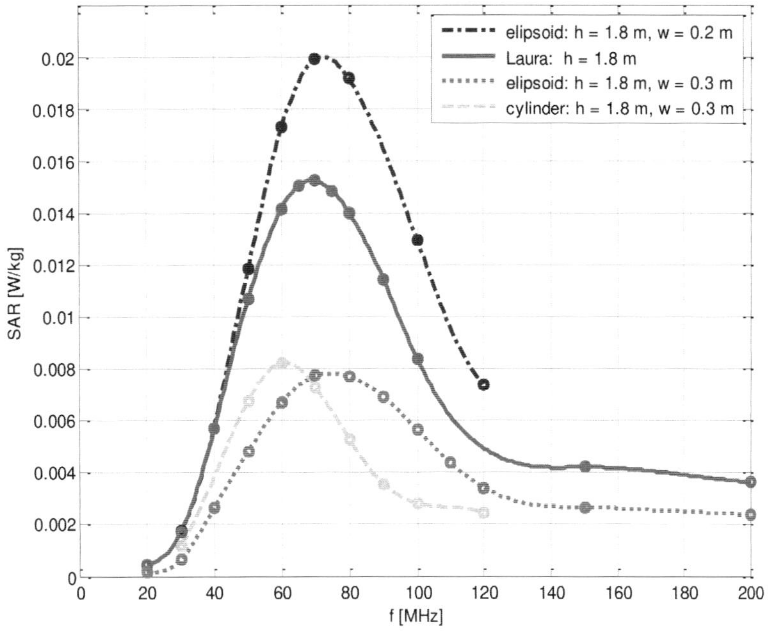

Figure 12. Frequency dependence of whole-body average SAR for the human body models of height $h = 1.8$ m: Laura model of woman body, ellipsoidal models of width $w = 0.2$ m and width $w = 0.3$ m and cylindrical model of width $w = 0.3$ m.

In figure 12 are shown frequency dependences of the whole-body average SAR for four models of the same height of $h = 1.8$ m: anthropomorphic woman model (Laura model), the ellipsoidal model of width $w = 0.2$ m and width $w = 0.3$ m, and cylindrical model of width $w = 0.3$ m. The resonance frequencies for ellipsoidal model and Laura model are about 70 MHz (69 MHz for Laura model, 72 MHz and 74 MHz for ellipsoidal model of width $w = 0.2$ m and $w = 0.3$ m, respectively), and resonance frequency for cylindrical model is 61 MHz. The resonance frequencies are mainly determined by the height of models, the width and shape of model is of less importance.

The total SAR at resonance frequency of Laura model is higher than of ellipsoidal and cylindrical model of width $w = 0.3$ m, but lower than resonance frequency of ellipsoidal model of width $w = 0.2$ m We suppose that this effect depends on width of model.

The values of resonance frequencies for different models resulted from simulations are summarized in figure 13, as well as the assumed relation according [9]. We can roughly say [10], that absorption reaches maximum (resonance frequency f_r), when height of model is equal to half of wavelength (λ):

$$f_{r0}(h) = \frac{c}{\lambda} = \frac{c}{2 \cdot h} \tag{3}$$

where λ is wavelength and c is speed of light.

The resonance frequency values for the ellipsoidal and cylindrical model (figure 13) are lower than that estimated according the relation (3). Therefore, we suggest that the more proper approximation of resonance frequency for these models would be according to corrected relation:

$$f_{r1}(h) = \frac{c}{2 \cdot h \cdot 1.2} \tag{4}$$

for the ellipsoidal model with constant height to width ratio of 3:1, or with the another correction constant of 2.1 for the cylindrical model ($h = w$):

$$f_{r2} = \frac{c}{2 \cdot h \cdot 2.1} \tag{5}$$

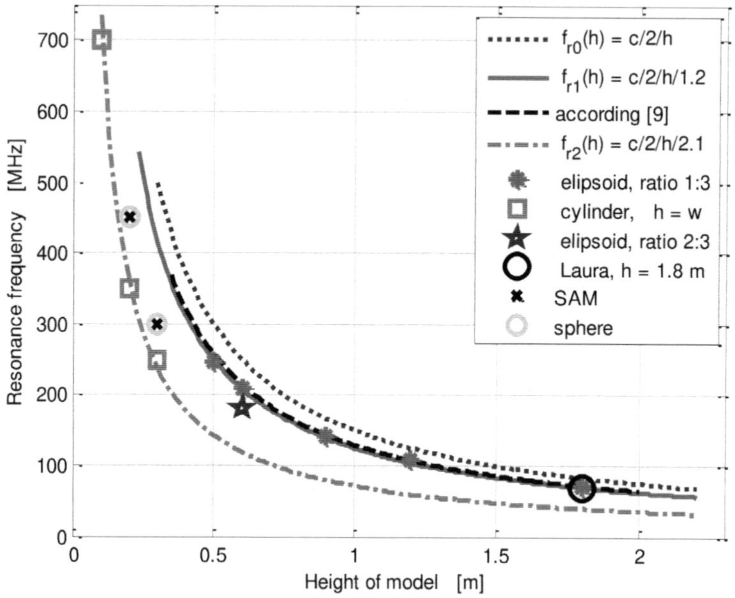

Figure 13. Resonance frequency dependence on the height h for different models of head and human body: the ellipsoidal body model (with the constant 3:1 ratio of height to width), cylindrical model and Laura model and different models of human head: spherical model, cylindrical and SAM model.

5. Conclusion

The dependencies of whole-body average SAR and resonance frequency on parameters of human body model have been investigated in this article. According to the simulation results we can conclude, that in the resonance frequency range the whole-body average SAR depends strongly on the human body size: the smaller is the model height, the higher is the whole-body average SAR.

We observed, that the values of resonance frequency are similar for the selected types of body models (ellipsoidal and anthropomorphic) of the same height. The simulation results show that the resonance frequency depends mainly on the model height: the smaller is the model height, the higher is the resonance frequency. The influence of electrical tissue properties on the whole-body average SAR was less pronounced (lower absorption for brain grey matter than for brain white matter).

The results of simulations show that even though the body resonance frequency is determined primarily by the body height, other parameters of body shape (correction constants for two model shapes are proposed) must be also included for more precise assessment.

Acknowledgments
This work was supported by the Ministry of Education of Slovak Republic under the grants VEGA No. 1/0055/10 and KEGA No. 3/7411/09.

References

[1] Ibrahiem A, Dale C, Tabbara W and Wiart J 2005 Analysis of the temperature increase linked to the power induced by RF source *Progress In Electromagnetics Research* **52** 23–46
[2] *Electromagnetic fields (300 Hz to 300 GHz)* Environmental Health Criteria 137 (Geneva: World Health Organization) 1993
[3] Hermann D M and Hossmann K A 1997 Neurological effects of microwave exposure related to mobile communication *J. Neurol. Sci.* **152** 1-14
[4] Krause C M, Pesonen M, Haarala B C and Hamalainen H 2007 Effects of pulsed and continuous wave 902 MHz mobile phone exposure on brain oscillatory activity during cognitive processing *Bioelectromagnetics* **28** 296-308
[5] Luria R, Eliyahu I, Hareuveny R, Margaliot M and Meiran N 2009 Cognitive effects of radiation emitted by cellular phones: The influence of exposure side and time *Bioelectromagnetics* **30** 198-204
[6] Hirata A, Kodera S, Wang J and Osamu Fujiwara O 2007 Dominant factors influencing whole-body average SAR due to far-field exposure in whole-body resonance frequency and GHz regions *Bioelectromagnetics* **28** 484-7
[7] CST Microwave Studio. [online], [15.06.2011] <http://www.cst.com/>
[8] Gabriel C and Gabriel S 1997 Compilation of the dielectric properties of body tissues at RF and microwave frequencies [online], [15.10.2007] <http://niremf.ifac.cnr.it/docs/DIELECTRIC/home.html>
[9] Durney C H, Massoudi H and Iskander M F 1986 *Radiofrequency radiation dosimetry handbook* (Salt Lake City: The University of Utah)
[10] Cocherová E, Šurda J, Púčik J and Štofanik V 2009 Dependence of the RF field absorption on the human body dimensions *Proc. of 19th Int. Conf. Radioelektronika 2009* Bratislava pp 327–329

Journal of Physics: Conference Series **329** (2011) 012034 doi:10.1088/1742-6596/329/1/012034

Impedance analysis of acupuncture points and pathways

Michal Teplan[1], Marek Kukučka[2], Alena Ondrejkovičová[3]

[1]Institute of Measurement Science, Slovak Academy of Sciences, Dúbravská cesta 9, 84104 Bratislava, Slovakia

[2]Department of radio electronics, Faculty of Electrical Engineering and Information TechnologySlovak University of Technology, Ilkovičova 3, 81219 Bratislava, Slovakia

[3]Liming, Mýtna 5, 81107 Bratislava, Slovakia

E-mail: michal.teplan@savba.sk

Abstract. Investigation of impedance characteristics of acupuncture points from acoustic to radio frequency range is addressed. Discernment and localization of acupuncture points in initial single subject study was unsuccessfully attempted by impedance map technique. Vector impedance analyses determined possible resonant zones in MHz region.

1. Introduction

In acupuncture and related fields it is assumed that there are special pathways in the body called meridians which are connected to main body systems, such as cardio-vascular, respiratory, digestive, etc. In this concept the pathways are exposed to the surface of human body in so called acupuncture points localized on skin surface. Although use of acupuncture is relatively well established in Western medicine as a complementary diagnostic and therapeutic tool, its physical and medical characterization is still largely unknown.

According to Zhang [1] within human body there is an invisible dissipative structure of EM field which in mainly composed of an interference patterns of standing waves in the resonance cavity of human body under the condition of permanent support of energy in an open system. This invisible structure may correspond to some extent to the mysterious acupuncture system and is closely related to different modalities of energetic medicine.

With noninvasive skin measurements typical values of impedance are around the range 100 k Ω and capacitive elements around fractions of μF. Some of the previous studies showed that electrical characteristics like impedance and capacitance at acupuncture points may show distinct values in comparison to values obtained from surrounding non-acupuncture points. Acupuncture sites were found to possess lower impedance and higher capacitance. However, some studies found no distinction from surrounding tissues. According to the review by Ahn et al. [2] preliminary evidence supports these findings however it is still impossible to determine whether acupuncture structures possess distinct electrical characteristics until better quality studies are performed.

Segments of human body may be modelled by electrical circuits. Complex electronic circuitry consists of conductors, capacitors and inductances. Resistive component reflects mainly proportion of liquids comprised in the tissue together with liquid properties of current conductance depending on concentration of different ions. Capacitive component reveals integrity of cell membranes. Phase angle is in bioelectrical impedance analysis used as an indicator for disease, hydration and nutritional status. Better overall health status should correspond with higher phase angle. Fukumoto [3] suggested a new approach for acupuncture point determination by constructing a ratio of imaginary and real part of complex impedance.

Within wide frequency range it may be beneficial to focus on the question whether resonant behavior may occur in systems of interest. Focus is directed to frequency windows with values of impedance modulus and/or phase angle different from values at surrounding frequencies. Basic resonance phenomena is often demonstrated on simple serial RLC circuits where inductance component is vitally needed in order to compensate capacitive part. Resonance should occur when phase angle vanishes. In biological systems inductance appears to play no significant role. However, also circuits with only resistive and capacitive components may possess resonant behaviour. Reichmanis et al. showed in [4] such an example: Infinite transmission line consisting of elements built from serial resistance with another resistance in parallel with capacitance (figure 7, model B) and terminated by its characteristic impedance.

Aim of the initial part of our study was to prepare methodology for several types of impedance measurements in order to study resonant behaviour of acupuncture system. Our investigation was focused on critical assessment of electrical properties of single acupuncture points and properties of meridians measured between two acupuncture points of the same meridian. In the first step we attempted to localize acupuncture points from impedance modulus maps. Then followed investigation of electrical properties of acupuncture system by vector impedance analysis in acoustic and radiofrequency range.

2. Methods

All data were measured from single healthy subject (corresponding author, male, 36 years old). Acupuncture points were chosen according to their accessibility – left arm, diversity of acupuncture pathways and significance regarded to therapeutic use in acupuncture medical system. The following points were used: Large intestine LI-4 and LI-11, heart HT-3 and HT-7, lungs LU-5and LU-7, pericardum PC-6 and PC-3, and triple warmer TW-3, TW-8, and TW-9. Experienced clinical acupuncturist localized all the points and afterwards they were labeled by marker and photographed. Control points were taken several centimeters apart from acupuncture points, namely in LI-4 and HT-3 area.

For the first part, impedance maps were measured with a grid of 64 brass telescopic needle electrodes (figure 1). Their telescopic tips maintained balanced pressure across measured area with side lenght 17.5 mm. Distance between neighbouring tips was 2.5 mm. Monopolar arrangement was applied with clamp reference electrode on the opposite arm (surface 6 cm^2). Frequency of probing current was 1 kHz. Unique recording system was controlled from Matlab environment via USB port. Prototype comprised microprocessor (Atmel ATmega16) with serial port and eight A/D converters with 10-bit resolution. The device measured voltage difference between active and passive electrodes based on spike detectors. Consequently, off-line transformation from voltage into impedance was performed based on interpolation of voltage data corresponding to known resistances.

For the second and the third experimental setup adhesive disposable Ag/AgCl ECG electrodes (Ambu Blue Sensor P) with measuring area 154 mm^2 and PUR sponge with highly conductive wet gel were used (figure 2). Electrode polarization effect was minimized by Ag/AgCl coating [5]. These active electrodes were placed on acupuncture point of the left arm.

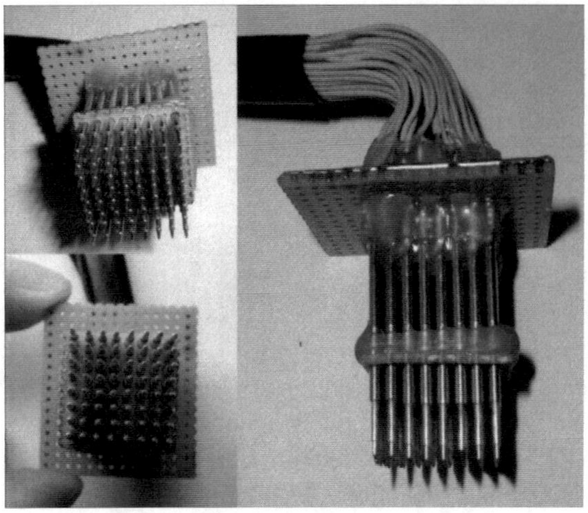

Figure 1: Brass needle electrodes with telescopic tips.

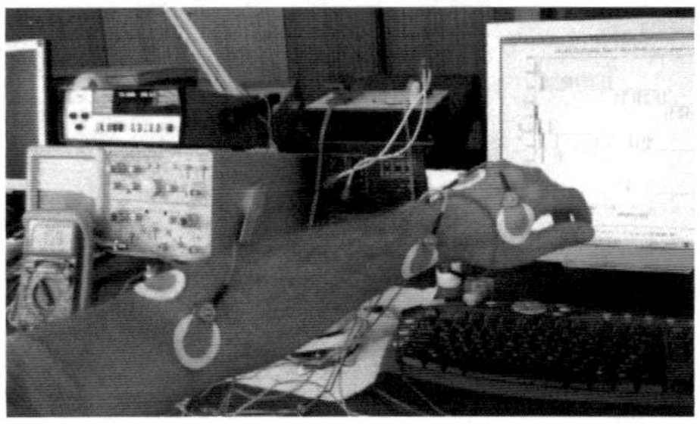

Figure 2: Adhesive disposable Ag/AgCl ECG electrodes.

3. Results and discussion

3.1. Impedance maps

We have found only one study depicting impedance/conductance maps of acupuncture point. Becker et al. [6] found higher conductance in the centre of acupuncture point surrounded by roughly circular equiconductance pattern (figure 3).

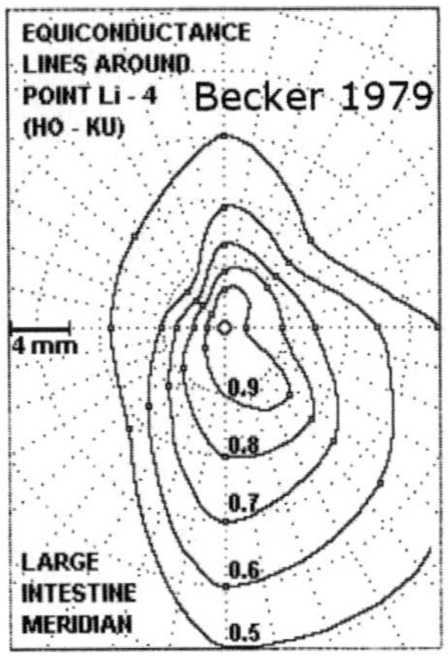

Figure 3: Sample of conductance map on acupuncture point LI-4 from study of Becker et al. [6].

Impedance maps from 11 acupuncture points did not consistently support any map structure with similar characteristics and pattern. In most of the cases distinctive minima surrounded by circular patterns were not found. In many maps there was no isolated depression in impedance modulus (figure 4). To show one of the most resembling map, there is local extreme in a form of minima at the upper left picture of figure 4 with certain indication of surrounding pattern. Control maps showed similar structures: irregular extremities spread across the measured skin surface, often with considerable variations within mm distance: Readings from neighbouring electrode sites differed up to the one order of magnitude. We found that repeatability of the same map was valid only under very restricted conditions – with immediate repetition and without removing the electrode grid from the skin. As a next step we plan to localize acupuncture points more precisely into the centre of the map and test different pressure levels of telescopic needles.

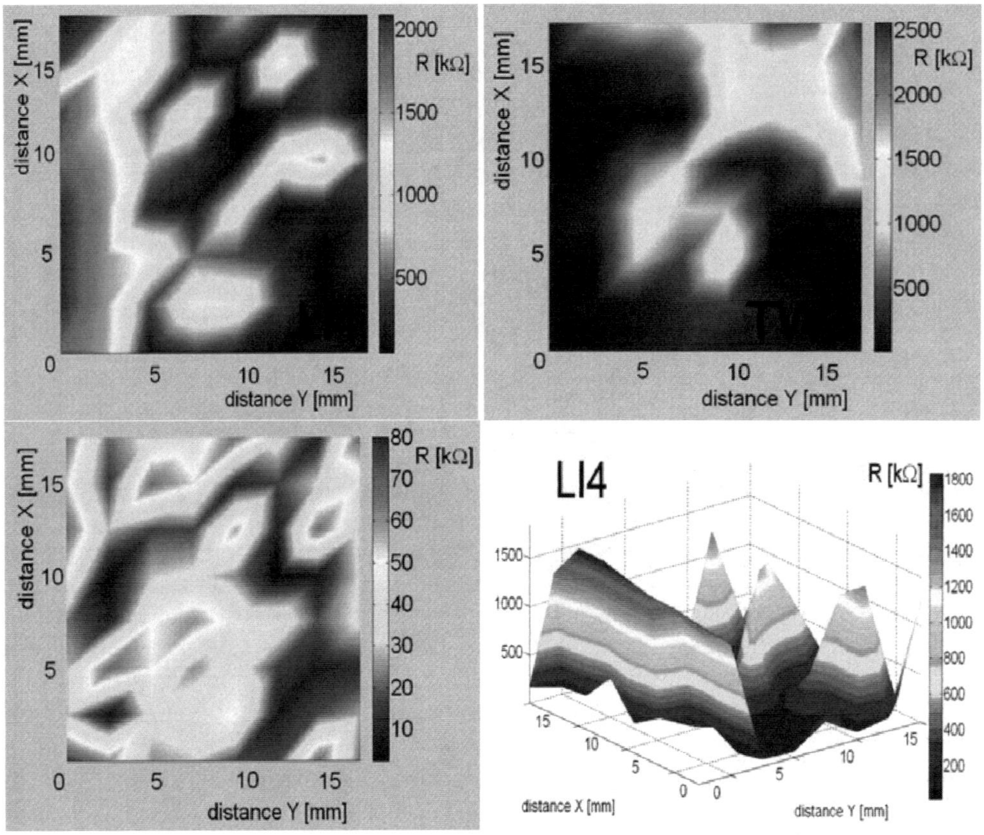

Figure 4: Sample of impedance maps of acupuncture points LI-4 and TW-3. Lower left: control point. Lower right: 3D plot.

3.2. Impedance in acoustic frequency range

For impedance measurement in acoustic frequency range Tesla BM 595 RLCG meter was used. Copper reference electrode in a shape of cylinder was placed under the right thigh. Supported frequency range was 100 – 20 000 Hz.

No distinction of acupuncture points from control points was obtained. On the contrary to expectations (e.g. [7]), control points had not higher impedance modulus (figure 5) in comparison to acupuncture points. While in impedance modulus all the curves were slightly decreasing, phase angle curves (figure 6) did not exhibit monotonous behaviour for most of the cases.

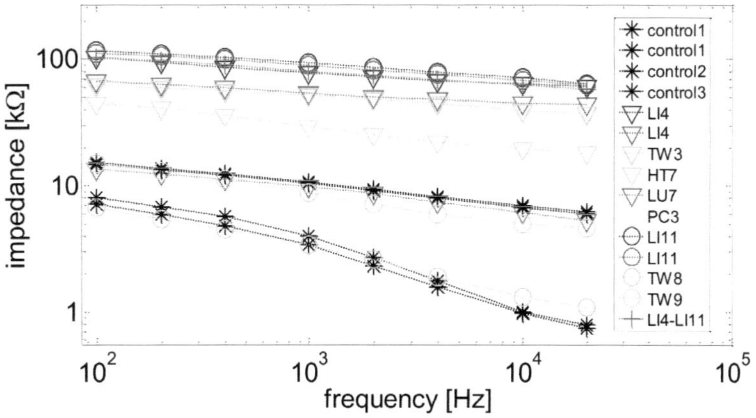

Figure 5: Impedance modulus dependance on frequency in acoustic frequency range. Black color holds for control points. Note the only meridian at the bottom of the legend.

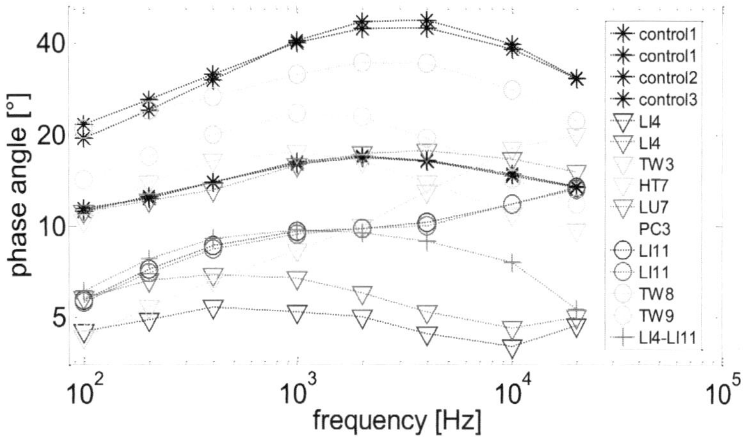

Figure 6: Phase angle in acoustic frequency range. Black color holds for control points. Note the only meridian at the bottom of the legend.

From obtained data it is possible to fit impedance |Z(f)| according to simple circuit models depicted in figure 7. Johng et al. [7] succesfully fitted their data according to model from scheme A of figure 7. Resulted estimates of one resistive and two capacitive components represented electrical elements based on skin layers: epidermis on the surface with dry tissue of stratum corneum (parallel R with C) together with serial R representing dermis. Their estimates agreed with distinction of acupuncture points as to be sites with smaller resistance and larger capacitance.

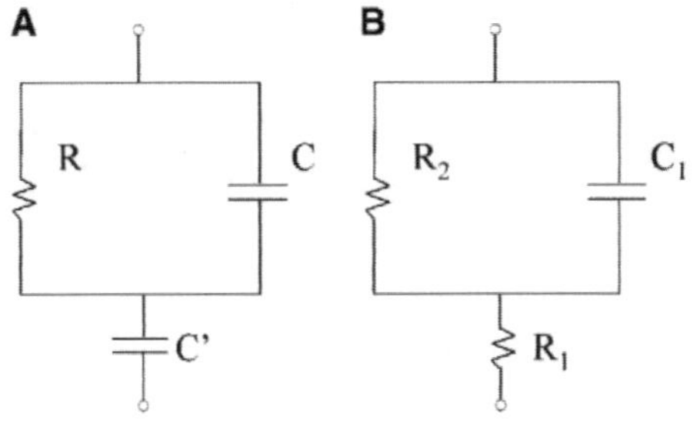

Figure 7: Electrical equivalent circuits from [2]: Model A was proposed by Johng et al. [7], model B by Rosendal [8].

Preliminary fitting of our data based on models from figure 7 was not appropriate, as the regression error was too large. In the next step we plan to build more complex model with more circuit elements. Also, relatively large surface of active electrode (1.5 cm^2) could cause that the largest current density flowed at electrodes' periphery and thus could avoid active points.

3.3. Impedance in radio frequency range

Depth of penetration into the skin is determined by AC frequency of measurement. Direct and low-frequency currents penetrates mainly only into the epidermis dominated by the dead stratum corneum. Moreover, at low frequencies (typically below several hundred kilohertz), the conductivity of the tissue is dominated by conduction in the electrolytes in the extracellular space. In order to investigate properties of deeper tissue structures and intracellular space as well, experimental setup with radio frequency impedance analyzer was prepared.

At radio frequencies, the tissue exhibits the beta dispersion, centered in the range 0.1 to 10 MHz, due to the charging of cell membranes through the intracellular and extracellular media. Above the beta dispersion, the cell membranes have negligible impedance, and the current passes through both the extracellular and intracellular media. On the other hand ionic flow is diminishing as relatively heavy ions are not able to adjust to rapid change of current direction.

Tomco TE1000 radio frequency impedance analyzer was applied with ECG active electrodes placed at acupuncture points and copper cylinder reference electrode hold in the right palm. The measurements were realized in the electromagnetically shielded room. Frequency from 0.5 – 150 MHz range was scanned with 1.5 MHz step.

The variations in impedance values across the whole frequency range were at the level of 1-2 orders of the magnitude (figure 8). Observed lower impedance zones around 50 and 70 MHz could point to the zone with improved information transmission in the studied tissues. Moreover, both zones bear another distinct property - zero phase angle (figure 9). These are the characteristics of resonant behaviour. Resonant properties may be demonstrated by either negative or positive impedance peaks. Different acupuncture subsystems differ from each other in resonant frequency around 50 MHz up to 10 MHz. A few another null angle zones appeared in the figure 8 while not binding to impedance extremities.

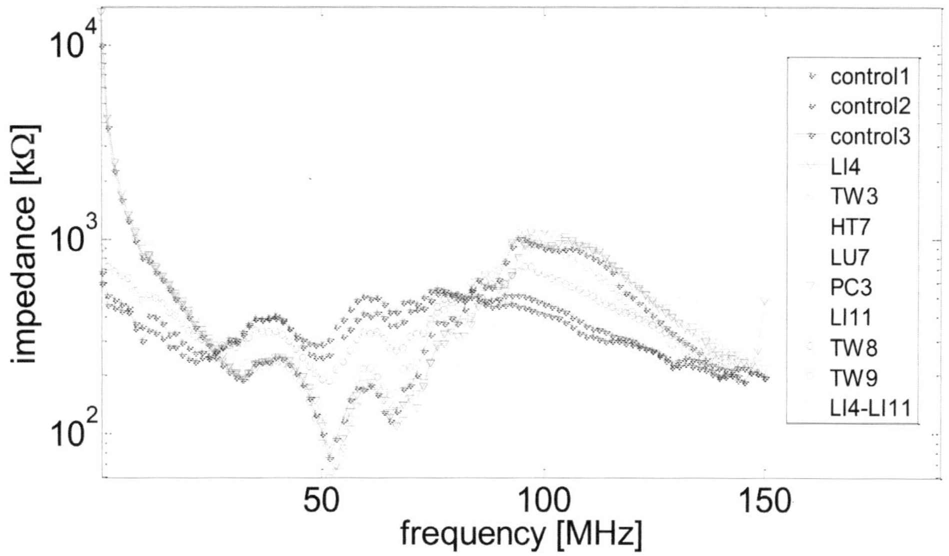

Figure 8: Impedance modulus for acupuncture (colored) and control (black) points. Note the only meridian at the bottom of the legend.

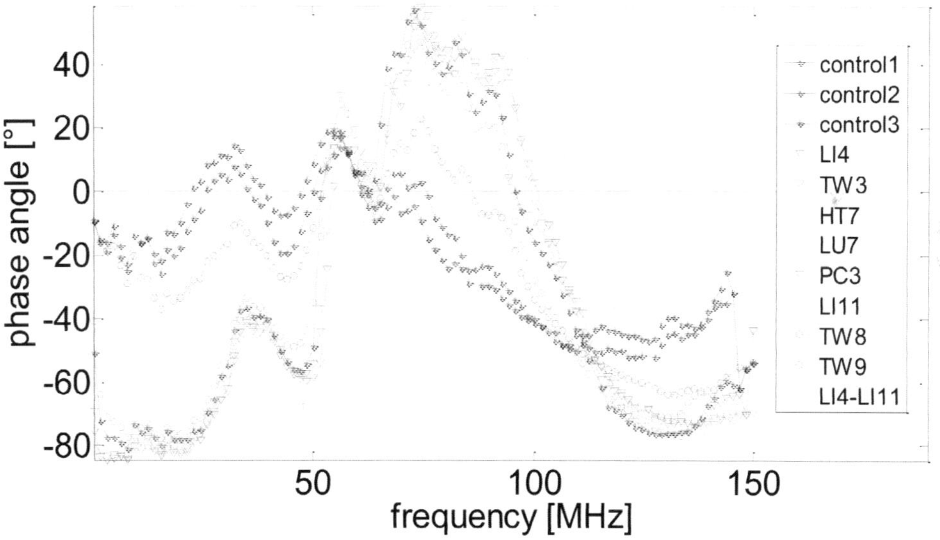

Figure 9: Phase angle for acupuncture (colored) and control (black) points. Note the only meridian at the bottom of the legend.

Acupuncture in comparison to control points seems to exhibit higher variations in impedance. Also, the main impedance minima at 50 MHz is for acupuncture points shifted to higher frequencies. The same holds for null angle characterizing resonant frequency around 50 MHz. The only acupuncture pathway measured between two acupuncture points exhibits distinct behaviour: sharper extremities and higher number of zero phase crossing.

Overall, the preliminary results require more data accumulation and further analysis. Experiences with measurement techniques confirm the fact that results may be heavily dependent on some of the following measurement characteristics: skin surface, presence of sweat glands, electrode geometry and polarizability, contact pressure, time fluctuation of electrical properties, etc.

Impedance analysis of acupuncture system deserves further attention along the need for improved understanding of physical mechanisms behind basic functioning of this relatively broadly used medical modality.

4. Acknowledgement

This work was supported by Slovak Grant Agency for Science (grant VEGA No 2/0210/10 and 1/0285/09).

References

[1] Zhang CL 2002 Skin resistance vs. body conductivity *Frontier Perspectives* 15-25

[2] Ahn AC, Colbert AP, Anderson BJ, Martinsen OG, Hammerschlag R, Cina S, Wayne PM and Langevin HM 2008 Electrical properties of acupuncture points and meridians: a systematic review *Bioelectromagnetics* **29** 4 245-256

[3] Fukumoto T and Akiyama H 2010 Acupuncture point position evaluating apparatus *US Patent* 7,818,054

[4] Reichmanis M, Marino AA and Becker RO 1977 Laplace plane analysis of transient impedance between acupuncture points Li-4 and Li-12 *IEEE Trans. Biomedical Engineering* **4** 402-405

[5] Ahn A C and Martinsen O. G. 2007 Electrical characterization of acupuncture points: technical issues and challenges *The journal of alternative and complementary medicine* **13** 8 817-824

[6] Becker RO, Reichmanis M, Marino AA and Spadaro JA 1976 Electrophysiological correlates of acupuncture points and meridians *Psychoenergetic Systems* **1** 106 195-212

[7] Johng HM, Cho JH, Shin HS, Sah KS, Koo, TH, Choi SY, Koo HS and Park MS 2002 Frequency dependence of impedances at the acupuncture point Quze (PC3) *Eng. Medicine and Biology Magazine IEEE* **21** 2 33-36

[8] Rosendal T 1944 Further Studies on the Conducting Properties of Human Skin to Direct and Alterting Current *Acta Physiologica Scandinavica* **8** 2-3 183-202

Journal of Physics: Conference Series **329** (2011) 012035 doi:10.1088/1742-6596/329/1/012035

Ablation Applicator for Destructive Hyperthermia Treatment

Tomas Vydra and Jan Vrba

CTU in Prague, Technická 2, Praha 6, 166 27, Czech Republic

vydratom@fel.cvut.cz

Abstract. This work is a preliminary research which is opening a field of examination of this complex clinic method (i.e. treatment using ablation applicators). Although it is established in many medical facilities its effects were not analyzed thoroughly and there are many unknown factors which can affect patients in clinical practice. We aim to bring new analysis methods which can simply and clearly determine how the tissue in the treated area will be affected and we want to inspect and overcome potential risks connected to this medical method.

1. Introduction

Although radiofrequency and microwave ablation is used widely in destructive hyperthermia, its precise effects on patients and a treated area are not well inspected. In this paper we focus on several applicators for radiofrequency hyperthermia (RF hyperthermia) and we analyze them in series of simulations (using SEMCAD X [1] and MATLAB [2] for results processing). Experiments in tissue samples were also conducted.

Radiofrequency ablation usually utilizes frequencies around 500 kHz (in comparison to microwave treatment with frequencies in range of GHz). Two electrodes (active and dispersive) are connected to a patient's body which closes an electrical circuit and once power is switched on electric currents start flowing through electrodes. Dispersive electrodes (there can be more than one) ensure that the electric current density is considerably lower than in the near vicinity of the active electrode (this is due to large connecting surface of dispersive electrodes to the patient's body). Electric current flowing through the tissue causes resistive losses which affect mainly the vicinity of an active electrode. These losses generate substantial heat which causes cell death (necrosis). [3]

In this paper we introduce four types of ablation applicators which can be used in practice. We analyze specific absorption rate (SAR) in the simplified model of a human body using SEMCAD simulations. Furthermore we inspect how the SAR distribution is affected when we change distance and position of dispersive electrodes. We also discuss obtained results considering its usage in clinic practice.

2. RF ablation system

The system layout is well shown in Fig. 1. It consists (as mentioned above) of two electrodes (dispersive electrode may have multiple pads), generator of high frequency voltage and additional controlling and measurement systems. Applicator (active electrode) is equipped with a handle for the operator which enables her to steer the tip of the electrode to the designated place for the treatment (CT and MRI results are used to locate the exact position of a tumor) and also (if the type of the electrode allows it) to eject or retract other structures from the inside of an applicator.

In further text electrode types used in this work are described.

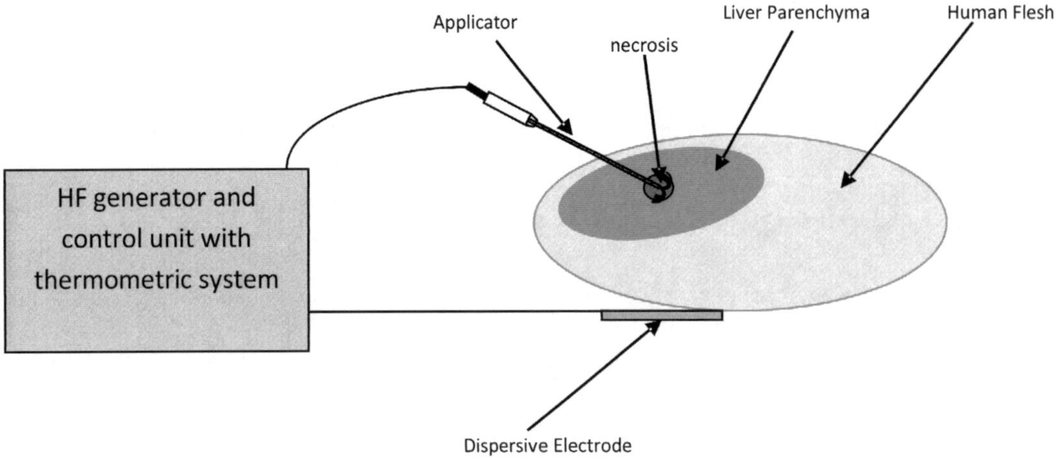

Fig. 1 System for RF ablation (of a liver in this specific case) layout

2.1. Active electrode

Due to the facts mentioned above an active electrode plays a crucial role in the RF ablation system. Its position (absolute and relative to a dispersive electrode) and shape can affect treatment procedure greatly. It is the reason why we inspect how the energy is absorbed if different electrodes are used or if their position is altered. Only through this careful analysis the best results can be obtained.

2.1.1. *A single-tip electrode.* A single-tip electrode can be seen in Fig. 2 a). This is the simplest electrode design which can be subsequently upgraded with additional systems (hollow tip with temperature control system – Fig. 2 b) etc.)
2.1.2. *An umbrella electrode.* Its usual design is shown in Fig. 2 c). Design of an electrode for radiofrequency ablation is restricted by the fact that it has to be inserted inside a patient's body with as little damage to healthy tissue as possible. This means that all the different electrodes must be retractable back inside its cannula. These limitations lead to designs as shown in Fig. 2 c), d) etc.
2.1.3. *A helical electrode.* Another design which allows ejecting electrode once a cannula is inside the body is a helical structure, see Fig. 2 d).

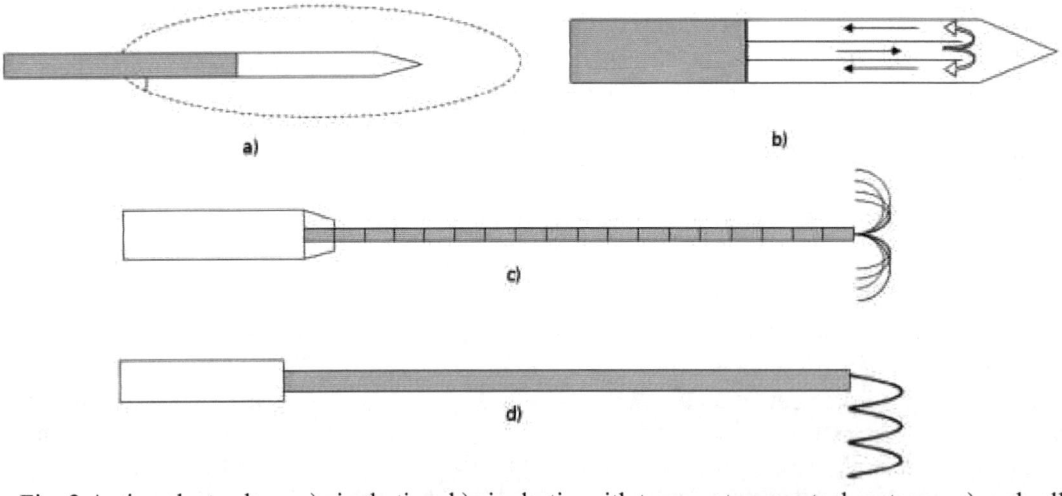

Fig. 2 Active electrodes – a) single tip – b) single-tip with temperature control system – c) umbrella – d) helix

2.2. Dispersive electrode

The other end of the system is a dispersive electrode. It can consist of several pads connected (usually using conductive gel) to the surface of a patient's body. The interface between skin and dispersive electrode has to be large to prevent any significant energy absorption near dispersive electrode (hence the name of the electrode – it disperses the electric current flowing through the treated area).

Dispersive electrodes are usually positioned on thighs where good contact with skin can be assured. Also no vital organs are near in case of any malfunction or unwanted local heating.

3. Models

For purposes of this work we used simulation software SEMCAD X which allows simple modelling. Four models where developed (two different umbrella applicators, one single-tip and one helical applicator). Models of the electrode tips can be seen in Fig. 3.

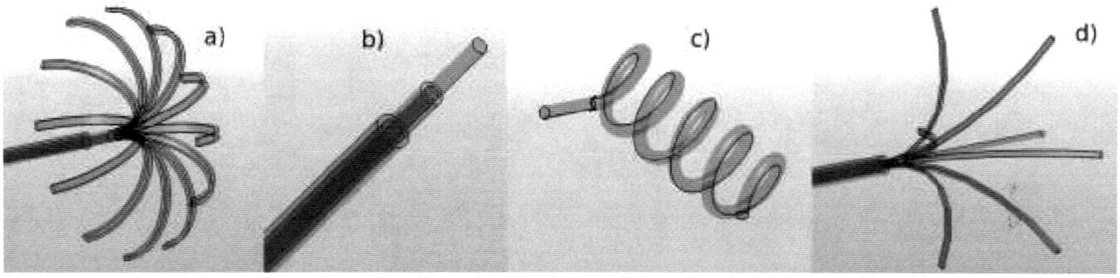

Fig. 3 Models of electrodes – a) umbrella – b) single-tip – c) helical – d) umbrella with central tip

Supply lines were included in models consisting of a conductive wire wrapped in dielectric isolator and inserted inside cannula. To keep models simple all metallic materials were simulated as PEC (Perfect Electric Conductor with no resistivity), cannula and isolator were simulated as dielectric material with no conductivity.

Model of human body was very simplified because effects of RF hyperthermia are limited to the near vicinity of the active electrode. Simulations with a more complex model of a human body (several organs were simulated with slightly different dielectric parameters) were conducted and no significant differences were found but the time required for a simulation is reduced significantly when the simplified model is used. Whole model for simulations with umbrella electrode with central tip is shown if Fig. 4.

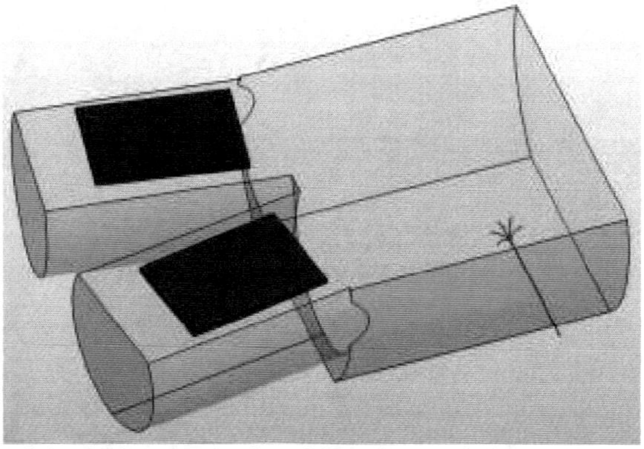

Fig. 4 The simplified model of a human body with active and dispersive electrodes inserted (black)

4. Simulations and results

Simulations were done in SEMCAD X and the results were processed using MATLAB. We inspected several things: how SAR differs for each type of electrodes, how SAR is influenced by the distance between active and passive electrodes, how SAR is influenced by a relative position of an active electrode to a dispersive electrode.

4.1. SAR for different types of electrodes

In this part of our work we conducted several simulations with electrodes presented above. It is evident that the surface of an electrode changes peak values greatly (larger an electrode is lower peak values are). That is why we normalized all results to the maximal value which was obtained in the case of a single-tip electrode (single-tip electrode has the smallest surface and thus the electric current density is the highest around it causing the highest losses in tissue). SAR in dB can be seen in Fig. 5.

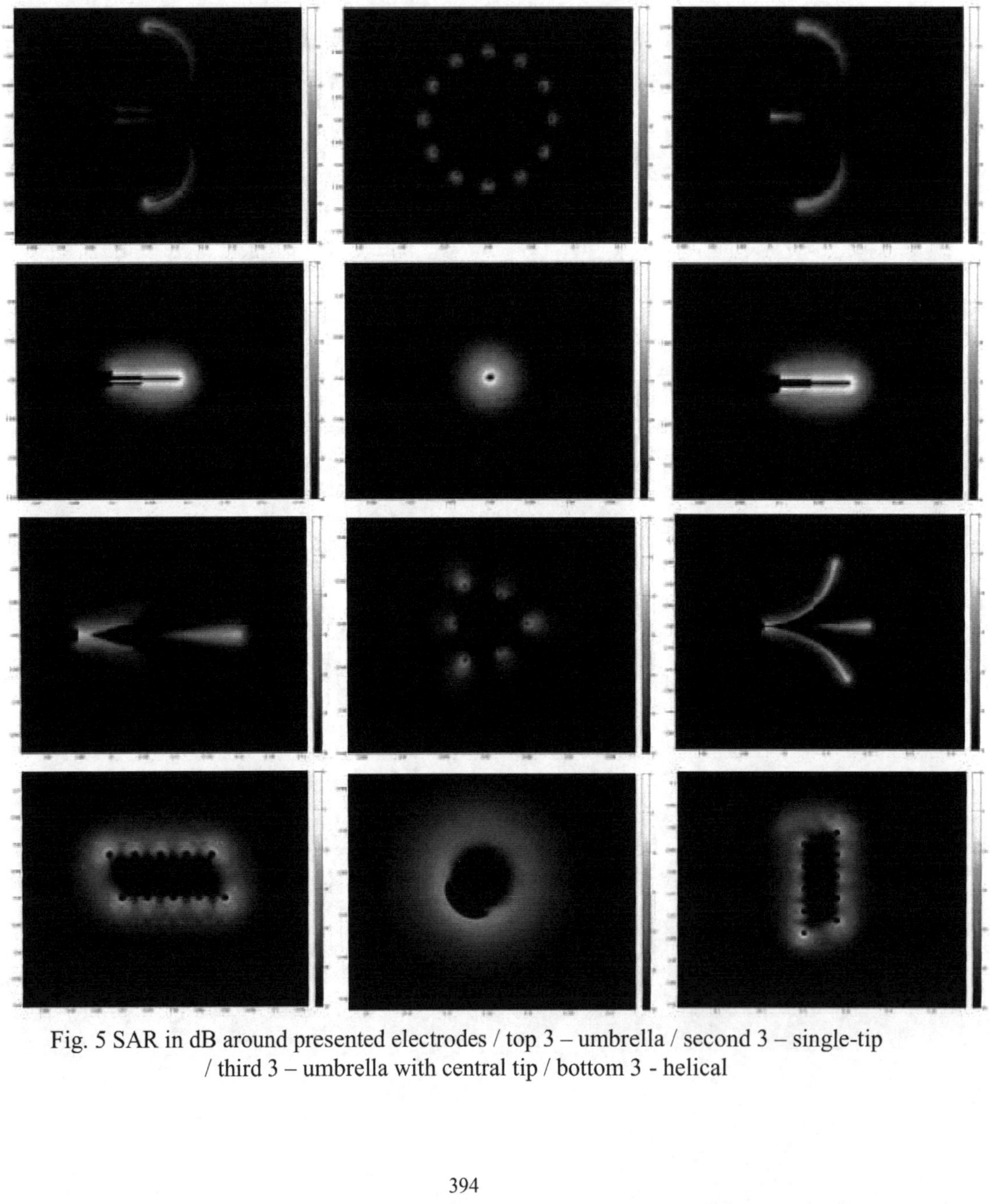

Fig. 5 SAR in dB around presented electrodes / top 3 – umbrella / second 3 – single-tip / third 3 – umbrella with central tip / bottom 3 - helical

From the results above it is evident that the energy is absorbed only very near to the electrodes. It can be also clearly seen that electrodes with larger surface reach lower SAR levels since currents flowing through nearby tissue distributed to larger volume. Specifically (single-tip electrode has maximum of SAR at 0 dB) SAR maximum is in case of umbrella electrode around -12 dB, in case of umbrella electrode with central tip it is around -4 dB and in case of helical electrode it is around -1.1 dB.

It is clear that in case of a single-tip electrode the effect of RF ablation is the most localized but can cause serious tissue damage if critical values of SAR are exceeded (intensive pain can occur in some cases [4]).

Usage of telescopic electrodes (i.e. umbrella, helical and other types) can prove to be problematic because, as seen especially in the case of helical electrode in Fig. 5, absorbed energy can be distributed mainly around the outer surface of the whole electrode causing some 'cold spots' inside the treated area [5]. This fact needs to be looked into since uneven treatment could cause regressions. Proper electrode design is, as mentioned above, unfortunately limited by the requirement of retractive electrode tips into a cannula.

4.2. Effects of various distance of electrodes

In this part of our work we wanted to find out how to augment SAR near the active electrode by moving dispersive electrodes (changing their distance to the active electrode). Fig. 6 depicts how the exposure changes when different distances are introduced to the system. As we can see in the left picture, absorbed energy is much higher if the electrodes are near each other. This can prove to provide much better results during a treatment procedure. On the other hand patient's safety must be considered and extremely high localized heating effects can cause excessive damage. Also peak values can appear on the edges of dispersive electrodes causing damage to the skin of a patient. According to our preliminary simulations current density is around ten times higher (+10 dB) and is still in safety zone but thorough testing would be required.

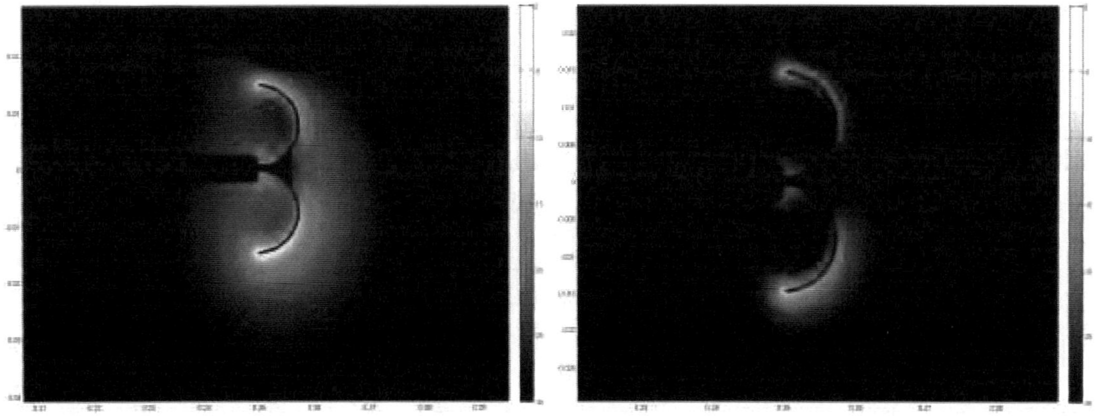

Fig. 6 SAR [dB] when (LEFT) dispersive electrodes were near the active electrode (distance around 10 cm) and (RIGHT) dispersive electrodes were far from the active electrode (distance around 30 cm)

4.3. Relative position – effect

Since the energy is primarily absorbed in very near vicinity of an active electrode and treatment effect is protruding through the tissue mainly by heat transfer we concluded (based on our simulations) that rotation of an active electrode does not affects treatment noticeably.

Results of our simulations for three different positions of an electrode are shown in Fig. 7. It is evident that the overall area of the treatment would be rather same in all cases (only rotated accordingly) thus allowing operator to choose the best position fitting particular tumor shape.

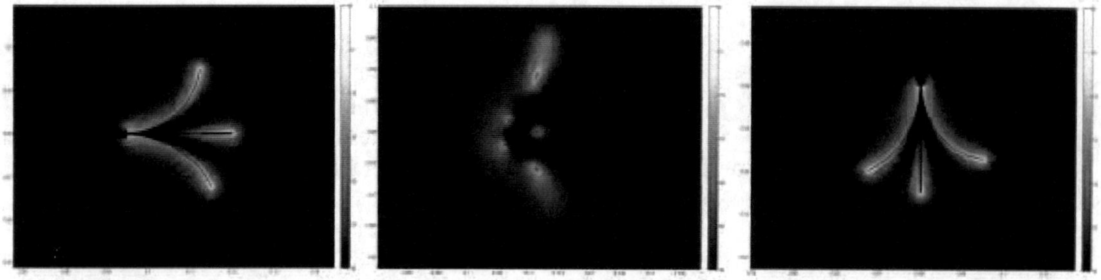

Fig. 7 SAR [dB] for three different orientations of an active electrode – from the left – 0°, 45° and 90°

5. Conclusion

From this work on RF ablation we conclude that there is very wide space for further research into the shape of active electrodes and possible unharmful positions of dispersive electrodes which would augment the efficiency of a treatment.

RF ablation is designed especially for small tumor treatment in the places that can be reached with difficulty by conventional methods or where standard procedures would pose too great risk (concerning heart, liver etc.). New and better telescopic electrodes can help overcome this limitation and expand area of the effective treatment and in some cases prevent excessive damage to the tissue.

In further development new designs of active electrodes should be presented. Their design will combine all the requirements (i.e. safety, retractable electrodes, as even as possible SAR etc.) while causing their strict specialization for various tumor types and shapes. Multiple advanced systems may be used [6].

References

[1] Online: http://www.speag.com/products/semcad/solutions/

[2] Online: http://www.mathworks.com/products/matlab/

[3] Chen C-C I, Miga M I, Galloway R L 2009 *IEEE TRANSACTIONS ON BIOMEDICAL ENGINEERING*. Optimizing Electrode Placement Using Finite-Element Models in Radiofrequency Ablation Treatment Planning. **56** 2 237-45

[4] Online: http://www.mayoclinic.org/radiofrequency-ablation/

[5] Ulucakli M E, Subedi Y, Ang I-C 2005 *Bioengineering Conference, 2005. Proceedings of the IEEE 31st Annual Northeast* Radiofrequency ablation of tumors 94-5

[6] Yang L et al. 2010 *Mechatronics, IEEE/ASME Transactions on* A Robotic System for Overlapping Radiofrequency Ablation in Large Tumor Treatment **15** 6 887-97

Disturbances of electrodynamic activity affect abortion in animals

M Nedbalova[1], A Jandova[2], and A Dohnalova[1]

[1]Institute of Physiology, First Faculty of Medicine, Charles University, Albertov 5, 128 00 Prague 2, Czech Republic
[2]Institute of Photonics and Electronics, v.v.i., Academy of Sciences of the Czech Republic, Chaberska 57, 182 51 Prague 8, Czech Republic

E-mail: martina.nedbalova@lf1.cuni.cz

Abstract. A specific kind of intracellular organelles, the mitochondria, is the place of metabolic energy production by oxidative mechanism. We used cell mediated immunity method for verification of the energy metabolism (ATP production). The antigen (immunological functional RNA) was obtained from blood of inbred laboratory mice strain C3H/H2K, infected with the lactate dehydrogenase elevating virus (LDV) and prepared by the high pressure gel chromatography (HPGC). We have studied the immunological adaptability of LDH viral antigen in 62 pigs (12 parents and 50 piglings). Exitus of piglings was in case of positive imunological response on LDV. The statement results from a comparison of the relative frequency of an incidence of identical findings in male piglets and sows and from identical findings in female piglets and pigs. The efficient elaboration and utilization of energy in cell may be damaged by the changes of energy production systems and also by long-term parasitary depletion of ATP energy. Biological activity is based not only on biochemical but also on biophysical mechanisms. Biophysical processes are also involved in the transfer of information and its processing for making decisions and providing control, which are important parts of biological activity. These experimental results were used for the same study in human.

1. Introduction

Research of infection by mice LDH virus (Riley virus) was going on in years 1965-1994 on Institute of Physiology of Universal Medicine Faculty in Prague. As a model was used mice strain C3H/H2K, rats, sewer rats, rabbits and dogs [1,2,3,4]. The model was extended by farm animals (cattle and pigs) [5,6]. The mode of the LDV transfer from mice to other animals was studied. The infection is transmited from mice to mice by means of saliva, sparmatic fluide and urine [7]. Statistical signifcant differences in cell-mediated immunity (CMI) answer to antigen obtained from blood of inbred laboratory mice strain C3H/H2K, infected with the LDV were discovered in cows, bulls and their calfs [8].

The CMI is important for the immunity of the organism to some viruses, moulds, and/or bacteria living inside cells and tumours, and it is also involved in a potential refusal of the transplant. The CMI was tested using the method of leukocyte adherence inhibition (LAI). An antigen as obtained from blood of inbred laboratory mice strain C3H/H2K infected with the LDV and prepared by the high pressure gel chromatography (HPGC) was used in the test. Pig males, sows and their progeny were used as a model, and their immunity response to the antigen applied was studied (immunologically active LDH-elevating virus RNA). For its survival, a cellular energy metabolism product, i.e. the ATP as formed prevailingly in mitochondria, is utilized by the given RNA. [9].

2. Materials

Preparation of Nonspecific Antigen – LDH Virus Antigen (Immunologically Functional Fraction from the Inbred Strain Mice Blood)

The immunologically functional fraction was obtained from the serum of inbred laboratory mice strains, e.g., C3H/H2K, infected with the virus (LDH) enhancing the lactate dehydrogenase iso-enzyme (NAD 1.1.1.27 Oxidoreductase). Plasma obtained from blood treated with heparin was subjected to at least doubled centrifugation for 20-30 minutes at 3-3.5 x 10^3 g, and the supernatant taken was clarified for 1-2 minutes at 9-11 x 10^3 g. The supernatant was subjected to further centrifugation for 50-70 minutes at 50-100 x 10^3g, and the sediment was suspended in buffer with pH of 7-7.4 consisting of tris-(hydroxymethyl)-aminomethane-NaCl [Tris-NaCl] and the disodium salt of ethylenediaminetetraacetic acid (EDTA) in a minimum volume of at most 1 ml. Equilibrium centrifugation in a saccharose density gradient separated the constituents, and the area corresponding to 32–38 % saccharose with the maximum of 35 % saccharose. This fraction was taken off and, after dilution with the above-mentioned buffer at the volume ratio of 1:0.5 to 1.5, was subjected to centrifugation at 50-100 x 10^3 g, for 50 to 70 minutes. Manipulation of blood plasma until the last sediment was obtained was carried out at 2 to 6°C.

The sediment obtained was dissolved in 1.5-3.0 ml of saline solution and used for high pressure gel chromatography. The fractions corresponding to molecular masses of 3 x 10^5 to 1.5 x 10^4 were immunologically active. The equilibrium centrifugation was carried out for a period of 180 to 210 minutes at 150 to 180 x 10^3 g. Separon Hema-300 glc®, which is a suspension copolymer of 2–hydroxyethyl-methacrylate with ethylenedimethacrylate with an exclusion limit of the column of 3 x 10^5, was used as a stationary phase for the high pressure gel chromatography [10]. Gel particles size was 32 to 40 μm. A differential refractometer and UV spectrophotometer with variable wavelength were used for detection. Saline solution was used as a mobile phase and the active fraction is captured at the wavelength of 340 nm [11].

3. Methods

3.1. LAI (Leukocyte Adherence Inhibition test)

Venous blood (8 ml) was taken from the patient and transferred into a test tube with 500 U of heparin, and incubated in a thermostat for 30 minutes at 37ºC. A Pasteur pipette was used to transfer the blood rich in white blood cells into a new test tube. The contents were diluted with a saline solution, medium to its original volume and centrifuged at 1500 - 2500 rpm.

The supernatant was poured off and the sediment was layered with 1 ml of distilled water, bubbled through a Pasteur pipette, and lysis of erythrocytes was achieved. Thereafter, dilution to volume with above mentioned solution according to an option was done. Thus, the lysis of erythrocytes was stopped. The contents were centrifuged once again for 10 minutes at 1500 - 2500 rpm. The rinsing solution was poured off, the sediment of white blood cells was diluted with a saline solution and diluted again so as to make available 4 x 10^6 cells per 1 ml for the test, as counted in the Bürker chamber.

Suspension with the white blood cells prepared according to the method described above was mixed with antigen at a ratio of 1 part of antigen to 4 parts of white blood cell float. 1 ml of the suspension with T lymphocytes sedimented in the test tubes from green Sial glass. After 60 minutes of sedimentation-adhesion process at 37°C number m of non-adherent cells was counted.

Cells, which didn´t meet the offered antigen, adhere steady to the test tube surface. The sensibilized cells don´t adhere.

3.2. The evaluation of LAI

The results of leukocyte adherence inhibition are expressed as a relative number N of non-adherent cells: $N = \frac{m}{M}$ x 100 (in %) or index of positivity: $\mathrm{IP} = \frac{N}{33}$

M is the number of cells in suspension before sedimentation-adhesion process

m is the number of non-adherent cells after 60 minutes incubation
33 is percent of non-adherent cells obtained continuously from healthy blood donors (1965-1995)
All examinations were practiced as a clinical-blind experiment (without the primary diagnosis) [12].

4. Results

The immunological response to LDV antigen was tested in pig, sow and their piglets. The results of examinations were expressed as index positivity (IP) values. The dependence of IP values in piglets on IP values in other sex parents was studied.

Table 1. Dependence of finding in pig male resp. female descendant on the finding in father resp. mother.

Father positive

	Descendant						
	male			female			descendant
Mother	positive	negative	total	positive	negative	total	total
positive	12	1	13	21	0	21	34
negative	1	17	18	17	4	21	39
total	13	18	31	38	4	42	73

The positive pig father had 91% of positive piglet daughters by transsexual transfer.

Table 2. Dependence of finding in pig male resp. female descendant on the finding in father resp. mother.

Mother positive

	Descendant						
	male			female			descendant
Father	positive	negative	total	positive	negative	total	total
positive	12	1	13	21	0	21	34
negative	10	2	12	1	11	12	24
total	22	3	25	22	11	33	58

The positive sow had 88% of positive piglet sons by transsexual transfer.

Table 3. Dependence of finding in pig male resp. female descendant on the finding in father resp. mother.

Father negative

	Descendant						
	male			female			descendant
Mother	positive	negative	total	positive	negative	total	total
positive	10	2	12	1	11	12	24
negative	0	20	20	2	23	25	45
total	10	22	32	3	34	37	69

The negative pig father had 92% of negative piglet daughters by transsexual transfer.

Table 4. Dependence of finding in pig male resp. female descendant on the finding in father resp. mother.

Mother negative

	Descendant						
	male			female			descendant
Father	positive	negative	total	positive	negative	total	total
positive	1	17	18	17	4	21	39
negative	0	20	20	2	23	25	45
total	1	37	38	19	27	46	84

The negative pig sow had 97% of negative piglet sons by transsexual transfer.

Table 5. Dependence of the finding in pig descendant on parent of the other sex.

	Father		Mother		
Female	positive	negative	Male	positive	negative
positive	38	3	positive	22	1
negative	4	34	negative	3	37
X.square	53.5		47.7		
significance	P< 0.0001		P < 0.0001		

Table 6. Mean values of IP in pig progeny, separately for positive and negative father and mother, and result of the test of mean values in progeny of positive and negative parent.

Descendant	Father		Significance	Mother		Significance
	positive	negative	t - test	positive	negative	t - test
Male	1.40	1.33	-	2.28	0.76	-
Female	2.33	0.78	P< 0.0001	1.84	1.43	P< 0.0001

Finally, the identical finding in immunological response to LDV antigen, obtained from blood of inbred laboratory mice strain C3H/H2K, infected with the lactate dehydrogenase elevating virus, in pig males and piglet daughters was 91%. The identical finding in sows and piglet sons was 94%. The dependence of finding on other sex parent is statistically high significant.

All these results became a base for research in human population.

5. Discussion

In one third of cases an increased content of the LDH-elevating virus RNA was ascertained in serum of experimental animals of a lower body mass (mice, rats/sewer rats, and dogs). Therefore, an immune response to the LDH virus antigen was studied in animals of a higher body mass (pigs). As known from literature, the LDH virus infection depends on the mice strain, age and gender and exhibits long-term parasitism on the cellular energy metabolism, which was proven specifically in the genealogical analysis of the frequency of LDH-elevating virus RNA incidence in parents and their progeny. The presence of the LDH-elevating virus RNA was studied in piglets (male or female) in dependence on the finding in their father or mother. It is evident from the results obtained that the identical result in the pig male and his piglet daughters was found in 91 %, while in case of a sow and her piglet sons the identical finding was in 94 %. If the increased LDH-elevating virus RNA occurred in both the parents, the positive progeny died without achieving a normal birth mass. If both the male and female organisms were burdened with the presence of a high amount of the LDH-elevating virus RNA (a high IP), the energetic metabolism of the pregnant sow is insufficient and it does not provide the fetus with a sufficient quantity of ATP through the placenta.

The ATP production (oxidative phosphorylation) is catalyzed by the ATP-synthase localized on the inner mitochondrial membrane. Nevertheless, if LDV is a parasite on the oxidative energy production system (i.e. on the mitochondrial production) of infected cells, the process monitored by the nonspecific antigen might be connected with disturbances of this system [13,14].

6. Conclusions

1) We studied a dependence of the LDH-elevating virus RNA incidence in piglets on the finding in a parent of an opposite gender than that of the offspring. A statistically significant 91% coincidence was found in a pig male and his female piglets. In a sow, the finding exhibited a 94% coincidence with that of male piglets.
2) If an increased LDH-elevating virus RNA occurred in both parents, their positive progeny either did not achieve the standard birth mass or they died.
3) The high content of LDH-elevating virus RNA in mitochondria affects the cellular energy metabolism.

7. References

[1] Motyčka K, Jandová A and Pezlarová J 1979 Studium přenosu infekce LDH virem z rodičů na potomstvo u myší kmene H. *Sb. lék.* **78** 176-80
[2] Motyčka K, Jandová A, Herzog P, Boštík J and Štěpánová I 1984 The lactic dehydrogenase virus (LDV) infection of C3H mice increases the number of spleen plaque forming cells (SPFC) above physiological values. *Physiol. Bohemoslov.* **33** 275
[3] Jandová A, Bendl J, Nedbalová M, Trojan S and Vávrová M 1994 Cell mediated immunity induced by the mouse LDH virus in rats of the Wistar strain. *Physiol. Res.* **43** 20
[4] Jandová A, Bendl J, Nedbalová M and Trojan S 1995 The Inhibition Adherence of Leukocytes in Relation to Specific Rabbit Antibody and to RNA Polymerase. *Phys. Res.* **44** 23P
[5] Říha J and Jandová A 1993 A preliminary study of lactate dehydrogenase virus (LDV) activity in breeding bulls. *Živočišná výroba* **38** 583-90
[6] Nedbalová M, Chválová S, Jandová A and Trojan S 1997 Immunological response of CD4 lymphocytes on LDV antigen in pigs and human parents and children. *Arch. Physiol. Biochem.* **105** 240
[7] Riley V 1974 Biological Contaminants and Scientific Misinterpretations. *Cancer Res.* **34** 1752-54
[8] Říha J, Jandová A and Polášek M 1993 The effects of lactate dehydrogenase virus (LDV) on reproductive performance and health of bovine females. *Živočišná výroba* **38** 223-38
[9] Jandová A, Motyčka K, Čoupek J, Sanitrák J, Heyberger K, Laurová L, Mejsnarová B, Novotná J, Pezlarová J, Macholda F, Šimonová J, Dostál C and Kubát Z 1979 "Disease" of the cellular energy system. *Sb. lék.* **81** 321-27
[10] Jandová A, Sanitrák J, Motyčka K, Pezlarová J and Čoupek J 1978 Isolation of immunoreactive components from experimental and human tissues and serums by high performance gel chromatography. *J. Chromatogr.* **146** 253-60
[11] Jandová A, Čoupek J, Sanitrák J, Motyčka K, Mach O and Pekárek J 1989 Method of preparation of immunologically functional fraction from the inbred strain mice blood. *Authors Certificate No. 259704*, Prague.
[12] Jandová A, Kobilková J, Čoupek J, Vejbora O, Postupa J and Pokorný J 1990 Method of determination of leukocyte adherence inhibition. *Authors Certificate No. 265601*, Prague.
[13] Jandová A, Nedbalová M and Trojan S 1997 LDH virus - parasite or energetic system of animals? *Arch. Physiol. Biochem.* **105** 239
[14] Jandová A, Pokorný J, Kobilková J, Trojan S, Nedbalová M, Dohnalová A, Čoček A, Mašata J, Holaj R, Tvrzická E, Zvolský P, Dvořáková M and Cifra M 2009 Mitochondrial Dysfunction. *Neural Netw. World* **4** 379-91

Neural network analysis of electrodynamic activity of yeast cells around 1 kHz

R Janca[1]
Czech Technical University in Prague, Faculty of Electrical Engineering, Prague, Czech Republic

E-mail: jancarad@fel.cvut.cz

Abstract. This paper deals with data analysis of electrodynamic activity of two mutants of yeast cells, cell cycle of which is synchronized and non-synchronized, respectively. We used data already published by Jelinek *et al.* and treat them with data mining method based on the multilayer neural network. Intersection of data mining and statistical distribution of the noise shows significant difference between synchronized and non-synchronized yeasts not only in total power, but also discrete frequencies.

1. Introduction

Electromagnetic activity and sensitivity of single cells has been extensively studied in the past [1]. Despite a number of indirect observations and theoretical models [2], electromagnetic activity of single cell has been measured only with limited success. However, advances in nanotechnology have led to reopening of this issue [3].

In recent paper by Jelinek *et al.* the evidence was shown for the electrical activity of yeast cells *Saccharomyces cerevisiae* in the frequency range 0.4 – 1.6 kHz [4]. However, this conclusion was based mainly on the mean electrical power in entire frequency range measured for two populations of cells in different phase of the cell cycle. One may have reservations about this method since it may be strongly influenced by the change in bio-chemical and electrical properties of the cells and the buffer due to metabolic activity of cells in certain phase of the cycle. In order to resolve whether this objection is valid, we employed artificial neural network in the analysis of data presented in [4]. In this paper we present results of the assay which revealed significant difference between two cellular cultures at individual frequencies. This difference, analysis proved, may not be attributed to technical artefact or chemical reason.

2. Materials and methods

We used data kindly provided by authors of Jelinek *et al.* [4]. In following we present a brief summary of measurement system and procedure. Detailed description is given in [4].

[1] Department of Circuit Theory, Faculty of Electrical Engineering, Czech Technical University in Prague. Technická 2, 166 27 Praha 6 – Dejvice, Czech Republic

2.1. Yeast cells

The β-tubulin *tub2-401* mutants of yeast cells *Saccharomyces cerevisiae* (strain CUY67 Mata tub2-401 ura3-52 ade2-101) were used for the measurement. The mutant cells were cultivated under the restrictive temperature of 14°C when microtubules cannot polymerize. Most cells interrupt their cell cycle before entering in to M-phase, because mitotic spindle cannot be assembled. The microtubules begin to polymerize after heating of the cells above the permissive temperature of 25°C, so the majority of cells assemble the spindle and start the mitosis synchronously (synchronized cells). The part of the mutant cells was cultivated at the temperature of 30°C. These cells were not synchronized (non-synchronized cells) and they were used as a reference.

2.2. Laboratory equipment

The measurement system is depicted in Figure 1. The important part of the system is the sensor followed by transform and power amplifiers placed in a triple electromagnetic shielded box which is housed inside a temperature stabilized chamber. The sensor is a sharp platinum wire with tip that is few micrometres above the bottom of the cuvette. Space between the bottom of the cuvette and the tip is comparable with the size of the yeast cell. Cells settle on the bottom of the cuvette, so the sensor can contact only one cell (see schematically in Figure 2). But a large part of the platinum wire is still immersed in the suspension, so the impedance of the suspension contributes to the background noise. The sensor is followed by the transform amplifier that is connected before the power amplifier. The gain of the amplifiers is around 110 dB at 1 kHz. Both amplifiers are supplied by batteries that are placed inside the shielded box. This helps to minimize spurious signals. The thermal box stabilizes temperature inside at 28±0.5°C with minimal hysteresis. Output of the power amplifier is connected to the spectrum analyser Agilent E4448A. Theoretical sensitivity of the system is $10^{-18} \div 10^{-19}$ W, that is the thermal noise level.

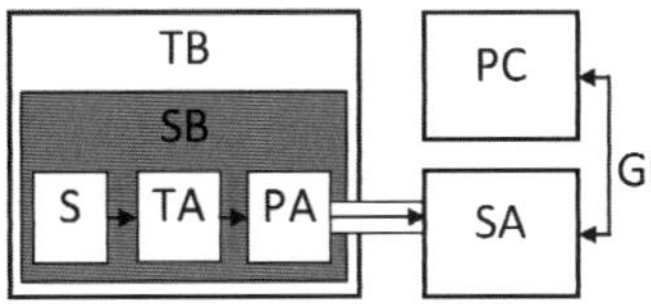

Figure 1. The schematics of the measurement system: S-sensor, TA-transform amplifier, PA- power amplifier, SA-spectral analyser, PC-personal computer.

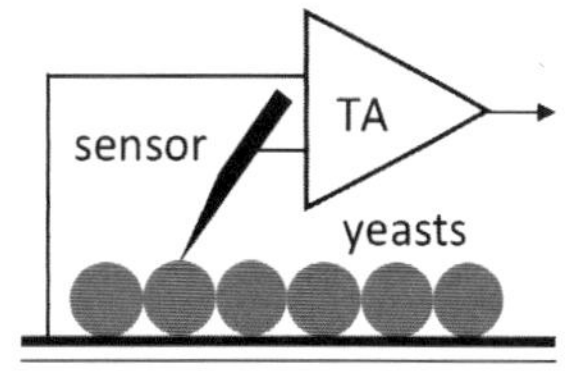

Figure 2. Detail scheme of the point sensor, bottom of cuvette, transform amplifier (TA) and sedimented yeasts.

2.3. Measurement

A suspension of the cells has optical density 4.5 (OD 600), which represents a concentration of about $2 \cdot 10^8$ cells per millilitre. The synchronized cells were cultivated under the 14°C and non-synchronized around 30°C. Before the measurement, a cuvette with the suspension was placed in 28°C water bath for 3 minutes. The sensor cuvette was then filled with the shaken suspension; volume was approximately 0.06 ml. The measurement of the electrical oscillation started immediately after filling. The spectral analyser measured in frequency range 400÷1601 Hz. Resolution and video bandwidths were set to 1 Hz, reference level was -80 dBm. Whole frequency band was scanned in two sweeps of 601 Hz; both sweeps and storing of data to the PC lasted 6 second. The 400 spectra were recorded during 40 minutes of the measurement.

3. Data analysis

The 49 sets of cells culture, 23 of which were synchronized and 26 were non-synchronized (as a reference), were measured during the experiment. Spectrum of the signal was recorded over time for

40 minutes. The spectrum had 1/f-noise character that is added to transfer function of the amplifiers [5]. The shape of the transfer function was estimated as a time-average spectra of a set, see Figure 3. The shape was depending on the impedance of measured suspension and it could be well approximated by polynomial of 3rd degree which corresponded with electrical stimulation of the amplifiers, an example is given in Figure 4. The spectra were independent on the transfer function after subtraction of its shape.

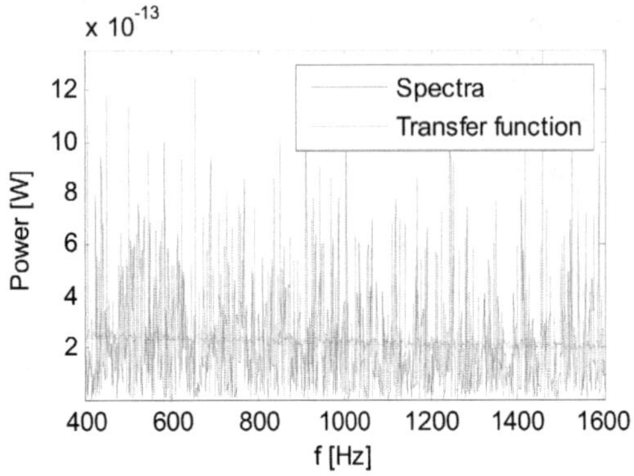

Figure 3. An example of the spectra and the transfer function estimated as a time-average spectra.

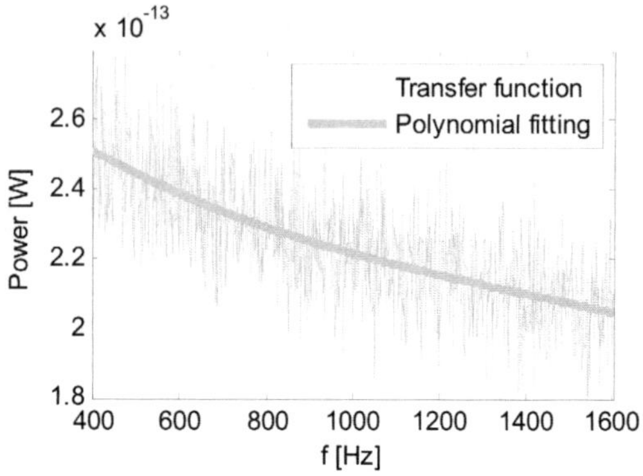

Figure 4. Transfer function estimated as a time-average spectra and its polynomial fitting.

3.1. Power analysis

Figure 5 shows normalized mean spectral power of all of synchronized and non-synchronized sets in time. Non-synchronized cells have uniform trend of the power emission in time. Synchronized cells have increased power between 5th and 20th minutes that correspond with the metaphase and anaphase. Average total power of the synchronized set was 7 % greater that of the non-synchronized. However, the growth and metabolic activity of yeasts during the mitotic process changes parameters of the suspension such as the impedance. The power of the thermal noise strongly depends on the impedance, so difference between sets do not have to be caused by the electrodynamic activity only.

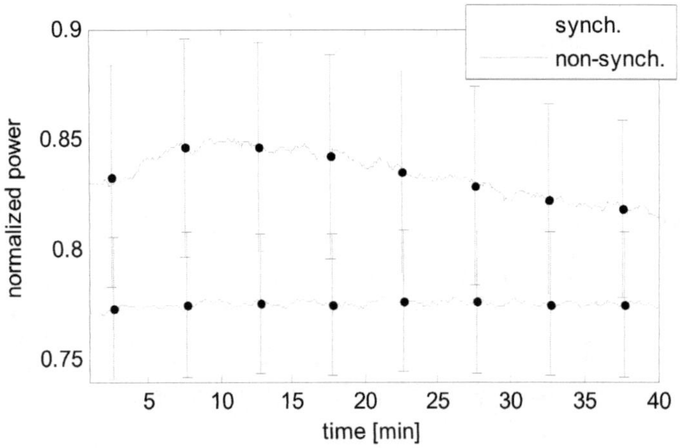

Figure 5. The normalized average spectral power of all cells sets in time. The power of the synchronized cells is increased between 5th and 20th minutes.

3.2. Frequencies analyses

The transfer function and thermal noise are directly dependent on the impedance of the suspension. A frequency analysis minimizes the mentioned dependency. The measured data have noise character without high spectral lines. Standard methods cannot find significant difference between discrete frequencies of the synchronized and non-synchronized sets. Therefore a method of the data mining using a neural network classifier was employed.

4. Data mining using a neural network

The method is based on following hypothesis:

- A multilayer neural network correctly classifies the spectra into two classes (synchronized, non-synchronized). The spectra do not contain information about an average spectral power. If previous clams are true, the information about difference in the spectra exists. Then weights of network's inputs show on important frequencies.

4.1. Multilayer neural network

A multilayer neural network (MLNN) is a complex classifier composed of simple classifiers called neurons; more details are in [6]. The McCulloch-Pitts neuron model is depicted in Figure 6. An input vector **x** of data is multiplied by a vector of the weight **w**. A sum of multiplied input values and thresholds θ is an argument of the transfer-function of the network. An output y of the neuron is a return value of the transfer-function *f()*. The index j is a neuron number; the variable n represents number of the input parameters. A first and second hidden layer use logarithmic-sigmoidal transfer function. The output layer uses linear function. The scheme of the topology1202-1801-300-2 of the MLNN is shown in Figure 7. A numerical designation of topology represents number of neurons in layers. The topology and transfer-function were chosen as the best of few variants [5].

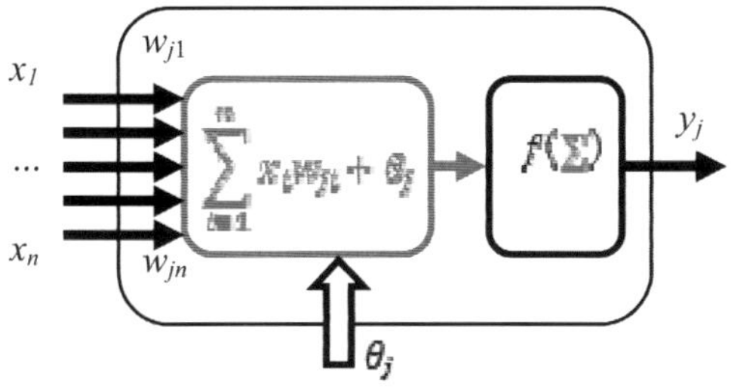

Figure 6. The McCulloch-Pitts model of the neuron.

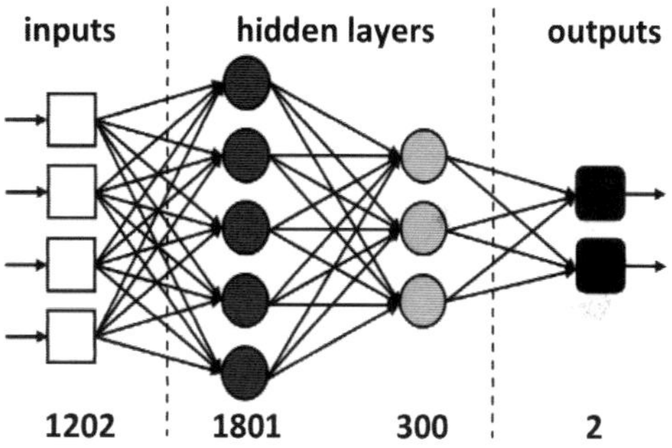

Figure 7. The scheme of the MLNN topology. Number of neurons in layers is below.

This neural network allows a transformation from 1202 to 2-dimensional space. The MLNN is realized in Matlab 7.9 using Neural Network Toolbox. A training algorithm adjusts neurons parameters (weights and thresholds) to minimize classification error on a training data set. The *learngdm* training-algorithm was used; it is a gradient descent method with momentum weight and bias learning function which prevents over-learning [6]. The training data contained input and target data. Trained MLNN is able to classify input test vectors of data into two classes (synchronized or non-synchronized). In other words, the neural network finds an optimal function that transforms data to known outcomes. The function can be very complex and cannot be determined through analytical methods. The network can be trained as a classifier which divides testing data into classes. A complex decision making strategy is composed from simple neuron functions.

4.2. Parameterizing of the data

Measured data had to be modified before using the MLNN. The polynomial shapes of amplifier's transfer function were estimated and subtracted from spectra. Spectral lines of each discrete frequency

Journal of Physics: Conference Series **329** (2011) 012037 doi:10.1088/1742-6596/329/1/012037

were normalized by maximal values of spectral lines throughout all sets. Duration of oscillations is expected less than 30 second. This time corresponded with 5 following measured spectra. Therefore 5 point moving average filter was used for highlight of oscillations and suppression of random high spectral lines in a frequency. The filtered spectra were powered to two in order to improve difference between high and low spectral lines. The parameterized spectra were used as input vectors, input of each corresponded with the frequencies. Additional parameters of the input vectors may be coefficients of the polynomial approximation of the polynomial shapes of amplifier's transfer function. These coefficients represented the average spectral powers. Using of additional coefficients only verified dependency classification on the average power. Each input vector had a target vector that contained two values representing class of the cells set. The synchronized set represented vector [+1, -1], the non-synchronized [-1, +1].

4.3. Data mining

The 2/3 of parameterized data was randomly separated to the training sets and the 1/3 to the test sets. The data were without additional power parameters. The MLLN was trained by the train sets with the targets vector representing the classes. A validating was proved by the test sets, but the output vector did not contain discrete values. The Equation (1) describes a rule of the assignment to the class. The vectors of the classes of the test sets are **X**; the **Y** is output vector of the trained net. If difference of the output vector **Y** has same sign like difference of the classes' vector **X**, the classification is correct (that is true); and conversely.

$$\operatorname{sign}(X(1)-X(2)) = \begin{cases} \operatorname{sign}(Y(1)-Y(2))\ldots \mathit{true} \\ -\operatorname{sign}(Y(1)-Y(2))\ldots \mathit{false} \end{cases} \tag{1}$$

A percentile of the true classification represented an ability of learning of the network. Inputs weights of the trained net have described strength of the input parameter for the classification. The strength **W** of the input was calculated as average value of the weights of the neurons in the first hidden layer; the *m* is a number of the neurons, the **w** is weights and the index *n* represents the input parameter; see equation (2).

$$W_n = \frac{1}{m}\sum_{i=1}^{m} w_{ni} \tag{2}$$

The similar method is used to pruning of a net that decrease number of neurons in the net; neurons with small weight are removed [7].

The described process was made 20 times for minimizing random errors. Greater number of repetitions does not significant improve the results. The final results were average of the cross validation.

5. Results

The neural network trained by data with the information about mean spectral power had a validation of the classification around 90 % which corresponded with the power analyses. Data without power parameters had less validation, but the value around 75 % was accepted. Importance of inputs parameters for classification success is represented by input weights. The input weights correspond with frequencies, see figure 8. The high weight shows on the significant frequencies that are the most important for data division into synchronized and non-synchronized. Consequently, values of weights directly related to information about differences between both sets.

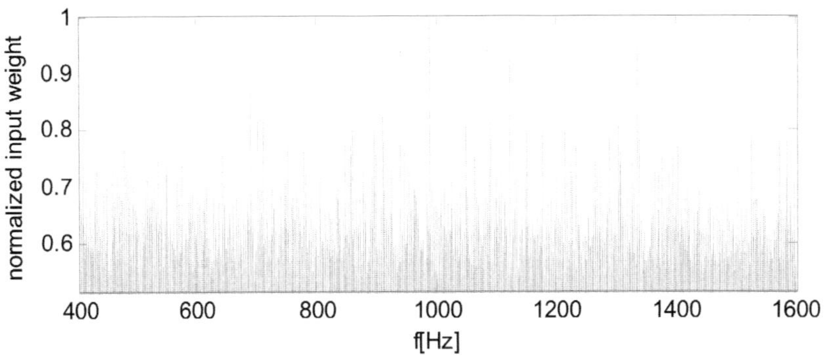

Figure 8. The normalized input weights as function of the frequency.

The probability density of noise was different in frequencies of measured spectra. The frequencies with the high weights had higher power of spectral lines than the frequencies with the small weight. This fact is described by a mean cumulative distribution function (CDF) of normalized noise of 10 frequencies with the highest and the lowest weights; see Figure 9.

The 90% percentiles ($k_{0.9}$) of the CDF show a significant difference between the discrete frequencies. The frequencies with the lowest percentiles correspond with the frequencies that have highest weights. The five most significant frequencies were found as an intersection of the 25 most important frequencies from the MLNN and CDF analyses. Frequencies of the highest weight and the lowest percentiles were 401, 702, 939, 1528 and 1573 Hz.

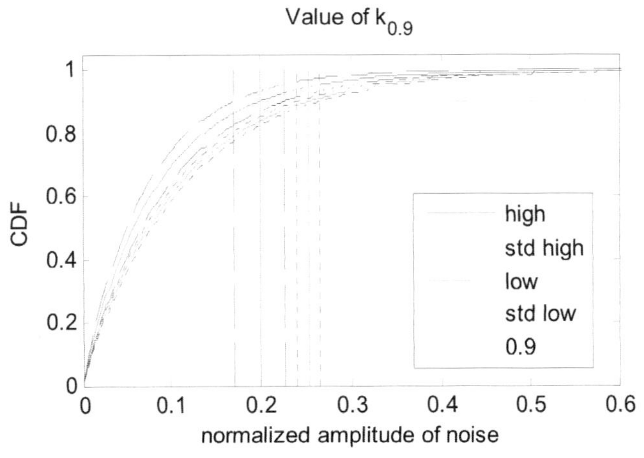

Figure 9. The mean cumulative distribution function (CDF) versus normalized noise amplitude of the 10 most important frequencies. The 90% percentiles ($k_{0.9}$) define a marker of the difference.

6. Conclusion

We used data kindly provided by authors of *Jelinek et al.* [4]. Process of the increasing electrodynamic activity of the yeasts in the metaphase and anaphase was described by Pohl in electrophoresis experiment [8]. A similar increasing of average power was measured on synchronized cells between 5th and 20th minutes of the measurement. The increasing was caused by a combination of an electrodynamic activity and thermal noise that is depend on an impedance of a cells suspension. A dynamic process of the mitosis changes a distribution and a shape of the cells under the sensor. It may cause changes of the impedance.

The frequency analyses using the data mining method and the CDF of noise shows on 401, 702, 939, 1528 and 1573 Hz. These frequencies were the most significant for classification into two classes (synchronized and non-synchronized). Same frequencies had distribution of noise with the greatest occurrence of high power spectral lines. Pelling measured – and Jelínek partially confirmed – the mechanical quasi-coherent oscillations on surface of yeast in acoustic frequency ranges for few seconds using atomic force microscope (AFM) [4, 9]. The mechanical oscillations of electrically polar structures may be one of the sources of the electromagnetic emission that was founded in discrete frequencies.

The results could be distorted by several aspects. The methods assumed independence of the oscillation on a temperature that is not completely fulfilled. The measured system stabilize temperature 28±0.5°C that could cause frequency shift of electrodynamic activity. Very weak signals like thermal noise were measured; so a part of the discovered oscillations may be caused by external noise. However influences of the aspects have not yet been observed.

Acknowledgement
I thank J. Pokorny for valuable discussions and for providing measurement data. This work has been supported by the grants IGA NT11460-4/2010, for which analysing method was primary developing.

Reference

[1] Cifra M, Fields J Z and Farhadi 2011 Electromagnetic cellular interactions *Progress in Biophysics and Molecular Biology* **105** 223

[2] Havelka D, Cifra M, Kučera O, Pokorný J and Vrba J 2011 High-frequency electric field and radiation characteristics of cellular microtubule network *Journal of Theoretical Biology* **286** 31

[3] Kučera O, Cifra M and Pokorný J 2010 Technical aspects of measurement of cellular electromagnetic activity *European Biophysics Journal* **39** 1465

[4] Jelínek F, Cifra M, Pokorný J, Vaniš J, Šimša J, Hašek J and Frýdlová I 2009 Measurement of electrical oscillations and mechanical vibrations of yeast cells membrane around 1 kHz *Electromagnetic Biology and Medicine* **28** 223

[5] Janča R 2010 Measurement of electromagnetic oscillations of cells (in Czech: Měrení elektromagentických oscilací buněk) *Faculty of Electrical Engineering, Czech Technical University in Prague*

[6] The MathWorks, Inc., © 2010 Neural Network Toolbox. *MATLAB documentation*

[7] Sakar A and Mammone R J 1993 Growing and pruning neural tree networks *IEEE Transactions on Computers* **42** 291

[8] Pohl H A 1980 Oscillating fields about growing cells *Int. J. Quant. Chem. Biol. Symp.* **7** 411

[9] Pelling A E, Sehati S, Gralla E B, Valentine J S and Gimzewski K 2004 Local Nanomechanical Motion of the Cell Wall of Sacharomyces cerevisiae *Science* **305** 1147

Journal of Physics: Conference Series **329** (2011) 012038 doi:10.1088/1742-6596/329/1/012038

Non invasive radiofrequency diagnostics of cancer. The Bioscanner —Trimprob technology and clinical applications

Clarbruno Vedruccio[1], Carla Ricci Vedruccio[2],

[1]Italy,clarbruno@tin.it ; web site: www.bioscanner.it
[2]Italy, drcarla@tin.it

Abstract. A new paper by Pokorny, Vedruccio, Cifra, Kucera, titled Cancer physics: Diagnostics based on damped cellular elasto-electrical vibrations in microtubules, recently available on Eur. Biophys. J., discloses the mechanism of active grown cancer tissues interaction with a Non- Linear Resonance Interaction (NLRI) Bioscanner Trimprob diagnostic device that is certified and ready to be used to investigate suspected cases of disease and cancer. This technology spreads early capabilities of cancer detection by means of low level radiofrequency oscillations in UHF band. The system is based on an unique and extremely innovative non- linear radiofrequency oscillator working on 462-465 MHz plus the harmonics. The diseased tissues suspected of cancer, are irradiated by means of a handy probe near field emission, while a spectrum analyzer placed in the far field detects by means of a small antenna, the oscillator interaction within the tissues. The Bioscanner is characterized by a high dynamic range, in the order of 30 or more decibel, and is useful for detection of small cancer agglomerates, if used by a well trained operator. At the resonance, the free running oscillator locks-in on the specific interaction frequency , in a sharp frequency window centered on 462 MHz; the resulting effect is evidenced by a deep decrease of the 462 MHz spectral line propagation in the far field around the oscillator probe. The NLRI provides a selective characterization, like a sort of a electronic biopsy response of biologic tissues in support of modern imaging diagnostics. Further to existing literature describing methods for cancer detections by means of electromagnetic fields this paper shows this innovative in vivo medical diagnostic equipment and some clinical applications.

1.Review of scientific literature

In the past century, a great number of researchers have given their contribution to the study of the interactions between biological matter and electromagnetic fields. Many investigated the dielectric properties of living matter. Some others analyzed the differences between a cancerous agglomerate of cells and homogenous or 'normal' tissues.
The period between the First and the Second World War spanned the early days of radio and electronics: vacuum tubes were the radio frequency oscillation generators, the spectrum ranged between a few kHz and 15 MHz. Measurements on biological materials were based on resistivity or impedance and instruments such as the Wheatstone bridge.
After the second world conflict, investigations on biological materials were extended into the microwave bands[1].

Among the pioneers in this field, there were H. Fricke[2] and S. Morse[3]. In 1926, in their paper entitled "*The electric capacity of tumors of the breast*", they reported that "*malignant tumors have a greater polarizability than normal breast tissues or benign tumors*". They carried out their experiments at low frequencies around 20 kHz. Tissues were cut into small blocks and placed in a conductivity cell for measurement. They claimed that measurements performed on tissues from locations other than the breast convinced them that the method was of general applicability and that in some cases the "*measurements may be made directly on the patient*". Following the publication of these results, Fricke published a paper in which he declared that "*It seems probable that the measurement of the capacity may provide a very practical method for diagnosing the malignancy of a tumor.*" These experiences are of a great importance to explain and clarify some aspects that arises in the common use of the Bioscanner/Trimprob device, and it is extremely interesting to read this paper in which the authors wrote: "*While the resistance of biologic tissues has been studied by many investigators, little attention has been directed to their capacity*". The term "capacity" is to be associated to the well known property of the tissues which is usually called its "polarization". Theoretically we assume two type of electric capacity, the first is the "static capacity" that is independent to the frequency of the alternating current, the second is the "polarization" type that depends upon the inter phases in the tissues and suggest that capacity might have a considerable biologic significance. The "polarization" capacity is related to the alternating current applied or irradiated to the tissue under test. In their paper, Fricke and Morse claim: It has been a constant surprise to find that *the capacity of malignant tumors of the breast is so consistently larger than that of normal tissues in the same location or of benign tumors as to make its estimation in any individual case clearly of diagnostic value.*

As above reported, these aspects are important to clarify the mechanism of the *non linear resonance interaction* applied to the diagnosis by means of this technology. It is known by the users, that the Trimprob works on three frequencies, and that the first is 462 MHz , while the others are the harmonics of the first ones.

Despite the frequency used for the analysis, but in accordance with the Fricke and Morse paper, the tissue capacity values have to be higher for the malignant tumors, lower for benign and much lower for healthy ones. The measured values are also greatly different in the order of four times greater for malignancy than for healthy tissues. In other words, we have to expect that a malignant cells agglomerate, that it is characterized by a high capacity, must have a non linear resonance interaction on the lower frequency of the harmonically related group emitted by the Bioscanner/Trimprob.
Differently, the benign pathologies, like benign prostate hypertrophy or breast fibromas, will not have the same capacity than a malignant tumor and of course, the non linear resonance interaction could be detected on a higher frequency.

2.Materials and Methods

The main feature of Trimprob apparatus is a cylindrical probe shown in fig. 1, within which a resonant cavity incorporates a transmission line tuned to the frequency of oscillation which is in the 65 cm wavelength band (462 MHz).

At the open end of this line there is a semiconductor with non-linear characteristics, which is activated by a nanosecond electromagnetic pulse. This transient provides an injection of electromagnetic energy into the tuned line, which performs a damped oscillation. This particular tunable amplifier-oscillator represents the core of the Trimprob diagnostic device. It possesses lock-in or synchronization characteristics and, because of its particular construction, it produces a harmonically related group of coherent electromagnetic waves. These oscillations are radiated as a beam through the "beam window" of the oscillator dome at the end of the probe, where it has been geometrically focused, and the beam is used to irradiate the diseased tissues.

The working principle can be explained by considering the equivalent circuit diagram of figure 2. The left part stands for the probe and the right part for the tested biological tissue, while the coupling is represented by (virtual) interrupted lines. Inside the probe, the transistor T activates an electric circuit, which has a natural frequency of oscillation f1 that is determined by self and capacity of this circuit. The current I passing through T is a *non-linear* function of the potential difference V. Actually, $I = -aV + bV^2 + gV^3$,where *a* defines a "negative resistance". It results from a positive feedback, mediated by magnetic coupling with the self of the first circuit. This non-linear system produces stationary oscillations of well-defined amplitude, but when the probe is brought close to the tested biological tissue, it becomes an "active oscillator" that interacts with a "passive oscillator".

Journal of Physics: Conference Series **329** (2011) 012038 doi:10.1088/1742-6596/329/1/012038

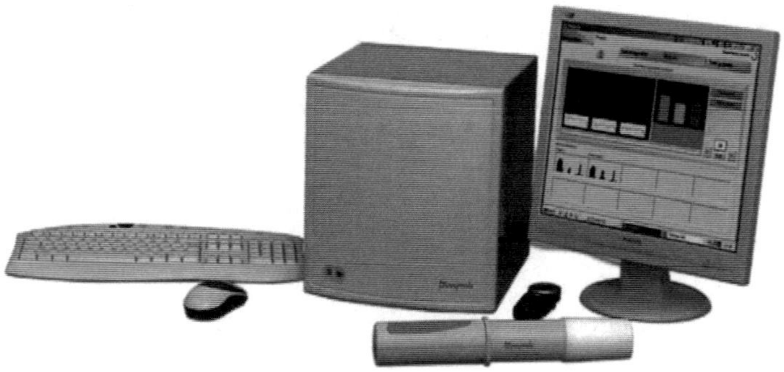

Fig. 1. The Trimprob equipment is composed by the Bioscanner probe and a computer based spectrum analyzer

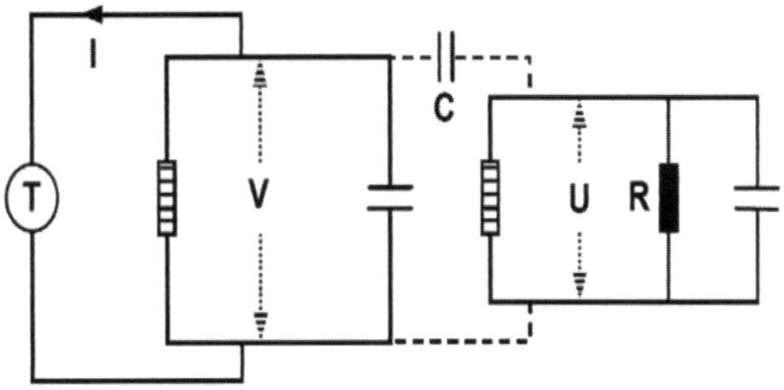

Fig. 2. Coupled active and passive oscillators equivalent electric circuit

Although the irradiated biological system contains various subsystems that could be set in forced oscillations, their mutual interactions are negligible. It is therefore sufficient to consider the effects of the active oscillator on one particular passive oscillator of given resonance frequency f_2.We can even imagine a circuit, where the self and capacity determine the frequency f_2 , while the resistance R defines energy absorption. The probe acts there like an "open capacity" and the tested biological tissue is subjected to the resulting electric field. This type of coupling is unusual. It involves a capacity C that increases when the probe approaches the tested tissue. Since this capacity facilitates the passage of high frequency currents, we can call this a *dynamic coupling*. All these features are taken into account by *two coupled differential equations*, describing the possible variations of the potential differences V and U. The detailed mathematical treatment is available on internet[1], but the basic ideas can be expressed in simple terms. Let us consider the particular case where the active oscillator is unperturbed (C = 0). The equation for V reduces then to the well-known *Van der Pol* equation, initially introduced to account for the possible actions of a triode. Even when the amplification coefficient *a* is very small, the rest-state (V = 0) will be unstable. The slightest perturbation will be amplified and the capacity will accumulate charges, but when they increase, there will be also a greater tendency towards discharging. The system will end up with a stationary harmonic oscillation of frequency f1 and given amplitude for the potential difference V. For larger values of *a*, higher *harmonics* will appear, since the equation for V contains terms that vary like V^2 and $V^{2.}$. This remains true when the active oscillator is coupled to a passive oscillator.

We can thus adopt a solution for V that accounts for the existence of oscillations at a fundamental frequency f and its harmonics, 2f and 3f. The value of f, as well as the amplitudes and phase factors of all these components can only be specified, when we take into account the fact that V produces forced oscillations for U and that this has an effects on V, because of C. The result can be summarized in the following way: the active oscillator is able to "feel" what happens inside the tested biological tissue, since *it has to transfer energy* to the passive oscillator to produce forced oscillations of the hidden entities. The active oscillator is also able to "tell" us how the passive oscillator is responding, since the amplitude of its own oscillations is strongly reduced when there is a large energy transfer. This is revealed, indeed, by a reduction of the amplitude of the emitted wave, displayed on the screen of the spectrum analyzer. The mathematical treatment reveals that the active oscillator draws more energy from the batteries when resonance is achieved, but its own energy is reduced, as if it had to make a "big effort". This mechanism is the essence of the *non-linear resonance interaction*[1,4,5].

Although the values of f^1 and f^2 are fixed, it is possible to achieve, or at least to approach, *ideal resonance* where the "dip" of a given spectral line is strongest, by changing the value of C through a modification of the distance between the probe and he tested tissue. The first spectral line is very sensitive to the existence of a resonance, when the negative resistance *a* is small, but a higher value will allow for a simultaneous search of resonance phenomena at the fundamental frequency f and its harmonics 2f, 3f, etc.

The effect of this interaction is easily detectable by means of a spectrum analyzer feed by a small antenna. At the resonance, on one or more of the spectral lines, two effects are detectable: the first is related to the transfer of an amount of radiofrequency from the generator probe to the diseased tissue, that absorbs a part of the signal on the proper frequency line (dynamic resonance), while the second effect it is related to the deformation of the electromagnetic pattern emitted by the probe, due to the interaction with a resonating diseased tissue, that produces in the "near field" a sort of parasitic resonating element able to deflect the waves in other spatial directions, in the same way that beam antennas for radio communications works.

The subject under test must be further from the probe than the "near field", and the same applies to the spectrum analyzer, which is a part of the system. Using this arrangement, it is possible to observe an effect that appears as absorption of one or more of the spectral lines radiated by the scanner. This is observed on the spectrum analyzer display, that transforms the received signal into a Fast Fourier Transform (FFT). These lines are specifically tuned to the types of tissues to be investigated. At the moment, three spectral lines are used: the first, corresponding to the wavelength, responds specifically to highly anisotropic states like micro-agglomerates of cancer cells; the second line responds to parenchyma (soft tissues) diseases; the third line responds to anomalies of the lymph and vascular system.

The interaction between a non-linear active oscillator and an ordinary (linear) passive oscillator leads to the peculiar phenomenon of "non-linear resonance interaction". A similar behavior is known as a *grid-dip meter* (g.d.m.). Initially, it contained a triode[6] that was associated with an oscillating circuit in such a way that it delivered a stationary oscillation at *one* particular, easily tunable frequency. The tunable active oscillator could be coupled by *magnetic induction* with another oscillating circuit, containing a real coil.

When such a grid-dip meter is tuned, so that its natural frequency is identical to the natural frequency of the passive oscillator, there will be a resonance. Since the active oscillator is transferring energy to the passive oscillator, the oscillating current passing through the coil of the active oscillator is reduced, and an ammeter, included in the grid circuit, will indicate this effect. At resonance, there appears a "grid-dip", but to avoid ambiguities, the active generator should produce no harmonics. When a spectrum analyzer is used to monitor the near field and primarily the far field emitted by the g.d.m. coil in the free space, while interacting with a tuned for resonance, passive L/C simple circuit, we can observe some interesting not commonly investigated effects.

Fig. 3A and 3B shown the necessary setup for this experiment: A Millen mod. 90651-A g.d.m. placed on a laboratory wooden table near a passive oscillator composed by an U shaped coil paralleled by a 30 pf variable air spaced capacitor. The circuit is tunable in frequency around the 140-170MHz band, that was used to facilitate the passive circuit realization as well as a proper coupling with the g.d.m. The passive oscillator U coil is placed in the near field of the g.d.m. test coil. At a distance of at least 50 cm, just outside the near field, another portable spectrum analyzer with a 1/8 wavelength rod antenna picks up the g.d.m. far field.

A slight tune of the g.d.m., to achieve the resonance with the passive circuit, is evidenced by a sharp dip of the ammeter current. This common and known effect represents the normal use of the instrument. At the same time, the far field received by the spectrum analyzer antenna shows a strong dip of the corresponding frequency line as evidenced in figs. 4-5;

The spectral line will drop the amplitude more than 20 decibel and could be in the order of 30 or more dB. In other words the frequency line will disappear from the display. Instead the near field detection will show a little attenuation of the spectral line in the order of few dB. This far field monitoring, to display the waves propagation of a passive oscillator interacting with an active one, was not previously reported in literature and represents the basis of the Trimprob operations.

The use of a g.d.m. not consent the cancer or other disease detection but it is used, scaled in frequency, for field modeling purposes and for other experiments and laboratory measurements, cause the magnetic coupling of the oscillators, although the propagation of the involved radiofrequency field is the same of the diagnostic device, that is not easily influenced by magnetic-coupled passive oscillators.

The EM cancer detector is different, since it allows for an *electric* and no magnetic coupling, by means of a quarter wavelength antenna, activating charged particles insidebiological tissues or other polarizable materials. Moreover, there are *harmonics,* that thespectrum analyzer allows for a distinction of possible resonance effects for anyone of the frequency components and could be considered like a sort of *'electric field capacity coupled grid dip meter'* provided of a far field detection. Both g.d.m. and Trimprob, are provided of synchronization capabilities1 that are evidenced by a loop locking of the active oscillator frequency respect the passive ones. Effect evidenced by the spectrum analyzer tracking capabilities that measures not only the amplitude, but also the precise frequency at the interaction resonance. It is astonishing observe the damping force opposite to frequency variations when the two oscillators are in their respective 'capture range'. To have diagnostic capabilities the irradiated radiofrequency by the probe has to be of few about ten milliwatt; or the interaction with the tissues will be no more evidenced cause excessive oscillator coupling and other saturation effects. A similar behavior is common with not well designed g.d.m., when these instruments are used to analyze the resonance of passive L/C oscillators, especially when the g.d.m. power is excessive. Instead, in the case of the Bioscanner, very low in level signals, in the order of microwatts could still interact with near the skin anomalies on 462 MHz, but a more sensitive spectrum analyzer is required, to display the far field. An experimental tunnel diode7 nonlinear oscillator probe was realized and laboratory tested by the author. This could represent a promising technology for a skin cancer like melanomas, detector, useful also for a low level e.m.f. interaction device with cells, in laboratory experiments.

The lock-in characteristic is also evidenced by the immediate synchronization in frequency of a couple of 'Bioscanner' probes when such a non-linear oscillators are in their respective 'capture range', that is about one wavelength wide. Greatest distances are possible with the aids of corner reflectors to focusing both the probe fields. The spectral far field line amplitude, due to the phase synchronization of the oscillators, is greater than for a single oscillator.

Fig 3 A Fig 3B

Experimental asset. The far-field spectrum analyzer is placed on the table about 50 cm.far from the g.d.m and the passive oscillator. A small antenna picks up the r.f. field. The author right hand is moving the L/C oscillator tuning to achieve a resonance with the grid dip meter: when the resonance is achieved, the spectral line on the display is immediately depressed (B).

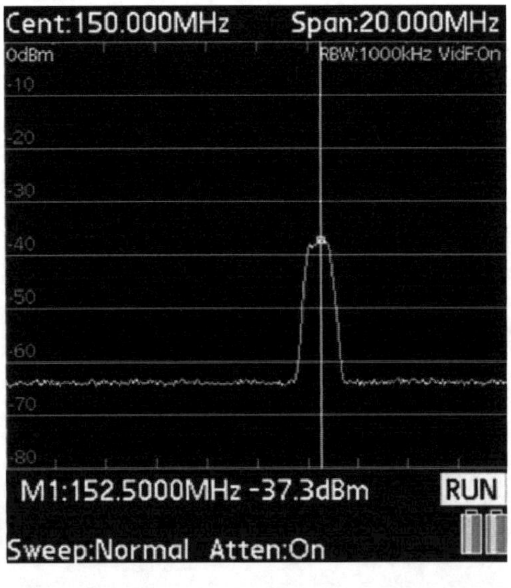

Fig. 4. The g.d.m. oscillator line out of resonance far at 152.5 Mhz.

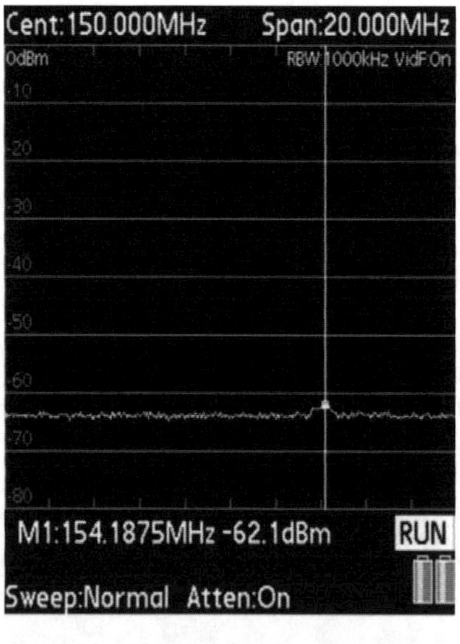

Fig. 5. Frequency resonance interaction, the field spectral line is depressed.

3.Opinions and implications

The present paper represents an integration of a published ones on 2010[20].

The first experiments, carried out by the author in the early days of the Bioscanner invention and development, as well as several clinical trials during the last years, have scientifically validated the efficacy of the described low level e.m.f. cancer detector in several body organs like breast[8], prostate[9-11], bladder[12,13], stomach-duodenum[14], thyroid[16,17], colon-rectum[18,21]. A new trial carried

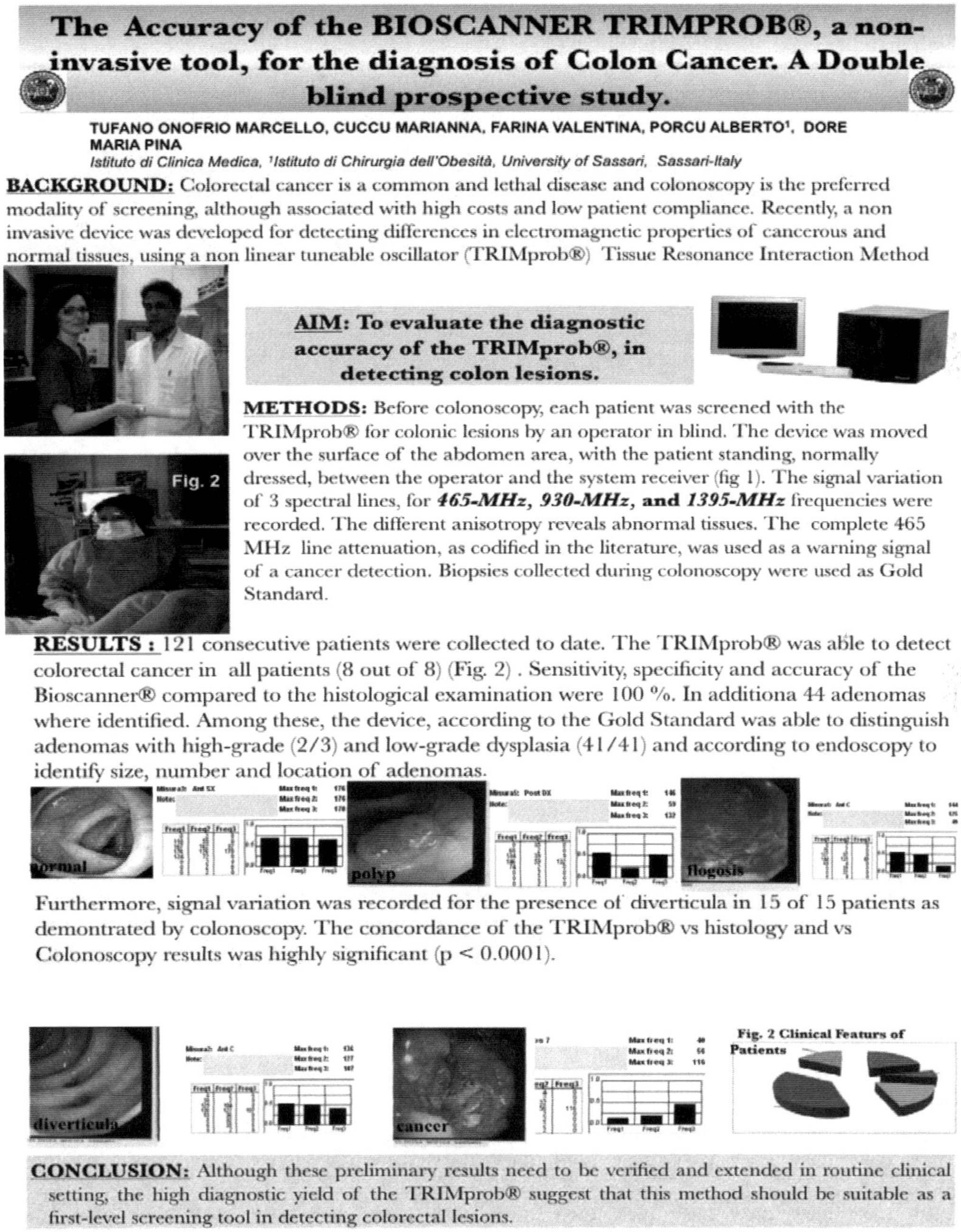

Fig.6.

out at the University of Sassari, Italy, by M. Tufano M.D. et al, was presented at the XVI Digestive Diseases National Congress (Gastro-enterology), March 2011, fig.6. The bioscanner spectrometry results demonstrate the elevate accuracy of the the method in colon cancer diagnostics: a sensitivity and specificity of 100% and the concordance of the bioscanner vs the histology and vs colonoscopy results was highly significant ($p < 0{,}0001$).

A new paper by Pokorny, Vedruccio, Cifra, Kucera, titled Cancer physics: Diagnostics based on damped cellular elasto-electrical vibrations in microtubules, recently available on Eur. Biophys. J., discloses the mechanism of active grown cancer tissues interaction with a Non- Linear Resonance Interaction (NLRI) Bioscanner. This paper represents an important bridge between the technological background of the Bioscanner and the biophysical behavior of cancer cells when irradiated by low level emf emission [24]. Trimprob clinical diagnostic accuracy as reported in table 1, that resumes the above mentioned clinical studies[19,20], spans several applications in the field of characterization of benign vs. malignant pathologies, prevention, screening capabilities and some other not disclosed here, possible applications.

In the last years was only possible to realize a not invasive diagnostic tool based on this technology, commercially named Trimprob, that was based on these researches, 'medical CE' certified, and quite diffused in Italy and abroad. The above mentioned results, still requires an important consideration: the cancer detection is possible, with the described device, only on the cited sharp frequency window centered on 462 MHz, no more than 8 MHz wide. Outside this range, the nonlinear resonance generator doesn't interact with the diseased tissues.

Table 1-Principal Clinical Trials Bioscanner® TRIMprob™ with pubblications

Organ	N° patients	Sensibility %	Specificity %	Positive Predictive Value (P.P.V)	Negative Predictive Value (N.P.V.)	Accuraty %
PROSTATE						
1- Trial by Prof. Scialpi[5] (Taranto)	43	94	92		96	93
2 – Dr. Bellorofonte C et Al.[9] (Milano)	757	95	43	94	90	
3 – Dr. Da Pozzo L et Al.[10] (Multicentre Clinical trials in five Univ.: HSRaffaele – Mi, Federico II NA, Careggi FI, S.Orsola-Malpighi BO, Cattinara Trieste)	188	80	51	44	84	60
4 – Prof. Tubaro A et Al.[11] (HSAndrea – Univ.Roma)	111	86	63	60	88	72
5 – Dr Gokce O et Al.[23] (Dip. Urologia –Univ. Istanbul)	148	76	61	40	89	
6- Prof. Cerulli, dr. Di Vaccaro et Al[22] (Univ.Umberto I° Roma)	782	(with PSA>4ng/ml) 84,1	(with PSA>4ng/ml) 67,8	(with PSA>4ng/ml) 61,7	(with PSA>4ng/ml) 83,8	
BLADDER						
Dr. Leucci, [13] (Osp Vito Fazzi – Lecce)	114	87,5 (Cohen's K = 0.77)	90,5 ($P < 0.001$)	83,3	91,1	89,5
BREAST						
Dr. Paganelli, dr. De Cicco C[8] (I.E.O. – Milano)	101	84	75		80	72
TYROID						
Prof. Sacco R et Al.[16] (Clinica Chir. –Univ Catanzaro)	51	100	100			100

STOMACH - DUODENUM						
1 – Prof Vedruccio C/ Prof. Mascia E et Al[14] (Clinica Chir. Osp MM- Taranto)	45	93	93	95	92	
2 - Prof. Sacco R et Al.[15] (Clinica Chir. – Univ Catanzaro)	43	100	100			100
RECTUM						
1- Prof. Leo, dr. Vannelli A et Al.[18] (INT – Milano) Preliminary Study	228	94	85	86	93	89
2 - Complete results[21]	442	94	84	92	88	90
	Tot. 3053					

References

[1] Meessen A. Working Principle of an EM Cancer Detector. Available on internet, http://www. meessnen.net/AMeessnen/EMcancerDet2.pdf; Institut de Physique, UniversitŽ Catholique de Louvain, Louvain-la-Neuvre, Belgium 2000.

[2] Fricke H. Phys Rev 21, 708 (1923); J Gen Physiol 9, 137-52 (1925); Phys Rev 1925; 26, 682-7.

[3] Fricke H, Morse A. Phys Rev 25, 361-367 (1925); J Gen Physiol 9, 153-167 (1925); The electric capacity of tumors of the breast. J Cancer Res 1926; 10: 340-76.

[4] Vedruccio C, Meessen A. EM cancer detection by means of non-linear resonance interaction. Proc.and Extended Papers book. PIERS 2004, Progress in Electromagnetics Research Simposium, Pisa Italy, March 28-31, 2004; 909-12.

[5] Vedruccio C, Meessen A. Nuove possibilità diagnostiche tramite onde elettromagnetiche. Fisica in Medicina, AIFM, 2004; 3: 225-30.

[6] Hund, A. High frequency measurements. McGraw-Hill, 1951.

[7] Manager, H.R.L. et al. Tunnel diode manual. General Electric, 1961.

[8] De Cicco C, Mariani L, Vedruccio C, *et al.* Clinical Application of Spectral Electromagnetic Interaction in Diagnosis of Breast Lesions. Results of a Pilot Study. Tumori 2006; 92(3): 207-12.

[9] Bellorofonte C, Vedruccio C, *et al.* Non-invasive detection of prostate cancer by electromagnetic interaction. Eur Urol 2005; 47: 29-37.

[10] Da Pozzo L. *et al.* Tissue Resonance Interaction Method for non Invasive Diagnosis of Prostate Cancer: a Multicenter Clinical Evaluation. BJU Int 2007; 100(5): 1055-9.

[11] Tubaro A, De Nunzio C, Trucchi A, *et al.* The Electromagnetic Detection of Prostatic Cancer: Evaluation of Diagnostic Accuracy. Urology 2008; 72(2): 340-4.

[12] Leucci G, Vedruccio C, *et al.* Studio Pilota per la Diagnosi del Carcinoma Vescicale mediante l'utilizzo del TRIMprob, (preliminary results), proc. of XI Congresso Nazionale AURO, Department of Urology, Ospedale Vito Fazzi, Lecce, Italy, 6-9 Oct 2004.

[13] Gervino G, *et al.* Diagnosis of Bladder Cancer at 465 MHz. Electromagn Biol Med 2007; 26(2): 119-34.

[14] Vedruccio C, Mascia E, Martines V. Ultra High Frequency and Microwave Non-linear Interaction Device for Cancer Detection and Tissue Characterization, a Military Research approach to prevent Health Diseases. International Review of the Armed Forces Medical Services (IRAFMS) 2005; 78(2): 121-6.

[15] Sacco R, Sammarco G, De Vinci R, *et al.* Relief of gastric cancer with an electromagnetic interaction system (TRIMprob) in outpatients. Chir Ital 2007; 59(6): 823-8.

[16] Lucisano AM, Innaro N, Pata F, *et al.* Diagnosis of Carcinoma in Multinodular Goiter by Electromagnetic Interactions. Preliminary Results. European Surgical Research 2006; 38: 129-32.

[17] Sacco R, Innaro N, Pata F, *et al.* Diagnosi Preoperatoria di Carcinoma Incidentale in Gozzo Multinodulare mediante Interazioni Elettromagnetiche. Chir Ital 2007; 59(2): 247-51.

[18] Vannelli A, Leo E, Battaglia L, *et al.* New technique for diagnosis of rectal cancer. Dis Colon Rectum 2009; 52(1): 162-6

[19] Vedruccio C, Ricci C. The Trimprob Non Linear Resonance Interaction for early cancer detection. In: Casciaro S, Samset E, Eds. Minimally Invasive Therapies and Novel Embedded Technologies. Lecce: Lupiensis Biomedical Publications, 2007:111-118.

[20] Vedruccio C, Polarizability of normal and cancerous tissues, a Radiofrequency Nonlinear Resonance Interaction non invasive diagnostic bioscanner "Trimprob" detector, ICEMS (eds.) "Fundamentals of bioelectromagnetics: physical, chemical, and biological mechanisms mediating the effects of low intensity, non-ionizing electromagnetic fields", edited by L. Giuliani, M. Soffritti , Eur. J. Oncol. Library, 2010; vol 5:177-185.

[21] Vannelli A, Battaglia L, Poiasina E and Leo E, Diagnosis of rectal cancer by Tissue Resonance Interaction Method, BMC Gastroenterology 2010, 10-45.
http://www.biomedcentral.com/1471-230X/10/45□doc\TRIM Rectal Cancer 2010.pdf

[22] Di Viccaro D, Perugia G, Cerulli C, Olivieri V, Bova J, Zanza C, Teodonio S, Liberti M. The accuracy of tissue resonance interaction method probe (Trimprob tm) in non-invasive diagnosis of prostatic cancer. Analysis of the results of 782 patient. Urologia. 2009;76 Suppl 15:1-3.

[23] O. Gokce, O. Sanli, A. Salmaslioglu, A. Tunaci, C. Ozsoy and F. Ozcan, Tissue Resonance Interaction Method (TRIMprob) has the potential to be used alongside the recognized tests in the screening protocols for prostate cancer, International Journal of Urology (Int J Urol.), 2009 Jun;16(6):580-3.

[24] Pokorny J, Vedruccio C, Cifra M and Kucera O, Cancer physics: diagnostics based on damped cellular elastoelectrical vibrations in microtubules, European Biophysics Journal, 2011; 40 (6): 747-759. http://www.springerlink.com/content/577757274043rk07/

AUTHOR INDEX